# Magnesium in Human Health and Disease

# Magnesium in Human Health and Disease

Editors

**Sara Castiglioni**
**Giovanna Farruggia**
**Concettina Cappadone**

MDPI • Basel • Beijing • Wuhan • Barcelona • Belgrade • Manchester • Tokyo • Cluj • Tianjin

*Editors*

Sara Castiglioni
Department of Biomedical and
Clinical Sciences Luigi Sacco
Università di Milano
Milan
Italy

Giovanna Farruggia
Department of Pharmacy and
Biotechnology
Università di Bologna
Bologna
Italy

Concettina Cappadone
Department of Pharmacy and
Biotechnology
Università di Bologna
Bologna
Italy

*Editorial Office*
MDPI
St. Alban-Anlage 66
4052 Basel, Switzerland

This is a reprint of articles from the Special Issue published online in the open access journal *Nutrients* (ISSN 2072-6643) (available at: www.mdpi.com/journal/nutrients/special_issues/Magnesium_Health).

For citation purposes, cite each article independently as indicated on the article page online and as indicated below:

LastName, A.A.; LastName, B.B.; LastName, C.C. Article Title. *Journal Name* **Year**, *Volume Number*, Page Range.

**ISBN 978-3-0365-1786-5 (Hbk)**
**ISBN 978-3-0365-1785-8 (PDF)**

© 2021 by the authors. Articles in this book are Open Access and distributed under the Creative Commons Attribution (CC BY) license, which allows users to download, copy and build upon published articles, as long as the author and publisher are properly credited, which ensures maximum dissemination and a wider impact of our publications.

The book as a whole is distributed by MDPI under the terms and conditions of the Creative Commons license CC BY-NC-ND.

# Contents

**About the Editors** . . . . . . . . . . . . . . . . . . . . . . . . . . . . . . . . . . . . . . . . . . . . . . . . . . . . . . . vii

**Sara Castiglioni**
Editorial of Special Issue "Magnesium in Human Health and Disease"
Reprinted from: *Nutrients* **2021**, *13*, 2490, doi:10.3390/nu13082490 . . . . . . . . . . . . . . . . . . 1

**Diana Fiorentini, Concettina Cappadone, Giovanna Farruggia and Cecilia Prata**
Magnesium: Biochemistry, Nutrition, Detection, and Social Impact of Diseases Linked to Its Deficiency
Reprinted from: *Nutrients* **2021**, *13*, 1136, doi:10.3390/nu13041136 . . . . . . . . . . . . . . . . . . 5

**Gabriele Piuri, Monica Zocchi, Matteo Della Porta, Valentina Ficara, Michele Manoni, Gian Vincenzo Zuccotti, Luciano Pinotti, Jeanette A. Maier and Roberta Cazzola**
Magnesium in Obesity, Metabolic Syndrome, and Type 2 Diabetes
Reprinted from: *Nutrients* **2021**, *13*, 320, doi:10.3390/nu13020320 . . . . . . . . . . . . . . . . . . . 49

**Mihnea-Alexandru Găman, Elena-Codruța Dobrică, Matei-Alexandru Cozma, Ninel-Iacobus Antonie, Ana Maria Alexandra Stănescu, Amelia Maria Găman and Camelia Cristina Diaconu**
Crosstalk of Magnesium and Serum Lipids in Dyslipidemia and Associated Disorders: A Systematic Review
Reprinted from: *Nutrients* **2021**, *13*, 1411, doi:10.3390/nu13051411 . . . . . . . . . . . . . . . . . . 65

**Serafi Cambray, Merce Ibarz, Marcelino Bermudez-Lopez, Manuel Marti-Antonio, Milica Bozic, Elvira Fernandez and Jose M. Valdivielso**
Magnesium Levels Modify the Effect of Lipid Parameters on Carotid Intima Media Thickness
Reprinted from: *Nutrients* **2020**, *12*, 2631, doi:10.3390/nu12092631 . . . . . . . . . . . . . . . . . . 95

**Flora O. Vanoni, Gregorio P. Milani, Carlo Agostoni, Giorgio Treglia, Pietro B. Faré, Pietro Camozzi, Sebastiano A. G. Lava, Mario G. Bianchetti and Simone Janett**
Magnesium Metabolism in Chronic Alcohol-Use Disorder: Meta-Analysis and Systematic Review
Reprinted from: *Nutrients* **2021**, *13*, 1959, doi:10.3390/nu13061959 . . . . . . . . . . . . . . . . . . 107

**Meng-Hua Tao and Kimberly G. Fulda**
Association of Magnesium Intake with Liver Fibrosis among Adults in the United States
Reprinted from: *Nutrients* **2021**, *13*, 142, doi:10.3390/nu13010142 . . . . . . . . . . . . . . . . . . . 119

**Balazs Odler, Andras T. Deak, Gudrun Pregartner, Regina Riedl, Jasmin Bozic, Christian Trummer, Anna Prenner, Lukas Söllinger, Marcell Krall, Lukas Höflechner, Carina Hebesberger, Matias S. Boxler, Andrea Berghold, Peter Schemmer, Stefan Pilz and Alexander R. Rosenkranz**
Hypomagnesemia Is a Risk Factor for Infections after Kidney Transplantation: A Retrospective Cohort Analysis
Reprinted from: *Nutrients* **2021**, *13*, 1296, doi:10.3390/nu13041296 . . . . . . . . . . . . . . . . . . 131

**Giuseppe Pietropaolo, Daniela Pugliese, Alessandro Armuzzi, Luisa Guidi, Antonio Gasbarrini, Gian Lodovico Rapaccini, Federica I. Wolf and Valentina Trapani**
Magnesium Absorption in Intestinal Cells: Evidence of Cross-Talk between EGF and TRPM6 and Novel Implications for Cetuximab Therapy
Reprinted from: *Nutrients* **2020**, *12*, 3277, doi:10.3390/nu12113277 . . . . . . . . . . . . . . . . . . 145

Julie Auwercx, Pierre Rybarczyk, Philippe Kischel, Isabelle Dhennin-Duthille, Denis Chatelain, Henri Sevestre, Isabelle Van Seuningen, Halima Ouadid-Ahidouch, Nicolas Jonckheere and Mathieu Gautier
$Mg^{2+}$ Transporters in Digestive Cancers
Reprinted from: *Nutrients* **2021**, *13*, 210, doi:10.3390/nu13010210 . . . . . . . . . . . . . . . . . . . **157**

Dominique Bayle, Cécile Coudy-Gandilhon, Marine Gueugneau, Sara Castiglioni, Monica Zocchi, Magdalena Maj-Zurawska, Adriana Palinska-Saadi, André Mazur, Daniel Béchet and Jeanette A. Maier
Magnesium Deficiency Alters Expression of Genes Critical for Muscle Magnesium Homeostasis and Physiology in Mice
Reprinted from: *Nutrients* **2021**, *13*, 2169, doi:10.3390/nu13072169 . . . . . . . . . . . . . . . . . . **179**

Monica Zocchi, Daniel Béchet, André Mazur, Jeanette A. Maier and Sara Castiglioni
Magnesium Influences Membrane Fusion during Myogenesis by Modulating Oxidative Stress in C2C12 Myoblasts
Reprinted from: *Nutrients* **2021**, *13*, 1049, doi:10.3390/nu13041049 . . . . . . . . . . . . . . . . . . **193**

Concettina Cappadone, Emil Malucelli, Maddalena Zini, Giovanna Farruggia, Giovanna Picone, Alessandra Gianoncelli, Andrea Notargiacomo, Michela Fratini, Carla Pignatti, Stefano Iotti and Claudio Stefanelli
Assessment and Imaging of Intracellular Magnesium in SaOS-2 Osteosarcoma Cells and Its Role in Proliferation
Reprinted from: *Nutrients* **2021**, *13*, 1376, doi:10.3390/nu13041376 . . . . . . . . . . . . . . . . . . **201**

Yutaka Shindo, Ryu Yamanaka, Kohji Hotta and Kotaro Oka
Inhibition of $Mg^{2+}$ Extrusion Attenuates Glutamate Excitotoxicity in Cultured Rat Hippocampal Neurons
Reprinted from: *Nutrients* **2020**, *12*, 2768, doi:10.3390/nu12092768 . . . . . . . . . . . . . . . . . . **215**

Jeanette A. Maier, Gisele Pickering, Elena Giacomoni, Alessandra Cazzaniga and Paolo Pellegrino
Headaches and Magnesium: Mechanisms, Bioavailability, Therapeutic Efficacy and Potential Advantage of Magnesium Pidolate
Reprinted from: *Nutrients* **2020**, *12*, 2660, doi:10.3390/nu12092660 . . . . . . . . . . . . . . . . . . **229**

Gisèle Pickering, André Mazur, Marion Trousselard, Przemyslaw Bienkowski, Natalia Yaltsewa, Mohamed Amessou, Lionel Noah and Etienne Pouteau
Magnesium Status and Stress: The Vicious Circle Concept Revisited
Reprinted from: *Nutrients* **2020**, *12*, 3672, doi:10.3390/nu12123672 . . . . . . . . . . . . . . . . . . **243**

Andrea Rosanoff, Rebecca B. Costello and Guy H. Johnson
Effectively Prescribing Oral Magnesium Therapy for Hypertension: A Categorized Systematic Review of 49 Clinical Trials
Reprinted from: *Nutrients* **2021**, *13*, 195, doi:10.3390/nu13010195 . . . . . . . . . . . . . . . . . . . **265**

Jiada Zhan, Taylor C. Wallace, Sarah J. Butts, Sisi Cao, Velarie Ansu, Lisa A. Spence, Connie M. Weaver and Nana Gletsu-Miller
Circulating Ionized Magnesium as a Measure of Supplement Bioavailability: Results from a Pilot Study for Randomized Clinical Trial
Reprinted from: *Nutrients* **2020**, *12*, 1245, doi:10.3390/nu12051245 . . . . . . . . . . . . . . . . . . **281**

**Svetlana Orlova, Galina Dikke, Gisele Pickering, Sofya Konchits, Kirill Starostin and Alina Bevz**
Magnesium Deficiency Questionnaire: A New Non-Invasive Magnesium Deficiency Screening Tool Developed Using Real-World Data from Four Observational Studies
Reprinted from: *Nutrients* **2020**, *12*, 2062, doi:10.3390/nu12072062 . . . . . . . . . . . . . . . . . . 293

**Hideki Mori, Jan Tack and Hidekazu Suzuki**
Magnesium Oxide in Constipation
Reprinted from: *Nutrients* **2021**, *13*, 421, doi:10.3390/nu13020421 . . . . . . . . . . . . . . . . . . 305

**Justyna Malinowska, Milena Małecka and Olga Ciepiela**
Variations in Magnesium Concentration Are Associated with Increased Mortality: Study in an Unselected Population of Hospitalized Patients
Reprinted from: *Nutrients* **2020**, *12*, 1836, doi:10.3390/nu12061836 . . . . . . . . . . . . . . . . . . 317

# About the Editors

**Sara Castiglioni**

Sara Castiglioni is Associate Professor of General Pathology at the Department of Biomedical and Clinical Sciences L. Sacco of University of Milan, Italy. With a degree in Biological Science (2003) and PhD in Molecular Medicine (2008), she has expertise in cellular and molecular biology. Her areas of research include osteogenic and myogenic differentiation, magnesium homeostasis, signal transduction, and cell microgravity.

**Giovanna Farruggia**

Giovanna Farruggia is an Assistant Professor at the Department of Pharmacy and Biotechnology of the University of Bologna. Her research mainly concerns the photochemical properties of new biosensors for intracellular cations, integrating spectroscopic, microscopic, and cytofluorimetric approaches. Her main interest is the application of such indicators in the study of magnesium cellular and tissue homeostasis under physiological and pathological conditions.

**Concettina Cappadone**

Concettina Cappadone is an Assistant Professor at the Department of Pharmacy and Biotechnology of the University of Bologna. The main topics of her research concern the control mechanisms of cell proliferation in cancer cells. She has been interested in magnesium cell homeostasis and its role in cell cycle progression, apoptosis and differentiation. In particular, she participated in the development of new analytical methods to quantify the intracellular magnesium content during the osteoblastic differentiation process of mesenchymal stem cells and osteosarcoma cells.

*Editorial*

# Editorial of Special Issue "Magnesium in Human Health and Disease"

Sara Castiglioni

Department of Biomedical and Clinical Sciences Luigi Sacco, Università di Milano, 20157 Milano, Italy; sara.castiglioni@unimi.it

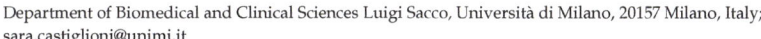

**Citation:** Castiglioni, S. Editorial of Special Issue "Magnesium in Human Health and Disease". *Nutrients* **2021**, *13*, 2490. https://doi.org/10.3390/nu13082490

Received: 16 July 2021
Accepted: 20 July 2021
Published: 21 July 2021

**Publisher's Note:** MDPI stays neutral with regard to jurisdictional claims in published maps and institutional affiliations.

**Copyright:** © 2021 by the author. Licensee MDPI, Basel, Switzerland. This article is an open access article distributed under the terms and conditions of the Creative Commons Attribution (CC BY) license (https://creativecommons.org/licenses/by/4.0/).

The fundamental role of magnesium in human health is extensively discussed in the review by Fiorentini and colleagues [1]. Magnesium acts both as a signalling element and a metabolite in cell physiology, and its homeostasis is regulated by the balance between intestinal absorption and renal excretion. Hypomagnesemia is probably the most underestimated electrolyte imbalance in Western countries. Apart from being caused by an insufficient dietary magnesium intake, magnesium deficiency is frequently associated with obesity, type 2 diabetes and metabolic syndrome [2]. The review of Piuri and colleagues offers interesting insights into the biochemical derangements occurring when magnesium is deficient, and how these altered metabolic pathways increase the risk of developing metabolic syndrome and type 2 diabetes in obese individuals. The data analysed in the review of Găman and collaborators point to an association of magnesium concentrations in the body with dyslipidemia and related disorders [3]. Another study found a relationship between magnesium, the lipid profile and atherosclerosis in patients with kidney diseases [4]. In particular, the carotid intima media is thicker when triglycerides or LDL levels are high and magnesemia is low. On the contrary, when magnesium levels are high, this effect disappears [4].

A correlation between the reduction in both total and ionised circulating magnesium and chronic alcohol use disorder was also demonstrated by performing a meta-analysis [5]. In particular, inappropriately high magnesium excretion was reported in hypomagnesemic patients with chronic alcohol use disorder [5]. The major consequence of alcoholic or non-alcoholic liver disease is the accumulation of the extracellular matrix within the liver, leading to the development of cirrhosis. A study was conducted to analyse the association between magnesium and calcium intake and liver fibrosis. While no association was found between significant fibrosis and calcium intake, some findings suggest that high total magnesium may reduce the risk of fibrosis [6].

A retrospective cohort study conducted on kidney transplant patients demonstrated that a magnesium-deficient status, defined as serum magnesium < 0.7 mmol/L, increases the risk of urinary tract infections and viral infections in the first year after transplantation [7].

Hypomagnesemia is often found in cancer patients in association with therapy with cetuximab. This is due to the fact that cetuximab, interfering with Epidermal Growth Factor (EGF) signalling, downregulates magnesium intestinal influx, inhibiting the TRPM6 (Transient Receptor Potential Cation Channel Subfamily M Member 6) magnesium channel [8].

Apart from TRPM6, which is the key channel that mediates magnesium influx in intestinal cells, other transporters are also responsible for maintaining magnesium homeostasis, such as Transient Receptor Potential Cation Channel Subfamily M Member 7 (TRPM7), Magnesium Transporter 1 (MagT1) and Cyclin and CBS Domain Divalent Metal Cation Transport Mediator 4 (CNNM4). Analysis of these transporters using Genotype-Tissue Expression (GTEx) and The Cancer Genome Atlas (TCGA) revealed their overexpression in digestive cancers. In particular, a positive correlation between TRPM7/MagT1/CNNM4 expression was found in esophageal adenocarcinoma and pancreatic ductal adenocarcinoma, and a correlation between TRPM7 and MagT1 expression was found in colorectal cancer [9]. Since in pancreatic ductal adenocarcinoma, the correlation between TRPM7/MagT1/CNNM4 expression is associated with a low survival, these proteins could be proposed as new biomarkers to predict life expectancy.

It is known that magnesium is essential for musculoskeletal health. A study on mice which received a magnesium-deficient diet demonstrated that even a mild magnesium deficiency is sufficient to alter the expression of several genes critical for muscle energy metabolism, muscle regeneration, proteostasis, mitochondrial dynamics and excitation–contraction coupling [10]. Magnesium also has a fundamental role during myogenesis. Non-physiological magnesium concentrations induce oxidative stress in myoblasts, and this oxidative stress is responsible for the inhibition of myoblast membrane fusion, thus impairing myogenesis [11]. Magnesium also affects bone growth and remodelling. Magnesium depletion inhibits the cell cycle progression and proliferation of SaOS-2 osteosarcoma cells. Magnesium is mainly confined at the plasma membrane in quiescent cells, and, when cells are stimulated to grow, magnesium moves toward the inner areas of the cell. In contrast, when SaOS-2 cells are cultured in magnesium deficiency conditions, the ion remains confined at the plasma membrane, even when the cells are stimulated to grow. This condition is associated with the inhibition of SaOS-2 proliferation [12].

Magnesium also has an important role in the nervous system. In particular, high magnesium in cerebrospinal fluid appears to enhance the neural functions, while low magnesium induces neuronal diseases. Shindo and colleagues demonstrated that high concentrations of glutamate induce excitotoxicity because, after a transient increase in intracellular magnesium due to its release from mitochondria, the cytosol magnesium concentration dramatically decreases. The authors showed that, by inhibiting the magnesium extrusion under toxic concentrations of glutamate, the excitotoxicity induced by glutamate is reduced, thus pointing out the importance of magnesium in the regulation of neuronal survival [13].

Magnesium deficiency is associated with mild and moderate tension-type headaches and migraines. A number of double-blind, randomised, placebo-controlled trials have demonstrated that magnesium supplementation is efficacious in relieving headaches. In particular, magnesium pidolate may have high bioavailability and good penetration at the intracellular level [14].

A large body of literature suggests a correlation between magnesium deficiency and stress. In particular, stress could induce magnesium deficiency, and, in turn, low magnesium levels could increase the susceptibility to stress, resulting in a magnesium and stress vicious circle [15].

Magnesium has also been proposed as therapy for hypertension. Rosanoff's review of 49 clinical trials shows that magnesium oral administration in association with treatment regimens in patients with partially controlled hypertension may be promising to control the blood pressure without increasing antihypertensive drugs [16].

Despite the fact that magnesium deficiency is very frequent in the population and is associated with altered levels of other electrolytes and various diseases, its diagnosis still represents a challenge because, to date, there is no gold standard to measure its concentration. The study of Zhan and colleagues showed that to assess magnesium bioavailability, circulating ionised magnesium might be a more sensitive measure of acute oral intake of magnesium than serum and urinary magnesium [17]. Orlova and co-workers proposed a magnesium deficiency questionnaire based on questions grouped into five categories: wellbeing, lifestyle, pregnancy, disease and medication, as a non-invasive assessment tool that could help to diagnose magnesium deficiency [18].

It is known that magnesium, particularly magnesium oxide, is clinically prescribed as a laxative. Compared to the newer drugs to treat constipation, magnesium is safe and of low cost, but daily use, particularly in patients with renal impairment, might lead to hypermagnesemia. Therefore, monitoring the magnesium concentration should be recommended [19].

Hypermagnesemia is an uncommon problem that can be caused by acute or chronic kidney diseases, hypothyroidism or corticoadrenal insufficiency and impacts on cell, tissue and organ functions, leading to many disorders. An analysis of serum magnesium concentrations in hospitalised patients revealed that either hypo- or hypermagnesemia was associated with an increased risk of in-hospital mortality, with dysmagnesemia being associated with severe diseases [20].

Overall, the 20 papers published in this Special Issue highlight that an adequate magnesium dietary intake, and maintenance of a correct magnesium homeostasis are essential for human health. To control magnesium availability, it would be advisable, in daily clinical practice, to include magnesium in the evaluation of blood ionograms.

**Funding:** The authors declare no funding.

**Conflicts of Interest:** The author declares no conflict of interest.

## References

1. Fiorentini, D.; Cappadone, C.; Farruggia, G.; Prata, C. Magnesium: Biochemistry, Nutrition, Detection, and Social Impact of Diseases Linked to Its Deficiency. *Nutrients* **2021**, *13*, 1136. [CrossRef] [PubMed]
2. Piuri, G.; Zocchi, M.; Della Porta, M.; Ficara, V.; Manoni, M.; Zuccotti, G.V.; Pinotti, L.; Maier, J.A.; Cazzola, R. Magnesium in Obesity, Metabolic Syndrome, and Type 2 Diabetes. *Nutrients* **2021**, *13*, 320. [CrossRef] [PubMed]
3. Găman, M.A.; Dobrică, E.C.; Cozma, M.A.; Antonie, N.I.; Stănescu, A.M.A.; Găman, A.M.; Diaconu, C.C. Crosstalk of magnesium and serum lipids in dyslipidemia and associated disorders: A systematic review. *Nutrients* **2021**, *13*, 1411. [CrossRef] [PubMed]
4. Cambray, S.; Ibarz, M.; Bermudez-Lopez, M.; Marti-Antonio, M.; Bozic, M.; Fernandez, E.; Valdivielso, J.M. Magnesium levels modify the effect of lipid parameters on carotid intima media thickness. *Nutrients* **2020**, *12*, 2631. [CrossRef]
5. Vanoni, F.O.; Milani, G.P.; Agostoni, C.; Treglia, G.; Faré, P.B.; Camozzi, P.; Lava, S.A.G.; Bianchetti, M.G.; Janett, S. Magnesium metabolism in chronic alcohol-use disorder: Meta-analysis and systematic review. *Nutrients* **2021**, *13*, 1959. [CrossRef]
6. Tao, M.H.; Fulda, K.G. Association of magnesium intake with liver fibrosis among adults in the United States. *Nutrients* **2021**, *13*, 142. [CrossRef] [PubMed]
7. Odler, B.; Deak, A.T.; Pregartner, G.; Riedl, R.; Bozic, J.; Trummer, C.; Prenner, A.; Söllinger, L.; Krall, M.; Höflechner, L.; et al. Hypomagnesemia is a risk factor for infections after kidney transplantation: A retrospective cohort analysis. *Nutrients* **2021**, *13*, 1296. [CrossRef] [PubMed]
8. Pietropaolo, G.; Pugliese, D.; Armuzzi, A.; Guidi, L.; Gasbarrini, A.; Lodovico Rapaccini, G.; Wolf, F.I.; Trapani, V. Magnesium absorption in intestinal cells: Evidence of cross-talk between egf and trpm6 and novel implications for cetuximab therapy. *Nutrients* **2020**, *12*, 3277. [CrossRef] [PubMed]
9. Auwercx, J.; Rybarczyk, P.; Kischel, P.; Dhennin-Duthille, I.; Chatelain, D.; Sevestre, H.; Van Seuningen, I.; Ouadid-Ahidouch, H.; Jonckheere, N.; Gautier, M. Mg2+ transporters in digestive cancers. *Nutrients* **2021**, *13*, 210. [CrossRef] [PubMed]
10. Bayle, D.; Coudy-Gandilhon, C.; Gueugneau, M.; Castiglioni, S.; Zocchi, M.; Maj-Zurawska, M.; Palinska-Saadi, A.; Mazur, A.; Béchet, D.; Maier, J.A. Magnesium Deficiency Alters Expression of Genes Critical for Muscle Magnesium Homeostasis and Physiology in Mice. *Nutrients* **2021**, *13*, 2169. [CrossRef] [PubMed]
11. Zocchi, M.; Béchet, D.; Mazur, A.; Maier, J.A.; Castiglioni, S. Magnesium Influences Membrane Fusion during Myogenesis by Modulating Oxidative Stress in C2C12 Myoblasts. *Nutrients* **2021**, *13*, 1049. [CrossRef] [PubMed]
12. Cappadone, C.; Malucelli, E.; Zini, M.; Farruggia, G.; Picone, G.; Gianoncelli, A.; Notargiacomo, A.; Fratini, M.; Pignatti, C.; Iotti, S.; et al. Assessment and imaging of intracellular magnesium in saos-2 osteosarcoma cells and its role in proliferation. *Nutrients* **2021**, *13*, 1376. [CrossRef] [PubMed]
13. Shindo, Y.; Yamanaka, R.; Hotta, K.; Oka, K. Inhibition of mg2+ extrusion attenuates glutamate excitotoxicity in cultured rat hippocampal neurons. *Nutrients* **2020**, *12*, 2768. [CrossRef] [PubMed]
14. Maier, J.A.; Pickering, G.; Giacomoni, E.; Cazzaniga, A.; Pellegrino, P. Headaches and magnesium: Mechanisms, bioavailability, therapeutic efficacy and potential advantage of magnesium pidolate. *Nutrients* **2020**, *12*, 2660. [CrossRef] [PubMed]
15. Pickering, G.; Mazur, A.; Trousselard, M.; Bienkowski, P.; Yaltsewa, N.; Amessou, M.; Noah, L.; Pouteau, E. Magnesium status and stress: The vicious circle concept revisited. *Nutrients* **2020**, *12*, 3672. [CrossRef] [PubMed]
16. Rosanoff, A.; Costello, R.B.; Johnson, G.H. Effectively prescribing oral magnesium therapy for hypertension: A categorized systematic review of 49 clinical trials. *Nutrients* **2021**, *13*, 195. [CrossRef] [PubMed]
17. Zhan, J.; Wallace, T.C.; Butts, S.J.; Cao, S.; Ansu, V.; Spence, L.A.; Weaver, C.M.; Gletsu-Miller, N. Circulating ionized magnesium as a measure of supplement bioavailability: Results from a pilot study for randomized clinical trial. *Nutrients* **2020**, *12*, 1245. [CrossRef] [PubMed]
18. Orlova, S.; Dikke, G.; Pickering, G.; Konchits, S.; Starostin, K.; Bevz, A. Magnesium deficiency questionnaire: A new non-invasive magnesium deficiency screening tool developed using real-world data from four observational studies. *Nutrients* **2020**, *12*, 2062. [CrossRef] [PubMed]

19. Mori, H.; Tack, J.; Suzuki, H. Magnesium oxide in constipation. *Nutrients* **2021**, *13*, 421. [CrossRef] [PubMed]
20. Malinowska, J.; Małecka, M.; Ciepiela, O. Variations in magnesium concentration are associated with increased mortality: Study in an unselected population of hospitalized patients. *Nutrients* **2020**, *12*, 1836. [CrossRef] [PubMed]

*Review*

# Magnesium: Biochemistry, Nutrition, Detection, and Social Impact of Diseases Linked to Its Deficiency

Diana Fiorentini †, Concettina Cappadone †, Giovanna Farruggia * and Cecilia Prata

Department of Pharmacy and Biotechnology, Alma Mater Studiorum—University of Bologna, 40126 Bologna, Italy; diana.fiorentini@unibo.it (D.F.); concettina.cappadone@unibo.it (C.C.); cecilia.prata@unibo.it (C.P.)
* Correspondence: giovanna.farruggia@unibo.it
† These authors contributed equally to this work.

**Abstract:** Magnesium plays an important role in many physiological functions. Habitually low intakes of magnesium and in general the deficiency of this micronutrient induce changes in biochemical pathways that can increase the risk of illness and, in particular, chronic degenerative diseases. The assessment of magnesium status is consequently of great importance, however, its evaluation is difficult. The measurement of serum magnesium concentration is the most commonly used and readily available method for assessing magnesium status, even if serum levels have no reliable correlation with total body magnesium levels or concentrations in specific tissues. Therefore, this review offers an overview of recent insights into magnesium from multiple perspectives. Starting from a biochemical point of view, it aims at highlighting the risk due to insufficient uptake (frequently due to the low content of magnesium in the modern western diet), at suggesting strategies to reach the recommended dietary reference values, and at focusing on the importance of detecting physiological or pathological levels of magnesium in various body districts, in order to counteract the social impact of diseases linked to magnesium deficiency.

**Keywords:** magnesium; nutrition; hypomagnesemia; mg deficiency; mg detection

**Citation:** Fiorentini, D.; Cappadone, C.; Farruggia, G.; Prata, C. Magnesium: Biochemistry, Nutrition, Detection, and Social Impact of Diseases Linked to Its Deficiency. *Nutrients* **2021**, *13*, 1136. https://doi.org/10.3390/nu13041136

Academic Editor: Mario Barbagallo

Received: 17 February 2021
Accepted: 26 March 2021
Published: 30 March 2021

**Publisher's Note:** MDPI stays neutral with regard to jurisdictional claims in published maps and institutional affiliations.

**Copyright:** © 2021 by the authors. Licensee MDPI, Basel, Switzerland. This article is an open access article distributed under the terms and conditions of the Creative Commons Attribution (CC BY) license (https://creativecommons.org/licenses/by/4.0/).

## 1. Introduction

Magnesium is the fourth most abundant element in the human body ($Ca^{2+}$ > $K^+$ > $Na^+$ > $Mg^{2+}$) and the second most abundant cation within the body's cells after potassium. The human body contains 760 mg of magnesium at birth and this quantity increases to 5 g at around 4–5 months [1,2]. The total $Mg^{2+}$ body amount varies between 20 and 28 g [2,3]. More than 99% of the total body $Mg^{2+}$ is located in the intracellular space, mainly stored in bone (50–65%), where, together with calcium and phosphorus, it participates in the constitution of the skeleton, but also muscle, soft tissues, and organs (34–39%), whereas less than 1–2% is present in blood and extracellular fluids [4,5]. Magnesium concentration within erythrocytes is three times higher than in plasma [6], where normal magnesium concentrations range between 0.75 and 0.95 millimoles (mmol)/L [7]. Up to 70% of all plasma $Mg^{2+}$ exists in the ionized (free) active form [8]. A serum magnesium level less than 1.7–1.8 mg/dL (0.75 mmol/L) is a condition defined as hypomagnesemia [9]. Magnesium levels superior to 2.07 mg/dL (0.85 mmol/L) are most likely linked to systemic adequate magnesium levels [10,11], as also reported by Razzaque, who in addition suggests to individuals with serum magnesium levels between 0.75 to 0.85 mmol/L to undergo further investigation to confirm body magnesium status [12]. In addition, a urinary excretion < 80 mg/*die* could indicate a risk of magnesium deficiency, because in this condition renal excretion decreases as a compensatory mechanism [13].

Analogously to calcium, body magnesium content is physiologically regulated through three main mechanisms: intestinal absorption, renal re-absorption/excretion, and exchange from the body pool of magnesium (i.e., bones). $Mg^{2+}$ stores are indeed tightly regulated via a balanced interplay between intestinal absorption and renal excretion under normal conditions. The elimination of magnesium by the kidneys increases of course when there is a magnesium surplus and can decreases to just 1 mEq of magnesium (~12 mg) in the urine during deficits. Despite renal conservation, magnesium can be derived from the bone (as well as muscles and internal organs) in order to preserve normal serum magnesium levels when intakes are low, as with calcium [14]. $Mg^{2+}$ insufficient levels have been documented in ill patients since the end of the last century [15]. Nevertheless, despite the well-recognized importance of magnesium, $Mg^{2+}$ availability is not generally determined and monitored in patients, therefore, magnesium has been called the "forgotten cation" [16,17]. Moreover, serum magnesium levels do not usually reflect the content of magnesium in different body districts. Therefore, a normal level of serum magnesium does not rule out magnesium deficiency [18]. In the last 20–30 years, a great number of epidemiological, clinical, and experimental research papers have shown that abnormalities of magnesium levels, such as hypomagnesemia and/or chronic magnesium deficiency, can result in disturbances in nearly every organ/body, contributing to or exacerbating pathological consequences and causing potentially fatal complications [19–21].

Magnesium subclinical deficiency is not uncommon among the general population [18]. Although kidneys limit urinary excretion of this mineral to avoid hypomagnesemia, habitual low intakes of magnesium or excessive losses, due to different causes and conditions, can lead to a magnesium subclinical deficiency. Early signs of magnesium deficiency include weakness, loss of appetite, fatigue, nausea, and vomiting. Afterwards, muscle contractions and cramps, numbness, tingling, personality changes, coronary spasms, abnormal heart rhythms, and seizures can occur when magnesium deficiency worsens [22]. Finally, severe magnesium deficiency can result in hypocalcemia or hypokalemia because mineral homeostasis is disrupted [23].

As previously cited, hypomagnesaemia is usually defined as serum magnesium concentration <0.75 mmol/L [21], nevertheless, there are various concerns about the use of this parameter as a marker of real magnesium content in cells/body [9].

This review focuses on the biochemical mechanisms underneath magnesium cellular functions, in order to elucidate the correlation with the potential risks associated to magnesium deficiency. Moreover, other important topics to be addressed are the nutritional strategies that enable the prevention of low magnesium intake and the methods to detect when magnesium content is truly physiological or pathological. Therefore, a comprehensive analysis of the data available on magnesium content and bioavailability (i.e., the fraction of an ingested nutrient that becomes available for use and storage in the body) is here reported together with a discussion about the available methods for the evaluation of magnesium content in the human organism.

Finally, a survey on the social impact of the principal diseases linked to magnesium deficiency is reported, focusing in particular on diabetes, osteoporosis, cardiovascular, and neurological diseases and cancer.

## 2. Biochemistry of Magnesium to Understand the Consequences of Its Deficiencies

Magnesium (atomic number 12, atomic mass 24.30 Da) is classed as an alkaline earth metal belonging to the second group of the periodic table of the elements. Like calcium, its oxidation state is 2+ and, owing to its strong reactivity, it frequently occurs as free cation $Mg^{2+}$ in aqueous solution or as the mineral part of a substantial variety of compounds, including chlorides, carbonates, and hydroxides rather than in a native metallic state.

$Mg^{2+}$ is mainly absorbed through the small intestine, although some is also taken up through the large intestine. There are two known transport systems for $Mg^{2+}$, a passive paracellular mechanism and transcellular transport by dedicated $Mg^{2+}$ channels and transporters. In particular, member 1 of solute carrier family 41 (SLC41A1), magnesium transporter 1 (MagT1), and transient receptor potential melastatin type 6 and 7 (TRPM6 and TRPM7) have been described. The role of these $Mg^{2+}$ transporters in the establishment of $Mg^{2+}$ homeostasis and the molecular mechanism of their action have been reviewed in detail in several recent papers present in the literature [2,24–32]. $Mg^{2+}$ homeostasis is maintained by the intestine, bone, and kidneys under hormonal control. Briefly, serum $Mg^{2+}$ is filtered by the renal glomeruli and then reabsorbed along the nephron, where the reabsorption pathways differ in each segment. Magnesium transport across cell membranes shows tissue variability and, among the body's tissues, is higher in the heart, liver, kidney, skeletal muscle, red cells, and brain. Thus, magnesium transport, the physiology of magnesium homeostasis, and metabolic activity of the cell are strictly correlated [19,23,25,33–35].

The level of renal excretion of $Mg^{2+}$ mainly depends on the serum $Mg^{2+}$ concentration. Blood magnesium concentrations are strictly regulated in order to maintain a normal range even if the dietary magnesium intakes are low or an excessive magnesium excretion occurs. While serum/plasma magnesium concentrations remain in the healthy range however, both bone and soft tissue intracellular magnesium concentrations may be depleted [7]. Unlike other ions, for which cells are observed to maintain transmembrane gradients, cellular and extracellular concentrations of free $Mg^{2+}$ are similar. $Mg^{2+}$ typical intracellular concentrations range from 10 to 30 mM. However, since most $Mg^{2+}$ is automatically associated with ribosomes, polynucleotides, ATP, and proteins upon its entry into the cells, its freely available concentration falls varies between 0.5 and 1.2 mM [36].

Approximately half of magnesium present in the body can be found in bone, 30% of which is exchangeable and functions as a pool to stabilize serum $Mg^{2+}$ concentration [37]. $Mg^{2+}$ is an integral part of apatite crystals, from which it is released in the course of bone resorption. The other half of magnesium is localized in soft tissue, with <1% present in the blood.

$Mg^{2+}$ is involved in practically every major metabolic and biochemical process within the cell and is responsible for numerous functions in the body, including bone development, neuromuscular function, signaling pathways, energy storage and transfer, glucose, lipid and protein metabolism, DNA and RNA stability, and cell proliferation.

Over 600 enzymes with $Mg^{2+}$ as a cofactor are currently listed by the enzymatic databases, while an additional 200 are listed in which $Mg^{2+}$ may act as an activator [38,39]. More specifically, it mainly interacts directly with the substrate, rather than acting as a real cofactor.

The involvement of magnesium in many cellular processes (Figure 1) is here briefly summarized and then detailed in the following paragraphs, explaining why habitually low intakes of magnesium induce changes in biochemical pathways that can lead to an increased risk of illness over time.

- The complex $MgATP^{2-}$ is required for the activity of many enzymes. In general, $Mg^{2+}$ acts as a cofactor in all reactions involving the utilization and transfer of ATP, including cellular responses to growth factors and cell proliferation, being thus implicated in virtually every process in the cells. $Mg^{2+}$ availability is a critical issue for carbohydrate metabolism, which may explain its role in diabetes mellitus type 2 [40];
- $Mg^{2+}$ is necessary for the correct structure and activity of DNA and RNA polymerases [41,42]. In addition, topoisomerases, helicases, exonucleases, and large groups of ATPases require $Mg^{2+}$ for their activity, therefore $Mg^{2+}$ is essential in DNA replication, RNA transcription, and protein formation, being thus involved in the control of cell proliferation. Moreover, $Mg^{2+}$ is crucial to maintain genomic and genetic stability, stabilizing the natural DNA conformation and acting as a cofactor for almost every enzyme involved in nucleotide and base excision repair and mismatch repair. Given these effects, low $Mg^{2+}$ availability can be involved in the development of cancer [2];
- Serum $Mg^{2+}$ concentrations are strongly related to bone metabolism; bone surface $Mg^{2+}$ is constantly exchanged with blood $Mg^{2+}$ [43,44]. Furthermore, $Mg^{2+}$ induces osteoblast proliferation [45] therefore, the consequences of $Mg^{2+}$ deficiency are accelerated bone loss and a decline in bone formation [46];
- $Mg^{2+}$ participates in controlling the activity of some ionic channels in many tissues. Its mechanism of action relies on either direct interaction with the channel, or an indirect

modification of channel function through other proteins (e.g., enzymes or G proteins), or via membrane surface charges and phospholipids [47]. Furthermore, $Mg^{2+}$ acts as a physiological $Ca^{2+}$ antagonist within cells, since it can compete with $Ca^{2+}$ for binding sites in proteins and $Ca^{2+}$ transporters [48]. These abilities are involved in the observed effect of magnesium on the cardiovascular system, muscle, and brain;
- Neuronal magnesium concentrations downregulate the excitability of the N-methyl-D-aspartate (NMDA) receptor, which is essential for excitatory synaptic transmission and neuronal plasticity in learning and memory [49]. Magnesium blocks the calcium channel in the NMDA receptor and must be removed for glutamatergic excitatory signaling. Low serum $Mg^{2+}$ levels increase NMDA receptor activity thus enhancing $Ca^{2+}$ and $Na^+$ influx and neuronal excitability. For these reasons, a deficiency of $Mg^{2+}$ has been hypothesized in many neurological disorders, such as migraine, chronic pain, epilepsy, Alzheimer's, Parkinson's, and stroke, as well as anxiety and depression [50].

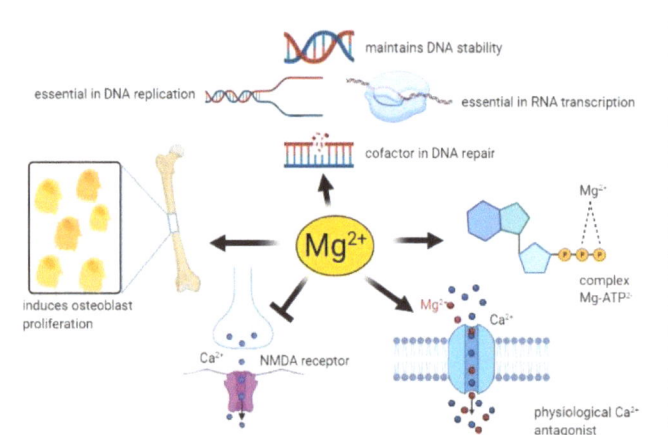

**Figure 1.** The biochemical involvement of magnesium in many cellular processes. This image is created with BioRender.com.

## 2.1. Magnesium as Enzymatic Cofactor

As stated above, magnesium is a cofactor for over 600 enzymes and an activator for an additional 200 enzymes [51]. Given the ability of $Mg^{2+}$ to bind inorganic phosphate, ATP, phosphocreatine, and other phosphometabolites make complex with magnesium, with important consequences for many metabolic reactions, especially those related to carbohydrate metabolism and cellular bioenergetics. The binding between ATP and $Mg^{2+}$ results in an adequate conformation that allow to weaken the terminal O–P bond of ATP, thereby facilitating the transfer of phosphate [14,52].

The paramount importance of magnesium in the glycolytic pathway [53] and mitochondrial synthesis of ATP [54] has been known for a long time. Many of the glycolytic enzymes are sensitive to magnesium, whose principal function is to facilitate the transfer of high energy phosphate. Thus, hexokinase, phosphofructokinase, phosphoglycerate kinase, and pyruvate kinase work in this manner, while aldolase and enolase require $Mg^{2+}$ for their stability and activity [53].

In mitochondria, the activity of three important dehydrogenases is dependent on $Mg^{2+}$. Isocitrate dehydrogenase is directly stimulated by the $Mg^{2+}$-isocitrate complex [55], α-ketoglutarate dehydrogenase complex by free $Mg^{2+}$ [56], pyruvate dehydrogenase is indirectly stimulated via the stimulatory effect of $Mg^{2+}$ on pyruvate dehydrogenase phosphatase, which dephosphorylates and thus activates the pyruvate dehydrogenase complex [57]. Furthermore, $Mg^{2+}$ has been shown to be the activator of ATP synthesis by mitochondrial $F_0/F_1$-ATPase [58]. Furthermore, it has been demonstrated that $Mg^{2+}$ concentration is low in the brain of patients affected by mitochondrial cytopathies and that supplementation with Coenzyme $Q_{10}$ has improved oxidative phosphorylation and the cytosolic magnesium level [59].

The enzyme creatine kinase, catalyzing the reversible reaction between creatine phosphate and ADP to form creatine and ATP, is strongly influenced by the concentration of free $Mg^{2+}$. This enzyme can synthesize ATP when the muscle or heart are subjected to a heavy workload and can be localized both in cytosol and mitochondrion [53].

In the liver, $Mg^{2+}$ is an important regulator of gluconeogenic enzymes, including glucose-6-phosphatase and phosphoenolpyruvate carboxykinase [40].

The fundamental role of $Mg^{2+}$ in glycolysis/gluconeogenesis, pentose phosphate pathway, and Krebs cycle was also demonstrated in a metabolomic analysis of the liver of rats fed a magnesium-deficient diet. Results of this study show that hepatic contents of glucose-6-phosphate, citric, fumaric, and malic acids were significantly reduced in animals fed a magnesium-deficient diet, and also fructose-6-phosphate and succinic acid were numerically lower. By contrast, the mRNA levels of the related enzymes, such as glucokinase, glucose-6-phosphatase, glucose-6-phosphate dehydrogenase, and phosphoenolpyruvate carboxykinase were not significantly different between control and treated animals. These data demonstrated that, in the liver, the metabolite content associated to glucose metabolism was altered by a magnesium deficiency [60], therefore a balanced magnesium status seems an important requisite for adequate carbohydrate metabolism [61].

In this regard, magnesium deficiency has been correlated to type 2 diabetes mellitus, metabolic syndrome, and insulin resistance [62]. As well as all other protein kinases, the tyrosine kinase activity of the β-subunit of the insulin receptor is dependent on magnesium concentration, therefore a magnesium deficiency may result in an impaired insulin signal. In other words, the lower the basal magnesium, the greater the amount of insulin required to metabolize the same glucose load, indicating decreased insulin sensitivity [62,63]. A recent study demonstrates that magnesium improves glucose consumption and glucose tolerance through two main mechanisms: stimulation of the GLUT4 gene expression and translocation of this glucose transporter to the plasma membrane, and suppression of the glucagon effect and gluconeogenesis pathway in the liver and muscle [64]. Table 1 summarizes the principal enzymes involved in carbohydrate metabolism which require magnesium for their proper action.

Table 1. Enzymes requiring magnesium.

| LOCALIZATION | ENZYME | Mg-ATP$^{2-}$ | FREE $Mg^{2+}$ | REF |
|---|---|---|---|---|
| Cytosol: glycolytic pathway | Hexokinase |  | - | [53] |
|  | Phosphofructokinase |  | - |  |
|  | Phosphoglycerate kinase |  | - |  |
|  | Pyruvate kinase |  | - |  |
|  | Aldolase | - |  |  |
|  | Enolase | - |  |  |
| Mitochondrion | Pyruvate dehydrogenase phosphatase | - |  | [65] |
|  | Isocitrate dehydrogenase | - |  | [55] |
|  | α-Ketoglutarate dehydrogenase | - |  | [56] |
|  | $F_0/F_1$-ATPase | - |  | [58] |
| Muscle cytosol/Heart mitochondrion | Creatine kinase |  | - | [53] |
| Liver, cytosol | Phosphoenolpyruvate carboxykinase | - |  | [40] |
|  | Glucose-6-phosphatase | - |  |  |
| β-subunit of the insulin receptor | Receptor tyrosine kinase activity |  | - | [62,63] |

## 2.2. Magnesium and Nucleic Acids

Divalent cations are widely known to affect the structure of duplex DNA. The possibility of hydrogen bond interactions between cations and DNA is greater for divalent ions, owing to their hydration properties. $Mg^{2+}$ ion attracts negatively-charged DNA phosphate groups and with its six-coordinated water molecules forms hydrogen bonds, locally reducing the DNA negative charge density and stabilizing its structure and natural conformation [66]. This magnesium effect may be named the 'protective effect'. However, when magnesium binds 'covalently' to DNA, it forms a bound coordination complex, causing a local distortion of the double helix, which may lead to the destruction of the cell. This effect occurs at higher concentrations of magnesium, therefore, the maintenance of the cellular $Mg^{2+}$ concentration within the physiological range is essential for DNA stability [66].

As stated above, $Mg^{2+}$ also has an important role in DNA repair mechanisms. Several enzymes of these systems depend on magnesium, some of them are common to different repair systems, such as DNA polymerase beta, DNA ligases, and DNA endonucleases, while others are specific to one repairing mechanism [67].

All DNA polymerases catalyze the same nucleotidyl-transfer reaction, forming and breaking phosphodiester bonds, and all DNA polymerases contain two-three conserved carboxylates in the catalytic center. The $Mg^{2+}$ ions neutralize the charge of the catalytic carboxylates and triphosphates of dNTP, thus facilitating the alignment of the substrates for the chemical reaction [68]. Moreover, other important nuclear enzyme activities involved in DNA replication depend on magnesium, like topoisomerases, helicases, and exonucleases. Thus, it may be concluded that magnesium is essential for DNA structure, duplication and repair, as well as maintaining genomic and genetic stability. Hence, magnesium deficiency could favor DNA mutations leading to the initiation of carcinogenesis [67].

As far as RNA is concerned, the role of $Mg^{2+}$ in restoring denatured tRNA molecules was reported in 1966 [69]. Later, the importance of $Mg^{2+}$ binding for the tRNA tertiary structure has been debated over the years, questioning the prevalent action of nonspecific diffuse binding of $Mg^{2+}$ (or other divalent cations) or a specific $Mg^{2+}$ interaction [2,70,71]. A very recent paper demonstrated that magnesium ions are required by tRNAPhe for proper recognition of UUC/UUU codons during ribosomal interactions with tRNA [72].

It has been recently demonstrated that the biological environment and biomolecules can stimulate RNA functions [73]. To this regard, amino acids and nucleotides are abundant cellular metabolites, and it has been shown that cellular concentrations of amino acid-chelated $Mg^{2+}$ stimulate RNA folding and catalysis [74]. The authors hypothesized that the amino acid-bound $Mg^{2+}$ ion may interact with the RNA, "sharing" the $Mg^{2+}$ ion with it, thus decreasing the folding free energy of RNA, stabilizing the RNA structure, and promoting RNA high catalytic activity [74]. Furthermore, the same authors demonstrated that also nucleotide diphosphate-chelated $Mg^{2+}$ promotes RNA catalysis much like amino acid-bound $Mg^{2+}$ [75]. The authors observed that the stimulatory effects of $Mg^{2+}$ diphosphate-containing metabolites is general for RNA and DNA enzymes [75].

Moreover, RNA synthesis require two $Mg^{2+}$ ions for the catalysis [68], thus magnesium is also involved in RNA transcription. The aspartate triad, three aspartate residues which reside at the center of the active site, is absolutely conserved among all DNA-dependent multi-subunit RNA polymerase. These amino acids chelate the first of two essential magnesium ions required for catalysis, the second ion is brought into the active center bound to the incoming nucleotide substrate [76].

Traduction is also highly dependent on intracellular magnesium concentration, it has been suggested that, upon growth factor binding to their receptors, $Mg^{2+}$ enters the cell and the increased cytosolic $Mg^{2+}$ level contributes to ribosomal activity and protein synthesis. Moreover, it should not be forgotten that the activities of receptor and non-receptors tyrosine kinases have an obligatory requirement for $Mg^{2+}$, thus involving magnesium in the signaling pathways of growth factors, such as VEGF, EGF, PDGF, and so on. Therefore, $Mg^{2+}$ is an important factor in controlling cell proliferation [77].

Given these considerations, it is reasonable to hypothesize the involvement of magnesium in tumor growth, owing to the ability of $Mg^{2+}$ to regulate several cancer-associated enzymes, particularly those involved in glycolysis—the preferred pathway used by neoplastic cells to produce energy—and DNA repair. On the other hand, low dietary $Mg^{2+}$ intake has been associated with the risk of several types of cancers, extensively reviewed by Castiglioni [78].

2.3. Magnesium and Bone Metabolism

It has been observed that bones belonging to magnesium-deficient animals are brittle and fragile, microfractures of the trabeculae can be detected and mechanical properties are severely impaired [79]. In general, all experimental data obtained from animal studies indicate that the reduced dietary intake of magnesium is a risk factor for osteoporosis through a variety of different mechanisms [80]. $Mg^{2+}$ increases the solubility of the minerals, which constitute the hydroxyapatite crystals, such as Pi and $Ca^{2+}$, thereby acting on crystal size and formation [81]. It has been demonstrated that osteoporotic women with a magnesium deficiency have larger and better organized crystals in trabecular bone than controls, making bone more susceptible to fractures [82].

Besides its direct effect on hydroxyapatite crystals, other mechanisms are involved in the structural role of magnesium and bone health. Of great importance is the complex interplay existing between magnesium and vitamin D: vitamin D stimulates intestinal magnesium absorption [83] and magnesium deficiency reduces the levels of $1,25(OH)_2D_3$, thus being implicated in magnesium-dependent vitamin D-resistant rickets [84]. In fact, magnesium is required for the activity of hepatic 25-hydroxylase and renal 1α-hydroxylase [85], both crucial to convert $25(OH)D$ into its biologically active form $1,25(OH)_2D_3$, and also to facilitate the transfer of vitamin D to target tissues through the vitamin D binding protein [86]. On the other hand, magnesium is also involved in the inactivation of vitamin D, being required for the activity of the renal 24-hydroxylase, to form $24,25(OH)_2D$ [87].

Another complex relationship occurs between magnesium and parathyroid hormone (PTH), and thus indirectly between magnesium and calcium. PTH is secreted by parathyroid glands in response to a low level of serum calcium and either high or low PTH levels can result in calcium dysregulation and bone disease. Particularly, a chronic sustained activation of the receptor by PTH, as observed in primary hyperparathyroidism, exerts catabolic effects on bone and leads to an enhanced bone turnover, resulting in bone loss and an increased fracture risk [88]. Physiological serum calcium level negatively regulates PTH secretion, but also the serum level of PTH and magnesium are co-dependent. Low levels of magnesium stimulate the secretion of PTH, but very low magnesium concentration inhibits PTH secretion [89]. Magnesium can also reduce PTH secretion at low calcium concentrations [90]. Interestingly, magnesium is also required for the sensitivity of the target organs to PTH signal [21] and impaired peripheral response to PTH leads to a low serum concentration of vitamin D [91].

Additionally, it has been demonstrated that magnesium deficiency in animal models induces a clinically inflammatory syndrome, characterized by leukocyte and macrophage activation, the release of inflammatory cytokines, and excessive production of free radicals. Since magnesium acts as a natural calcium antagonist, the molecular basis for inflammatory response could be the result of modulation of intracellular calcium concentration [92]. Many studies have demonstrated that in humans, a moderate or subclinical magnesium deficiency can induce chronic low-grade inflammation or exacerbate inflammatory stress caused by other factors [93]. This low-grade inflammation increases the secretion of pro-inflammatory cytokines, which stimulate the resorption of bone by the induction of the differentiation of osteoclasts from their precursors [94]. The ability of $Mg^{2+}$ to decrease the inflammatory response and oxidative stress, as well as improving lung inflammation, possibly by inhibiting IL-6 pathway, NF-κB pathway, and L-type calcium channels [95], has raised the hypothesis of a possible magnesium supplementation in the prevention and treatment of COVID-19 patients, as suggested in the recent papers by Tang [96] and Iotti [97].

### 2.4. Magnesium, Calcium, and Cardiovascular System

As previously mentioned, intricate interactions between magnesium and calcium exist, and it has long been known that calcium intake affects magnesium retention and vice versa. As stated above, the $Mg^{2+}/Ca^{2+}$ ratio is very important for the activity of $Ca^{2+}$-ATPases and other $Ca^{2+}$ transporting proteins [48], thus small changes in the $Mg^{2+}$ availability within the cell may cause perturbed $Ca^{2+}$ signaling.

Magnesium plays an important role in the cardiovascular system, influencing myocardial metabolism, $Ca^{2+}$ homeostasis, and endothelium-dependent vasodilation. It also acts as an antihypertensive, antidysrhythmic, anti-inflammatory, and anticoagulant agent. In myocardium, the opening of L-type $Ca^{2+}$ channels produces a long-lasting $Ca^{2+}$ current, corresponding to the second phase of the cardiac action potential. $Mg^{2+}$ inhibits these channels, preventing $Ca^{2+}$ overload and cell toxicity and thus exerting a myocardial protective effect [98]. Two general mechanisms could explain how $Mg^{2+}$ regulates $Ca^{2+}$ fluxes through L-type channels: alteration of ion permeation and/or modulation of channel gating properties. L-type $Ca^{2+}$ channel gating is, in turn, regulated by membrane potential, cytosolic $Ca^{2+}$ concentration, and channel phosphorylation. It has been demonstrated that the effect of $Mg^{2+}$ is dependent on the channel's phosphorylation state, since phosphatase treatment decreases the inhibitory effect of $Mg^{2+}$ [99]. Furthermore, $Mg^{2+}$ is necessary for $Na^+/K^+$-ATPase, which is responsible for the active transport of $K^+$ intracellularly during the action potential duration. $Mg^{2+}$ is also involved in regulating the $K^+$ influx through different $K^+$ channels [100]. The modulation of cardiac action potential can explain the antidysrhythmic action of $Mg^{2+}$: Its infusion provokes the slowing of atrioventricular nodal conduction, and also determines the prolongation of PR interval and QRS duration in the electrocardiogram [101]. The effect of $Mg^{2+}$ on cardiomyocytes also depends on other mechanisms, including the ability of $Mg^{2+}$ to compete with $Ca^{2+}$ for binding sites in proteins, such as calmodulin, troponin C, and parvalbumin [48], to act as substrate in a complex with ATP for cardiac $Ca^{2+}$-ATPases, and to alter the affinity of $Na^+$-$Ca^{2+}$ exchanger [2]. In summary, tight regulation of $Mg^{2+}$ concentration in myocytes is necessary for optimal cardiac function, indeed hypomagnesemia can impact physiological activity, leading to cardiovascular diseases [102].

A relationship between $Mg^{2+}$ levels and blood pressure has been established [2,21,103]. Many different mechanisms are involved in the $Mg^{2+}$ vasodilation effect, among which the ability of $Mg^{2+}$ to act as a natural calcium channel blocker [104] and to upregulate the endothelial nitric oxide synthase, thus increasing nitric oxide (NO) release [102]. Additionally, $Mg^{2+}$ is able to increase the production of prostacyclin ($PGI_2$), a platelet inhibiting factor [2], it is involved in a step of the synthesis of prostaglandin $E_1$ ($PGE_1$), a vasodilator and platelet inhibitor agent [104], and owing to its anti-inflammatory role, $Mg^{2+}$ results in an improved lipid profile, reduced free oxygen radicals, and improved endothelial function [2].

Finally, the antagonistic action of $Mg^{2+}$ on calcium channels and $Ca^{2+}$-binding proteins has been related to muscle cramps and spasms observed as recurrent symptoms in patients with hypomagnesia [105]. In this regard, conflicting results have been published about magnesium supplementation in clinically cramp prophylaxis [106].

### 2.5. Magnesium, Calcium, and Brain

One of the main neurological functions of magnesium is due to magnesium's interaction with the N-methyl-D-aspartate (NMDA) receptor. NMDA receptors are activated upon glutamate binding and mediate the influx of $Ca^{2+}$ and $Na^+$ ions and the efflux of $K^+$ ions. Glutamate is the major excitatory neurotransmitter in the brain, acting by binding to various transporters and receptors, among which are the cation channels AMPA (2-amino-3-(3-hydroxy-5-methyl-isoxazol-4-yl) propanoic acid) and NMDA. The NMDA receptor family includes a wide variety of receptor subtypes, both diheteromers and triheteromers, each with distinct biophysical, pharmacological, and signaling properties and different locations between brain regions [107]. Going into the detail of $Mg^{2+}$ mechanism of action, glutamate from presynaptic neuron binds both AMPA and NMDA receptors on the postsynaptic neuron, but at a normal membrane potential, $Mg^{2+}$ ions block NMDA receptors, allowing only AMPA receptor activation to occur. When the membrane potential rises, NMDA receptors are unlocked, and they facilitate the cation influx upon glutamate binding. When $Mg^{2+}$ concentration is reduced, less NMDA channels are blocked and this increased excitatory postsynaptic potential causes hyperexcitability of the neurons, which can lead to oxidative stress and neuronal cell death [2]. Another mechanism involving $Mg^{2+}$ in neuron hyperexcitability is its ability to regulate the function of the inhibitory GABA receptors [108]. In case of low $Mg^{2+}$ concentration, the membrane potential will be higher, thus relieving $Mg^{2+}$ block of the NMDA receptors and contributing to hyperexcitability of the neurons.

Noteworthy, abnormal glutamatergic neurotransmission has been implicated in many neurological and psychiatric disorders, including migraine, chronic pain, epilepsy, Alzheimer's, Parkinson's, depression, and anxiety [109].

Furthermore, a NMDA receptor-dependent mechanism is involved in the $Mg^{2+}$ dependent enhancement of the activity of nitric oxide synthases, causing the NO release, which has multiple functions in the brain, such as vasodilation, regulation of gene transcription, and neurotransmitter release [110]. $Mg^{2+}$ also increases the expression and secretion of calcitonin gene-related peptide (CGRP), which has a vasodilatory effect [111], while $Mg^{2+}$ deficiency increases the release of substance P, a neuroinflammatory agent, which stimulates the secretion of inflammatory mediators [112]. Hence, $Mg^{2+}$ has also a role in the regulation of neuropeptides release and oxidative stress, contributing to the maintenance of a healthy neurological function.

### 3. Nutritional Strategies to Avoid Magnesium Deficiencies

Magnesium is an essential nutrient for living organisms therefore it must be supplied regularly from our diet to reach the recommended intake, preventing deficiency. Consequently, it is important not only to identify the possible sources of magnesium, but also to assess the bioavailability and factors that can influence its absorption and elimination.

*3.1. Recommended Intake and Categories of People That Risk Inadequate Magnesium Intake*

Intake recommendations for magnesium and other nutrients have been provided by the World Health Organization (WHO) and the Food and Agriculture Organization (FAO), by the American National Academy of Medicine (NAM), previously called the Institute of Medicine (IoM), and by the European Food Safety Agency (EFSA). According to the development of scientific knowledge about the roles played by nutrients in human health, the Food and Nutrition Board at the National Academy, with the Health Canada partnership, has updated what used to be known as Recommended Dietary Allowances (RDAs) and renamed a new version of these guidelines as "Dietary Reference Intakes" (DRIs) [3,113,114]. Analogously, the European Food Safety Agency (EFSA) provides Dietary Reference Values (DRVs) [115,116]. LARN ("Livelli di Assunzione di Riferimento di Nutrienti ed energia per la popolazione italiana" corresponding to "Recommended Levels of Nutrients and Energy Intakes") are the last version of the Italian DRVs, more recently released in 2014 by the Italian Society of Human Nutrition and periodically updated by the Commission of the Human Nutrition Society (SINU) and by the Ministry of Agricultural, Food, and Forestry Policies (CREA), in line with the EFSA technical reports [16,117–119].

These values, which vary according to sex and specific-age ranges, can be used to identify nutrient intakes that are relevant for diet planning in individuals as well as in the general population and include:

- The population reference intakes (PRIs), which refer to the level of nutrient intake that is adequate for the majority of people in a population group;

- The average requirements (ARs), which refer to the intake level that is adequate to meet the physiological requirements of 50% of healthy individuals. This parameter is usually taken into consideration not only to assess the nutrient intakes of groups of people and to plan nutritionally adequate diets for them but also to assess the nutrient intakes of individuals.

In case of insufficient scientific evidence to estimate the AR and/or PRI, adequate intake (AI) is established by estimating the intake of an apparently healthy population group that is assumed to have adequate intake.

- Adequate intake (AI), therefore, refers to the intake assumed to ensure nutritional adequacy;
- Tolerable upper intake level (UL): Maximum daily intake which is considered to be safe/without adverse health effects on the totality of the considered population.

The current intake recommendations for magnesium are reported in Table 2.

**Table 2.** Magnesium intake recommendations expressed in terms of: Population Reference Intake (PRI), Average Requirement (AR), Recommended Dietary Allowance (RDAs)—Dietary Reference Intakes (DRIs), Dietary Reference Values (DRVs)—Adequate Intake (AI), "Livelli di Assunzione di Riferimento di Nutrienti ed energia per la popolazione italiana" (LARN) and **tolerable** Upper intake Level (UL).

| Life Stage | PRI (mg) | AR (mg) | UL * (mg) | RDA-DRI (mg) | DRV-AI (mg) | LARN (mg) |
|---|---|---|---|---|---|---|
| Birth to 6 months | - |  | Nd | 30 |  |  |
| Infants 7–12 months | 80 | Nd | Nd | 75 | 80 | 80 |
| Children 1–3 years | 80 | 65 | 250 | 80 | 170 | 80 |
| Children 4–6 years | 100 | 85 | 250 | 130 | 230 | 100 |
| Children 7–10 years | 150 | 130 | 250 | 240 | 230 | 150 |
| Teen boys 11–18 years | 240 | 170–200 | 250 | 410 | 300 | 240 |
| Teen girls 11–18 years | 240 | 170–200 | 250 | 360 | 250 | 240 |
| Men | 240 | 170 | 250 | 400–420 | 350 | 240 |
| Women | 240 | 170 | 250 | 310–320 | 300 | 240 |
| Pregnant | 240 | 170 | 250 | 350–400 | 300 | 240 |
| Breastfeeding | 240 | 170 | 250 | 310–360 | 300 | 240 |

* the UL value refers to the magnesium taken in pharmaceutical or supplement form, in addition to magnesium content already present in the diet.

Magnesium adequate intake for infants from birth to 12 months is determined by considering the mean intake of magnesium in healthy, breastfed infants, with added solid foods during the first 7–12 months of life.

The gradual transition from an exclusively milk-based diet to one including a different range of family foods that occurs during the 6–24 months of life, requires a consumption of a healthy and balanced diet. Although an adequate intake of micronutrients is clearly critical during this sensitive period of growth and development [120], insufficient intake of some micronutrients are observed also in industrialized countries. Regarding magnesium, recommendations on infant requirements from WHO/FAO, American National Academy of Medicine, and EFSA were based on estimations of intake [116,121,122]. There is insufficient information on either magnesium or phosphorus to establish a UL for infants and young children (0–3 years old).

Considering all the evidence available from prospective observational studies and balance studies, the EFSA Panel, in the last version of Scientific Opinion on Dietary Reference Values (DRV) for magnesium (2015) [116], decided to set an AI based on observed intakes in nine European Union countries (Italy, Finland, France, Germany, Ireland, Latvia, the Netherlands, Sweden, and the UK). The panel proposed to set AIs according to sex, for adults of all ages. Considering the distribution of observed average intakes, the panel proposed AI values according to sex and ages, as reported in Table 2.

There are several international guidelines that give advice for the general population in order to maintain a healthy status therefore, when specifying or describing an advisory amount of macro-micronutrients it is important to indicate what system and/or updated sources has been used because the values are similar but not always the same. For example, magnesium requirements (RDA) for adults (18–29 years) in Japan are 340–370 and 270–290 mg/die for male and female, respectively [115,123].

During pregnancy and lactation, the correct uptake of magnesium is particularly important as evidenced by Durlach J. [124]. Nevertheless, since pregnancy causes only a small increase in magnesium requirement, which is probably satisfied by adaptive physiological mechanisms, the EFSA panel considers the same AI for nonpregnant and pregnant women. Data from the Cochrane Database of Systematic Reviews suggest that there is not enough high-quality evidence to show that dietary magnesium supplementation during pregnancy is beneficial, as deeply analyzed and reported by Makrides and Colleagues [125].

Analogously, taking into consideration that 25 mg/day is secreted with breast milk during the first six months of exclusive breastfeeding and that there is the strategy adaptation of magnesium metabolism, at the level of both absorption and elimination, the panel considers the same AI both for non-lactating and lactating women. The Italian LARN are in agreement with these considerations, although the values are lower than those suggested by the EFSA.

Athletes are recommended to consume higher amounts of potassium and magnesium. In particular, 420 mg/die for male and 320 mg/die for female, considering 19–50 years as an age range [18]. Usually, renal elimination involves approximately 100 mg of $Mg^{2+}$ per die, whereas the losses via sweat are generally low. However, during intense exercise, these losses can increase considerably. Moreover, since $Mg^{2+}$ activates the enzymes involved in protein synthesis, it is involved in ATP metabolism and $Mg^{2+}$ serum levels decrease with exercise, magnesium supplementation could therefore improve energy metabolism and ATP availability. Magnesium supplementation generally does not affect an athlete's performance, unless there is a deficiency state [126–130], although this result seems to be disproved by a recent double-blind study carried out by Reno A.M. and colleagues [131]. This double-blind study, although with several criticisms (e.g., small number of subjects and absence of initial magnesium status detection), shows that magnesium supplementation (vs. Pla) significantly reduced muscle soreness and improved perceived recovery. As far as the role of magnesium against skeletal muscle cramps is concerned, a recent update of a Cochrane Review that assessed evaluating magnesium for exercise-associated muscle cramps or disease-state-associated muscle cramps was found. Moreover, magnesium supplementation did not provide clinically meaningful cramp prophylaxis to older adults experiencing skeletal muscle cramps. Further research is also needed to also evaluate the protective effect of magnesium [106].

Besides athletes, the following groups of people suffer more frequently from magnesium deficiency:

- Older people absorb less magnesium from the gut and lose more magnesium because of an increased renal excretion. Chronic magnesium deficiency is indeed common in the elderly, usually due to a decrease both in diet assumption and intestinal absorption, and it is probably exacerbated by estrogen deficit, which occurs in aging women and men and cause hypermagnesuria [132]. In a very recent and comprehensive review [34], Lo Piano and colleagues highlight the risk and consequences of the reduce intake and absorption of magnesium by elderly people;
- People affected by gastrointestinal diseases with consequent general malabsorption, such as Crohn's disease [117,133–140], inflammatory bowel diseases [135,138,140,141], and celiac disease [142–150]. In particular, besides the absorption inefficiency due to celiac disease, a gluten free-diet was found to be poor in fiber and micronutrients,

such as magnesium [151,152]. Therefore, people suffering from celiac disease are a typical example of subjects particularly susceptible to magnesium deficiency as they are simultaneously exposed to two risk factors;
- People affected by type 2 diabetes, although it still remains unclear if magnesium deficiency represents a cause or a consequences of this pathology [3,21,135,153–156];
- People who used to drink alcohol/alcoholics or are affected by long-term alcoholism [3,157–160] and are therefore affected by intestinal malabsorption. Spirits (such as brandy, cognac, gin, rum, vodka, and whisky) do not contain significant traces of magnesium. Moreover, alcohol consumption, such as wine and beer during meals, is acceptable and is also included in the Mediterranean food pyramid (2–4 units/day), however, despite beer and wine having magnesium levels that range from 30–250 mg/L and fermented apple ciders ranging from 10–50 mg/L, such beverages cannot be considered as a reliable source of magnesium because they cause magnesiuresis and can have a laxative effect, with consequent problems on bioavailability and absorption. Ethanol is indeed magnesiuretic by causing proximal tubular dysfunction and increasing urinary magnesium loss, and its effect is rapid and common in people with an already negative magnesium balance [6,161];
- People under treatment with drugs (e.g., diuretics, proton pump inhibitors, tacrolimus, an immunosuppressor, chemotherapeutic agents, and some phosphate-based drugs) [6].

However, it is important to point out that most apparently healthy people risk an insufficient magnesium intake due to a decreased presence of this metal in the modern Western diet characterized by a wide use of demineralized water, processed foods, and agricultural practices that use soil deficient in magnesium for growing food [162–164], as discussed in the next paragraph and reported for the Spanish population, where about 75% of the population revealed intakes below 80% of the national and European recommended daily intakes [165]. Accordingly, data on people's dietary habits still reveal that intakes of magnesium are lower than the recommended amounts either in the United States or in Europe. Epidemiological studies have shown that people consuming Western-type diets introduce an insufficient amount of micronutrients and in particular, a quantity of magnesium that is <30–50% of the RDA. Accordingly, the magnesium dietary intakes in the United States have been decreasing over the last 100 years from about 500 mg/day to 175–225 mg/day [21], and a general similar decrease in magnesium daily uptake in people fed a Western diet is reported in a recent and interesting review by Cazzola and Colleagues [164].

*3.2. Magnesium Food Content and Bioavailability*

Magnesium is considered widely distributed in foods, although the amount of magnesium contained in food is influenced by various factors including the soil and water used to irrigate, fertilizers, conservation, and also refining, processing, and cooking methods. In general, seeds, legumes, nuts (almonds, cashews, Brazil nuts, and peanuts), whole grain breads, and cereals (brown rice, millet), some fruits, and cocoa are considered good sources of magnesium. Nevertheless, acidic, light, and sandy soil is usually deficient in magnesium content. Moreover, agricultural techniques, such the use of potassium and ammonium at high concentration in fertilizers lead to magnesium depletion in food [1] and a recent meta-analysis on the effects of magnesium fertilization has been recently published [166].

Green leafy vegetables are frequently counted among the food rich in magnesium according to the hypothesis that chlorophyll-bound magnesium may represent important nutritional sources of magnesium. This hypothesis relies on what is known about iron, which is similarly bound in the porphyrin ring of heme, and is absorbed to a greater extent than non-heme iron. This concept is incorrect for many reasons: the acidic pH of the gastric juice induces a fast and irreversible degradation of the chlorophylls to their corresponding pheophytins and the theoretical amount of chlorophyll-bound magnesium presents in chlorophyll a is 2.72% and chlorophyll b is 2.68% of total mass. In leafy green vegetables, such as lettuce and spinach, chlorophyll-bound magnesium represents 2.5% to 10.5% of total magnesium, whereas other common green vegetables, pulses, and fruits contain <1% chlorophyll-bound magnesium. Bohn and colleagues in the conclusion section of a paper stated that "chlorophyll bound magnesium contributes a small and nutritionally insignificant part of total magnesium intake in industrialized countries" [167].

As previously mentioned, some methods of food processing, such as boiling vegetables and refining grains with the consequent removal of germ and bran, cause a substantially lower magnesium content. The loss of magnesium during food refining is considerable: white flour (−82%), polished rice (−83%), starch (−97%), and white sugar (−99%). Since 1968, a 20% decrease of magnesium content in wheat has occurred, probably due to acidic soil, yield dilution, and unbalanced crop fertilization (high levels of nitrogen, phosphorus, and potassium) [168]. The hydrosphere (i.e., sea and oceans) is the most abundant source of biologically available magnesium (about 55 mmol/L). Unrefined sea salt is indeed rich in magnesium, which represents approximately 12% of sodium mass, although refined salt, usually present in food and added for cooking either at industrial or domestic levels, lacks this mineral [6,18]. Therefore, the Western diet, characterized by easy-to-cook meals and fast food such as refined and processed food, junk food, and the near absence of legumes and seeds, predisposes apparently healthy people to magnesium deficiency.

It is important to point out that the quantification of the nutrient content in foods must be critically analyzed because nutrient bioavailability and the amount of nutrients in food portions should also be taken into consideration. Intrinsic and extrinsic factors can indeed notably affect the bioavailability of nutrients present in food- and non-food sources of nutrients [169]. Moreover, the real potential intake of the nutrient by the assumption of a determined food in a healthy and balanced diet needs indeed to be considered.

Consequently, these and other considerations must be taken into account during the consultation of nutritional tables. Selected nutritional sources of magnesium are listed in Table 3. According to data from 13 dietary surveys in nine European Union (EU) countries before Brexit, dietary intake of magnesium was estimated by EFSA, considering food consumption data from the EFSA Comprehensive European Food Consumption Database and composition data from the EFSA Food Composition Database (https://www.efsa.europa.eu/en/microstrategy/food-composition-data, accessed on February 2021)). CREA provides a list arranged by decreasing nutrient content (https://www.crea.gov.it/ accessed on February 2021), similarly to EFSA. The U.S. Department of Agriculture's (USDA's) Food Data Central (https://fdc.nal.usda.gov/ accessed on February 2021) lists the nutrient content of many foods and presents a comprehensive list of foods containing magnesium according to measure/portion.

Pseudo cereal and whole-grain wheat, oat, and millet were shown to be great sources of magnesium even if the cooking methods influence the real magnesium assumption per portion. For example, 100 g of wholemeal pasta cooked in water contain 42 mg of magnesium. The introduction of unrefined whole grains, nuts, legumes, and unrefined dark chocolate in the daily diet is useful to reaching a satisfactory amount of magnesium, because they represent good dietary sources of magnesium [18]. Among fruit, a high content of magnesium is found in dried apricot and dried bananas even if the normal serving of dried fruit (30 g) contains a similar amount of magnesium to a serving (100–150 g) of some fresh fruit (e.g., avocado, blackberries, prickly pears, chokecherries) [170,171].

According to the U.S. Department of Agriculture's National Nutrient Database, Magnesium content in cocoa is at significant levels (2–4 mg/g dry powder). Therefore, a 40 g portion of 70–80%-cocoa dark chocolate would contain ≈40 mg of magnesium, enough to satisfy about ~10% of the recommended daily allowance (300–400 mg magnesium/day in adults) [172].

**Table 3.** Magnesium content in Food according to the EFSA Comprehensive European Food Consumption Database (CREA) and U.S. Department of Agriculture, Agricultural Research Service—Food Data Central.

| Food | EFSA (mg/100 g) | CREA (mg/100 g) | USDA (mg/Measure) | Measure and Weight |
|---|---|---|---|---|
| Wheat/Cereal bran | 451 | 550 | 354 | 1 cup, 50 g |
| Pumpkin and squash seed, dried | 429 | 592 | 764 | 1 cup, 46 g |
| Cocoa powder | 545 | 499 | 29 | 1 ts [1], 6 g |
| Sunflower seeds dried | 346 | n.a [2] | 173 | 1 cup, 130 g |
| Wheat germ | 276 | 255 | 275 | 1 cup, 115 g |
| Amaranth flour | 266 | 266 | 476 | 1 cup, 193 g |
| Cashews dried | 258 | 260 | 352 | 1 cup, 137 g |
| Sweet, dried almonds | 251 | 264 | 386 | 1 cup, 143 g |
| Peanuts, roasted | 229 | 175 | 260 | 1 cup, 146 g |
| Quinoa | n.a | 189 | 335 | 1 cup, 170 g |
| Pecans | 168 | 121 | 132 | 1 cup, 109 g |
| Hazelnuts, dried | 163 | 163 | 187 | 1 cup, 187 g |
| Beans, dried | 158 | 170 | 258 | 1 cup, 184 g |
| Walnuts, dried | 150 | 158 | 185 | 1 cup, 169 g |
| Chickpeas, dried | 150 | 131 | 158 | 1 cup, 100 g |
| Pistachios, dried | 147 | 160 | 149 | 1 cup, 123 g |
| Millet, shelled | 136 | 160 | 228 | 1 cup, 200 g |
| Wheat flour, hard | 136 | 120 | 164 | 1 cup, 120 g |
| Oat flour | 131 | n.a [2] | 150 | 1 cup, 169 g |
| Buckwheat flour, whole-groats | 121 | 231 | 301 | 1 cup, 120 g |
| Macadamia | 115 | 120 | 156 | 1 cup, 132 g |
| Wholemeal pasta | 111 | 101 | 95 | 1 cup, 90 g |
| Lentils, dried | 101 | 83.1 | 113 | 1 cup, 100 g |

[1] Ts: Teaspoon; [2] n.a: Not available data.

It is used to say that approximately 300 mg are ingested daily in the diet, however there are several factors that hinder or facilitate magnesium availability. Unfortunately, the bioavailability studies present in the literature cover a wide range of $Mg^{2+}$ loading administration (i.e., from <100 to >1000 mg/day) and observed different periods of time. Moreover, other important variables, such as the age of subjects (infants—adults), their physical condition, and the proximity of magnesium administration to meals and different meal matrices, did not allow for a comparison of results, leading to confusing and apparently conflicting results. Obviously, systematic studies comparing $Mg^{2+}$ absorption efficiency between magnesium-depleted and -saturated subjects were not possible due to ethical reasons.

Approximately 30% to 40% of dietary magnesium consumption is usually absorbed by the body. However, variables that can facilitate or obstruct magnesium absorption are here discussed and schematized in Figure 2.

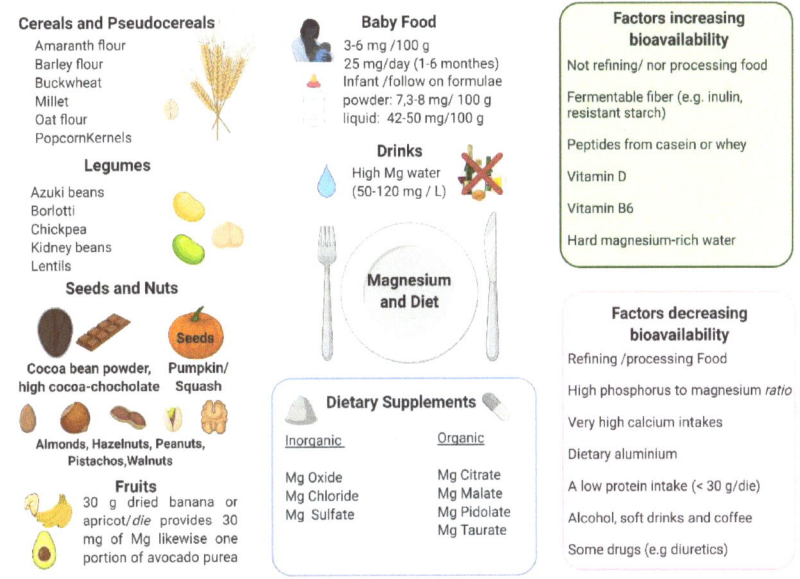

**Figure 2. Magnesium and Diet.** Main sources of magnesium, magnesium supplements, and factors that increase or decrease magnesium bioavailability are schematized. This image is created with BioRender.com.

In general, foods containing dietary non-fermentable fiber have indeed a high content of magnesium, nevertheless the bioavailability is low, analogously to iron [173]. By contrast, fermentable low- or indigestible carbohydrates (e.g., inulin, oligosaccharides, resistant starch, mannitol, and lactulose) enhance $Mg^{2+}$ uptake [4].

Among the compounds that can influence magnesium absorption, there are:

- Phytates and oxalates present in foods rich in fiber can decrease the absorption of magnesium because of metal chelation. Nevertheless, the decrease of magnesium absorption caused by phytate and cellulose is usually compensated by an increased magnesium intake due to high magnesium concentrations in phytate- and cellulose-rich products [4,174–177];
- Phosphorus: high luminal concentrations of phosphates can reduce magnesium absorption, mainly because of salt formation [178]. A major source of phosphorus is represented by soft drinks: the consumption of these beverages, typically rich in phosphoric acid, has been significantly rising in the last quarter of a century. The increase in dietary phosphate is also linked to phosphate additives, present in many food items but mainly processed meats [18]. Dairy and in particular cheese have a very high phosphorus/magnesium ratio. For example, cheddar cheese has a phosphorus/magnesium ratio of ~18 and a calcium/magnesium ratio of ~26.66. On the contrary, pumpkin seeds have a phosphorus/magnesium ratio of 0.35 and a calcium/magnesium ratio of 0.21 [18];
- Very high calcium intakes can reduce the absorption of magnesium, in particular, magnesium bioavailability decreases when calcium intake is over 10 mg/kg/day [18]. Increasing evidence suggests that the optimal serum magnesium/calcium ratio is 0.4 and if it is in the range 0.36–0.28, it is considered too low. This ratio is a more practical and sensitive of magnesium status and/or turnover, than the serum magnesium level alone [12];
- Dietary aluminum may contribute to a magnesium deficit by means of an approximately 5-fold reduction of its absorption, of 41% of its retention, and by causing a reduction of magnesium in the bone. Since aluminum is widespread in modern day

society (such as in cookware, deodorants, over the counter and prescription drugs, powder, baked products, and others), this could represent an important contributor to magnesium deficiency [18];
- Peptides from casein or whey could bind magnesium, which may promote absorption, analogously to other divalent cations [179]. A low protein intake (<30 g/die) could negatively influence the absorption of magnesium, however, other studies showed that magnesium use was not affected by the level of protein intake [180];
- Vitamin D seems to have a favorable role on $Mg^{2+}$ absorption [14,85,181,182] and $Mg^{2+}$ is important for vitamin D activation and inactivation [85];
- Vitamin B6 collaborates with magnesium in many enzyme systems and increases the accumulation of intracellular magnesium; a vitamin B6-deficient diet can lead to a negative magnesium balance via increased magnesium excretion [183,184];
- High doses of zinc can interfere with magnesium. Nielsen et al. reported that an intake of 53 mg zinc/day (4-fold higher than LARN) over 90 days can decrease magnesium balance [185];
- As for beverages, magnesium levels are decreased by excess ethanol, soft drinks, and coffee intake [186];
- Some drugs negatively affect the state of magnesium, in particular diuretics, insulin, and digitalis [23].

In turn, the magnesium content can influence the bioavailability of other nutrients: Magnesium deficiency can cause hypocalcemia [124]; Durlach suggested that the optimal dietary calcium:magnesium ratio is close to 2:1. Moreover, as previously reported, magnesium deficiency can alter the responses to vitamin D; individuals with hypocalcemia and magnesium deficiency become resistant to pharmacological doses of 1,25-dihydroxyvitamin D (active form of vitamin D), as detailed in paragraph 2.1 and recently reviewed in depth in order to highlight the role of magnesium in counteracting Covid-19 infection [85,97,187].

The magnesium content of tap/bottled water can be a significant contributor to the intake of this mineral. Tap, mineral, and bottled waters can indeed be sources of magnesium, but the amount of magnesium in water varies by source and brand (ranging from 1 mg/L to more than 120 mg/L) [188–190]. The magnesium in drinking water could indeed be an interesting option in order to meet the organism's magnesium necessities, as it is highly bioavailable [191]. A study by Sabatier and colleagues demonstrated a higher magnesium bioavailability when $Mg^{2+}$ rich mineral water was consumed during a meal [192].

Although magnesium-rich water could provide up to 30% of daily RDA, magnesium may be almost absent in soft water: the magnesium content of water tends to be ignored at the moment of purchase. Indeed, when the recommendations on the type of water for human consumption are reviewed, it is customary to downplay the importance of the magnesium it contains. The European regulations on public drinking water refer to the content of magnesium. However, as for natural mineral water, the Codex standard does not refer to magnesium content. European regulations on bottled drinking water indicate that the label can mention that it is rich in magnesium if it contains more than 50 mg/L of this mineral [*Directive 2009/54/EC of the European Parliament and of the Council of 18 June 2009 on the exploitation and marketing of natural mineral waters*].

As for the presence of magnesium in breast milk, concentrations vary over a wide range (15–64 mg/L) with a median value of 31 mg/L and 75% of reported mean concentrations below 35 mg/L [193,194].

Breastfed infants (6–24 months old) need complementary foods to satisfy 50% of their requirements for micronutrients, including magnesium, which is among those micronutrients particularly important during the sensitive period of rapid growth and development from weaning to 2 years old, and every effort should be made to ensure their adequacy in the diet [121,195].

As for baby food, in the EFSA table, it is reported that infant formulae powder contains 42 mg/100 g and 7.3 mg/100 g if considered a liquid and a slightly higher concentration is found for the follow-on formula: 8 and 50 mg/100 g for liquid and powder, respectively. In a recent study about the comparison between infants fed commercially prepared baby food and non-consumers of prepared baby food, a higher magnesium intake in the first group was reported [195].

In developed countries, children are frequently overfed and undernourished, which means "Even though children may consume an excess of energy, they may not be meeting all of their micronutrient needs" [18]. For example, more than a quarter of obese and non-obese youth do not satisfy adequate intakes of magnesium (27% and 29%, respectively), as reported by Gillis [196].

Magnesium is often added to some breakfast cereals and other fortified foods [181,197], in order to counteract deficiencies. Foods providing ≥20% of 420 mg of magnesium are considered to be high sources of this fundamental micronutrients for adults and children aged 4 years and older, as reported by the U.S. Food and Drug Administration (FDA) [198,199]. Foods providing lower percentages also contribute to a healthy diet.

### 3.3. Nutritional and Health Claims for Magnesium

The claim **"a source of vitamins and/or minerals"** and any other claim with the same meaning for the consumer, may only be attributed to a product that contains at least a significant amount of the micronutrients, as defined in the Annex to Directive 90/496/EEC or according to Article 7 of Regulation (EC) No. 1925/2006 of the European Parliament and of the Council of 20 December 2006 on "the addition of vitamins and minerals and of certain other substances to foods"; this value for magnesium is 300 mg. Therefore, a claim that a food **is high in** magnesium and any claim likely to have the same meaning for the consumer, is authorized only if the product contains **at least twice the value** of 'source of magnesium' i.e., a Magnesium content ≥112 mg supplied by 100 g or 100 mL or per a single portion package, according to the Annex to Directive 90/496/EEC that report "As a rule, 15% of the recommended allowance specified in this Annex (375 mg for magnesium) is supplied by 100 g or 100 mL or per package if the package contains only a single portion should be taken into consideration in deciding what constitutes a significant amount".

A claim stating that **the content** in one or more nutrients, other than vitamins and minerals, **has been increased,** and any claim endowed with the same meaning for the consumer, is only allowed if the product meets the conditions for the claim 'source of' and **the increase in content is at least 30% compared to a similar product.**

According to European Register on Nutrition and Health Claims [116], the following claims about magnesium have been authorized: Art, 13(1). The claim may be used only for food which is at least a source of magnesium as referred to in the claim "a source of" magnesium as listed in the Annex to Regulation (EC) No 1924/2006.

- "Magnesium contributes to a reduction of tiredness and fatigue";
- "Magnesium contributes to electrolyte balance";
- "Magnesium contributes to normal energy-yielding metabolism";
- "Magnesium contributes to normal functioning of the nervous system";
- "Magnesium contributes to normal muscle function";
- "Magnesium contributes to normal protein synthesis";
- "Magnesium contributes to normal psychological function";
- "Magnesium contributes to the maintenance of normal bones";
- "Magnesium contributes to the maintenance of normal teeth";
- "Magnesium has a role in the process of cell division".

### 3.4. Dietary Supplements of Magnesium

Magnesium supplements are available in a variety of formulations, including inorganic salt (e.g., magnesium oxide, chloride, sulfate) and organic compounds (e.g., citrate, malate, pidolate, taurate). Magnesium absorption from different kinds of supplements is not the same, nevertheless, the results obtained in the available studies in humans are hardly comparable due to the differences among the study designs. The $Mg^{2+}$ load administered varied widely among studies (from <100 to >1000 mg/d), despite the age of subjects (infants to adults), their physical condition, or the proximity of meals to administration. In addition, the absorption depends on the magnesium status of the subjects therefore, bioavailability studies are hampered by the fact that test persons do not automatically have the same Mg status when starting the study. Moreover, due to the absence of simple, rapid, and accurate laboratory tests to measure total body magnesium stores or to evaluate its distribution in different compartments, the assessment of magnesium status after its administration is not easy to achieve. The assessment of magnesium blood level provides information about acute changes in magnesium, but blood levels are not a good marker for magnesium status and do not correlate with levels in tissue pools [200]. As a result, the data often appear confusing and conflicting. However, the amount of $Mg^{2+}$ uptake is resulted to be dependent on the ingested dose [201]. For example, when dietary $Mg^{2+}$ intake is low, the relative absorption rate can reach 80%, whereas it is reduced to 20% in $Mg^{2+}$ abundance state [33]. In general, $Mg^{2+}$ is absorbed as an ion [202,203]. Soluble forms of magnesium are more absorbed in the gut than less soluble forms [204,205]. In a recent review, Schuchardt and Hahn reported that the relative $Mg^{2+}$ bioavailability is higher if the mineral is ingested in multiple low doses during the day rather than a single assumption of a high amount of $Mg^{2+}$ [4]. Small studies showed that magnesium in the aspartate, chloride, citrate, and lactate salt is absorbed almost completely and is more bioavailable than magnesium oxide and magnesium sulfate [204,206]. In general, it has been suggested that absorption of organic magnesium salts is better than the absorption of inorganic compounds, whereas other studies did not find differences between salt formulations [4,200,207,208]. Unabsorbed magnesium salts in general cause diarrhea and laxative effects due to the osmotic activity in the intestine and colon and the stimulation of gastric motility [58].

Although magnesium oxide can indeed be accompanied by diarrhea with the potential to reduce magnesium absorption, it is present in mineral supplements as reviewed by Hillier [209].

Intravenous magnesium sulfate has been used as a tocolytic agent and to reduce pre-term eclampsia [210], nevertheless since magnesium chloride and sulfate have both similar and proper effects, choosing magnesium chloride seems preferable because of its more interesting clinical and pharmacological effects and its lower tissue toxicity as compared to magnesium sulfate, as reported in the review by Durlach et al. [211].

A daily supplement of 200 mg of chelated magnesium (citrate, lactate) is suggested to be likely safe, adequate, and sufficient to significantly increase serum magnesium concentration in a fasting non-hemolyzed serum sample to levels >0.85 mmol/L but <1.1 mmol/L. A steady state is usually achieved in 20–40 weeks of supplementation and is dependent on the dose [6].

Supplements may contain mixed salt of magnesium citrate and malate, with a magnesium content of 12–15%, a mixed salt formulation suggested to be used in food supplements that are intended to provide up to 300–540 mg/day magnesium. The EFSA panel concluded that it is a source from which magnesium is bioavailable, but the extent of its bioavailability still remains unclear [212].

Ates and colleagues [200] recently studied in a mouse model the distribution of different organic magnesium compounds to tissues, evaluating also the effects of different administration doses. Moreover, they evaluated the potential differences between the organic acid–bounded compounds (magnesium citrate and magnesium malate) and the amino acid–bounded compounds (magnesium acetyl taurate and magnesium glycinate), in terms of bioavailability. Magnesium acetyl taurate increases brain magnesium levels independently of dose. This observation may be due to the presence of a taurine transport system in capillary endothelial cells of the blood–brain barrier [213] that allow a rapid absorption rate even in small doses of magnesium taurate. The saturated capacity of the taurine transport system acts as a rate limiting boundary [214]. On the other hand, muscle magnesium levels increased only after a high-dose administration (405 mg/70 kg).

Another salt frequently present in supplements is magnesium pidolate. Its dissociation constant is similar to that one of the inorganic salts, therefore it is highly dissociated at physiological pH [215]. Farruggia et al. suggested that magnesium pidolate is unable to maintain the normal intracellular magnesium content in a osteoblast model when used at the concentration of 1 mM, which corresponds to the normal magnesium level in the serum [216].

The beneficial effects of magnesium supplementation appeared to be more pronounced in the elderly and alcoholics, but were not particularly apparent in athletes and physically active individuals [217]. Further research on long-term administration of different magnesium compounds and their effect on other tissues are needed.

In a placebo controlled randomized trial in adolescent girls, showing that magnesium oxide daily supplementation significantly increased BMD (bone mineral density) in one year [218]. A strong association between severe hypomagnesemia and increased risk of fractures was reported in a recent prospective cohort study that takes into consideration 2245 men over a 25-year period [219].

As deeply discussed in Sections 2 and 5, the deficiency of magnesium is not related to a healthy status, however an indiscriminate over-treatment leading to significant hypermagnesemia must be avoided, in order to avert the risk of diseases related to the toxic effects of this mineral. Diseases associated with magnesium deficiency and toxicity are summarized in Table 4.

Table 4. Diseases associated with magnesium deficiency and toxicity.

| Magnesium Deficiency | Magnesium Toxicity |
| --- | --- |
| Hypocalcemia, hypokalemia | Diarrhea, nausea and vomiting |
| Osteoporosis | Muscle weakness |
| Cardiovascular disorders | Low blood pressure |
| Neurological disorders | Loss of deep tendon reflexes |
| Diabetes | Sinoatrial or atrioventricular node blocks |
| Tumors | Respiratory paralysis |
| Covid-19 | Cardiac arrest |

If the intake slightly exceeds the daily requirement, absorption of magnesium from the gut is reduced and its active renal secretion in the urine can exceed 100% of the filtered load [6]. Although an excess of magnesium from food does not represent a health risk in healthy individuals because the kidneys eliminate excess amounts in the urine [220], high doses of magnesium from dietary supplements, drugs, or other sources can often cause not only diarrhea accompanied by nausea and abdominal cramping but also the onset of diseases [221]. As previously reported, magnesium formulations most commonly associated with diarrhea, include magnesium carbonate, chloride, gluconate, and oxide [222].

Toxic hypermagnesemia is only observed at oral magnesium doses higher than 2500 mg, i.e., doses exceeding more than 10 times the UL.

## 4. Methods to Evaluate Magnesium Status

Since the importance of magnesium in human (and animal) health has been understood (the first review reporting the interplay between magnesium and health was published in 1965) [223], two questions have been posed: which is the correct sample reflecting the magnesium status and which is the important fraction of this element? In other words, is it better to consider the ionized free $Mg^{2+}$ or its total amount composed by both the free ion and the fraction bound to cellular and extracellular elements [224,225]? The two questions are strictly interconnected and it is probably impossible to give a single answer because it depends on several aspects, as clearly stated by the literature [226–228] and resumed in the previous sections of this review.

Obviously, the easiest samples to obtain are those derived from urine or blood. Urine is relatively simple to collect, but its magnesium content is heavily affected by several factors, such as hormones or drugs, and by the complex homeostasis between the dietary intake and the mobilization mainly from bones and/or muscle [9,203,220,226,227]. Age and gender also affect urinary excretion [229–232]. Furthermore, the more reliable urine sample seems to be the 24 h collected samples [227,229,231,232], but often the urine sample consists of the first urine in the morning [233,234]. For all these reasons, urine magnesium level seems poorly correlated with the magnesium status of the body, even if it could be an important component of the "metallomic signature" in severe pathologies, such as pancreatic cancer [90].

Blood samples could consist of serum, plasma, or the corpusculated part, i.e., erythrocytes, peripheral blood mononuclear cells (PBMC), and platelets. However, several papers claimed that serum magnesium does not give an appropriated estimation of total body magnesium, being, as previously stated, around 0.3–1% of total magnesium [4,5,227]. However, the corpusculated blood constituents also represent a similarly small fraction of total magnesium, corresponding to 0.5% in the erythrocytes and an even lower fraction in PBMC or platelets. Therefore, 99% of magnesium mainly resides in bones, muscle, and soft tissues, as reported in the previous sections [4,5,227].

Tissue magnesium could represent the more reliable sample to assess, but its withdrawal could obviously be highly invasive. The analysis of fixed sublingual epithelial cells by energy dispersive x-ray microanalysis (EXA tm) is claimed to be a valid alternative to monitor magnesium status in physiological or pathological status. These cells could be easily collected by gentle scratching the sublingual tissue, then fixed on a carbon slide with cytology fixative [235–237].

Magnesium in the cytosol, is mainly bound to ATP, or to other phosphorylated molecules, such as phosphocreatine or inorganic phosphate. For this reason, an indirect evaluation of ionized magnesium in organs like the brain or muscle could be assessed by phosphorus magnetic resonance spectroscopy (31P-MRS). This technique enables the indirect detection of the cytosolic amount of magnesium by the shift of ATP resonance frequencies, occurring when this nucleotide is bound to $Mg^{2+}$, and this chemical shift is a function of free $Mg^{2+}$ concentration. This powerful technique, even if limited to the detection of this fraction, can be applied to measure free $Mg^{2+}$ in vivo in normal [238–240] or pathological subjects [241–244], in several tissues, such as the brain or muscle. Recently, the application of proton NMR has been proposed for ionized magnesium evaluation in skeletal muscle [245] and, by means of clinical NMR instruments, in human plasma samples [246]. However, these techniques need very expensive and complex equipment and require very sophisticated analytical software [247], features that have confined them to very specialized laboratories.

*4.1. Atomic Absorption Spectroscopy*

To assess magnesium in biological samples, atomic absorption spectroscopy (AAS) is probably the oldest and the most widespread technique [248–250]. It has the important pro that could be applied to all kinds of biological sample [251–253], but its major cons are that sample preparation (usually acidic extracts), instrument calibration, and analysis are time expensive. Furthermore, Flame AAS raises two problems: safety as it needs dangerous gases (a mixture of air/acetylene) to burn the samples, and the sample size (millions of cells or grams of tissue). These cons are partly reduced by inductively coupled plasma-atomic emission spectrometry (ICP-AES), which allows the simultaneous multi-element analysis of small biological samples [250,253–255]. All these techniques are destructive and need specialized personnel to be performed. Finally, they do not allow researchers to discriminate between the free and bound form of the ion analyzed.

*4.2. Ion Selective Electrodes*

$Mg^{2+}$ can be measured potentiometrically using an ion selective electrode that, together with a reference electrode, forms an electrochemical system [3,256]. Ionized $Mg^{2+}$ can be measured in whole blood [257], serum or plasma [258], or in cells as erythrocytes [251]. As in the case of $Ca^{2+}$, several interferences, such as changes in pH, due to the loss of $CO_2$, can influence the complex balance of $Mg^{2+}$ in a serum [227,259,260]. The main disadvantages of this technique are the lack of specificity of the electrodes [261] and the rather long reaction times. However, it has significant improving potential and is spreading more and more in practice, owing to accumulating evidence regarding the presence of ionized $Mg^{2+}$ in different clinical situations, a parameter believed, by some authors, to be more important than total magnesium content [3]. A development of this technique is represented by ion-selective microelectrodes, which can be applied to whole alive cells. In this case, the cells must be impaled on the microelectrodes, to allow for the measurement of the cytosolic concentration of $Mg^{2+}$. A quite exhaustive discussion of this technique is presented in [256].

*4.3. Optical Sensors*

Optical chemosensors for magnesium determination represent an important and growing field of application due to their good selectivity, sensitivity, and simplicity in preparation. Several fluorimetric and colorimetric assays are proposed however, the application development seems to be more focused on fluorimetric ones than to colorimetric, maybe for the higher selectivity and sensitivity of the former [262].

4.3.1. Colorimetric or Enzymatic Assay

Colorimetric assays are commercially available and are based on the direct binding of magnesium to a chromophore, such as calmagite [262–265] or xylidyl-blue [266] or taking advantage from the enzyme which activity strictly depends on $Mg^{2+}$. In the assay based on glycerol kinase [267], the product of the reaction (glycerol-3–phosphate) is oxidized to dihydroxyacetone phosphate and $H_2O_2$ by using glycerophosphate oxidase. Then, a peroxidase utilizes the produced $H_2O_2$ to reduce a chromogenic substrate, whose formation is proportional to magnesium concentration. Several commercial kits are available and built on patented reactives.

4.3.2. Fluorescent Chemosensors

In the recent past, there has been a strong impulse in the synthesis and characterization of fluorescent dyes specific for magnesium, aiming to clarify not only the magnesium content in cells or body fluids, but also to evaluate the flows of this ion both between the intra- and extra-cellular environment and at an intracellular level [268–270]. Most of the commercial dyes are designed to have good specificity for the free ionized $Mg^{2+}$, like the widely-used Mag-Fura-2, Mag-Indo-1, Mag-Fluo-3, and do not allow the detection of magnesium compartmentalization. Furthermore, they often show poor selectivity against other ions.

Several new molecules have been proposed in the past two decades, such as the family of dyes having a coumarin structure equipped with a charged β-diketone as a binding site for $Mg^{2+}$ [271–273] or that based on alkoxystyryl-functionalized BODIPY fluorophores decorated with a 4-oxo-4H-quinolizine-3-carboxylic acid metal binding moiety [274]. Recently, strategies to obtain dyes capable of being addressed in specific organelles have also been proposed [275,276].

The diazacrown-hydroxyquinoline based dyes, the so-called DCHQ family, are instead specifically designed to monitor total magnesium content in cells or in cellular lysates [277–279]. The dye DCHQ-5, in particular, is useful in monitoring total magnesium content in whole viable cells analyzed by flow cytometric assays or in cellular lysates subjected to fluorimetric assays, starting from very small samples, up to $5 \times 10^4$ cell/mL [280–282].

Other dyes have been synthetized and proposed and are waiting for a widespread confirmation of their biological application, as those proposed by Yadav [283].

*4.4. Element Bioimaging*

Chemical imaging is quite a new field of study and allows for the detection of elements with high sensitivity and spatial resolution, with the further advantage that it allows for the simultaneous evaluation of several elements or molecules [283,284]. Chemical imaging techniques could provide a detailed map of elements and molecules within the cell at a nanometric/micrometric scale, opening a new window in understanding cellular functions; a detailed overview of these techniques and of the methods of samples preparations are reported in the very recent review of Decelle and colleagues [284]. In particular, metals are good subjects for x-ray fluorescence (XRF) microscopy, which exploits the excitation of the core electrons of atoms, thus generating the emission of x-rays characteristic of the elements in the sample. However, these techniques request x-ray sources, and even if laboratory sources are improving their performances in an impressive way [285], the difference in the results obtained by laboratory equipment and the synchrotron radiation sources are still noticeable [270,286].

As reported previously, the first application of x-ray detection of magnesium was proposed by Hagney and colleagues in 1995 [235]. From then, x-ray biological applications became a reality, even if they still remain a niche technique. However, it is important to report that these techniques allow researchers to simultaneously detect magnesium and other trace elements in several tissues. Clinical application of analytical scanning electron microscopy (ASEM) using computerized elemental x-ray analysis on sublingual epithelial cells was used for the simultaneous detection of magnesium and potassium in diabetes patients treated with magnesium supplementation [287]; energy dispersive x-ray spectroscopy coupled to scanning electron microscopy enables researchers to quantify differences in elemental composition in muscle biopsies of patients with peripheral artery disease (PAD), finding significative difference particularly in calcium, magnesium, and sulfur content, while C, O, K, and Na do not change within the different samples [288].

By using synchrotron sources, characterized by extremely high brilliance, a really detailed reconstruction of the map of different elements and of course of magnesium is possible, even if a routine application to clinical specimens remains obviously difficult. By this application on single cells analysis, for example, it was possible to highlight the different distribution of intracellular magnesium in cells sensitive and resistant to doxorubicin, in cells deprived of extracellular magnesium [265,289], or in cells in which different magnesium channels are modulated, as reported [265,282]. Finally, this powerful technique has recently allowed researchers to dissect the initial step of mineral deposition during osteogenic commitment, opening a window on magnesium involvement in these events [286].

## 5. Magnesium Deficiency and High Social Impact Diseases

Over the last 30 years several, experimental, clinical, and epidemiological studies have shown that chronic magnesium deficiency is associated with and/or amplifies many major diseases [6]. Most of them are well known "social pathologies" such as diabetes, osteoporosis, and cardiovascular diseases, with a significant impact on the lives of the people affected and their families, but also on the community's economy and social life.

The disease's social impact can be defined as "the nexus between biological event, its perception by patient and practitioner, and the collective effort to make cognitive and policy sense out of those perceptions" [290]. Nevertheless, a comprehensive and real picture of the disease's social impact must consider both the direct and indirect costs on the economic system. The direct costs represent the value of resources used to prevent, detect, and treat a health impairment or its effects. The indirect costs are those for employed individuals and, in the case of disabled patients, their caregivers, including the value of production lost to society due to absence from work, reduced ability, and death of productive people [291]. In fact, each pathology brings with it multiple effects, involving concentric circles of subjects, ranging from the patient directly involved, to their relational networks, to the life worlds in which it operates.

Growing scientific evidence supports the view that low intakes of magnesium could induce changes in biochemical signaling pathways, increasing the risk of illness over time. Among a few works that focus on the social impact of magnesium deficiency, a recent study is noteworthy. It asserts that subclinical magnesium deficiency increases the risk of numerous types of cardiovascular disease, burdens nations around the world with incalculable healthcare costs and suffering, and should be considered a public health crisis [18]. In this context it is important to reiterate that the acute hypomagnesemia shows clear clinical features (severe cramps, nystagmus, cardiac arrhythmias, etc.), and is easily detectable. On the contrary, subclinical or chronic magnesium deficiency is often underestimated because it reflects reduced levels of magnesium within cells and bone, not extracellular magnesium [6].

Magnesium levels should be routinely measured not only in critically ill patients but more generally in people at risk of chronic hypomagnesemia, considering that it is inexpensive to diagnose and easy to treat. This approach would permit the prevention of the onset of high social impact diseases and eventually to improve their outcome, preserving considerable resources for the whole community, such as the great savings that could be obtained by reducing the incidence and mortality due to diabetes. In fact, this illness imposes a considerable burden on society, consisting of higher medical costs, lost productivity, premature mortality, and intangible costs in the form of reduced quality of life. It has been reported that health expenses in the U.S. for diabetes increased by 26% from 2012 to 2017, passing from $245 billion to $327 billion, respectively [292]. A similar great benefit in terms of social impact could be obtained by reducing the incidence of neurological disorders, as they represent the third most common cause of disability and premature death in the EU and their burden and prevalence will increase according to the progressive ageing of the population [293].

The following section focuses on five high social impact diseases in which magnesium deficiency appears to be involved: diabetes mellitus, osteoporosis, cardiovascular diseases, cancer, and neurological disorders.

*5.1. Diabetes Mellitus*

It is well known that magnesium acts as an insulin sensitizer by inducing autophosphorylation of insulin receptors and regulating tyrosine kinase activity on these receptors [63,153,294,295]. In addition, magnesium may directly affect the activity of the glucose transporter 4 (GLUT4) and help to regulate glucose uptake into the cell [21]. Consequently, diets with higher amounts of magnesium are related to a significantly lower risk of diabetes [296]. Several studies report that a reduced intracellular magnesium level can lead to increased insulin resistance [93,297]. The incidence of hypomagnesemia in patients with type 2 diabetes is wide, ranging from 13.5–47.7% [298].

A 100 mg/day increase in total magnesium intake is reported to decrease the risk of diabetes by a statistically significant 15% [299]. Moreover, a meta-analysis of eight prospective cohort studies, involving 271,869 men and women over 4 to 18 years, showed a significant inverse association between magnesium intake from food and risk of type 2 diabetes; the relative risk reduction was 23% when the highest to lowest intakes were compared [300]. According to this, Dong et al. reported a meta-analysis of prospective cohort studies of magnesium intake and risk of type 2 diabetes included 13 studies with a total of 536,318 participants and 24,516 cases of diabetes. It was demonstrated that the magnesium intake is inversely associated with risk of contracting the disease in a dose-response manner [301]. The same conclusion was drawn from a prospective study on high risk population, involving 2582 community-dwelling participants followed up for 7 years [302]. Moreover, in a randomized controlled trial involving 116 adults with prediabetes and hypomagnesemia, the reduction of plasma glucose levels and the improvement of the glycemic status by oral magnesium supplementation were also demonstrated [303]. Additionally, a very recent trial sequential analysis confirmed that magnesium intake has an inverse dose-response association with type 2 diabetes incidence, and magnesium supplementation appears to be advisable in terms of glucose parameters in high-risk individuals [304].

Interestingly, some studies documented that hypomagnesemia could have an impact on many dysfunctions indicated in the pathophysiology of diabetes, such as diabetic nephropathy, poor lipid profile, and high risk of atherosclerosis, even indicating hypomagnesemia as a marker [103,305].

### 5.2. Osteoporosis

The most common bone disease in humans is osteoporosis, which represents a major public health problem and is more common in Caucasians, women, and older people [306].

It is well accepted that magnesium deficiency might represent a risk factor for osteoporosis [80,91]. Both dietary intake and supplementation of magnesium were investigated in relation to osteoporosis and risk of fractures in humans. Early works examining the effect of oral supplementation of magnesium in postmenopausal women evidenced a significant increase in BMD (bone mineral density), but the little number of enrolled subjects limited the conclusions that could be drawn [307,308]. According to one short-term study, 290 mg/day of elemental magnesium for 30 days in 20 postmenopausal women with osteoporosis counteract bone turnover and thus decreased bone loss compared with placebo [309]. Other investigations found a positive association between dietary magnesium, BMD, and lower risk of osteoporosis, suggesting that increasing magnesium intakes from food or supplements might increase BMD in postmenopausal and elderly patients [310,311].

A recent meta-analysis evidenced a positive slightly significant correlation between magnesium intake and BMD only for the femoral neck and total hip, but not for the lumbar spine [312].

Fractures and in particular osteoporotic fractures are widespread causes of disability and morbidity, especially among the aging population, and increase the burden on health systems [306]. The prevention of fractures and the evaluation of putative risk factors could be very important for the public health: serum magnesium, which may have predictive or causal relevance to the risk of fractures, could help to personalize preventive and therapeutic interventions [219]. Although several studies evidenced a positive correlation between BMD and magnesium intake, the relation to fracture outcomes is yet unclear. A prospective cohort study on 73,684 postmenopausal women showed that a lower magnesium intake is linked to decreased bone density in the hip and whole body. However this does not relate to an increase of fracture risks [313]. On the other hand, data from a large perspective study [314] and cross-sectional analysis [315] showed that by satisfying the recommended magnesium intake, the risk of fractures is lower. Accordingly, a strong association between low serum magnesium and increased risk of fractures was reported in a prospective cohort study of 2245 middle-aged Caucasian men over a 25-year period [219].

### 5.3. Cardiovascular Diseases

Increasing evidence from epidemiological studies, randomized controlled trials, and meta-analyses has shown inverse relationship between magnesium intake and cardiovascular disorders (CVD) [316]. Indeed, high magnesium intake is related to lower probability of major CV risk factors (such as hypertension and diabetes), stroke, and total CVD. In addition, a reduced risk of ischemic and coronary heart disease is related to higher levels of circulating magnesium [317].

It is well known that hypertension is an important risk factor for heart disease and stroke. As stated by A. Rosanoff, "Magnesium status has a direct effect upon the relaxation capability of vascular smooth muscle cells and the regulation of the cellular placement of other cations important to blood pressure—cellular sodium: potassium ratio and intracellular calcium. As a result, nutritional magnesium has both direct and indirect impacts on the regulation of blood pressure and therefore on the occurrence of hypertension" [318]. Early studies have shown that a magnesium deficiency could impact blood pressure, leading to hypertension. Oral magnesium supplementation may exert a moderate antihypertensive effect [319]. Afterwards, a meta-analysis of 12 clinical trials found that magnesium supplementation for 8–26 weeks in 545 hypertensive subjects obtained only a slight reduction in diastolic blood pressure with magnesium supplementation, ranging from nearly 243 to 973 mg/day [320]. Next, Kass et al. analyzed 22 studies with 1173 normotensive and hypertensive adults concluding that magnesium supplements for 3–24 weeks reduced both systolic and diastolic blood pressure, albeit to a small extent [321]. Other authors have pooled six prospective cohort studies including 20,119 cases and 180,566 participants. They found a statistically significant inverse association between dietary magnesium and hypertension risk without apparent evidence of heterogeneity between studies. The range of dietary magnesium intake among the included studies was 96–425 mg/day, and the follow-up ranged from 4 to 15 years [322]. Additionally, a meta-analysis on 11 randomized controlled trials counting 543 participants with preclinical or non-communicable diseases who were monitored for a range of 1–6 months, showed that the group supplemented with oral magnesium had a considerably greater decrease in blood pressure. An average reduction of 4.18 mmHg in systolic blood pressure and 2.27 mmHg in diastolic blood pressure was found after magnesium supplementation [323].

Magnesium deficiency reduces cardiac Na-K-ATPase, determining greater levels of sodium and calcium and lower levels of magnesium and potassium in the heart. Consequently, the vasoconstriction in the coronary arteries increases, inducing coronary artery spasms, heart attack, and cardiac arrhythmia [18]. Higher magnesium serum levels were significantly linked to a lower risk of CVD, as shown by a systematic review and meta-analysis of prospective studies, involving 313,041 individuals with 11,995 cardiovascular diseases, 7534 ischemic heart diseases, and 2686 fatal ischemic heart disease. Moreover, higher dietary magnesium intakes (up to approximately 250 mg/day) were correlated with a substantially lower risk of ischemic heart disease caused by a lowered blood supply to the heart muscle. Circulating serum magnesium (per 0.2 mmol/L increment) was associated with a 30% lower risk of CVD and trends toward lesser risks of ischemic heart disease and fatal ischemic heart disease [324]. In a monocentric, controlled, double-blind study, 79 patients with severe chronic heart failure under optimal medical cardiovascular treatment were randomized to receive either magnesium orotate or placebo. The two groups were similar in demographic data, duration of heart failure, and pre- and concomitant treatment. The survival rate was 75.7% compared to 51.6% under placebo, after 1 year of treatment. Clinical symptoms improved in 38.5% of patients under magnesium orotate, whereas they worsened in 56.3% of patients under placebo [325].

Additionally, magnesium has a well-established role in the management of torsade de pointes, a repetitive polymorphous ventricular tachycardia with prolongation of QT interval of the electrocardiogram. The guideline of the American Heart Association and the American College of Cardiology recommends intravenous administration of magnesium and potassium for the prevention and treatment of torsade de pointes, and tachycardia [326,327].

Low magnesium levels can also enhance endothelial cell dysfunction, potentially increasing the risk of atherosclerosis and thrombosis, stimulating a proatherogenic phenotype in endothelial cells [328]. The Atherosclerosis Risk in Communities study evaluated heart disease risk factors and concentrations of serum magnesium in a cohort of 14,232 white and African American men and women aged 45 to 64 years at baseline. Over an average of 12 years of follow-up, individuals with a normal physiologic range of serum magnesium (at least 0.88 mmol/L) had a 38% reduced risk of sudden cardiac death in comparison with individuals with 0.75 mmol/L or less. Nevertheless, dietary magnesium intakes did not show any risk of sudden cardiac death [329].

In an updated meta-analysis involving more than 400,000 adults from different cohorts, who were followed for 5 to 28 years, the summary estimate comparing individuals at the higher versus the lowest categories of dietary magnesium intake demonstrated a protection of 14% against the risk of CVD death. Additional assessment of the subtypes of CVD death indicated that dietary magnesium intake was inversely and significantly associated with a lower risk for heart failure and sudden cardiac death. Further dose–response analysis showed a protection of 25% in women for the increment of 100 mg/day of magnesium intake [322]. Another prospective population study of 7664 adults aged 20 to 75 years without cardiovascular disease verified the protective action of magnesium in this context: it was found that low urinary magnesium excretion levels (an indicator for low dietary magnesium intake) were related to a superior risk of ischemic heart disease over a median follow-up period of 10.5 years [330].

### 5.4. Cancer

Hypomagnesemia is also a common medical problem that contributes to the morbidity of cancer patients. Cancer is the leading cause of death worldwide; over 1.7 million people were diagnosed with cancer and over 600,000 deaths have resulted from this disease in 2018 alone [331,332].

The effects of diet in cancer metabolism are certainly an area of popular interest. A recent review highlights the mechanisms underlying magnesium disturbances due to cancer and/or its treatment [333]. Hypomagnesemia can be due to these physio-pathological mechanisms: (i) decreased intake, (ii) transcellular shift, (iii) gastrointestinal losses, and (iv) kidney losses. Moreover, cancer patients are at risk for opportunistic infections, cardiovascular complications, and are treated with classes of medications that cause or emphasize hypomagnesemia, like platinum-based chemotherapy, anti-EGFR monoclonal antibodies, and human epidermal growth factor receptor-2 target inhibitors (HER2) [334].

Several epidemiologic studies demonstrated that a diet poor in magnesium increases the risk of developing cancer, evidencing its importance in the field of hematology and oncology. Being an enzyme cofactor involved in the DNA repair mechanisms, magnesium plays a major role in maintaining genomic stability and fidelity, modulating cell cycle progression, cell proliferation, differentiation, and apoptosis. Thus, magnesium deficiency could affects these systems, leading to DNA mutations, which may result in tumorigenesis and in both the risk and prognosis of cancers [78,335]. Moreover, a protective effect of magnesium against chemical carcinogenesis has been recently reported [27].

Some studies have focused on the effect of dietary magnesium on breast cancer prognosis, suggesting that higher dietary intake is inversely associated with mortality among breast cancer patients [336]. The effect of magnesium intake on breast cancer risk has been explored, both directly and indirectly via its effect on inflammatory markers C-reactive protein and interleukin-6 [337].

Liu et al. in a recent review evidenced that magnesium supplementation can protect the liver and reduce the morbidity and mortality associated also with liver cancer. Furthermore, the risk of cancer metastasis to the liver increases in cancer patients with magnesium deficiency [338]. According to this, an in vitro study has shown that magnesium cantharidate has an inhibitory effect on human hepatoma SMMC-7721 cell proliferation by blocking the MAPK signaling pathway [339]. Moreover, magnesium administration can increase the expression of protein phosphatase magnesium dependent 1A (PPM1a), blocking TGF-β signaling by dephosphorylating of p-Smad2/3, and thus preventing the transcription of specific genes necessary for hepatocellular cancer growth [340].

The association between magnesium and calcium intake and colorectal cancer (CRC) recurrence and all-cause mortality was also reported. It has been observed that 25(OH)D3 and magnesium may work synergistically in decreasing the risk of all-cause mortality in these patients [341]. Higher concentrations of 25-hydroxyvitamin D3 at diagnosis are associated with a lower mortality risk in CRC patients. This is expected given the crucial roles of magnesium in several biochemical processes involved in the synthesis and metabolism of vitamin D [85,342]. In addition, in a meta-analysis that involved 3 case-control studies of colorectal adenomas and six prospective cohort studies of carcinomas, every 100 mg (4.11 mmol)/d increase in magnesium intake was associated with a 13% lower risk of colorectal adenomas and 12% lower risk of colorectal tumors [343]. Moreover, epidemiological studies have linked a magnesium deficiency with high Ca:Mg intake ratios to a higher incidence of colon cancer and mortality [342]. It has been proposed that this kind of magnesium deficiency increases intracellular calcium levels in part by increasing TRPM7 expression and unblocking the gating effect of magnesium on intracellular calcium entry. Increased intracellular calcium levels promote reactive oxygen species (ROS) generation and magnesium deficiency likely blunts cell-associated antioxidant capacity to further promote oxidative stress. This study also sheds some insight on the epidemiological findings that link high Ca:Mg ratios with increased incidence of cancer [344–346] and increased mortality among colon cancer patients [347].

Observational studies evidenced that elevated magnesium content in drinking water is linked with a reduced risk of esophageal cancer and decreased mortality due to prostate and ovarian cancers. Higher dietary intake of magnesium decreases the risk beyond of above-mentioned colorectal cancer, also of pancreatic cancer and lung cancer [348–353].

Although most of the literature regards solid tumors, hypomagnesemia has also been correlated with a higher viral load of the Epstein Barr virus, a virus associated with a multitude of hematologic malignancies. Studies of patients with a rare primary immunodeficiency known as XMEN disease (x-linked immunodeficiency with magnesium defect, Epstein–Barr virus (EBV) infection, and Neoplasia disease) elucidated the role of magnesium in the immune system. These patients have a mutation in the MAGT1 gene, which codes for a magnesium transporter. The mutation leads to impaired T cell activation and an increased risk of developing hematologic malignancies. Furthermore, magnesium replacement may increase the immune system's ability to target and destroy cancer cells through this mechanism highlighted in patients with XMEN [27]. On the other hand, a very recent study has redefined MagT1 as a non-catalytic subunit of the oligosaccharyltransferase complex that facilitates Asparagine (N)-linked glycosylation of specific substrates. The authors proposed updating XMEN to "X-linked MAGT1 deficiency with increased susceptibility to EBV-infection and N-linked glycosylation defect". However, the precise mechanism by which MAGT1 is involved in the homeostasis of magnesium and how this affects the glycosylation defect requires further investigation [354].

Moreover, a very recent work assessed the disturbance of electrolyte in leukemia. In particular, a significantly higher concentration of calcium and a lower content of magnesium in the serum and whole blood of Acute leukemia children were found, as compared to healthy subjects. Furthermore, magnesium is replaced by calcium and harmful metals (As, Cd, and Pb) which results in its deficiency, producing physiological disorders, which may be involved in acute leukemia. The level of magnesium in normal children had the range of 150–279% than AL patients [355]. This finding is consistent with other previously reported data, which indicates an association between insufficiency of magnesium and development of malignant disorders [224,356–359]. These studies highlight that a diet enriched with magnesium can decrease the incidence of cancers and the possibility that hypomagnesemia is associated with poor outcomes in cancer patients undergoing treatment.

*5.5. Neurological Diseases*

Neurological diseases are a substantial and wide spreading health burden worldwide, as shown in the Global Burden of Diseases (GBD) Study 2016. They represent the third most common cause of disability and premature death in the EU and their prevalence will presumably increase with the progressive ageing of the European population [293].

Numerous studies report the involvement of magnesium in these pathologies, the recurrent deficiency in the patients and the effectiveness of dietary integration [103,360,361]. The mechanisms by which magnesium can modulate these disorders are multiple and not fully understood. However, variation in the excitability of the central nervous system, spontaneous neuronal depolarization, and abnormal mitochondria functioning have been connected to most of them. Since glutamate is the most abundant excitatory neurotransmitter, it is often linked to etiology, prevention, and treatment of neuropathology [362,363]. For this reason, magnesium has been a potential strategy for neurological diseases mainly due to its negative modulation of the glutamatergic N-methyl-D-aspartate (NMDA) receptor. Furthermore, magnesium is a key metabolic factor in mitochondrial functioning, lowering membrane permeability and consequently reducing the possibility of spontaneous neuronal depression due to hyperexcitability [50].

A very exhaustive review describes "The Role of Magnesium in Neurological Disorders", summarizing the recent literature on the role played by magnesium in counteracting the onset and co-treating the most frequent neurological diseases: chronic pain, migraine, stroke, epilepsy, Alzheimer's, and Parkinson's, as well as the commonly comorbid conditions of anxiety and depression. The authors claim that "despite to a great number of publications in this field the amount of quality data on the association of magnesium with various neurological disorders differs greatly." Nevertheless, compelling evidence is reported about the role of magnesium in migraine and depression and for counteracting chronic pain conditions and in anxiety as well [50].

From the social impact point of view, it is worth noting that a migraine is a debilitating brain disorder with serious social and financial consequences for the individual and society. The economic impact of headache disorders is enormous in EU countries, with an annual cost of 111 billion Euros. A total of 93% of the costs are indirect and attributable to reduced productivity rather than absenteeism [364]. The serum level of magnesium in migraine patients is frequently lower than healthy subjects. Oral magnesium supplementation is prescribed for prophylaxis while intravenous magnesium administration is routinely suggested for acute migraine. The American Academy of Neurology has revealed the effectiveness of oral magnesium usage in migraine prevention [365]. The efficacy of magnesium in acute migraine treatment was confirmed by different studies [366–368].

Depression is a frequent and debilitating disorder that affects almost 11% of adults older than 60 and 18.8% of those younger than 60. Depression is linked to inadequate quality of life with severe impairments and is often associated with other comorbid disorders, such as anxiety and chronic pain. Interestingly, magnesium plays a role in many pathways involved in the pathophysiology of depression and it is important for the activity of several enzymes, hormones, and neurotransmitters [157,369]. Low magnesium status has been associated with increased depressive symptoms in several different age groups and ethnic populations [370,371]. Recently, it has been reported that there is a significant association between very low magnesium intake and depression, especially in younger adults [372]. Magnesium supplementation has been associated with the improvements of symptoms linked to major depression, premenstrual condition, postpartum depression, and chronic fatigue syndrome [373,374]. A recent open-label randomized trial with 126 adults comparing 248 mg of magnesium to a placebo over six weeks, showed a significant improvement of depression scores within the magnesium group within the first two weeks of treatment [375,376].

Epilepsy is a disease that affects 50 million people worldwide, characterized by seizures occurrence. Seizure activity has been strongly linked to excessive glutamatergic neurotransmission thus, magnesium could also modulate the excitotoxicity connected to epilepsy [377]. In fact, it is well known that severe hypomagnesaemia, itself, can cause seizure activity [378]. Interestingly, it has been reported that pre-eclampsia and eclampsia, conditions associated with symptomatic seizures, improved after magnesium supplementation [50].

Stroke is a cerebrovascular disease characterized by symptoms such as slurred speech, paralysis/numbness, and difficulty walking. A recent publication on stroke reviewed multiple meta-analyses and reported a dose-dependent protective effect of magnesium against stroke. Most of the meta-analyses reviewed found that each 100 mg/day increment of dietary magnesium intake provided between 2% and 13% protection against total stroke. Another updated meta-analysis, including 40 prospective cohort studies, found a 22%

protection against the risk of stroke when comparing people with the highest to the lowest categories of dietary magnesium intake [322].

Alzheimer's and Parkinson's diseases (AD and PD) represent two aging disease of neurodegenerative character with higher social impact. The cost burden of these pathologies in European countries rises year by year, and by 2050 it will be almost two times higher in comparison with the year of 2010, estimated to reach 357 billion Euros [293,379].

AD is characterized by profound synapse loss and impairments of learning and memory. Excitotoxicity, neuroinflammation, and mitochondrial dysfunction have all been implicated in Alzheimer's disease, thus, hypomagnesaemia could further hinder neuronal activity [380]. The level of magnesium in a diet is critical to support synaptic plasticity, and the decline in hippocampal synaptic connections has been associated with impaired memory [381]. Recent findings in animal studies are encouraging and provide novel insights into the neuroprotective effects of magnesium. Magnesium treatment, in fact, at an early stage may decrease the risk of cognitive decline in AD [382]. This coincides with earlier studies proving that the increase in the concentration of magnesium in the extracellular fluid results in a permanent increase in synaptic plasticity of hippocampal neurons cultured in vitro and improves learning and memory in rats [383]. Moreover, recent research suggests that ionized magnesium, cerebral spinal fluid magnesium, hair magnesium, plasma magnesium, and red blood cell magnesium concentrations are significantly reduced in AD patients compared to healthy and medical controls [21,384]. Nevertheless, the exact role of magnesium in AD pathogenesis remains unclear.

Parkinson's disease is a common neurodegenerative disease that occurs in the *substantia nigra* and *striatum*. The exact cause of its pathological changes is still not very clear, although genetic, aging, and oxidative stress have been suggested to be linked to it. It has been shown that the concentration of magnesium in the cortex, white matter, basal ganglia, and brainstem of the PD brain is low [385,386]. However, the association between circulating magnesium and PD is still ambiguous and controversial. Human research of magnesium concentrations in PD is severely lacking, despite growing evidence implicating magnesium in animal studies [356]. The latest published study on magnesium and PD was a multicentered hospital-based case-control study that examined the dietary intake of metals in patients who were found to be within six years of onset for PD. The study found that higher magnesium concentrations were associated with a reduced risk of PD [387].

Furthermore, the involvement of magnesium in Attention-Deficit Hyperactivity Disorder (ADHD), a serious neurodevelopmental condition characterized by inattention, hyperactivity, and impulsivity, has been reported. The estimated prevalence of ADHD is between 5% and 7% in schoolchildren worldwide. Frequently, learning disorders are associated with this disease and these impairments can influence children's quality of life and impose substantial costs on their family, health-care services, and educational systems worldwide [388]. It is well accepted that magnesium might be useful as a therapeutic agent in the treatment of ADHD because it has been reported that the serum magnesium level in ADHD children was lower than the controls [389,390]. Moreover, magnesium supplementation (alone or in combination with vitamins or other metals) significantly improved ADHD symptoms [391,392]. Magnesium supplementation along with standard treatment ameliorated inattention, hyperactivity, impulsivity, opposition, and conceptual level in children with ADHD. A very recent paper assessed that magnesium and vitamin D supplementation in children with ADHD disorder was effective on conduct problems, social problems, and anxiety/shy scores compared with placebo intake [381,388].

## 6. Conclusions

This multifaceted analysis of the importance of magnesium for maintaining a good state of health, starting from the tuning role played by this element at cellular level, revealed the importance of disseminating dietary strategies that satisfy the recommended daily value. Moreover, it is fundamental to have reliable and minimally invasive methods either to promptly identify magnesium deficiency in various body districts or to accurately monitor the efficacy of supplements to prevent and counteract diseases that correlate with magnesium deficiency. Indeed, magnesium has to be considered as a real metabolite instead of a simple electrolyte and its deficiency has a great impact on different physiological functions.

Data from many studies indicate that in about 60% of adults, magnesium intakes from the diet is insufficient and that subclinical magnesium deficiency is a widely diffused condition in the western population. Hence, more attention should be paid to the preventive role of magnesium for social pathologies, encouraging a more adequate dietary intake of the cation and supplementations. As extensively described above, magnesium is found in a wide variety of non-refined foods and is among the less expensive available supplements. Moreover, magnesium trials have shown that magnesium supplements are well tolerated and generally improve multiple markers of disease status.

**Author Contributions:** Conceptualization, D.F., G.F., C.P., and C.C.; writing—original draft preparation D.F., G.F., C.P., and C.C.; writing—review and editing, D.F., G.F., C.P., and C.C.; supervision, C.P.; project administration, G.F. All authors have read and agreed to the published version of the manuscript.

**Funding:** This research received no external funding.

**Institutional Review Board Statement:** Not applicable.

**Informed Consent Statement:** Not applicable.

**Data Availability Statement:** Not applicable.

**Acknowledgments:** Authors would like to express deep gratitude to Stefano Iotti for his valuable and constructive suggestions during the pre-submission revision of the manuscript. The authors wish to also acknowledge Tullia Maraldi (University of Modena and Reggio Emilia, Modena, Italy) for the license to use BioRender.com for creating the images.

**Conflicts of Interest:** The authors declare no conflict of interest.

## References

1. Romani, A.M. Cellular magnesium homeostasis. *Arch. Biochem. Biophys.* **2011**, *512*, 1–23. [CrossRef] [PubMed]
2. De Baaij, J.H.F.; Hoenderop, J.G.J.; Bindels, R.J.M. Magnesium in Man: Implications for Health and Disease. *Physiol. Rev.* **2015**, *95*, 1–46. [CrossRef]
3. Saris, N.E.L.; Mervaala, E.; Karppanen, H.; Khawaja, J.A.; Lewenstam, A. Magnesium: An update on physiological, clinical and analytical aspects. *Clin. Chim. Acta* **2000**, *294*, 1–26. [CrossRef]
4. Schuchardt, J.P.; Hahn, A. Intestinal Absorption and Factors Influencing Bioavailability of Magnesium-An Update. *Curr. Nutr. Food Sci.* **2017**, *13*, 260–278. [CrossRef] [PubMed]
5. Konrad, M.; Schlingmann, K.P.; Gudermann, T. Insights into the molecular nature of magnesium homeostasis. *Am. J. Physiol. Physiol.* **2004**, *286*, F599–F605. [CrossRef]
6. Ismail, A.A.A.; Ismail, Y.; Ismail, A.A. Chronic magnesium deficiency and human disease; time for reappraisal? *QJM* **2018**, *111*, 759–763. [CrossRef]
7. Elin, R.J. Assessment of magnesium status for diagnosis and therapy. *Magnes. Res.* **2010**, *23*, 194–198.
8. Reddi, A.S.; Reddi, A.S. Disorders of Magnesium: Hypomagnesemia. In *Fluid, Electrolyte and Acid-Base Disorders*; Springer: Berlin/Heidelberg, Germany, 2018.
9. Witkowski, M.; Hubert, J.; Mazur, A. Methods of assessment of magnesium status in humans: A systematic review. *Magnes. Res.* **2011**, *24*, 163–180. [CrossRef]
10. Nielsen, F.H. Guidance for the determination of status indicators and dietary requirements for magnesium. *Magnes. Res.* **2016**, *29*, 154–160. [CrossRef]
11. Costello, R.B.; Elin, R.J.; Rosanoff, A.; Wallace, T.C.; Guerrero-Romero, F.; Hruby, A.; Lutsey, P.L.; Nielsen, F.H.; Rodriguez-Moran, M.; Song, Y.; et al. Perspective: The Case for an Evidence-Based Reference Interval for Serum Magnesium: The Time Has Come. *Adv. Nutr.* **2016**, *7*, 977–993. [CrossRef]
12. Razzaque, M.S. Magnesium: Are we consuming enough? *Nutrients* **2018**, *10*, 1863. [CrossRef]
13. Costello, R.; Wallace, T.; Rosanoff, A. Nutrient Information: Magnesium. *Adv. Nutr. Int. Rev. J.* **2016**, *7*, 199–201. [CrossRef]
14. Rude, R.K.; Gruber, H.E. Magnesium deficiency and osteoporosis: Animal and human observations. *J. Nutr. Biochem.* **2004**, *15*, 710–716. [CrossRef]
15. Whang, R. Frequency of hypomagnesemia and hypermagnesemia. Requested vs routine. *JAMA* **1990**, *263*, 3063–3064. [CrossRef]

16. Rosanoff, A.; Weaver, C.M.; Rude, R.K. Suboptimal magnesium status in the United States: Are the health consequences underestimated? *Nutr. Rev.* **2012**, *70*, 153–164. [CrossRef]
17. Khalil, S.I. Magnesium the forgotten cation. *Int. J. Cardiol.* **1999**, *68*, 133–135.
18. DiNicolantonio, J.J.; O'Keefe, J.H.; Wilson, W. Subclinical magnesium deficiency: A principal driver of cardiovascular disease and a public health crisis. *Open Heart* **2018**, *5*, e000668. [CrossRef] [PubMed]
19. Martin, K.J.; González, E.A.; Slatopolsky, E. Clinical Consequences and Management of Hypomagnesemia. *J. Am. Soc. Nephrol.* **2008**, *20*, 2291–2295. [CrossRef] [PubMed]
20. Van Laecke, S. Hypomagnesemia and hypermagnesemia. *Acta Clin. Belgica Int. J. Clin. Lab. Med.* **2019**, *74*, 41–47. [CrossRef]
21. Gröber, U.; Schmidt, J.; Kisters, K. Magnesium in Prevention and Therapy. *Nutrients* **2015**, *7*, 8199–8226. [CrossRef] [PubMed]
22. Jahnen-Dechent, W.; Ketteler, M. Magnesium basics. *Clin. Kidney J.* **2012**, *5*, i3–i14. [CrossRef]
23. Gröber, U. Magnesium and Drugs. *Int. J. Mol. Sci.* **2019**, *20*, 2094. [CrossRef] [PubMed]
24. Schäffers, O.J.M.; Hoenderop, J.G.J.; Bindels, R.J.M.; De Baaij, J.H.F. The rise and fall of novel renal magnesium transporters. *Am. J. Physiol. Physiol.* **2018**, *314*, F1027–F1033. [CrossRef]
25. Auwercx, J.; Rybarczyk, P.; Kischel, P.; Dhennin-Duthille, I.; Chatelain, D.; Sevestre, H.; Van Seuningen, I.; Ouadid-Ahidouch, H.; Jonckheere, N.; Gautier, M. $Mg^{2+}$ transporters in digestive cancers. *Nutrients* **2021**, *13*, 210. [CrossRef] [PubMed]
26. Zou, Z.-G.; Rios, F.J.; Montezano, A.C.; Touyz, R.M. TRPM7, Magnesium, and Signaling. *Int. J. Mol. Sci.* **2019**, *20*, 1877. [CrossRef]
27. Gile, J.; Ruan, G.; Abeykoon, J.; McMahon, M.M.; Witzig, T. Magnesium: The overlooked electrolyte in blood cancers? *Blood Rev.* **2020**, *44*, 100676. [CrossRef] [PubMed]
28. Huang, Y.; Jin, F.; Funato, Y.; Xu, Z.; Zhu, W.; Wang, J.; Sun, M.; Zhao, Y.; Yu, Y.; Miki, H.; et al. Structural basis for the $Mg^{2+}$ recognition and regulation of the CorC $Mg^{2+}$ transporter. *Sci. Adv.* **2021**, *7*, eabe6140. [CrossRef]
29. Giménez-Mascarell, P.; Schirrmacher, C.E.; Martínez-Cruz, L.A.; Müller, D. Novel aspects of renal magnesium homeostasis. *Front. Pediatr.* **2018**, *6*, 77. [CrossRef]
30. Blaine, J.; Chonchol, M.; Levi, M. Renal Control of Calcium, Phosphate, and Magnesium Homeostasis. *Clin. J. Am. Soc. Nephrol.* **2014**, *10*, 1257–1272. [CrossRef]
31. Wang, J.; Um, P.; Dickerman, B.A.; Liu, J. Zinc, Magnesium, Selenium and Depression: A Review of the Evidence, Potential Mechanisms and Implications. *Nutrients* **2018**, *10*, 584. [CrossRef]
32. Dolati, S.; Rikhtegar, R.; Mehdizadeh, A.; Yousefi, M. The Role of Magnesium in Pathophysiology and Migraine Treatment. *Biol. Trace Element Res.* **2019**, *196*, 375–383. [CrossRef] [PubMed]
33. Quamme, G.A. Recent developments in intestinal magnesium absorption. *Curr. Opin. Gastroenterol.* **2008**, *24*, 230–235. [CrossRef] [PubMed]
34. Piano, F.L.; Corsonello, A.; Corica, F. Magnesium and elderly patient: The explored paths and the ones to be explored: A review. *Magnes. Res.* **2019**, *32*, 1–15.
35. Sun, Y.; Sukumaran, P.; Singh, B.B. Magnesium-Induced Cell Survival Is Dependent on TRPM7 Expression and Function. *Mol. Neurobiol.* **2020**, *57*, 528–538. [CrossRef]
36. Ebel, H.; Günther, T.; Günther, H.E.T. Magnesium Metabolism: A Review. *Clin. Chem. Lab. Med.* **1980**, *18*, 257–270. [CrossRef]
37. Seo, J.W.; Park, T.J. Magnesium Metabolism. *Electrolytes Blood Press.* **2008**, *6*, 86–95. [CrossRef]
38. Bairoch, A. The ENZYME database in 2000. *Nucleic Acids Res.* **2000**, *28*, 304–305. [CrossRef]
39. Caspi, R.; Altman, T.; Dreher, K.; Fulcher, C.A.; Subhraveti, P.; Keseler, I.M.; Kothari, A.; Krummenacker, M.; Latendresse, M.; Mueller, L.A.; et al. The MetaCyc database of metabolic pathways and enzymes and the BioCyc collection of pathway/genome databases. *Nucleic Acids Res.* **2012**, *40*, D742–D753. [CrossRef]
40. Feng, J.; Wang, H.; Jing, Z.; Wang, Y.; Cheng, Y.; Wang, W.; Sun, W. Role of Magnesium in Type 2 Diabetes Mellitus. *Biol. Trace Element Res.* **2020**, *196*, 74–85. [CrossRef]
41. Brautigam, C.A.; Steitz, T.A. Structural and functional insights provided by crystal structures of DNA polymerases and their substrate complexes. *Curr. Opin. Struct. Biol.* **1998**, *8*, 54–63. [CrossRef]
42. Chul Suh, W.; Leirmo, S.; Thomas Record, M. Roles of $Mg^{2+}$ in the Mechanism of Formation and Dissociation of Open Complexes between Escherichia coli RNA Polymerase and the λPR Promoter: Kinetic Evidence for a Second Open Complex Requiring $Mg^{2+}$. *Biochemistry* **1992**, *31*, 7815–7825.
43. Alfrey, A.C.; Miller, N.L.; Trow, R. Effect of Age and Magnesium Depletion on Bone Magnesium Pools in Rats. *J. Clin. Investig.* **1974**, *54*, 1074–1081. [CrossRef] [PubMed]
44. Mammoli, F.; Castiglioni, S.; Parenti, S.; Cappadone, C.; Farruggia, G.; Iotti, S.; Davalli, P.; Maier, J.A.; Grande, A.; Frassineti, C. Magnesium Is a Key Regulator of the Balance between Osteoclast and Osteoblast Differentiation in the Presence of Vitamin D3. *Int. J. Mol. Sci.* **2019**, *20*, 385. [CrossRef]
45. Lu, W.-C.; Pringa, E.; Chou, L. Effect of magnesium on the osteogenesis of normal human osteoblasts. *Magnes. Res.* **2017**, *30*, 42–52. [CrossRef]
46. Zofkova, I.; Davis, M.; Blahos, J. Trace Elements Have Beneficial, as Well as Detrimental Effects on Bone Homeostasis. *Physiol. Res.* **2017**, *66*, 391–402. [CrossRef]
47. Mubagwa, K.; Gwanyanya, A.; Zakharov, S.; Macianskiene, R. Regulation of cation channels in cardiac and smooth muscle cells by intracellular magnesium. *Arch. Biochem. Biophys.* **2007**, *458*, 73–89. [CrossRef]
48. Iseri, L.T.; French, J.H. Magnesium: Nature's physiologic calcium blocker. *Am. Heart J.* **1984**, *108*, 188–193. [CrossRef]

49. Paoletti, P.; Bellone, C.; Zhou, Q. NMDA receptor subunit diversity: Impact on receptor properties, synaptic plasticity and disease. *Nat. Rev. Neurosci.* **2013**, *14*, 383–400. [CrossRef]
50. Kirkland, A.E.; Sarlo, G.L.; Holton, K.F. The Role of Magnesium in Neurological Disorders. *Nutrients* **2018**, *10*, 730. [CrossRef]
51. Caspi, R.; Billington, R.; Keseler, I.M.; Kothari, A.; Krummenacker, M.; Midford, P.E.; Ong, W.K.; Paley, S.; Subhraveti, P.; Karp, P.D. The MetaCyc database of metabolic pathways and enzymes—A 2019 update. *Nucleic Acids Res.* **2020**, *48*, D445–D453. [CrossRef]
52. Sanders, G.T.; Huijgen, H.J.; Sanders, R. Magnesium in Disease: A Review with Special Emphasis on the Serum Ionized Magnesium. *Clin. Chem. Lab. Med.* **1999**, *37*, 1011–1033. [CrossRef]
53. Garfinkel, L.; Garfinkel, D. Magnesium regulation of the glycolytic pathway and the enzymes involved. *Magnesium* **1985**, *4*, 60–72.
54. Gomez Puyou, A.; Ayala, G.; Muller, U.; Tuena de Gomez Puyou, M. Regulation of the synthesis and hydrolysis of ATP by mitochondrial ATPase. Role of $Mg^{2+}$. *J. Biol. Chem.* **1983**, *258*, 13680–13684. [CrossRef]
55. Willson, V.J.C.; Tipton, K.F. The Activation of Ox-Brain NAD+-Dependent Isocitrate Dehydrogenase by Magnesium Ions. *JBIC J. Biol. Inorg. Chem.* **1981**, *113*, 477–483. [CrossRef] [PubMed]
56. Panov, A.; Scarpa, A. Independent Modulation of the Activity of α-Ketoglutarate Dehydrogenase Complex by $Ca^{2+}$ and $Mg^{2+}$. *Biochemtry* **1996**, *35*, 427–432. [CrossRef] [PubMed]
57. Thomas, A.P.; Diggle, T.A.; Denton, R.M. Sensitivity of pyruvate dehydrogenase phosphate phosphatase to magnesium ions. Similar effects of spermine and insulin. *Biochem. J.* **1986**, *238*, 83–91. [CrossRef]
58. Galkin, M.A.; Syroeshkin, A.V. Kinetic mechanism of ATP synthesis catalyzed by mitochondrial Fo x F1-ATPase. *Biochemtry* **1999**, *64*, 1176–1185.
59. Barbiroli, B.; Iotti, S.; Cortelli, P.; Martinelli, P.; Lodi, R.; Carelli, V.; Montagna, P. Low Brain Intracellular Free Magnesium in Mitochondrial Cytopathies. *Br. J. Pharmacol.* **1999**, *19*, 528–532. [CrossRef]
60. Shigematsu, M.; Nakagawa, R.; Tomonaga, S.; Funaba, M.; Matsui, T. Fluctuations in metabolite content in the liver of magnesium-deficient rats. *Br. J. Nutr.* **2016**, *116*, 1694–1699. [CrossRef]
61. Mooren, F.C. Magnesium and disturbances in carbohydrate metabolism. *Diabetes Obes. Metab.* **2015**, *17*, 813–823. [CrossRef] [PubMed]
62. Barbagallo, M.; Dominguez, L.J. Magnesium metabolism in type 2 diabetes mellitus, metabolic syndrome and insulin resistance. *Arch. Biochem. Biophys.* **2007**, *458*, 40–47. [CrossRef] [PubMed]
63. Kostov, K. Effects of Magnesium Deficiency on Mechanisms of Insulin Resistance in Type 2 Diabetes: Focusing on the Processes of Insulin Secretion and Signaling. *Int. J. Mol. Sci.* **2019**, *20*, 1351. [CrossRef] [PubMed]
64. Sohrabipour, S.; Sharifi, M.R.; Talebi, A.; Soltani, N. Effect of magnesium sulfate administration to improve insulin resistance in type 2 diabetes animal model: Using the hyperinsulinemic-euglycemic clamp technique. *Fundam. Clin. Pharmacol.* **2018**, *32*, 603–616. [CrossRef] [PubMed]
65. Bohn, T.; Davidsson, L.; Walczyk, T.; Hurrell, R.F. Fractional magnesium absorption is significantly lower in human subjects from a meal served with an oxalate-rich vegetable, spinach, as compared with a meal served with kale, a vegetable with a low oxalate content. *Br. J. Nutr.* **2004**, *91*, 601–606. [CrossRef]
66. Anastassopoulou, J.; Theophanides, T. Magnesium-DNA interactions and the possible relation of magnesium to carcinogenesis. Irradiation and free radicals. *Crit. Rev. Oncol.* **2002**, *42*, 79–91. [CrossRef]
67. Wolf, F.; Maier, J.; Nasulewicz, A.; Feillet-Coudray, C.; Simonacci, M.; Mazur, A.; Cittadini, A. Magnesium and neoplasia: From carcinogenesis to tumor growth and progression or treatment. *Arch. Biochem. Biophys.* **2007**, *458*, 24–32. [CrossRef]
68. Yang, W. An overview of Y-family DNA polymerases and a case study of human DNA polymerase π. *Biochemistry* **2014**, *53*, 2793–2803. [CrossRef] [PubMed]
69. Lindahl, T.; Adams, A.; Fresco, J.R. Renaturation of transfer ribonucleic acids through site binding of magnesium. *Proc. Natl. Acad. Sci. USA* **1966**, *55*, 941–948. [CrossRef]
70. Misra, V.K.; Draper, D.E. The linkage between magnesium binding and RNA folding 1 1Edited by B. Honig. *J. Mol. Biol.* **2002**, *317*, 507–521. [CrossRef]
71. Tan, Z.J.; Chen, S.J. Importance of diffuse metal ion binding to RNA. *Met. Ions Life Sci.* **2011**, *9*, 101–124. [PubMed]
72. Fandilolu, P.M.; Kamble, A.S.; Dound, A.S.; Sonawane, K.D. Role of Wybutosine and $Mg^{2+}$ Ions in Modulating the Structure and Function of tRNAPhe: A Molecular Dynamics Study. *ACS Omega* **2019**, *4*, 21327–21339. [CrossRef]
73. Strulson, C.A.; Boyer, J.A.; Whitman, E.E.; Bevilacqua, P.C. Molecular crowders and cosolutes promote folding cooperativity of RNA under physiological ionic conditions. *RNA* **2014**, *20*, 331–347. [CrossRef]
74. Yamagami, R.; Bingaman, J.L.; Frankel, E.A.; Bevilacqua, P.C. Cellular conditions of weakly chelated magnesium ions strongly promote RNA stability and catalysis. *Nat. Commun.* **2018**, *9*, 2149. [CrossRef]
75. Yamagami, R.; Huang, R.; Bevilacqua, P.C. Cellular Concentrations of Nucleotide Diphosphate-Chelated Magnesium Ions Accelerate Catalysis by RNA and DNA Enzymes. *Biochemtry* **2019**, *58*, 3971–3979. [CrossRef] [PubMed]
76. Forrest, D. Unusual relatives of the multisubunit RNA polymerase. *Biochem. Soc. Trans.* **2019**, *47*, 219–228. [CrossRef]
77. Rubin, H. The membrane, magnesium, mitosis (MMM) model of cell proliferation control. *Magnes. Res.* **2005**, *18*, 268–274.
78. Castiglioni, S.; Maier, J.A. Magnesium and cancer: A dangerous liason. *Magnes. Res.* **2011**, *24*, 92–100. [CrossRef] [PubMed]
79. Boskey, A.L.; Rimnac, C.M.; Bansal, M.; Federman, M.; Lian, J.; Boyan, B.D. Effect of short-term hypomagnesemia on the chemical and mechanical properties of rat bone. *J. Orthop. Res.* **1992**, *10*, 774–783. [CrossRef]

80. Castiglioni, S.; Cazzaniga, A.; Albisetti, W.; Maier, J.A.M. Magnesium and Osteoporosis: Current State of Knowledge and Future Research Directions. *Nutrients* **2013**, *5*, 3022–3033. [CrossRef]
81. Salimi, M.H.; Heughebaert, J.C.; Nancollas, G.H. Crystal growth of calcium phosphates in the presence of magnesium ions. *Langmuir* **1985**, *1*, 119–122. [CrossRef]
82. Cohen, L.; Kitzes, R. Infrared spectroscopy and magnesium content of bone mineral in osteoporotic women. *ISR J. Med. Sci.* **1981**, *17*, 1123–1125. [PubMed]
83. Swaminathan, R. Magnesium Metabolism and its Disorders. *Clin. Biochem. Rev.* **2003**, *24*, 47–66. [PubMed]
84. Ozsoylu, S.; Hanioğlu, N. Serum magnesium levels in children with vitamin D deficiency rickets. *Turk. J. Pediatr.* **1977**, *19*, 89–96.
85. Uwitonze, A.M.; Razzaque, M.S. Role of Magnesium in Vitamin D Activation and Function. *J. Am. Osteopat. Assoc.* **2018**, *118*, 181–189. [CrossRef]
86. Reddy, V.; Sivakumar, B. Magnesium-dependent vitamin-D-resistant rickets. *Lancet* **1974**, *303*, 963–965. [CrossRef]
87. Erem, S.; Atfi, A.; Razzaque, M.S. Anabolic effects of vitamin D and magnesium in aging bone. *J. Steroid Biochem. Mol. Biol.* **2019**, *193*, 105400. [CrossRef]
88. Lewiecki, E.M.; Miller, P.D. Skeletal Effects of Primary Hyperparathyroidism: Bone Mineral Density and Fracture Risk. *J. Clin. Densitom.* **2013**, *16*, 28–32. [CrossRef]
89. Vetter, T.; Lohse, M.J. Magnesium and the parathyroid. *Curr. Opin. Nephrol. Hypertens.* **2002**, *11*, 403–410. [CrossRef] [PubMed]
90. Rodríguez-Ortiz, M.E.; Canalejo, A.; Herencia, C.; Martínez-Moreno, J.M.; Peralta-Ramírez, A.; Perez-Martinez, P.; Navarro-González, J.F.; Rodríguez, M.; Peter, M.; Gundlach, K.; et al. Magnesium modulates parathyroid hormone secretion and upregulates parathyroid receptor expression at moderately low calcium concentration. *Nephrol. Dial. Transplant.* **2014**, *29*, 282–289. [CrossRef]
91. Rude, R.K.; Singer, F.R.; Gruber, H.E. Skeletal and Hormonal Effects of Magnesium Deficiency. *J. Am. Coll. Nutr.* **2009**, *28*, 131–141. [CrossRef]
92. Mazur, A.; Maier, J.A.; Rock, E.; Gueux, E.; Nowacki, W.; Rayssiguier, Y. Magnesium and the inflammatory response: Potential physiopathological implications. *Arch. Biochem. Biophys.* **2007**, *458*, 48–56. [CrossRef] [PubMed]
93. Nielsen, F.H. Magnesium deficiency and increased inflammation: Current perspectives. *J. Inflamm. Res.* **2018**, *11*, 25–34. [CrossRef] [PubMed]
94. Klein, G.L. The Role of Calcium in Inflammation-Associated Bone Resorption. *Biomolecules* **2018**, *8*, 69. [CrossRef]
95. Güzel, A.; Doğan, E.; Türkçü, G.; Kuyumcu, M.; Kaplan, İ.; Çelik, F.; Yıldırım, Z.B. Dexmedetomidine and Magnesium Sulfate: A Good Combination Treatment for Acute Lung Injury? *J. Investig. Surg.* **2019**, *32*, 331–342. [CrossRef]
96. Tang, C.-F.; Ding, H.; Jiao, R.-Q.; Wu, X.-X.; Kong, L.-D. Possibility of magnesium supplementation for supportive treatment in patients with COVID-19. *Eur. J. Pharmacol.* **2020**, *886*, 173546. [CrossRef] [PubMed]
97. Iotti, S.; Wolf, F.; Mazur, A.; Maier, J.A. The COVID-19 pandemic: Is there a role for magnesium? Hypotheses and perspectives. *Magnes. Res.* **2020**, *33*, 21–27. [CrossRef]
98. White, R.E.; Hartzell, H.C. Effects of intracellular free magnesium on calcium current in isolated cardiac myocytes. *Science* **1988**, *239*, 778–780. [CrossRef] [PubMed]
99. Wang, M.; Tashiro, M.; Berlin, J.R. Regulation of L-type calcium current by intracellular magnesium in rat cardiac myocytes. *J. Physiol.* **2004**, *555*, 383–396. [CrossRef]
100. Chakraborti, S.; Chakraborti, T.; Mandal, M.; Mandal, A.; Das, S.; Ghosh, S. Protective role of magnesium in cardiovascular diseases: A review. *Mol. Cell. Biochem.* **2002**, *238*, 163–179. [CrossRef] [PubMed]
101. Rasmussen, H.S.; Thomsen, P.E.B. The electrophysiological effects of intravenous magnesium on human sinus node, atrioventricular node, atrium, and ventricle. *Clin. Cardiol.* **1989**, *12*, 85–90. [CrossRef] [PubMed]
102. Severino, P.; Netti, L.; Mariani, M.V.; Maraone, A.; D'Amato, A.; Scarpati, R.; Infusino, F.; Pucci, M.; LaValle, C.; Maestrini, V.; et al. Prevention of Cardiovascular Disease: Screening for Magnesium Deficiency. *Cardiol. Res. Pract.* **2019**, *2019*, 4874921. [CrossRef] [PubMed]
103. Al Alawi, A.M.; Majoni, S.W.; Falhammar, H. Magnesium and Human Health: Perspectives and Research Directions. *Int. J. Endocrinol.* **2018**, *2018*, 1–17. [CrossRef]
104. Houston, M. The Role of Magnesium in Hypertension and Cardiovascular Disease. *J. Clin. Hypertens.* **2011**, *13*, 843–847. [CrossRef] [PubMed]
105. Bilbey, D.L.; Prabhakaran, V.M. Muscle cramps and magnesium deficiency: Case reports. *Can. Fam. Physician Med. Fam. Can.* **1996**, *42*, 1348–1351.
106. Garrison, S.R.; Korownyk, C.S.; Kolber, M.R.; Allan, G.M.; Musini, V.M.; Sekhon, R.K.; Dugré, N. Magnesium for skeletal muscle cramps. *Cochrane Database Syst. Rev.* **2020**. [CrossRef]
107. Stroebel, D.; Casado, M.; Paoletti, P. Triheteromeric NMDA receptors: From structure to synaptic physiology. *Curr. Opin. Physiol.* **2018**, *2*, 1–12. [CrossRef]
108. Möykkynen, T.; Uusi-Oukari, M.; Heikkilä, J.; Lovinger, D.M.; Lüddens, H.; Korpi, E.R. Magnesium potentiation of the function of native and recombinant GABAA receptors. *Neuroreport* **2001**, *12*, 2175–2179. [CrossRef]
109. Olloquequi, J.; Cornejo-Córdova, E.; Verdaguer, E.; Soriano, F.X.; Binvignat, O.; Auladell, C.; Camins, A. Excitotoxicity in the pathogenesis of neurological and psychiatric disorders: Therapeutic implications. *J. Psychopharmacol.* **2018**, *32*, 265–275. [CrossRef]

110. Steinert, J.R.; Postlethwaite, M.; Jordan, M.D.; Chernova, T.; Robinson, S.W.; Forsythe, I.D. NMDAR-mediated EPSCs are maintained and accelerate in time course during maturation of mouse and rat auditory brainstem in vitro. *J. Physiol.* **2010**, *588*, 447–463. [CrossRef] [PubMed]
111. Bigal, M.E.; Walter, S.; Rapoport, A.M. Calcitonin Gene-Related Peptide (CGRP) and Migraine Current Understanding and State of Development. *Headache J. Head Face Pain* **2013**, *53*, 1230–1244. [CrossRef]
112. Weglicki, W.B. Hypomagnesemia and Inflammation: Clinical and Basic Aspects. *Annu. Rev. Nutr.* **2012**, *32*, 55–71. [CrossRef] [PubMed]
113. Medeiros, D.M. Dietary Reference Intakes: The Essential Guide to Nutrient Requirements. *Am. J. Clin. Nutr.* **2007**, *85*, 924. [CrossRef]
114. Institute of Medicine (US) Standing Committee on the Scientific Evaluation of Dietary Reference Intakes. *Dietary Reference Intakes for Calcium, Phosphorus, Magnesium, Vitamin D, and Fluoride*; National Academies Press: Washington, DC, USA, 1997.
115. Price, M.; Preedy, V. Dietary Reference Values. In *Metabolism and Pathophysiology of Bariatric Surgery*; Elsevier BV: Amsterdam, The Netherlands, 2017; pp. 399–417.
116. EFSA Scientific Panel NDA. Scientific Opinion on Dietary Reference Values for magnesium. *EFSA J.* **2015**, *13*, 4186.
117. Cioffi, I.; Imperatore, N.; Di Vincenzo, O.; Pagano, M.C.; Santarpia, L.; Pellegrini, L.; Testa, A.; Marra, M.; Contaldo, F.; Castiglione, F.; et al. Evaluation of nutritional adequacy in adult patients with Crohn's disease: A cross-sectional study. *Eur. J. Nutr.* **2020**, *59*, 3647–3658. [CrossRef]
118. Di Riferimento, L.L.D.A. *Di Nutrienti ed Energia per la Popolazione Italiana*; Doc. di Sintesi per XXXV Congr.; Società Italiana di Nutrizione Umana (SINU): Milano, Italy, 2012.
119. Dietary Reference Values for nutrients Summary report. *EFSA Support. Publ.* **2017**, *14*, e15121. [CrossRef]
120. Tedstone, A.; Dunce, N.; Aviles, M.; Shetty, P.; Daniels, L. Effectiveness of interventions to promote healthy feeding in infants under one year of age. *Natl. Inst. Heal. Res.* **2018**, *61*, 1012–1021. [CrossRef]
121. O'Neill, L.M.; Dwyer, J.T.; Bailey, R.L.; Reidy, K.C.; Saavedra, J.M. Harmonizing Micronutrient Intake Reference Ranges for Dietary Guidance and Menu Planning in Complementary Feeding. *Curr. Dev. Nutr.* **2020**, *4*, nzaa017. [CrossRef]
122. Shergill-Bonner, R. Micronutrients. *Paediatr. Child Health.* **2017**, *27*, 357–362. [CrossRef]
123. Melby, M.K.; Utsugi, M.; Miyoshi, M.; Watanabe, S. Overview of nutrition reference and dietary recommendations in Japan: Application to nutrition policy in Asian countries. *Asia Pac. J. Clin. Nutr.* **2008**, *17* (Suppl. S2), 394–398. [PubMed]
124. Durlach, J.; Pagès, N.; Bac, P.; Bara, M.; Guiet-Bara, A. New data on the importance of gestational Mg deficiency. *Magnes. Res.* **2004**, *17*, 116–125. [CrossRef] [PubMed]
125. Makrides, M.; Crosby, D.D.; Shepherd, E.; Crowther, C.A. Magnesium supplementation in pregnancy. *Cochrane Database Syst. Rev.* **2014**, *2014*, CD000937. [CrossRef]
126. Lukaski, H.C. Magnesium, zinc, and chromium nutriture and physical activity. *Am. J. Clin. Nutr.* **2000**, *72*, 585S–593S. [CrossRef] [PubMed]
127. Zhang, Y.; Xun, P.; Wang, R.; Mao, L.; He, K. Can Magnesium Enhance Exercise Performance? *Nutrients* **2017**, *9*, 946. [CrossRef] [PubMed]
128. Nielsen, F.H.; Lukaski, H.C. Update on the relationship between magnesium and exercise. *Magnes. Res.* **2006**, *19*, 180.
129. Setaro, L.; Santos-Silva, P.R.; Nakano, E.Y.; Sales, C.H.; Nunes, N.; Greve, J.M.; Colli, C. Magnesium status and the physical performance of volleyball players: Effects of magnesium supplementation. *J. Sports Sci.* **2013**, *32*, 438–445. [CrossRef] [PubMed]
130. Casoni, I.; Guglielmini, C.; Graziano, L.; Reali, M.; Mazzotta, D.; Abbasciano, V. Changes of Magnesium Concentrations in Endurance Athletes. *Int. J. Sports Med.* **1990**, *11*, 234–237. [CrossRef]
131. Reno, A.M.; Green, M.; Killen, L.G.; O'Neal, E.K.; Pritchett, K.; Hanson, Z. Effects of Magnesium Supplementation on Muscle Soreness and Performance. *J. Strength Cond. Res.* **2020**. [CrossRef] [PubMed]
132. Ismail, A.A.A.; Ismail, Y.; Ismail, A.A. Clinical assessment of magnesium status in the adult: An overview. In *Magnesium in Human Health and Disease*; Springer: Berlin/Heidelberg, Germany, 2013; ISBN 9781627030441.
133. Filippi, J.; Al-Jaouni, R.; Wiroth, J.B.; Hébuterne, X.; Schneider, S.M. Nutritional deficiencies in patients with Crohn's disease in remission. *Inflamm. Bowel Dis.* **2006**, *12*, 185–191. [CrossRef] [PubMed]
134. Van Langenberg, D.; Della Gatta, P.; Warmington, S.A.; Kidgell, D.J.; Gibson, P.R.; Russell, A.P. Objectively measured muscle fatigue in Crohn's disease: Correlation with self-reported fatigue and associated factors for clinical application. *J. Crohn's Coliti* **2014**, *8*, 137–146. [CrossRef]
135. Naser, S.A. Domino effect of hypomagnesemia on the innate immunity of Crohn's disease patients. *World J. Diabetes* **2014**, *5*, 527–535. [CrossRef]
136. Habtezion, A.; Silverberg, M.S.; Parkes, R.; Mikolainis, S.; Steinhart, A.H. Risk Factors for Low Bone Density in Crohn's Disease. *Inflamm. Bowel Dis.* **2002**, *8*, 87–92. [CrossRef]
137. Mukai, A.; Yamamoto, S.; Matsumura, K. Hypocalcemia secondary to hypomagnesemia in a patient with Crohn's disease. *Clin. J. Gastroenterol.* **2015**, *8*, 22–25. [CrossRef]
138. Taylor, L.; Almutairdi, A.; Shommu, N.; Fedorak, R.; Ghosh, S.; Reimer, R.A.; Panaccione, R.; Raman, M. Cross-Sectional Analysis of Overall Dietary Intake and Mediterranean Dietary Pattern in Patients with Crohn's Disease. *Nutrients* **2018**, *10*, 1761. [CrossRef]

139. Pierote, N.R.; Braz, A.F.; Barros, S.L.; Neto, J.M.M.; Parente, J.M.L.; Silva, M.D.C.M.; Beserra, M.S.; Soares, N.R.M.; Marreiro, D.N.; Nogueira, N.D.N. Effect of mineral status and glucocorticoid use on bone mineral density in patients with Crohn's disease. *Nutrients* **2018**, *48*, 13–17. [CrossRef] [PubMed]
140. Kruis, W.; Phuong Nguyen, G. Iron Deficiency, Zinc, Magnesium, Vitamin Deficiencies in Crohn's Disease: Substitute or Not? *Dig. Dis.* **2016**, *34*, 105–111. [CrossRef] [PubMed]
141. Weisshof, R.; Chermesh, I. Micronutrient deficiencies in inflammatory bowel disease. *Curr. Opin. Clin. Nutr. Metab. Care* **2015**, *18*, 576–581. [CrossRef] [PubMed]
142. Balamtekin, N.; Aksoy, Ç.; Baysoy, G.; Uslu, N.; Demir, H.; Köksal, G.; Saltık-Temizel, İ.N.; Özen, H.; Gürakan, F.; Yüce, A. Is compliance with gluten-free diet sufficient? Diet composition of celiac patients. *Turk. J. Pediatr.* **2015**, *57*, 374. [PubMed]
143. Caruso, R.; Pallone, F.; Stasi, E.; Romeo, S.; Monteleone, G. Appropriate nutrient supplementation in celiac disease. *Ann. Med.* **2013**, *45*, 522–531. [CrossRef] [PubMed]
144. Kupper, C. Dietary guidelines and implementation for celiac disease. *Gastroenterology* **2005**, *128*, S121–S127. [CrossRef] [PubMed]
145. Zanchi, C.; Di Leo, G.; Ronfani, L.; Martelossi, S.; Not, T.; Ventura, A. Bone Metabolism in Celiac Disease. *J. Pediatr.* **2008**, *153*, 262–265. [CrossRef]
146. Martin, J.; Geisel, T.; Maresch, C.; Krieger, K.; Stein, J. Inadequate Nutrient Intake in Patients with Celiac Disease: Results from a German Dietary Survey. *Digestion* **2013**, *87*, 240–246. [CrossRef]
147. Fernández, C.B.; Varela-Moreiras, G.; Úbeda, N.; Alonso-Aperte, E. Nutritional Status in Spanish Children and Adolescents with Celiac Disease on a Gluten Free Diet Compared to Non-Celiac Disease Controls. *Nutrients* **2019**, *11*, 2329. [CrossRef] [PubMed]
148. Di Nardo, G.; Villa, M.P.; Conti, L.; Ranucci, G.; Pacchiarotti, C.; Principessa, L.; Raucci, U.; Parisi, P. Nutritional Deficiencies in Children with Celiac Disease Resulting from a Gluten-Free Diet: A Systematic Review. *Nutrients* **2019**, *11*, 1588. [CrossRef]
149. González, T.; Larretxi, I.; Vitoria, J.C.; Castaño, J.; Simón, E.; Churruca, I.; Navarro, V.; Lasa, A. Celiac Male's Gluten-Free Diet Profile: Comparison to that of the Control Population and Celiac Women. *Nutrients* **2018**, *10*, 1713. [CrossRef] [PubMed]
150. Babio, N.; Alcázar, M.; Castillejo, G.; Recasens, M.; Martínez-Cerezo, F.; Gutiérrez-Pensado, V.; Masip, G.; Vaqué, C.; Vila-Martí, A.; Torres-Moreno, M.; et al. Patients With Celiac Disease Reported Higher Consumption of Added Sugar and Total Fat Than Healthy Individuals. *J. Pediatr. Gastroenterol. Nutr.* **2017**, *64*, 63–69. [CrossRef]
151. Vici, G.; Belli, L.; Biondi, M.; Polzonetti, V. Gluten free diet and nutrient deficiencies: A review. *Clin. Nutr.* **2016**, *35*, 1236–1241. [CrossRef]
152. Bascuñán, K.A.; Vespa, M.C.; Araya, M. Celiac disease: Understanding the gluten-free diet. *Eur. J. Nutr.* **2017**, *56*, 449–459. [CrossRef]
153. Barbagallo, M. Magnesium and type 2 diabetes. *World J. Diabetes* **2015**, *6*, 1152–1157. [CrossRef] [PubMed]
154. Chaudhary, D.P.; Sharma, R.; Bansal, D.D. Implications of Magnesium Deficiency in Type 2 Diabetes: A Review. *Biol. Trace Element Res.* **2009**, *134*, 119–129. [CrossRef]
155. Lopez-Ridaura, R.; Willett, W.C.; Rimm, E.B.; Liu, S.; Stampfer, M.J.; Manson, J.E.; Hu, F.B. Magnesium intake and risk of type 2 diabetes in men and women. *Diabetes Care* **2003**, *27*, 134–140. [CrossRef] [PubMed]
156. Ramadass, S.; Basu, S.; Srinivasan, A. SERUM magnesium levels as an indicator of status of Diabetes Mellitus type 2. *Diabetes Metab. Syndr. Clin. Res. Rev.* **2015**, *9*, 42–45. [CrossRef]
157. Serefko, A.; Szopa, A.; Poleszak, E. Magnesium and depression. *Magnes. Res.* **2016**, *29*, 112–119. [CrossRef]
158. Long, S.; Romani, A.M. Role of Cellular Magnesium in Human Diseases. *Austin J. Nutr. Food Sci.* **2014**, *2*, 1051.
159. Prior, P.L.; Vaz, M.J.; Ramos, A.C.; Galduróz, J.C.F. Influence of Microelement Concentration on the Intensity of Alcohol Withdrawal Syndrome. *Alcohol Alcohol.* **2015**, *50*, 152–156. [CrossRef]
160. Grochowski, C.; Blicharska, E.; Baj, J.; Mierzwińska, A.; Brzozowska, K.; Forma, A.; Maciejewski, R. Serum iron, Magnesium, Copper, and Manganese Levels in Alcoholism: A Systematic Review. *Molecules* **2019**, *24*, 1361. [CrossRef]
161. Ismail, A.A.; Ismail, N.A. Magnesium: A Mineral Essential for Health Yet Generally Underestimated or Even Ignored. *J. Nutr. Food Sci.* **2016**, *6*, 4. [CrossRef]
162. Marles, R.J. Mineral nutrient composition of vegetables, fruits and grains: The context of reports of apparent historical declines. *J. Food Compos. Anal.* **2017**, *56*, 93–103. [CrossRef]
163. Mayer, A. Historical changes in the mineral content of fruits and vegetables. *Br. Food J.* **1997**, *99*, 207–211. [CrossRef]
164. Cazzola, R.; Della Porta, M.; Manoni, M.; Iotti, S.; Pinotti, L.; Maier, J.A. Going to the roots of reduced magnesium dietary intake: A tradeoff between climate changes and sources. *Heliyon* **2020**, *6*, e05390. [CrossRef] [PubMed]
165. Olza, J.; Aranceta-Bartrina, J.; Gonzalez-Gross, M.; Ortega, R.M.; Serra-Majem, L.; Varela-Moreiras, G.; Gil, A. Reported Dietary Intake, Disparity between the Reported Consumption and the Level Needed for Adequacy and Food Sources of Calcium, Phosphorus, Magnesium and Vitamin D in the Spanish Population: Findings from the ANIBES Study. *Nutrients* **2017**, *9*, 168. [CrossRef] [PubMed]
166. Wang, Z.; Hassan, M.U.; Nadeem, F.; Wu, L.; Zhang, F.; Li, X. Magnesium Fertilization Improves Crop Yield in Most Production Systems: A Meta-Analysis. *Front. Plant Sci.* **2020**, *10*, 1727. [CrossRef]
167. Bohn, T.; Walczyk, T.; Leisibach, S.; Hurrell, R. Chlorophyll-bound Magnesium in Commonly Consumed Vegetables and Fruits: Relevance to Magnesium Nutrition. *J. Food Sci.* **2006**, *69*, S347–S350. [CrossRef]
168. Guo, W.; Nazim, H.; Liang, Z.; Yang, D. Magnesium deficiency in plants: An urgent problem. *Crop. J.* **2016**, *4*, 83–91. [CrossRef]

169. Melse-Boonstra, A. Bioavailability of Micronutrients From Nutrient-Dense Whole Foods: Zooming in on Dairy, Vegetables, and Fruits. *Front. Nutr.* **2020**, *7*, 101. [CrossRef]
170. Roe, M.; Bell, S.; Oseredczuk, M.; Christensen, T.; Westenbrink, S.; Pakkala, H.; Presser, K.; Finglas, P. Updated food composition database for nutrient intake. *EFSA Support. Publ.* **2013**, *10*, 355E. [CrossRef]
171. Departamento de Agricultura de Estados Unidos (USDA). FoodData Central. 2019. Available online: https://fdc.nal.usda.gov/ (accessed on 10 February 2021).
172. Ellam, S.; Williamson, G. Cocoa and Human Health. *Annu. Rev. Nutr.* **2013**, *33*, 105–128. [CrossRef]
173. Brink, E.J.; Beynen, A.C. Nutrition and magnesium absorption: A review. *Prog. Food Nutr. Sci.* **1992**, *16*, 125–162.
174. Schlemmer, U.; Frølich, W.; Prieto, R.M.; Grases, F. Phytate in foods and significance for humans: Food sources, intake, processing, bioavailability, protective role and analysis. *Mol. Nutr. Food Res.* **2009**, *53*, S330–S375. [CrossRef] [PubMed]
175. Lopez, H.W.; Krespine, V.; Guy, C.; Messager, A.; Demigne, C.; Remesy, C. Prolonged Fermentation of Whole Wheat Sourdough Reduces Phytate Level and Increases Soluble Magnesium. *J. Agric. Food Chem.* **2001**, *49*, 2657–2662. [CrossRef]
176. Lopez, H.W.; Leenhardt, F.; Coudray, C.; Remesy, C. Minerals and phytic acid interactions: Is it a real problem for human nutrition? *Int. J. Food Sci. Technol.* **2002**, *37*, 727–739. [CrossRef]
177. Gibson, R.; Dahdouh, S.; Grande, F.; Najera, S.; Fialon, M.; Vincent, A.; King, J.; Bailey, K.; Raboy, V.; Charrondiere, U.R. New phytate data collection: Implications for nutrient reference intakes for minerals, programmes and policies. *Ann. Nutr. Metab.* **2017**, *71*, 209–210.
178. Severo, J.S.; Morais, J.B.S.; De Freitas, T.E.C.; Cruz, K.J.C.; De Oliveira, A.R.S.; Poltronieri, F.; Marreiro, D.D.N. Metabolic and nutritional aspects of magnesium. *Nutr. Clin. Diet. Hosp.* **2015**, *35*, 67–74.
179. Vegarud, G.E.; Langsrud, T.; Svenning, C. Mineral-binding milk proteins and peptides; occurrence, biochemical and technological characteristics. *Br. J. Nutr.* **2000**, *84*, 91–98. [CrossRef]
180. Kitano, T.; Esashi, T.; Azami, S. Effect of protein intake on mineral (calcium, magnesium, and phosphorus) balance in Japanese males. *J. Nutr. Sci. Vitaminol.* **1988**, *34*, 387–398. [CrossRef] [PubMed]
181. Whiting, S.J.; Kohrt, W.M.; Warren, M.P.; Kraenzlin, M.I.; Bonjour, J.-P. Food fortification for bone health in adulthood: A scoping review. *Eur. J. Clin. Nutr.* **2016**, *70*, 1099–1105. [CrossRef]
182. Deng, X.; Song, Y.; Manson, J.E.; Signorello, L.B.; Zhang, S.M.; Shrubsole, M.J.; Ness, R.M.; Seidner, D.L.; Dai, Q. Magnesium, vitamin D status and mortality: Results from US National Health and Nutrition Examination Survey (NHANES) 2001 to 2006 and NHANES III. *BMC Med.* **2013**, *11*, 187. [CrossRef] [PubMed]
183. Bieńkowski, P. Commentary on: Pouteau et al. Superiority of magnesium and vitamin B6 over magnesium alone on severe stress in healthy adults with low magnessemia: A randomized, single-blind clinical trial. *PLoS ONE* **2018**, *13*, e0208454. [CrossRef]
184. Pouteau, E.; Kabir-Ahmadi, M.; Noah, L.; Mazur, A.; Dye, L.; Hellhammer, J.; Pickering, G.; DuBray, C. Superiority of combined magnesium (MG) and vitamin B6 (VITB6) supplementation over magnesium alone on severe stress in adults with low magnesemia: A randomised, single blind trial. *Clin. Nutr.* **2018**, *37*, S289–S290. [CrossRef]
185. Nielsen, F.H.; Milne, D.B. A moderately high intake compared to a low intake of zinc depresses magnesium balance and alters indices of bone turnover in postmenopausal women. *Eur. J. Clin. Nutr.* **2004**, *58*, 703–710. [CrossRef] [PubMed]
186. Johnson, S. The multifaceted and widespread pathology of magnesium deficiency. *Med. Hypotheses* **2001**, *56*, 163–170. [CrossRef]
187. Wallace, T.C. Combating COVID-19 and Building Immune Resilience: A Potential Role for Magnesium Nutrition? *J. Am. Coll. Nutr.* **2020**, *39*, 685–693. [CrossRef]
188. Takahashi, Y.; Imaizumi, Y. Hardness in Drinking Water. *Eisei Kagaku* **1988**, *34*, 475–479. [CrossRef]
189. Van Der Aa, M. Classification of mineral water types and comparison with drinking water standards. *Environ. Earth Sci.* **2003**, *44*, 554–563. [CrossRef]
190. Maraver, F.; Vitoria, I.; Ferreira-Pêgo, C.; Armijo, F.; Salas-Salvadó, J. Magnesium in tap and bottled mineral water in Spain and its contribution to nutritional recommendations. *Nutr. Hosp.* **2015**, *31*, 2297–2312. [PubMed]
191. Verhas, M.; De La Guéronnière, V.; Grognet, J.-M.; Paternot, J.; Hermanne, A.; Winkel, P.V.D.; Gheldof, R.; Martin, P.; Fantino, M.; Rayssiguier, Y. Magnesium bioavailability from mineral water. A study in adult men. *Eur. J. Clin. Nutr.* **2002**, *56*, 442–447. [CrossRef]
192. Sabatier, M.; Arnaud, M.J.; Kastenmayer, P.; Rytz, A.; Barclay, D.V. Meal effect on magnesium bioavailability from mineral water in healthy women. *Am. J. Clin. Nutr.* **2002**, *75*, 65–71. [CrossRef] [PubMed]
193. Dórea, J.G. Magnesium in Human Milk. *J. Am. Coll. Nutr.* **2000**, *19*, 210–219. [CrossRef]
194. Yang, Z.; Huffman, S.L. Review of fortified food and beverage products for pregnant and lactating women and their impact on nutritional status. *Matern. Child Nutr.* **2011**, *7*, 19–43. [CrossRef]
195. Reidy, K.C.; Bailey, R.L.; Deming, D.M.; O'Neill, L.; Carr, B.T.; Lesniauskas, R.; Johnson, W. Food Consumption Patterns and Micronutrient Density of Complementary Foods Consumed by Infants Fed Commercially Prepared Baby Foods. *Nutr. Today* **2018**, *53*, 68–78. [CrossRef]
196. Gillis, L.; Gillis, A. Nutrient Inadequacy in Obese and Non-Obese Youth. *Can. J. Diet. Pract. Res.* **2005**, *66*, 237–242. [CrossRef]
197. Poitevin, E. Determination of calcium, copper, iron, magnesium, manganese, potassium, phosphorus, sodium, and zinc in fortified food products by microwave digestion and inductively coupled plasma-optical emission spectrometry: Single-laboratory validation and ring tri. *J. AOAC Int.* **2012**, *95*, 177–185. [CrossRef]

198. Food and Drug Administration. Food Labeling: Revision of the Nutrition and Supplement Facts Labels. Final rule. *Fed. Regist.* **2016**, *81*, 33741–37999.
199. Food and Drug Administration. Food Labeling: Serving Sizes of Foods That Can Reasonably Be Consumed at One Eating Occasion; Dual-Column Labeling; Updating, Modifying, and Establishing Certain Reference Amounts Customarily Consumed; Serving Size for Breath Mints; and Technical Amendmen. *Fed. Regist.* **2016**, *81*, 34000–34047.
200. Ates, M.; Kizildag, S.; Yuksel, O.; Hosgorler, F.; Yuce, Z.; Guvendi, G.; Kandis, S.; Karakilic, A.; Koc, B.; Uysal, N. Dose-Dependent Absorption Profile of Different Magnesium Compounds. *Biol. Trace Elem. Res.* **2019**, *192*, 244–251. [CrossRef]
201. Schweigel, M.; Martens, H. Magnesium transport in the gastrointestinal tract. *Front. Biosci.* **2000**, *3*, D666–D677. [CrossRef]
202. Vormann, J. Magnesium: Nutrition and metabolism. *Mol. Aspects Med.* **2003**, *24*, 27–37. [CrossRef]
203. Vormann, J. Magnesium: Nutrition and Homoeostasis. *AIMS Public Health* **2016**, *3*, 329–340. [CrossRef]
204. Ranade, V.V.; Somberg, J.C. Bioavailability and Pharmacokinetics of Magnesium After Administration of Magnesium Salts to Humans. *Am. J. Ther.* **2001**, *8*, 345–357. [CrossRef]
205. Cosaro, E.; Bonafini, S.; Montagnana, M.; Danese, E.; Trettene, M.; Minuz, P.; Delva, P.; Fava, C. Effects of magnesium supplements on blood pressure, endothelial function and metabolic parameters in healthy young men with a family history of metabolic syndrome. *Nutr. Metab. Cardiovasc. Dis.* **2014**, *24*, 1213–1220. [CrossRef] [PubMed]
206. Walker, A.F.; Marakis, G.; Christie, S.; Byng, M. Mg citrate found more bioavailable than other Mg preparations in a randomised, double-blind study. *Magnes. Res.* **2003**, *16*, 183–191. [PubMed]
207. Uysal, N.; Kizildag, S.; Yuce, Z.; Guvendi, G.; Kandis, S.; Koc, B.; Karakilic, A.; Camsari, U.M.; Ates, M. Timeline (Bioavailability) of Magnesium Compounds in Hours: Which Magnesium Compound Works Best? *Biol. Trace Elem. Res.* **2019**, *187*, 128–136. [CrossRef]
208. Coudray, C.; Rambeau, M.; Feillet-Coudray, C.; Gueux, E.; Tressol, J.C.; Mazur, A.; Rayssiguier, Y. Study of magnesium bioavailability from ten organic and inorganic Mg salts in Mg-depleted rats using a stable isotope approach. *Magnes. Res.* **2005**, *18*, 215–223.
209. Hillier, K. Magnesium Oxide. In *xPharm: The Comprehensive Pharmacology Reference*; Elsevier BV: Amsterdam, The Netherlands, 2007.
210. Hunter, L.A.; Gibbins, K.J. Magnesium Sulfate: Past, Present, and Future. *J. Midwifery Women's Heal.* **2011**, *56*, 566–574. [CrossRef]
211. Durlach, J.; Guiet-Bara, A.; Pagès, N.; Bac, P.; Bara, M. Magnesium chloride or magnesium sulfate: A genuine question. *Magnes. Res.* **2005**, *18*, 187–192. [PubMed]
212. EFSA Panel on Nutrition, Novel Foods and Food Allergens (NDA); Turck, D.; Castenmiller, J.; De Henauw, S.; Hirsch-Ernst, K.I.; Kearney, J.; Knutsen, H.K.; Maciuk, A.; Mangelsdorf, I.; McArdle, H.J.; et al. Magnesium citrate malate as a source of magnesium added for nutritional purposes to food supplements. *EFSA J.* **2018**, *16*, e05484. [CrossRef] [PubMed]
213. Tamai, I.; Senmaru, M.; Terasaki, T.; Tsuji, A. Na$^+$- and Cl$^-$-Dependent transport of taurine at the blood-brain barrier. *Biochem. Pharmacol.* **1995**, *50*, 1783–1793. [CrossRef]
214. Tsuji, A.; Tamai, I. Sodium- and chloride-dependent transport of taurine at the blood-brain barrier. *Single Mol. Single Cell Seq.* **1996**, *403*, 385–391.
215. Covington, A.K.; Danish, E.Y. Measurement of Magnesium Stability Constants of Biologically Relevant Ligands by Simultaneous Use of pH and Ion-Selective Electrodes. *J. Solut. Chem.* **2009**, *38*, 1449–1462. [CrossRef]
216. Farruggia, G.; Castiglioni, S.; Sargenti, A.; Marraccini, C.; Cazzaniga, A.; Merolle, L.; Iotti, S.; Cappadone, C.; Maier, J.A.M. Effects of supplementation with different Mg salts in cells: Is there a clue? *Magnes. Res.* **2014**, *27*, 25–34. [CrossRef]
217. Wang, R.; Chen, C.; Liu, W.; Zhou, T.; Xun, P.; He, K.; Chen, P. The effect of magnesium supplementation on muscle fitness: A meta-analysis and systematic review. *Magnes. Res.* **2017**, *30*, 120–132. [CrossRef] [PubMed]
218. Carpenter, T.O.; DeLucia, M.C.; Zhang, J.H.; Bejnerowicz, G.; Tartamella, L.; Dziura, J.; Petersen, K.F.; Befroy, D.; Cohen, D. A Randomized Controlled Study of Effects of Dietary Magnesium Oxide Supplementation on Bone Mineral Content in Healthy Girls. *J. Clin. Endocrinol. Metab.* **2006**, *91*, 4866–4872. [CrossRef]
219. Kunutsor, S.K.; Whitehouse, M.R.; Blom, A.W.; Laukkanen, J.A. Low serum magnesium levels are associated with increased risk of fractures: A long-term prospective cohort study. *Eur. J. Epidemiol.* **2017**, *32*, 593–603. [CrossRef]
220. Musso, C.G. Magnesium metabolism in health and disease. *Int. Urol. Nephrol.* **2009**, *41*, 357–362. [CrossRef]
221. Nanduri, A.; Saleem, S.; Khalaf, M. Severe hypermagnesemia. *Chest* **2020**, *158*, A1016. [CrossRef]
222. National Institutes of Health NIH Magnesium—Health Professional Fact Sheet. *Fact Sheet Healyh Prof.* **2018**. Available online: https://ods.od.nih.gov/factsheets/Magnesium-HealthProfessional/ (accessed on 10 February 2021).
223. Moodie, E. Modern Trends in Animal Health and HUSBANDRY Hypocalcaemia and Hypomagnesaemia. *Br. Vet. J.* **1965**, *121*, 338–349. [CrossRef]
224. Murphy, E. Mysteries of Magnesium Homeostasis. *Circ. Res.* **2000**, *86*, 245–248. [CrossRef] [PubMed]
225. Romani, A.M.P. Magnesium in health and disease. *Met. Ions Life Sci.* **2013**, *2013*, 49–79.
226. Glasdam, S.M.; Glasdam, S.; Peters, G.H. The Importance of Magnesium in the Human Body: A Systematic Literature Review. *Adv. Clin. Chem.* **2016**, *73*, 169–193. [PubMed]
227. Workinger, J.L.; Doyle, R.P.; Bortz, J. Challenges in the Diagnosis of Magnesium Status. *Nutrients* **2018**, *10*, 1202. [CrossRef]
228. Ismail, Y.; Ismail, A.A.; Ismail, A.A.A. The underestimated problem of using serum magnesium measurements to exclude magnesium deficiency in adults; A health warning is needed for "normal" results. *Clin. Chem. Lab. Med.* **2010**, *48*, 323–327. [CrossRef]

229. Rylander, R.; Remer, T.; Berkemeyer, S.; Vormann, J. Acid-Base Status Affects Renal Magnesium Losses in Healthy, Elderly Persons. *J. Nutr.* **2006**, *136*, 2374–2377. [CrossRef]
230. Löwik, M.R.; Van Dokkum, W.; Kistemaker, C.; Schaafsma, G.; Ockhuizen, T. Body composition, health status and urinary magnesium excretion among elderly people (Dutch Nutrition Surveillance System). *Magnes. Res.* **1993**, *6*, 223–232. [PubMed]
231. Ware, E.B.; Smith, J.A.; Zhao, W.; Ganesvoort, R.T.; Curhan, G.C.; Pollak, M.; Mount, D.B.; Turner, S.T.; Chen, G.; Shah, R.J.; et al. Genome-wide Association Study of 24-Hour Urinary Excretion of Calcium, Magnesium, and Uric Acid. *Mayo Clin. Proc. Innov. Qual. Outcomes* **2019**, *3*, 448–460. [CrossRef]
232. Gant, C.M.; Soedamah-Muthu, S.S.; Binnenmars, S.H.; Bakker, S.J.L.; Navis, G.; Laverman, G.D. Higher Dietary Magnesium Intake and Higher Magnesium Status Are Associated with Lower Prevalence of Coronary Heart Disease in Patients with Type 2 Diabetes. *Nutrients* **2018**, *10*, 307. [CrossRef]
233. Xu, B.; Sun, J.; Deng, X.; Huang, X.; Sun, W.; Xu, Y.; Xu, M.; Lu, J.; Bi, Y. Low Serum Magnesium Level Is Associated with Microalbuminuria in Chinese Diabetic Patients. *Int. J. Endocrinol.* **2013**, *2013*, 1–6. [CrossRef]
234. Suliburska, J.; Bogdański, P.; Szulińska, M.; Pupek-Musialik, D. Short-Term Effects of Sibutramine on Mineral Status and Selected Biochemical Parameters in Obese Women. *Biol. Trace Element Res.* **2012**, *149*, 163–170. [CrossRef] [PubMed]
235. Haigney, M.C.; Silver, B.; Tanglao, E.; Silverman, H.S.; Hill, J.D.; Shapiro, E.; Gerstenblith, G.; Schulman, S.P. Noninvasive Measurement of Tissue Magnesium and Correlation With Cardiac Levels. *Circulation* **1995**, *92*, 2190–2197. [CrossRef]
236. Shechter, M.; Sharir, M.; Paul, M.J.; James, L.; Burton, F.; Noel, S.C.; Merz, B. Oral magnesium therapy improves endothelial function in patients with coronary artery disease. *Circulation* **2000**, *102*, 2353–2358. [CrossRef]
237. Silver, B.B. Development of Cellular Magnesium Nano-Analysis in Treatment of Clinical Magnesium Deficiency. *J. Am. Coll. Nutr.* **2004**, *23*, 732S–737S. [CrossRef] [PubMed]
238. Cameron, D.; Welch, A.A.; Adelnia, F.; Bergeron, C.M.; Reiter, D.A.; Dominguez, L.J.; Brennan, N.A.; Fishbein, K.W.; Spencer, R.G.; Ferrucci, L. Age and Muscle Function Are More Closely Associated With Intracellular Magnesium, as Assessed by 31P Magnetic Resonance Spectroscopy, Than With Serum Magnesium. *Front. Physiol.* **2019**, *10*, 1454. [CrossRef]
239. Iotti, S.; Malucelli, E. In vivo assessment of $Mg^{2+}$ in human brain and skeletal muscle by 31P-MRS. *Magnes. Res.* **2008**, *21*, 157–162.
240. McCully, K.K.; Turner, T.N.; Langley, J.; Zhao, Q. The reproducibility of measurements of intramuscular magnesium concentrations and muscle oxidative capacity using 31P MRS. *Dyn. Med.* **2009**, *8*, 5. [CrossRef]
241. Malucelli, E.; Lodi, R.; Martinuzzi, A.; Tonon, C.; Barbiroli, B.; Iotti, S. Free $Mg^{2+}$ concentration in the calf muscle of glycogen phosphorylase and phosphofructokinase deficiency patients assessed in different metabolic conditions by 31P MRS. *Dyn. Med.* **2005**, *4*, 7. [CrossRef]
242. Pironi, L.; Malucelli, E.; Guidetti, M.; Lanzoni, E.; Farruggia, G.; Pinna, A.D.; Barbiroli, B.; Iotti, S. The complex relationship between magnesium and serum parathyroid hormone: A study in patients with chronic intestinal failure. *Magnes. Res.* **2009**, *22*, 37–43. [CrossRef]
243. Nelander, M.; Weis, J.; Bergman, L.; Larsson, A.; Wikstrom, A.K.; Wikstrom, J. Cerebral magnesium levels in preeclampsia; A phosphorus magnetic resonance spectroscopy study. *Am. J. Hypertens.* **2017**, *30*, 667–672. [CrossRef]
244. Mairiang, E.; Hanpanich, P.; Sriboonlue, P. In vivo 31P-MRS assessment of muscle-pH, cytolsolic-[$Mg^{2+}$] and phosphorylation potential after supplementing hypokaliuric renal stone patients with potassium and magnesium salts. *Magn. Reson. Imaging* **2004**, *22*, 715–719. [CrossRef]
245. Reyngoudt, H.; Kolkovsky, A.L.L.; Carlier, P.G. Free intramuscular $Mg^{2+}$ concentration calculated using both 31P and 1H NMRS-based pH in the skeletal muscle of Duchenne muscular dystrophy patients. *NMR Biomed.* **2019**, *32*, e4115. [CrossRef] [PubMed]
246. Schutten, J.C.; Gomes-Neto, A.W.; Navis, G.; Gansevoort, R.T.; Dullaart, R.P.F.; Kootstra-Ros, J.E.; Danel, R.M.; Goorman, F.; Gans, R.O.B.; De Borst, M.H.; et al. Lower Plasma Magnesium, Measured by Nuclear Magnetic Resonance Spectroscopy, is Associated with Increased Risk of Developing Type 2 Diabetes Mellitus in Women: Results from a Dutch Prospective Cohort Study. *J. Clin. Med.* **2019**, *8*, 169. [CrossRef]
247. Lutz, N.W.; Bernard, M. Multiparametric quantification of heterogeneity of metal ion concentrations, as demonstrated for [$Mg^{2+}$] by way of 31P MRS. *J. Magn. Reson.* **2018**, *294*, 71–82. [CrossRef]
248. López-Pedrouso, M.; Lorenzo, J.M.; Zapata, C.; Franco, D. Proteins and amino acids. In *Innovative Thermal and Non-Thermal Processing*; Barba, F.J., Saraiba, J.M.A., Cravotto, G., Lorenzo, J.M., Eds.; Elsevier Inc.: Amsterdam, The Netherlands, 2019; pp. 139–168. ISBN 9781469816593.
249. Christian, G.D. Medicine, trace elements, and atomic absorption spectroscopy. *Anal. Chem.* **1969**, *41*, 24A–40A. [CrossRef]
250. DiPietro, E.; Bashor, M.; Stroud, P.; Smarr, B.; Burgess, B.; Turner, W.; Neese, J. Comparison of an inductively coupled plasma-atomic emission spectrometry method for the determination of calcium, magnesium, sodium, potassium, copper and zinc with atomic absorption spectroscopy and flame photometry methods. *Sci. Total Environ.* **1988**, *74*, 249–262. [CrossRef]
251. Uğurlu, V.; Binay, Ç.; Şimşek, E.; Bal, C. Cellular Trace Element Changes in Type 1 Diabetes Patients. *J. Clin. Res. Pediatr. Endocrinol.* **2016**, *8*, 180–186. [CrossRef]
252. Millart, H.; Durlach, V.; Durlach, J. Red blood cell magnesium concentrations: Analytical problems and significance. *Magnes. Res.* **1995**, *8*, 65–76.
253. Tashiro, M.; Inoue, H.; Konishi, M. Magnesium Homeostasis in Cardiac Myocytes of Mg-Deficient Rats. *PLoS ONE* **2013**, *8*, e73171. [CrossRef] [PubMed]

254. Schilling, K.; Larner, F.; Saad, A.; Roberts, R.; Kocher, H.M.; Blyuss, O.; Halliday, A.N.; Crnogorac-Jurcevic, T. Urine metallomics signature as an indicator of pancreatic cancer. *Metallomics* **2020**, *12*, 752–757. [CrossRef] [PubMed]
255. Ma, J.; Yan, L.; Guo, T.; Yang, S.; Liu, Y.; Xie, Q.; Ni, D.; Wang, J. Association between Serum Essential Metal Elements and the Risk of Schizophrenia in China. *Sci. Rep.* **2020**, *10*, 10875. [CrossRef] [PubMed]
256. Günzel, D.; Schlue, W.-R. Determination of [$Mg^{2+}$]i—An update on the use of $Mg^{2+}$-selective electrodes. *BioMetals* **2002**, *15*, 237–249. [CrossRef]
257. Kamochi, M.; Aibara, K.; Nakata, K.; Murakami, M.; Nandate, K.; Sakamoto, H.; Sata, T.; Shigematsu, A. Profound ionized hypomagnesemia induced by therapeutic plasma exchange in liver failure patients. *Transfusion* **2002**, *42*, 1598–1602. [CrossRef]
258. Fu, C.-Y.; Chen, S.-J.; Cai, N.-H.; Liu, Z.-H.; Zhang, M.; Wang, P.-C.; Zhao, J.-N. Increased risk of post-stroke epilepsy in Chinese patients with a TRPM6 polymorphism. *Neurol. Res.* **2019**, *41*, 378–383. [CrossRef]
259. Ordak, M.; Maj-Zurawska, M.; Matsumoto, H.; Bujalska-Zadrozny, M.; Kieres-Salomonski, I.; Nasierowski, T.; Muszynska, E.; Wojnar, M. Ionized magnesium in plasma and erythrocytes for the assessment of low magnesium status in alcohol dependent patients. *Drug Alcohol Depend.* **2017**, *178*, 271–276. [CrossRef]
260. International Federation of Clinica Ben Rayana; Burnett, R.W.; Covington, A.K.; D'Orazio, P.; Fogh-Andersen, N.; Jacobs, E.; Külpmann, W.R.; Kuwa, K.; Larsson, L.; Lewenstam, A.; et al. IFCC Guideline for sampling, measuring and reporting ionized magnesium in plasma. *Clin. Chem. Lab. Med.* **2008**, *46*, 21–26. [CrossRef]
261. Maj-Żurawska, M.; Lewenstam, A. Selectivity coefficients of ion-selective magnesium electrodes used for simultaneous determination of magnesium and calcium ions. *Talanta* **2011**, *87*, 295–301. [CrossRef]
262. Lvova, L.; Gonçalves, C.G.; Di Natale, C.; Legin, A.; Kirsanov, D.; Paolesse, R. Recent advances in magnesium assessment: From single selective sensors to multisensory approach. *Talanta* **2018**, *179*, 430–441. [CrossRef] [PubMed]
263. Lindstrom, F.; Diehl, H. Indicator for the Titration of Calcium Plus Magnesium with (Ethylenedinitrilo)tetraacetate. *Anal. Chem.* **1960**, *32*, 1123–1127. [CrossRef]
264. Abernethy, M.H.; Fowler, R.T. Micellar improvement of the calmagite complexometric measurement of magnesium in plasma. *Clin. Chem.* **1982**, *28*, 520–522. [CrossRef] [PubMed]
265. Malucelli, E.; Procopio, A.; Fratini, M.; Gianoncelli, A.; Notargiacomo, A.; Merolle, L.; Sargenti, A.; Castiglioni, S.; Cappadone, C.; Farruggia, G.; et al. Single cell versus large population analysis: Cell variability in elemental intracellular concentration and distribution. *Anal. Bioanal. Chem.* **2018**, *410*, 337–348. [CrossRef]
266. Chromý, V.; Svoboda, V.; Štěpánová, I. Spectrophotometric determination of magnesium in biological fluids with xylidyl blue II. *Biochem. Med.* **1973**, *7*, 208–217. [CrossRef]
267. Wimmer, M.C.; Artiss, J.D.; Zak, B. A kinetic colorimetric procedure for quantifying magnesium in serum. *Clin. Chem.* **1986**, *32*, 629–632. [CrossRef] [PubMed]
268. Trapani, V.; Schweigel-Röntgen, M.; Cittadini, A.; Wolf, F.I. Intracellular Magnesium Detection by Fluorescent Indicators. *Methods Enzymol.* **2012**, *505*, 421–444. [CrossRef] [PubMed]
269. Liu, M.; Yu, X.; Li, M.; Liao, N.; Bi, A.; Jiang, Y.; Liu, S.; Gong, Z.; Zeng, W. Fluorescent probes for the detection of magnesium ions ($Mg^{2+}$): From design to application. *RSC Adv.* **2018**, *8*, 12573–12587. [CrossRef]
270. Picone, G.; Cappadone, C.; Farruggia, G.; Malucelli, E.; Iotti, S. The assessment of intracellular magnesium: Different strategies to answer different questions. *Magnes. Res.* **2020**, *33*, 1–11. [CrossRef] [PubMed]
271. Suzuki, Y.; Komatsu, H.; Ikeda, T.; Saito, N.; Araki, K.; Citterio, D.; Hisamoto, H.; Kitamura, Y.; Kubota, T.; Nakagawa, J.; et al. Design and Synthesis of $Mg^{2+}$-Selective Fluoroionophores Based on a Coumarin Derivative and Application for $Mg^{2+}$ Measurement in a Living Cell. *Anal. Chem.* **2002**, *74*, 1423–1428. [CrossRef] [PubMed]
272. Komatsu, H.; Iwasawa, N.; Citterio, D.; Suzuki, Y.; Kubota, T.; Tokuno, K.; Kitamura, Y.; Oka, K.; Suzuki, K. Design and Synthesis of Highly Sensitive and Selective Fluorescein-Derived Magnesium Fluorescent Probes and Application to Intracellular 3D $Mg^{2+}$ Imaging. *J. Am. Chem. Soc.* **2004**, *126*, 16353–16360. [CrossRef]
273. Suzuki, Y.; Yokoyama, K. Development of Functional Fluorescent Molecular Probes for the Detection of Biological Substances. *Biosensors* **2015**, *5*, 337–363. [CrossRef] [PubMed]
274. Lin, Q.; Buccella, D. Highly selective, red emitting BODIPY-based fluorescent indicators for intracellular $Mg^{2+}$ imaging. *J. Mater. Chem. B* **2018**, *6*, 7247–7256. [CrossRef]
275. Gruskos, J.J.; Zhang, G.; Buccella, D. Visualizing Compartmentalized Cellular $Mg^{2+}$ on Demand with Small-Molecule Fluorescent Sensors. *J. Am. Chem. Soc.* **2016**, *138*, 14639–14649. [CrossRef] [PubMed]
276. Fujii, T.; Shindo, Y.; Hotta, K.; Citterio, D.; Nishiyama, S.; Suzuki, K.; Oka, K. Design and synthesis of a FlAsH-type $Mg^{2+}$ fluorescent probe for specific protein labeling. *J. Am. Chem. Soc.* **2014**, *136*, 2374–2381. [CrossRef]
277. Farruggia, G.; Iotti, S.; Prodi, L.; Montalti, M.; Zaccheroni, N.; Savage, P.B.; Trapani, V.; Sale, P.; Wolf, F.I. 8-Hydroxyquinoline derivatives as fluorescent sensors for magnesium in living cells. *J. Am. Chem. Soc.* **2006**, *128*, 344–350. [CrossRef]
278. Farruggia, G.; Iotti, S.; Prodi, L.; Zaccheroni, N.; Montalti, M.; Savage, P.B.; Andreani, G.; Trapani, V.; Wolf, F.I. A Simple Spectrofluorometric Assay to Measure Total Intracellular Magnesium by a Hydroxyquinoline Derivative. *J. Fluoresc.* **2009**, *19*, 11–19. [CrossRef]
279. Farruggia, G.; Iotti, S.; Lombardo, M.; Marraccini, C.; Petruzziello, D.; Prodi, L.; Sgarzi, M.; Trombini, C.; Zaccheroni, N. Microwave Assisted Synthesis of a Small Library of Substituted N,N′-Bis((8-hydroxy-7-quinolinyl)methyl)-1,10-diaza-18-crown-6 Ethers. *J. Org. Chem.* **2010**, *75*, 6275–6278. [CrossRef]

280. Sargenti, A.; Farruggia, G.; Zaccheroni, N.; Marraccini, C.; Sgarzi, M.; Cappadone, C.; Malucelli, E.; Procopio, A.; Prodi, L.; Lombardo, M.; et al. Synthesis of a highly $Mg^{2+}$-selective fluorescent probe and its application to quantifying and imaging total intracellular magnesium. *Nat. Protoc.* **2017**, *12*, 461–471. [CrossRef] [PubMed]
281. Sargenti, A.; Farruggia, G.; Malucelli, E.; Cappadone, C.; Merolle, L.; Marraccini, C.; Andreani, G.; Prodi, L.; Zaccheroni, N.; Sgarzi, M.; et al. A novel fluorescent chemosensor allows the assessment of intracellular total magnesium in small samples. *Analyst* **2014**, *139*, 1201–1207. [CrossRef] [PubMed]
282. Merolle, L.; Sponder, G.; Sargenti, A.; Mastrototaro, L.; Cappadone, C.; Farruggia, G.; Procopio, A.; Malucelli, E.; Parisse, P.; Gianoncelli, A.; et al. Overexpression of the mitochondrial Mg channel MRS2 increases total cellular Mg concentration and influences sensitivity to apoptosis. *Metallomics* **2018**, *10*, 917–928. [CrossRef] [PubMed]
283. Yadav, N.; Kumar, R.; Singh, A.K.; Mohiyuddin, S.; Gopinath, P. Systematic approach of chromone skeleton for detecting $Mg^{2+}$ ion: Applications for sustainable cytotoxicity and cell imaging possibilities. *Spectrochim. Acta Part A Mol. Biomol. Spectrosc.* **2020**, *235*, 118290. [CrossRef] [PubMed]
284. Decelle, J.; Veronesi, G.; Gallet, B.; Stryhanyuk, H.; Benettoni, P.; Schmidt, M.; Tucoulou, R.; Passarelli, M.; Bohic, S.; Clode, P.; et al. Subcellular Chemical Imaging: New Avenues in Cell Biology. *Trends Cell Biol.* **2020**, *30*, 173–188. [CrossRef]
285. De Santis, S.; Sotgiu, G.; Crescenzi, A.; Taffon, C.; Felici, A.C.; Orsini, M. On the chemical composition of psammoma bodies microcalcifications in thyroid cancer tissues. *J. Pharm. Biomed. Anal.* **2020**, *190*, 113534. [CrossRef] [PubMed]
286. Picone, G.; Cappadone, C.; Pasini, A.; Lovecchio, J.; Cortesi, M.; Farruggia, G.; Lombardo, M.; Gianoncelli, A.; Mancini, L.; Ralf, H.M.; et al. Analysis of Intracellular Magnesium and Mineral Depositions during Osteogenic Commitment of 3D Cultured Saos2 Cells. *Int. J. Mol. Sci.* **2020**, *21*, 2368. [CrossRef] [PubMed]
287. Zghoul, N.; Alam-Eldin, N.; Mak, I.T.; Silver, B.; Weglicki, W.B. Hypomagnesemia in diabetes patients: Comparison of serum and intracellular measurement of responses to magnesium supplementation and its role in inflammation. *Diabetes Metab. Syndr. Obes. Targets Ther.* **2018**, *11*, 389–400. [CrossRef]
288. Becker, R.A.; Cluff, K.; Duraisamy, N.; Casale, G.P.; Pipinos, I.I. Analysis of ischemic muscle in patients with peripheral artery disease using X-ray spectroscopy. *J. Surg. Res.* **2017**, *220*, 79–87. [CrossRef]
289. Malucelli, E.; Iotti, S.; Gianoncelli, A.; Fratini, M.; Merolle, L.; Notargiacomo, A.; Marraccini, C.; Sargenti, A.; Cappadone, C.; Farruggia, G.; et al. Quantitative Chemical Imaging of the Intracellular Spatial Distribution of Fundamental Elements and Light Metals in Single Cells. *Anal. Chem.* **2014**, *86*, 5108–5115. [CrossRef]
290. Hughes, D. Chapter 49 Cultural Influences on Medical Knowledge. In *Handbook of the Philosophy of Medicine*; Schramme, T., Edwards, S., Eds.; Springer: Dordrecht, The Netherlands, 2017; pp. 1–18.
291. Pradelli, L.; Ghetti, G. A general model for the estimation of societal costs of lost production and informal care in Italy. *Farmeconomia. Health Econ. Ther. Pathw.* **2017**, *18*, A365.
292. Yang, W.; Dall, T.M.; Beronjia, K.; Lin, J.; Semilla, A.P.; Chakrabarti, R.; Hogan, P.F.; Petersen, M.P. Economic costs of diabetes in the U.S. in 2017. *Diabetes Care* **2018**, *41*, 917–928.
293. Deuschl, G.; Beghi, E.; Fazekas, F.; Varga, T.; Christoforidi, K.A.; Sipido, E.; Bassetti, C.L.; Vos, T.; Feigin, V.L. The burden of neurological diseases in Europe: An analysis for the Global Burden of Disease Study 2017. *Lancet Public Health* **2020**, *5*, e551–e567. [CrossRef]
294. Cruz, K.J.C.; De Oliveira, A.R.S.; Pinto, D.P.; Morais, J.B.S.; Lima, F.D.S.; Colli, C.; Torres-Leal, F.L.; Marreiro, D.D.N. Influence of Magnesium on Insulin Resistance in Obese Women. *Biol. Trace Element Res.* **2014**, *160*, 305–310. [CrossRef] [PubMed]
295. Günther, T. The biochemical function of $Mg^{2+}$ in insulin secretion, insulin signal transduction and insulin resistance. *Magnes. Res.* **2010**, *23*, 5–18. [CrossRef]
296. Castellanos-Gutiérrez, A.; Sánchez-Pimienta, T.G.; Carriquiry, A.; Da Costa, T.H.M.; Ariza, A.C. Higher dietary magnesium intake is associated with lower body mass index, waist circumference and serum glucose in Mexican adults. *Nutr. J.* **2018**, *17*, 114. [CrossRef] [PubMed]
297. Rodríguez-Morán, M.; Mendía, L.E.S.; Galván, G.Z.; Guerrero-Romero, F. The role of magnesium in type 2 diabetes: A brief based-clinical review. *Magnes. Res.* **2011**, *24*, 156–162. [CrossRef] [PubMed]
298. Palmer, B.F.; Clegg, D.J. Electrolyte and Acid–Base Disturbances in Patients with Diabetes Mellitus. *N. Engl. J. Med.* **2015**, *373*, 548–559. [CrossRef] [PubMed]
299. Larsson, S.C.; Wolk, A. Magnesium intake and risk of type 2 diabetes: A meta-analysis. *J. Intern. Med.* **2007**, *262*, 208–214. [CrossRef]
300. Schulze, M.B.; Schulz, M.; Heidemann, C.; Schienkiewitz, A.; Hoffmann, K.; Boeing, H. Fiber and magnesium intake and incidence of type 2 diabetes: A prospective study and meta-analysis. *Arch. Intern. Med.* **2007**, *167*, 956–965. [CrossRef]
301. Dong, J.Y.; Xun, P.; He, K.; Qin, L.Q. Magnesium intake and risk of type 2 diabetes meta-analysis of prospective cohort studies. *Diabetes Care* **2011**, *34*, 2116–2122. [CrossRef] [PubMed]
302. Hruby, A.; Meigs, J.B.; O'Donnell, C.J.; Jacques, P.F.; McKeown, N.M. Higher Magnesium Intake Reduces Risk of Impaired Glucose and Insulin Metabolism and Progression From Prediabetes to Diabetes in Middle-Aged Americans. *Diabetes Care* **2013**, *37*, 419–427. [CrossRef] [PubMed]
303. Guerreroromero, F.; Simentalmendia, L.E.; Hernández-Ronquillo, G.; Rodriguezmoran, M. Oral magnesium supplementation improves glycaemic status in subjects with prediabetes and hypomagnesaemia: A double-blind placebo-controlled randomized trial. *Diabetes Metab.* **2015**, *41*, 202–207. [CrossRef]

304. Zhao, B.; Deng, H.; Li, B.; Chen, L.; Zou, F.; Hu, L.; Wei, Y.; Zhang, W. Association of magnesium consumption with type 2 diabetes and glucose metabolism: A systematic review and pooled study with trial sequential analysis. *Diabetes Metab. Res. Rev.* **2020**, *36*, e3243. [CrossRef] [PubMed]
305. Lin, C.-C.; Huang, Y.-L. Chromium, zinc and magnesium status in type 1 diabetes. *Curr. Opin. Clin. Nutr. Metab. Care* **2015**, *18*, 588–592. [CrossRef]
306. Sozen, T.; Ozisik, L.; Basaran, N.C. An overview and management of osteoporosis. *Eur. J. Rheumatol.* **2017**, *4*, 46–56. [CrossRef]
307. Abraham, G.E.; Grewal, H. A total dietary program emphasizing magnesium instead of calcium. Effect on the mineral density of calcaneous bone in postmenopausal women on hormonal therapy. *J. Reprod. Med.* **1990**, *35*, 503–507.
308. Stendig-Lindberg, G.; Tepper, R.; Leichter, I. Trabecular bone density in a two year controlled trial of peroral magnesium in osteoporosis. *Magnes. Res.* **1993**, *6*, 155–163.
309. Aydın, H.; Deyneli, O.; Yavuz, D.; Gozu, H.; Mutlu, N.; Kaygusuz, I.; Akalın, S.; Kaygusuz, I. Short-Term Oral Magnesium Supplementation Suppresses Bone Turnover in Postmenopausal Osteoporotic Women. *Biol. Trace Element Res.* **2009**, *133*, 136–143. [CrossRef]
310. Tucker, K.L.; Hannan, M.T.; Chen, H.; Cupples, L.A.; Wilson, P.W.; Kiel, D.P. Potassium, magnesium, and fruit and vegetable intakes are associated with greater bone mineral density in elderly men and women. *Am. J. Clin. Nutr.* **1999**, *69*, 727–736. [CrossRef]
311. Mederle, O.A.; Balas, M.; Ioanoviciu, S.D.; Gurban, C.-V.; Tudor, A.; Borza, C. Correlations between bone turnover markers, serum magnesium and bone mass density in postmenopausal osteoporosis. *Clin. Interv. Aging* **2018**, *13*, 1383–1389. [CrossRef]
312. Farsinejad-Marj, M.; Saneei, P.; Esmaillzadeh, A. Dietary magnesium intake, bone mineral density and risk of fracture: A systematic review and meta-analysis. *Osteoporos. Int.* **2016**, *27*, 1389–1399. [CrossRef]
313. Orchard, T.S.; Larson, J.C.; Alghothani, N.; Bout-Tabaku, S.; Cauley, J.A.; Chen, Z.; Lacroix, A.Z.; Wactawski-Wende, J.; Jackson, R.D. Magnesium intake, bone mineral density, and fractures: Results from the Women's Health Initiative Observational Study. *Am. J. Clin. Nutr.* **2014**, *99*, 926–933. [CrossRef] [PubMed]
314. Veronese, N.; Stubbs, B.; Solmi, M.; Noale, M.; Vaona, A.; Demurtas, J.; Maggi, S. Dietary magnesium intake and fracture risk: Data from a large prospective study. *Br. J. Nutr.* **2017**, *117*, 1570–1576. [CrossRef] [PubMed]
315. Welch, A.A.; Skinner, J.; Hickson, M. Dietary Magnesium May Be Protective for Aging of Bone and Skeletal Muscle in Middle and Younger Older Age Men and Women: Cross-Sectional Findings from the UK Biobank Cohort. *Nutrients* **2017**, *9*, 1189. [CrossRef] [PubMed]
316. Da Cunha, M.M.L.; Trepout, S.; Messaoudi, C.; Wu, T.-D.; Ortega, R.; Guerquin-Kern, J.-L.; Marco, S. Overview of chemical imaging methods to address biological questions. *Micron* **2016**, *84*, 23–36. [CrossRef] [PubMed]
317. Rosique-Esteban, N.; Guasch-Ferré, M.; Hernández-Alonso, P.; Salas-Salvadó, J. Dietary Magnesium and Cardiovascular Disease: A Review with Emphasis in Epidemiological Studies. *Nutrients* **2018**, *10*, 168. [CrossRef]
318. Rosanoff, A. Magnesium and hypertension. *Clin. Calcium* **2005**, *15*, 255–260.
319. Touyz, R.M.; Milne, F.J.; Reinach, S.G. Intracellular $Mg^{2+}$, $Ca^{2+}$, $Na^{2+}$ and $K^+$ in platelets and erythrocytes of essential hypertension patients: Relation to blood pressure. *Clin. Exp. Hypertens. Part A Theory Pract.* **1992**, *14*, 1189–1209. [CrossRef] [PubMed]
320. Dickinson, H.O.; Nicolson, D.; Campbell, F.; Cook, J.V.; Beyer, F.R.; Ford, G.A.; Mason, J. Magnesium supplementation for the management of primary hypertension in adults. *Cochrane Database Syst. Rev.* **2006**, *2006*, CD004640. [CrossRef]
321. Kass, L.S.; Weekes, J.; Carpenter, L.W. Effect of magnesium supplementation on blood pressure: A meta-analysis. *Eur. J. Clin. Nutr.* **2012**, *66*, 411–418. [CrossRef]
322. Fang, X.; Han, H.; Li, M.; Liang, C.; Fan, Z.; Aaseth, J.; He, J.; Montgomery, S.; Cao, Y. Dose-Response Relationship between Dietary Magnesium Intake and Risk of Type 2 Diabetes Mellitus: A Systematic Review and Meta-Regression Analysis of Prospective Cohort Studies. *Nutrients* **2016**, *8*, 739. [CrossRef] [PubMed]
323. Dibaba, D.T.; Xun, P.; Song, Y.; Rosanoff, A.; Shechter, M.; He, K. The effect of magnesium supplementation on blood pressure in individuals with insulin resistance, prediabetes, or noncommunicable chronic diseases: A meta-analysis of randomized controlled trials. *Am. J. Clin. Nutr.* **2017**, *106*, 921–929. [CrossRef] [PubMed]
324. Del Gobbo, L.C.; Imamura, F.; Wu, J.H.Y.; Otto, M.C.D.O.; Chiuve, S.E.; Mozaffarian, D. Circulating and dietary magnesium and risk of cardiovascular disease: A systematic review and meta-analysis of prospective studies. *Am. J. Clin. Nutr.* **2013**, *98*, 160–173. [CrossRef] [PubMed]
325. Stepura, O.B.; Martynow, A.I. Magnesium orotate in severe congestive heart failure (MACH). *Int. J. Cardiol.* **2009**, *131*, 293–295. [CrossRef] [PubMed]
326. Vierling, W.; Liebscher, D.H.; Micke, O.; Von Ehrlich, B.; Kisters, K. Magnesium deficiency and therapy in cardiac arrhythmias: Recommendations of the German Society for Magnesium Research. *DMW—Dtsch. Med. Wochenschr.* **2013**, *138*, 1165–1171.
327. Drew, B.J.; Ackerman, M.J.; Funk, M.; Gibler, W.B.; Kligfield, P.; Menon, V.; Philippides, G.J.; Roden, D.M.; Zareba, W. Prevention of Torsade de Pointes in Hospital Settings. *Circulationa* **2010**, *121*, 1047–1060. [CrossRef]
328. Ferrè, S.; Baldoli, E.; Leidi, M.; Maier, J.A. Magnesium deficiency promotes a pro-atherogenic phenotype in cultured human endothelial cells via activation of NFkB. *Biochim. Biophys. Acta Mol. Basis Dis.* **2010**, *1802*, 952–958. [CrossRef]
329. Peacock, J.M.; Ohira, T.; Post, W.; Sotoodehnia, N.; Rosamond, W.; Folsom, A.R. Serum magnesium and risk of sudden cardiac death in the Atherosclerosis Risk in Communities (ARIC) Study. *Am. Heart J.* **2010**, *160*, 464–470. [CrossRef]

330. Joosten, M.M.; Gansevoort, R.T.; Mukamal, K.J.; Van Der Harst, P.; Geleijnse, J.M.; Feskens, E.J.M.; Navis, G.; Bakker, S.J.L.; The PREVEND Study Group. Urinary and plasma magnesium and risk of ischemic heart disease. *Am. J. Clin. Nutr.* **2013**, *97*, 1299–1306. [CrossRef]
331. Cronin, K.A.; Lake, A.J.; Scott, S.; Sherman, R.L.; Noone, A.M.; Howlader, N.; Henley, S.J.; Anderson, R.N.; Firth, A.U.; Ma, J.; et al. Annual Report to the Nation on the Status of Cancer, part I: National cancer statistics. *Cancer* **2018**, *124*, 2785–2800. [CrossRef]
332. Workeneh, B.T.; Uppal, N.N.; Jhaveri, K.D.; Rondon-Berrios, H. Hypomagnesemia in the Cancer Patient. *Kidney360* **2021**, *2*, 154–166. [CrossRef]
333. Gray, A.; Dang, B.N.; Moore, T.B.; Clemens, R.; Pressman, P. A review of nutrition and dietary interventions in oncology. *SAGE Open Med.* **2020**, *8*, 2050312120926877. [CrossRef] [PubMed]
334. Hsieh, M.-C.; Wu, C.-F.; Chen, C.-W.; Shi, C.-S.; Huang, W.-S.; Kuan, F.-C. Hypomagnesemia and clinical benefits of anti-EGFR monoclonal antibodies in wild-type KRAS metastatic colorectal cancer: A systematic review and meta-analysis. *Sci. Rep.* **2018**, *8*, 2047. [CrossRef] [PubMed]
335. Blaszczyk, U.; Duda-Chodak, A. Magnesium: Its role in nutrition and carcinogenesis. *Rocz. Państwowego Zakładu Hig.* **2013**, *64*, 3.
336. Tao, M.; Dai, Q.; Millen, A.E.; Nie, J.; Edge, S.B.; Trevisan, M.; Shields, P.G.; Freudenheim, J. Abstract 884: Associations of intakes of magnesium and calcium and survival among women with breast cancer: Results from Western New York Exposures and Breast Cancer (WEB) Study. *Epidemiology* **2015**, *75*, 884. [CrossRef]
337. Huang, W.-Q.; Long, W.-Q.; Mo, X.-F.; Zhang, N.-Q.; Luo, H.; Lin, F.-Y.; Huang, J.; Zhang, C.-X. Direct and indirect associations between dietary magnesium intake and breast cancer risk. *Sci. Rep.* **2019**, *9*, 5764. [CrossRef]
338. Liu, M.; Yang, H.; Mao, Y. Magnesium and liver disease. *Ann. Transl. Med.* **2019**, *7*, 578. [CrossRef]
339. Liu, Y.; Li, X.; Zou, Q.; Liu, L.; Zhu, X.; Jia, Q.; Wang, L.; Yan, R. Inhibitory effect of magnesium cantharidate on human hepatoma SMMC-7721 cell proliferation by blocking MAPK signaling pathway. *Chin. J. Cell. Mol. Immunol.* **2017**, *33*, 347–351.
340. Liu, Y.; Xu, Y.; Ma, H.; Wang, B.; Xu, L.; Zhang, H.; Song, X.; Gao, L.; Liang, X.; Ma, C. Hepatitis B virus X protein amplifies TGF-ß promotion on HCC motility through down-regulating PPM1a. *Oncotarget* **2016**, *7*, 33125. [CrossRef] [PubMed]
341. Wesselink, E.; Kok, D.E.; Bours, M.J.L.; De Wilt, J.H.W.; Van Baar, H.; Van Zutphen, M.; Geijsen, A.M.J.R.; Keulen, E.T.P.; Hansson, B.M.E.; Ouweland, J.V.D.; et al. Vitamin D, magnesium, calcium, and their interaction in relation to colorectal cancer recurrence and all-cause mortality. *Am. J. Clin. Nutr.* **2020**, *111*, 1007–1017. [CrossRef]
342. Dai, Q.; Shu, X.-O.; Deng, X.; Xiang, Y.-B.; Li, H.; Yang, G.; Shrubsole, M.J.; Ji, B.; Cai, H.; Chow, W.-H.; et al. Modifying effect of calcium/magnesium intake ratio and mortality: A population-based cohort study. *BMJ Open* **2013**, *3*, e002111. [CrossRef]
343. Wark, P.A.; Lau, R.; Norat, T.; Kampman, E. Magnesium intake and colorectal tumor risk: A case-control study and meta-analysis. *Am. J. Clin. Nutr.* **2012**, *96*, 622–631. [CrossRef]
344. Sun, Y.; Selvaraj, S.; Varma, A.; Derry, S.; Sahmoun, A.E.; Singh, B.B. Increase in Serum $Ca^{2+}/Mg^{2+}$ Ratio Promotes Proliferation of Prostate Cancer Cells by Activating TRPM7 Channels. *J. Biol. Chem.* **2013**, *288*, 255–263. [CrossRef] [PubMed]
345. Steck, S.E.; Omofuma, O.O.; Su, L.J.; Maise, A.A.; Woloszynska-Read, A.; Johnson, C.S.; Zhang, H.; Bensen, J.T.; Fontham, E.T.H.; Mohler, J.L.; et al. Calcium, magnesium, and whole-milk intakes and high-aggressive prostate cancer in the North Carolina–Louisiana Prostate Cancer Project (PCaP). *Am. J. Clin. Nutr.* **2018**, *107*, 799–807. [CrossRef] [PubMed]
346. Sahmoun, A.E.; Singh, B.B. Does a higher ratio of serum calcium to magnesium increase the risk for postmenopausal breast cancer? *Med. Hypotheses* **2010**, *75*, 315–318. [CrossRef]
347. Kumar, G.; Chatterjee, P.K.; Madankumar, S.; Mehdi, S.F.; Xue, X.; Metz, C.N. Magnesium deficiency with high calcium-to-magnesium ratio promotes a metastatic phenotype in the CT26 colon cancer cell line. *Magnes. Res.* **2020**, *33*, 68–85. [CrossRef] [PubMed]
348. Ma, E.; Sasazuki, S.; Inoue, M.; Iwasaki, M.; Sawada, N.; Takachi, R.; Tsugane, S.; Members of the JPHC Study Group. High Dietary Intake of Magnesium May Decrease Risk of Colorectal Cancer in Japanese Men. *J. Nutr.* **2010**, *140*, 779–785. [CrossRef] [PubMed]
349. Folsom, A.R.; Hong, C.-P. Magnesium Intake and Reduced Risk of Colon Cancer in a Prospective Study of Women. *Am. J. Epidemiol.* **2005**, *163*, 232–235. [CrossRef] [PubMed]
350. Brandt, P.A.V.D.; Smits, K.M.; Goldbohm, R.A.; Weijenberg, M.P. Magnesium intake and colorectal cancer risk in the Netherlands Cohort Study. *Br. J. Cancer* **2007**, *96*, 510–513. [CrossRef] [PubMed]
351. Mahabir, S.; Wei, Q.; Barrera, S.L.; Dong, Y.Q.; Etzel, C.J.; Spitz, M.R.; Forman, M.R. Dietary magnesium and DNA repair capacity as risk factors for lung cancer. *Carcinogenesis* **2008**, *29*, 949–956. [CrossRef] [PubMed]
352. Jansen, R.J.; Robinson, D.P.; Stolzenberg-Solomon, R.Z.; Bamlet, W.R.; De Andrade, M.; Oberg, A.L.; Rabe, K.G.; Anderson, K.E.; Olson, J.E.; Sinha, R.; et al. Nutrients from Fruit and Vegetable Consumption Reduce the Risk of Pancreatic Cancer. *J. Gastrointest. Cancer* **2012**, *44*, 152–161. [CrossRef] [PubMed]
353. Dibaba, D.; Xun, P.; Yokota, K.; White, E.; He, K. Magnesium intake and incidence of pancreatic cancer: The VITamins and Lifestyle study. *Br. J. Cancer* **2015**, *113*, 1615–1621. [CrossRef] [PubMed]
354. Ravell, J.C.; Chauvin, S.D.; He, T.; Lenardo, M. An Update on XMEN Disease. *J. Clin. Immunol.* **2020**, *40*, 671–681. [CrossRef] [PubMed]
355. Afridi, H.I.; Kazi, T.G.; Talpur, F.N. Correlation of Calcium and Magnesium Levels in the Biological Samples of Different Types of Acute Leukemia Children. *Biol. Trace Element Res.* **2018**, *186*, 395–406. [CrossRef] [PubMed]

356. Slahin, G.; Ertem, U.; Duru, F.; Birgen, D.; Yuuksek, N. High Prevelance of Chronic Magnesium Deficiency in T Cell Lymphoblastic Leukemia and Chronic Zinc Deficiency in Children with Acute Lymphoblastic Leukemia and Malignant Lymphoma. *Leuk. Lymphoma* **2000**, *39*, 555–562. [CrossRef] [PubMed]
357. Sikora, P.; Borzęcka, H.; Kołłątaj, B.; Majewski, M.; Wieczorkiewicz-Płaza, A.; Zajączkowska, M. The diagnosis of familial hypomagnesemia with hypercalciuria and nephrocalcinosis in a girl with acute lymphoblastic leukemia—Case report. *Pol. Merkur. Lek.* **2006**, *118*, 430–432.
358. Canbolat, O.; Kavutcu, M.; Durak, I. Magnesium contents of leukemic lymphocytes. *BioMetals* **1994**, *7*, 313–315. [CrossRef] [PubMed]
359. Atkinson, S.A.; Halton, J.M.; Bradley, C.; Wu, B.; Barr, R.D. Bone and mineral abnormalities in childhood acute lymphoblastic leukemia: Influence of disease, drugs and nutrition. *Int. J. Cancer* **1998**, *8*, 35–39. [CrossRef]
360. Volpe, S.L. Magnesium in Disease Prevention and Overall Health. *Adv. Nutr.* **2013**, *4*, 378S–383S. [CrossRef] [PubMed]
361. Pickering, G.; Mazur, A.; Trousselard, M.; Bienkowski, P.; Yaltsewa, N.; Amessou, M.; Noah, L.; Pouteau, E. Magnesium Status and Stress: The Vicious Circle Concept Revisited. *Nutrients* **2020**, *12*, 3672. [CrossRef] [PubMed]
362. Goadsby, P.J.; Holland, P.R.; Martins-Oliveira, M.; Hoffmann, J.; Schankin, C.; Akerman, S. Pathophysiology of Migraine: A Disorder of Sensory Processing. *Physiol. Rev.* **2017**, *97*, 553–622. [CrossRef] [PubMed]
363. Hoffmann, J.; Charles, A. Glutamate and Its Receptors as Therapeutic Targets for Migraine. *Neurotherapeutics* **2018**, *15*, 361–370. [CrossRef] [PubMed]
364. Linde, M.; Gustavsson, A.; Stovner, L.J.; Steiner, T.J.; Barré, J.; Katsarava, Z.; Lainez, J.M.; Lampl, C.; Lantéri-Minet, M.; Rastenyte, D.; et al. The cost of headache disorders in Europe: The Eurolight project. *Eur. J. Neurol.* **2011**, *19*, 703–711. [CrossRef] [PubMed]
365. Nattagh-Eshtivani, E.; Sani, M.A.; Dahri, M.; Ghalichi, F.; Ghavami, A.; Arjang, P.; Tarighat-Esfanjani, A. The role of nutrients in the pathogenesis and treatment of migraine headaches: Review. *Biomed. Pharmacother.* **2018**, *102*, 317–325. [CrossRef]
366. Lodi, R.; Iotti, S.; Cortelli, P.; Pierangeli, G.; Cevoli, S.; Clementi, V.; Soriani, S.; Montagna, P.; Barbiroli, B. Deficient energy metabolism is associated with low free magnesium in the brains of patients with migraine and cluster headache. *Brain Res. Bull.* **2001**, *54*, 437–441. [CrossRef]
367. Choi, H.; Parmar, N. The use of intravenous magnesium sulphate for acute migraine: Meta-analysis of randomized controlled trials. *Eur. J. Emerg. Med.* **2014**, *21*, 2–9. [CrossRef]
368. Chiu, H.-Y.; Yeh, T.-H.; Huang, Y.-C.; Chen, P.-Y. Effects of Intravenous and Oral Magnesium on Reducing Migraine: A Meta-analysis of Randomized Controlled Trials. *Pain Physician* **2016**, *19*, E97–E112.
369. Serefko, A.; Szopa, A.; Wlaź, P.; Nowak, G.; Radziwoń-Zaleska, M.; Skalski, M.; Poleszak, E. Magnesium in depression. *Pharmacol. Rep.* **2013**, *65*, 547–554. [CrossRef]
370. Barragán-Rodríguez, L.; Rodríguez-Morán, M.; Guerrero-Romero, F. Efficacy and safety of oral magnesium supplementation in the treatment of depression in the elderly with type 2 diabetes: A randomized, equivalent trial. *Magnes. Res.* **2008**, *21*, 218–223.
371. Jacka, F.N.; Maes, M.; Pasco, J.A.; Williams, L.J.; Berk, M. Nutrient intakes and the common mental disorders in women. *J. Affect. Disord.* **2012**, *141*, 79–85. [CrossRef] [PubMed]
372. Tarleton, E.K.; Littenberg, B. Magnesium intake and depression in adults. *J. Am. Board Fam. Med.* **2015**, *28*, 249–256. [CrossRef] [PubMed]
373. Eby, G.A.; Eby, K.L. Rapid recovery from major depression using magnesium treatment. *Med. Hypotheses* **2006**, *67*, 362–370. [CrossRef] [PubMed]
374. Parazzini, F.; Di Martino, M.; Pellegrino, P. Magnesium in the gynaecological practice: A literature review. *Magnes. Res.* **2017**, *30*, 1–7. [PubMed]
375. Tarleton, E.K.; Littenberg, B.; MacLean, C.D.; Kennedy, A.G.; Daley, C. Role of magnesium supplementation in the treatment of depression: A randomized clinical trial. *PLoS ONE* **2017**, *12*, e0180067. [CrossRef] [PubMed]
376. Li, B.; Lv, J.; Wang, W.; Zhang, D. Dietary magnesium and calcium intake and risk of depression in the general population: A meta-analysis. *Aust. N. Z. J. Psychiatry* **2017**, *51*, 219–229. [CrossRef]
377. Barkerhaliski, M.L.; White, H.S. Glutamatergic Mechanisms Associated with Seizures and Epilepsy. *Cold Spring Harb. Perspect. Med.* **2015**, *5*, a022863. [CrossRef]
378. Chen, B.B.; Prasad, C.; Kobrzynski, M.; Campbell, C.; Filler, G. Seizures Related to Hypomagnesemia: A Case Series and Review of the Literature. *Child Neurol. Open* **2016**, *3*, 2329048X16674834. [CrossRef]
379. Maresova, P.; Klimova, B.; Novotny, M.; Kuca, K. Alzheimer's and Parkinson's Diseases: Expected Economic Impact on Europe-A Call for a Uniform European Strategy. *J. Alzheimer's Dis.* **2016**, *54*, 1123–1133. [CrossRef]
380. Zádori, D.; Veres, G.; Szalárdy, L.; Klivényi, P.; Vécsei, L. Alzheimer's Disease: Recent Concepts on the Relation of Mitochondrial Disturbances, Excitotoxicity, Neuroinflammation, and Kynurenines. *J. Alzheimer's Dis.* **2018**, *62*, 523–547. [CrossRef]
381. Fan, X.; Wheatley, E.G.; Villeda, S.A. Mechanisms of Hippocampal Aging and the Potential for Rejuvenation. *Annu. Rev. Neurosci.* **2017**, *40*, 251–272. [CrossRef] [PubMed]
382. Xu, Z.-P.; Li, L.; Bao, J.; Wang, Z.-H.; Zeng, J.; Liu, E.-J.; Li, X.-G.; Huang, R.-X.; Gao, D.; Li, M.-Z.; et al. Magnesium Protects Cognitive Functions and Synaptic Plasticity in Streptozotocin-Induced Sporadic Alzheimer's Model. *PLoS ONE* **2014**, *9*, e108645. [CrossRef]
383. Slutsky, I.; Abumaria, N.; Wu, L.-J.; Huang, C.; Zhang, L.; Li, B.; Zhao, X.; Govindarajan, A.; Zhao, M.-G.; Zhuo, M.; et al. Enhancement of Learning and Memory by Elevating Brain Magnesium. *Neuron* **2010**, *65*, 165–177. [CrossRef]

384. Veronese, N.; Zurlo, A.; Solmi, M.; Luchini, C.; Trevisan, C.; Bano, G.; Manzato, E.; Sergi, G.; Rylander, R. Magnesium Status in Alzheimer's Disease: A Systematic Review. *Am. J. Alzheimers. Dis. Other Demen.* **2016**, *31*, 208–213. [CrossRef] [PubMed]
385. Xue, W.; You, J.; Su, Y.; Wang, Q. The Effect of Magnesium Deficiency on Neurological Disorders: A Narrative Review Article. *Iran. J. Public Health* **2019**, *48*, 379–387. [CrossRef]
386. Barbiroli, B.; Martinelli, P.; Patuelli, A.; Lodi, R.; Iotti, S.; Cortelli, P.; Montagna, P. Phosphorus magnetic resonance spectroscopy in multiple system atrophy and Parkinson's disease. *Mov. Disord.* **1999**, *14*, 430–435. [CrossRef]
387. Miyake, Y.; Tanaka, K.; Fukushima, W.; Sasaki, S.; Kiyohara, C.; Tsuboi, Y.; Yamada, T.; Oeda, T.; Miki, T.; Kawamura, N.; et al. Dietary intake of metals and risk of Parkinson's disease: A case-control study in Japan. *J. Neurol. Sci.* **2011**, *306*, 98–102. [CrossRef] [PubMed]
388. Hemamy, M.; Heidari-Beni, M.; Askari, G.; Karahmadi, M.; Maracy, M. Effect of Vitamin D and Magnesium Supplementation on Behavior Problems in Children with Attention-Deficit Hyperactivity Disorder. *Int. J. Prev. Med.* **2020**, *11*, 4. [CrossRef] [PubMed]
389. Archana, E.; Pai, P.; Prabhu, B.K.; Shenoy, R.P.; Prabhu, K.; Rao, A. Altered Biochemical Parameters in Saliva of Pediatric Attention Deficit Hyperactivity Disorder. *Neurochem. Res.* **2012**, *37*, 330–334. [CrossRef]
390. Mahmoud, M.M.; El-Mazary, A.-A.M.; Maher, R.M.; Saber, M.M. Zinc, ferritin, magnesium and copper in a group of Egyptian children with attention deficit hyperactivity disorder. *Ital. J. Pediatr.* **2011**, *37*, 60. [CrossRef] [PubMed]
391. Mousain-Bosc, M.; Roche, M.; Polge, A.; Pradal-Prat, D.; Rapin, J.; Bali, J.P. Improvement of neurobehavioral disorders in children supplemented with magnesium-vitamin B6: I. Attention deficit hyperactivity disorders. *Magnes. Res.* **2006**, *19*, 46–52. [PubMed]
392. El Baza, F.; AlShahawi, H.A.; Zahra, S.; AbdelHakim, R.A. Magnesium supplementation in children with attention deficit hyperactivity disorder. *Egypt. J. Med. Hum. Genet.* **2016**, *17*, 63–70. [CrossRef]

*Review*

# Magnesium in Obesity, Metabolic Syndrome, and Type 2 Diabetes

Gabriele Piuri [1,†], Monica Zocchi [1,†], Matteo Della Porta [1], Valentina Ficara [1], Michele Manoni [2], Gian Vincenzo Zuccotti [1,3], Luciano Pinotti [2], Jeanette A. Maier [1] and Roberta Cazzola [1,*]

[1] Department of Biomedical and Clinical Sciences "L. Sacco", Università di Milano, 20157 Milan, Italy; gabriele.piuri@unimi.it (G.P.); monica.zocchi@unimi.it (M.Z.); matteo.dellaporta@unimi.it (M.D.P.); valentina.ficara@unimi.t (V.F.); gianvincenzo.zuccotti@unimi.it (G.V.Z.); jeanette.maier@unimi.it (J.A.M.)
[2] Department of Health, Animal Science and Food Safety, Università di Milano, 20133 Milan, Italy; michele.manoni@unimi.it (M.M.); luciano.pinotti@unimi.it (L.P.)
[3] Department of Pediatrics, Ospedale dei Bambini, 2154 Milan, Italy
* Correspondence: roberta.cazzola@unimi.it
† These authors contributed equally to this work.

**Abstract:** Magnesium ($Mg^{2+}$) deficiency is probably the most underestimated electrolyte imbalance in Western countries. It is frequent in obese patients, subjects with type-2 diabetes and metabolic syndrome, both in adulthood and in childhood. This narrative review aims to offer insights into the pathophysiological mechanisms linking $Mg^{2+}$ deficiency with obesity and the risk of developing metabolic syndrome and type 2 diabetes. Literature highlights critical issues about the treatment of $Mg^{2+}$ deficiency, such as the lack of a clear definition of $Mg^{2+}$ nutritional status, the use of different $Mg^{2+}$ salts and dosage and the different duration of the $Mg^{2+}$ supplementation. Despite the lack of agreement, an appropriate dietary pattern, including the right intake of $Mg^{2+}$, improves metabolic syndrome by reducing blood pressure, hyperglycemia, and hypertriglyceridemia. This occurs through the modulation of gene expression and proteomic profile as well as through a positive influence on the composition of the intestinal microbiota and the metabolism of vitamins B1 and D.

**Keywords:** magnesium; magnesium deficiency; magnesium supplementation; obesity; metabolic syndrome; type 2 diabetes; gut microbiota; vitamin D

**Citation:** Piuri, G.; Zocchi, M.; Della Porta, M.; Ficara, V.; Manoni, M.; Zuccotti, G.V.; Pinotti, L.; Maier, J.A.; Cazzola, R. Magnesium in Obesity, Metabolic Syndrome, and Type 2 Diabetes. *Nutrients* **2021**, *13*, 320. https://doi.org/10.3390/nu13020320

Academic Editor: Mario Barbagallo
Received: 30 November 2020
Accepted: 19 January 2021
Published: 22 January 2021

**Publisher's Note:** MDPI stays neutral with regard to jurisdictional claims in published maps and institutional affiliations.

**Copyright:** © 2021 by the authors. Licensee MDPI, Basel, Switzerland. This article is an open access article distributed under the terms and conditions of the Creative Commons Attribution (CC BY) license (https://creativecommons.org/licenses/by/4.0/).

## 1. Introduction

Magnesium ($Mg^{2+}$) is the second most abundant intracellular cation and the fourth most abundant cation of the human body. Almost all the body $Mg^{2+}$ is found in the bones (about 60%) and soft tissues (about 40%), while <1% is in the blood. It is a cofactor of hundreds of enzymatic reactions, acting both on the enzymes as a structural or catalytic component and on the substrates. An example of $Mg^{2+}$ bioactive activity is given by the reactions involving the complex Mg-ATP, which is an essential cofactor of kinases. For this reason, $Mg^{2+}$ is a rate-limiting factor for many enzymes involved in carbohydrate and energy metabolism. Furthermore, $Mg^{2+}$ is essential in the intermediary metabolism for the synthesis of the macromolecules [1]. Other vital $Mg^{2+}$-dependent functions are muscle contraction and relaxation, normal neurological function, and release of neurotransmitters [2].

At the cellular level, $Mg^{2+}$ homeostasis is fine-tuned by the coordinated activity of membrane channels and transporters. Some of them are ubiquitously expressed, such as transient receptor potential melastatin (TRPM) 7, $Mg^{2+}$ transporter 1 (MagT1) and solute carrier family 41 member 1 (SLC41A1). Others are tissue-specific, such as TRPM6, expressed in the kidney and the colon, cyclin and CBS domain divalent metal cation transport mediator cyclin M2 (CNNM2), expressed in the kidney, and CNNM4, expressed in the colon [3].

Obesity, metabolic syndrome, and type 2 diabetes mellitus are three interrelated conditions that share a series of pathophysiological mechanisms attributable to "low-grade" systemic inflammation [4]. $Mg^{2+}$ deficit is frequent in obese subjects [3] and is a highly prevalent condition in patients with diabetes or metabolic syndrome. Moreover, it increases the risk of developing type-2 diabetes [5]. Besides, $Mg^{2+}$ depletion can promote chronic inflammation both directly [6–8] and indirectly by modifying the intestinal microbiota [9].

This review aims to offer insights into the pathophysiological mechanisms linking $Mg^{2+}$ deficiency with obesity and the risk of developing metabolic syndrome and type 2 diabetes (Figure 1).

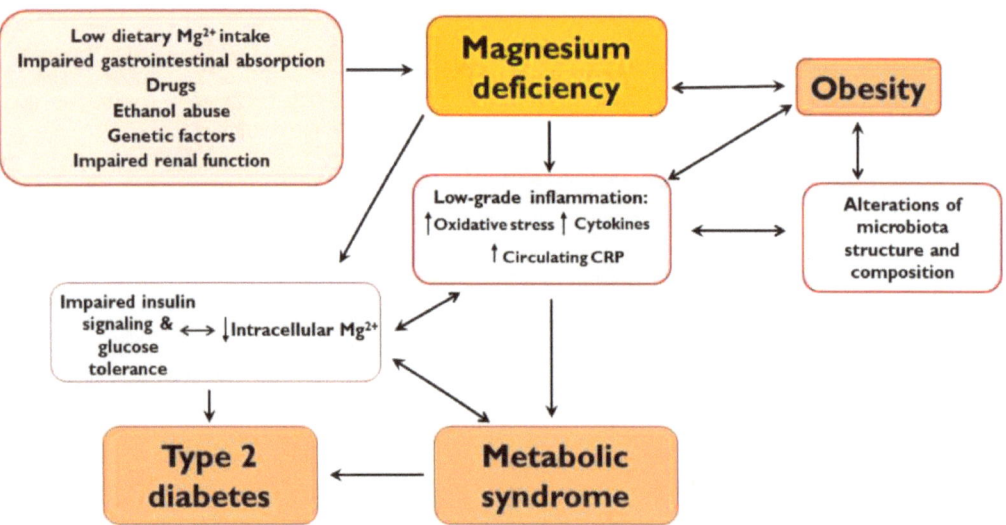

**Figure 1.** Physio-pathological mechanisms of magnesium deficiency in obesity, metabolic syndrome, and type 2 diabetes.

## 2. $Mg^{2+}$ Deficiency

Among all the lab tests most frequently used to evaluate $Mg^{2+}$ status in routine clinical practice is magnesemia because it is feasible and inexpensive [1,10]. However, magnesemia does not correlate with tissue pools because serum $Mg^{2+}$ is just a tiny percentage of the intracellular/total body $Mg^{2+}$ content [2]. This is one of the reasons why $Mg^{2+}$ deficiency is the most underestimated electrolyte imbalance in Western countries, where a significantly high risk of latent hypomagnesemia occurs [11,12]. Based on distribution patterns of $Mg^{2+}$ in the blood, the reference range for serum $Mg^{2+}$ concentration is 0.75–0.95 mmol/L [13–16] and hypomagnesemia is generally defined as serum $Mg^{2+}$ level lower than 0.7 mmol/L [3,17]. Recently, a panel of experts proposed that urinary $Mg^{2+}$ secretion should also be considered. Specifically, a magnesemia lower than 0.82 mmol/L with $Mg^2$ urinary excretion of 40–80 mg/die should be considered indicative of $Mg^{2+}$ deficiency [13]. Since serum $Mg^{2+}$ content is only 1% of total $Mg^{2+}$ in the body and is not representative for global intracellular $Mg^{2+}$ status, $Mg^{2+}$ deficiency may be underestimated and persist latently for years [16]. Subclinical hypomagnesemia is responsible for a variety of clinical manifestations that are non-specific and can overlap with symptoms of other electrolyte imbalances [18]. Some of these symptoms are depression, fatigue, muscle spasms and arrhythmias. Furthermore, a chronic low-$Mg^{2+}$ status has been associated with an increased risk of chronic non-transmissible diseases, among which osteoporosis and sarcopenia [19–21]. Severe $Mg^{2+}$ depletion, defined by serum $Mg^{2+}$ concentration below 0.3–0.4 mmol/L, may lead to cardiac arrhythmias, tetany and seizures [3].

There are several causes of hypomagnesemia and one of the most relevant is an insufficient dietary intake. In fact, several studies show that the majority of the population in Europe and North America consumes less than the recommended daily allowance

(RDA) of $Mg^{2+}$, i.e., approximately 420 $Mg^{2+}$ for adult males and 320 $Mg^{2+}$ for adult females [22–24]. This deficit mainly derives from the Western-style diet (WD) that often contains only 30–50% of $Mg^{2+}$ RDA. Indeed, the WD is based on massive consumption of processed foods, demineralized water and low amounts of vegetables and legumes, often grown in $Mg^{2+}$-poor soil [18]. Hypomagnesemia may also be a consequence of pre-existing pathological conditions. For example, $Mg^{2+}$ depletion is frequent in subjects affected by impaired gastrointestinal absorption caused by celiac disease [25], inflammatory bowel diseases [26–28] or in the presence of colon cancer, gastric bypass and other minor gastrointestinal disorders [29]. Additional causes of $Mg^{2+}$ deficit are type 1 diabetes mellitus, renal disorders and hydro-electrolyte imbalances [30]. Hypomagnesemia is also associated, through different molecular mechanisms, with the frequent use of several medications such as diuretics (furosemide, thiazide), epidermal growth factor receptor inhibitors (cetuximab), calcineurin inhibitors (cyclosporine A), cisplatin and some antimicrobials (rapamycin, aminoglycosides antibiotics, pentamidine, foscaret, amphotericin B). It is also interesting to highlight that the wide use of proton pump inhibitors (PPI–omeprazole, pantoprazole, esomeprazole), which is generally considered safe, induces hypomagnesemia in 13% of the cases but the underlying mechanism is still unknown [31]. Moreover, ethanol abuse results in $Mg^{2+}$ deficiency [32,33].

A low-$Mg^{2+}$ status may also have genetic origins and derive from mutations of genes such as TRPM6, CLDN16-19 (claudin 16 and 19), KCNA1 (potassium voltage-gated channel subfamily A member 1), CNNM2 [34]. These mutations result in a severe hypomagnesemia accompanied by calcium wasting, renal failure, seizures and mental retardation [3]. Finally, from a physiological point of view, $Mg^{2+}$ deficit may be observed after intensive sport activities with an increase in sweating, in healthy postmenopausal women [35] or during lactation [30]. Moreover, $Mg^{2+}$ status is generally impaired in older people [36].

Considering the focus of this work, it is important to underline that a moderate or subclinical $Mg^{2+}$ deficiency induces a chronic low-grade inflammation sustained by the release of inflammatory cytokines and production of free radicals, which exacerbate a pre-existing inflammatory status [7]. For this reason $Mg^{2+}$ depletion is considered a risk factor for pathological conditions characterized by chronic inflammation, such as hypertension and cardiovascular disorders but also metabolic syndrome and diabetes [29,37,38].

## 3. $Mg^{2+}$ and Obesity

Obesity and its comorbidities, including metabolic syndrome and type 2 diabetes, are a relevant medical problem worldwide. Obesity is the result of unhealthy diets, high in calories, but poor in essential nutrients. As a consequence, obese subjects are often $Mg^{2+}$ deficient [39]. Indeed, the National Health and Nutrition Examination Survey (NHANES) 3 study underlines that $Mg^{2+}$ deficit is more prevalent in subjects with body mass index (BMI) in the obese range than in the normal American population [40,41]. Analogously, $Mg^{2+}$ intake is impaired in 35% of French individuals with BMI > 35 kg/m$^2$ [42]. The 30-year longitudinal CARDIA study, performed on more than 5000 subjects, indicates that $Mg^{2+}$ intake is inversely associated with the incidence of obesity and with the levels of C reactive protein [43]. Besides, in animal models of diet-induced obesity $Mg^{2+}$ supplementation prevents the accumulation of adipose tissue [44] and human studies report an inverse association between $Mg^{2+}$ intake and markers of adiposity, such as BMI and waist circumference [45–47].

In obese subjects, most of the energy of the diet derives from refined grains and simple sugars and, consequently, their hepatic glucose catabolism is very active. Several key enzymes of glucose oxidation pathways are $Mg^{2+}$-dependent and $Mg^{2+}$ is necessary also for the activation of vitamin B1 into thiamine diphosphate (TDP) that is another critical coenzyme of oxidative metabolism. Importantly, TDP-dependent enzymes require $Mg^{2+}$ to reach optimal activation [48]. Therefore, low intracellular concentrations of $Mg^{2+}$ and/or TDP may alter the oxidative metabolism of glucose. In the liver, a decrease of the activity of the $Mg^{2+}$- and TDP-dependent enzyme pyruvate dehydrogenase may divert glucose metabolism into the oxidative phase of the pentose phosphate pathway, thus generating an excess of NADPH [48]. NADPH provides essential redox potential for synthetic pathways, including fatty acid biosynthesis, thus promoting an increased synthesis of triglycerides and very low-density lipoprotein and, consequently, a higher triglyceride storage in adipocytes that increases the extent of obesity and the risk of obesity co-morbidities such as dyslipidemia, metabolic syndrome and type 2 diabetes [49–51].

Moreover, obese subjects are often deficient also in vitamin D [49,52] both in the presence and in the absence of type 2 diabetes [53], and $Mg^{2+}$ is essential also for vitamin D synthesis and activation [54]. A randomized controlled trial suggests that optimal $Mg^{2+}$ status may be fundamental for optimizing vitamin D status [55]. Because of its role in the renin-angiotensin system and its immunomodulatory properties, vitamin D deficiency is identified as a potential risk factor in cardiometabolic disorders, including insulin resistance, metabolic syndrome and cardiovascular diseases [56]. Moreover, chronic latent $Mg^{2+}$ deficiency and/or Vitamin D deficiency predispose non-diabetic obese subjects to an increased risk of cardiometabolic diseases. Meanwhile, maintaining a normal $Mg^{2+}$ status improves the beneficial effect of Vitamin D on cardiometabolic risk indicators [57]. Interestingly, an interventional study performed on healthy women showed a significant increase in serum concentration of $Mg^{2+}$ in obese but not in non-obese subjects after vitamin D intramuscular injection, probably caused by increased $Mg^{2+}$ renal retention induced by vitamin D and emphasized by baseline $Mg^{2+}$ deficiency of the obese subjects [58].

## 4. $Mg^{2+}$ in Metabolic Syndrome

Obesity and metabolic syndrome (MetS) are both characterized by excessive accumulation of body fat. However, while obesity only implies the accumulation of excess body fat, metabolic syndrome is a disorder of accumulation and use of energy, promoted by low-grade systemic inflammation, and resulting in central adiposity, hypertension, dyslipidemia, or insulin resistance. Many studies have found a positive correlation between low dietary $Mg^{2+}$ intake and MetS risk independently from other risk factors such as age, gender, BMI, race, educational attainment, marital status, smoking, alcohol intake, exercise, energy intake, percentage of calories from saturated fat, use of an antihypertensive or lipid medication [59–64]. Dibaba et al. showed in the last meta-analysis available that the dietary $Mg^{2+}$ intake is inversely associated with the prevalence of MetS [63]. A recent cross-sectional analysis performed in a large Chinese population reports an inverse correlation between dietary $Mg^{2+}$ intake and the prevalence of MetS [65]. In more than 11.000 middle-aged and older women high dietary $Mg^{2+}$ intake lowers systemic inflammation and the risk of the MetS [59]. An interesting Serbian study shows a positive association between chronic exposure to insufficient $Mg^{2+}$ in drinking municipalities water and the prevalence of hypertension and MetS [66].

MetS exponentially increases the risk of developing type 2 diabetes, cardiovascular disease and, in general, morbidity and mortality. Proper $Mg^{2+}$ intake reduces cardiometabolic risk and is associated with a reduced hazard of cardiovascular disease, diabetes, and all-cause mortality [67–70]. Likewise, higher levels of circulating $Mg^{2+}$ are associated with a lower risk of cardiovascular disease, mainly coronary artery disease [71].

Low chronic $Mg^{2+}$ dietary intake leads to serum and intracellular $Mg^{2+}$ deficiency. This is particularly evident in obese people with MetS, in elderly subjects and non-white people with insulin resistance [72–74].

$Mg^{2+}$ is a natural calcium ($Ca^{2+}$) antagonist, and its metabolic effect needs to be discussed according to $Ca^{2+}$ concentration. A recent meta-analysis suggests that high $Ca^{2+}$ dietary intake reduces the risk of MetS [75]. Other experimental data suggest that a higher $Ca^{2+}/Mg^{2+}$ intracellular ratio, induced by a diet high in $Ca^{2+}$ and low in $Mg^{2+}$, may lead to hypertension, insulin resistance, and MetS [76]. Accordingly, subjects who meet the recommended daily allowance for both $Mg^{2+}$ and $Ca^{2+}$ have reduced risk of MetS [76]. $Mg^{2+}$ and $Ca^{2+}$ work together to regulate the metabolic response of overweight and obese subjects, and an unbalanced $Ca^{2+}/Mg^{2+}$ ratio maximizes the effect of their single deficiency. The optimal $Ca^{2+}/Mg^{2+}$ ratio leads to the best-decreased risk of MetS [77,78].

## 5. $Mg^{2+}$ in Type 2 Diabetes

Type 2 diabetes (T2D) is often associated with altered $Mg^{2+}$ homeostasis and $Mg^{2+}$ intake is inversely associated with the risk of T2D in a dose-response manner [79,80]. Epidemiologic studies have shown a high prevalence of hypomagnesemia in subjects with T2D [81,82]. $Mg^{2+}$ depletion in patients with T2D is mainly caused by a low intake and an increased urinary loss of $Mg^{2+}$, probably resulting from impaired renal function [82]. Moreover, recent findings demonstrate that hypomagnesemia is strongly associated with the progression of T2D [83]. In particular, if it is true that insulin regulates $Mg^{2+}$ homeostasis, at the same time $Mg^{2+}$ is also a significant determinant of post-receptor insulin signaling. The influence of $Mg^{2+}$ on glucose metabolism, insulin sensitivity, and insulin action could explain the negative association between $Mg^{2+}$ intake and T2D incidence [82,84–87] (Figure 2). To better understand this issue, it is worth recalling that insulin secretion is started by a $Ca2+$ influx that is competitively inhibited by extracellular $Mg^{2+}$ and, consequently, insulinemia is inversely correlated with magnesemia. Circulating glucose is easily taken from cells β through the glucose transporter 2 (GLUT2), and then converted in glucose-6-phosphate (G6P) by glucokinase (GK). The oxidation of G6P in glycolysis determines an increase in the ATP/ADP ratio leading to the closure of ATP-sensitive $K^+$ channels (KATP channels) and, consequently, to the depolarization of the membrane, followed by the opening of voltage-dependent $Ca^{2+}$ channels [88]. The increase in intracellular concentrations of $Ca^{2+}$ triggers the fusion of insulin-containing granules with the membrane and the subsequent release of their content. The molecular mechanisms by which $Mg^{2+}$ contributes to insulin resistance are mostly unrevealed. However, it is accepted that $Mg^{2+}$ deficiency has a significant impact on insulin secretion and may contribute to dysfunction of pancreatic beta cells in T2D [89]. This depends on the key roles played by $Mg^{2+}$ in the glucose-dependent signaling inducing insulin release. The activities of GK and many glycolytic enzymes depend on Mg-ATP complex, thus, a low intracellular $Mg^{2+}$ concentration results in decreased ATP level in the cells. In addition, the closure of KATP channels depends on ATP binding to the Kir6.2 subunit, while the opening of these channels depends on Mg-ATP binding to the SUR1 subunit. The reduction in the intracellular levels of both ATP and Mg-ATP deranges the fine regulation of KATP channels. This leads to an increase in the basal secretion of insulin and induces hyperinsulinemia, thus contributing to a chronic exposure of cells to insulin and to the development of insulin resistance fostered also by the concomitant low grade inflammation [89]. Moreover, the prolonged hyperinsulinemia typical of insulin resistance induces an increase in renal excretion of $Mg^{2+}$, thus perpetuating a vicious cycle [90]. In addition, we recall that physiological concentrations of insulin and glucose stimulate $Mg^{2+}$ transport, thus increasing intracellular $Mg^{2+}$ content. It is noteworthy that low intracellular $Mg^{2+}$ impairs cell responsiveness to insulin, because low intracellular $Mg^{2+}$ alters the tyrosine-kinase activity of the insulin receptor (INSR), leading to the development of post-receptor insulin resistance and decreased cellular glucose utilization [89,91]. In particular, $Mg^{2+}$ and Mg-ATP complex are key regulators of the PI3K/Akt kinase pathway downstream to the INSR. This pathway starts with INSR auto-phosphorylation, which triggers the downstream kinase cascade. Insulin receptor substrate (IRS) mainly activates phosphatidylinositol-4,5-bisphosphate-3-kinase (PI3K), which generates the second messenger phosphatidylinositol-3,4,5-triphosphate (PIP3). PIP3 activates 3-phosphoinositide dependent protein kinase-1 (PDK1), which activates Akt. Akt regulates the metabolic actions of insulin, including glucose uptake by GLUT4 mobilization in skeletal muscle and adipose tissue, glycogen and protein synthesis and lipogenesis. For this reason, the lower is the basal intracellular $Mg^{2+}$ concentration, the higher is the amount of insulin required to metabolize the same glucose load, indicating decreased insulin sensitivity [89,91]. All these data underline that insulin action is strictly dependent on the intracellular $Mg^{2+}$ concentration.

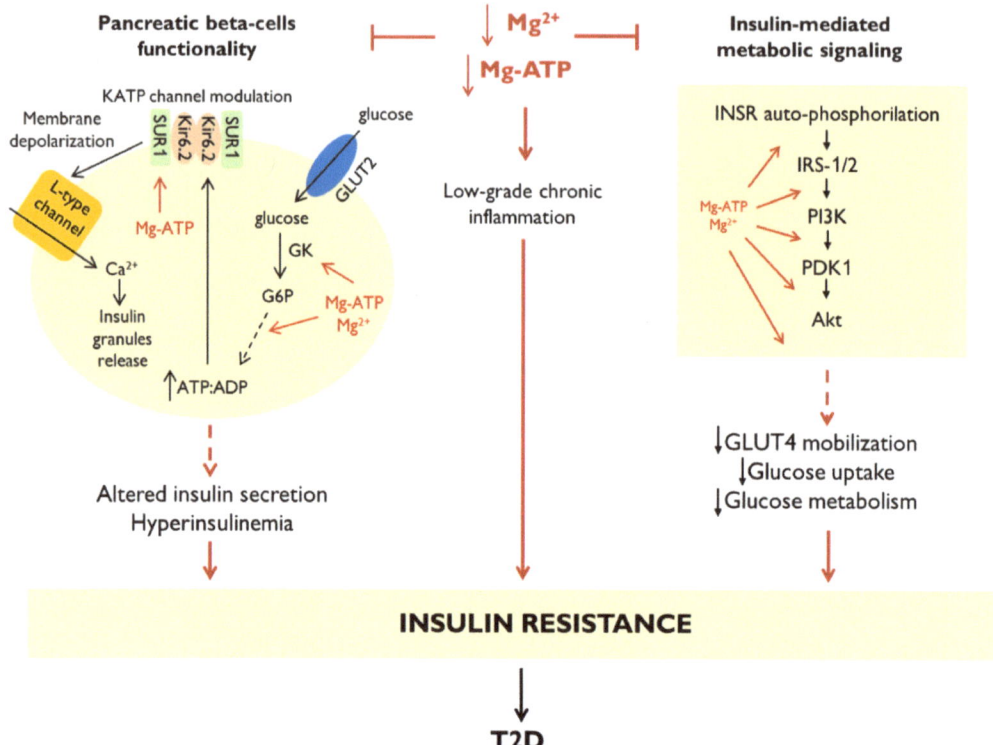

**Figure 2.** Links between $Mg^{2+}$ and insulin signaling. For details, please see the text.

$Mg^{2+}$ deficiency can also contribute to T2D through the modulation of $Na^+/K^+$-ATPase that is crucial for maintaining the membrane potential and low cytoplasmic sodium concentration. $Mg^{2+}$ ions drive the conformational change of the sodium pump whose dysfunction has been correlated to T2D [92,93]. Moreover, some single nucleotide polymorphisms in the TRPM6 gene are associated with an increased risk of developing T2D because TRPM6 cannot be activated by insulin in the presence of these mutations [94].

The observation that several pharmacological treatments for diabetes, such as metformin, appear to increase $Mg^{2+}$ levels further supports this assumption and suggests a substantial interdependence between $Mg^{2+}$ deficiency and the development of insulin resistance and T2D. $Mg^{2+}$ deficiency may not be a secondary consequence of T2D, but it may contribute to insulin resistance and altered glucose tolerance, thereby leading to T2D [82].

Few studies have discussed the relationship between hypomagnesemia and metabolic disorders in childhood and adolescence. $Mg^{2+}$ deficiency in obese children may be secondary to decreased dietary $Mg^{2+}$ intake. Obese children show lower serum $Mg^{2+}$ levels than the normal-weight control group. In obese children and adolescents, $Mg^{2+}$ blood concentration is inversely correlated with the degree of obesity and is related to an unfavorable serum lipid profile and higher systemic blood pressure than healthy controls [95,96]. The association between $Mg^{2+}$ deficiency and insulin resistance has been described also in childhood [97]. $Mg^{2+}$ supplementation or increased intake of $Mg^{2+}$-rich foods to correct its deficiency may represent an essential and inexpensive tool in preventing T2D in obese children.

## 6. $Mg^{2+}$ and Gut Microbiota

The gut microbiota is a complex microbial ecosystem, symbiotic with humans, that plays a crucial role in a series of pathophysiological processes. In healthy subjects the gut microbiota is rich in microbial species and, through its genes and metabolites (i.e., short-chain fatty acids, amino acid derivatives, secondary bile acid), it acts as an immunologic and metabolic organ [98,99]. By contrast, obesity and related metabolic disorders, such as MetS and T2D, determine profound functional and compositional alterations in the intestinal microbiota, collectively referred to as dysbiosis [100].

Little is known about $Mg^{2+}$ deficiency and gut microbiota in humans, while some data are available in animal models. A 6-week $Mg^{2+}$-deficient diet in rodents altered the gut microbiota and was associated with anxiety-like behavior [101]. In particular, $Mg^{2+}$ deficiency may mediate an imbalance of the microbiota–gut–brain axis, which contributes to the development of depressive-like behavior [102]. It should be pointed out that obesity increases the risk of depression and depression was found to be predictive of developing obesity [103]. Moreover, epidemiological data have demonstrated that obesity is an important risk factor for the development of gastroesophageal reflux disease [104] and PPI used for the treatment of such disease, may lead to $Mg^{2+}$ deficiency also through the involvement of the gut microbiome [105]. As previously mentioned, $Mg^{2+}$ deficiency is a nutritional disorder connected to a low-grade, latent chronic inflammatory state. Interestingly, in $Mg^{2+}$-deficient mice, changes in intestinal bifidobacteria levels are associated with an inflammatory response, thus creating an effective link between $Mg^{2+}$ status, gut microbiota and inflammation [106].

It is now widely accepted that an altered gut microbiota composition participates in systemic low-grade inflammation [107–110]. An analysis of patients with different glucose tolerance suggests that both structure and diversity of gut microbiota are altered in the presence of impaired glucose regulation and T2D [111,112]. However, Thingholm et al. compared the microbiota composition of obese versus lean subjects and obese versus obese with T2D. The authors observed that microbiome diversity and functionality were significantly reduced in obese compared to lean subjects, while only modest differences emerged when comparing the microbiome of obese versus obese with T2D [113]. Therefore, the development of obesity-associated T2D could be related to a progressive disruption of the gut microbiome. Gut microbiota manipulation through dietary adjustment has become an important research direction in T2D prevention and therapy. In this perspective, $Mg^{2+}$ supplementation might help in remodeling the microbiota. Indeed, $Mg^{2+}$ supplementation in obese subjects with and without T2D affects microbial composition and functional potential [113]. Moreover, dietary supplementation with a multi-mineral functional food derived from seaweed and seawater, rich in bioactive $Mg^{2+}$ and other trace elements, significantly enhances the gut microbial diversity in adult male rats [114].

To conclude, an adequate $Mg^{2+}$ dietary intake could positively affect the composition of the intestinal microbiota and, consequently, the host metabolism, thus helping in preventing metabolic alterations associated with the development of MetS and TD2. However, the path for clarifying the impact of $Mg^{2+}$ in this emerging field of research is still long.

## 7. Dietary $Mg^{2+}$

The intakes of food rich in $Mg^{2+}$, including whole grains, nuts and seeds, legumes, and dark-green vegetables, were associated with a lower incidence of obesity, T2D and MetS [43]. Therefore, correcting unhealthy diets is a priority to meet the daily-recommended requirement for $Mg^{2+}$. However, because of agronomic and environmental factors as well as food processing, $Mg^{2+}$ content in fruits and vegetables dropped in the last 50 years [115] and it might be necessary to supplement it. This is an approach that has been proven beneficial in T2D and MetS (Figure 2). The daily administration of 250 mg of elemental $Mg^{2+}$ for three months improves glycemic control in T2D subjects as demonstrated by the significant reduction of glycated hemoglobin, insulin levels, C-peptide, and Homeostatic Model Assessment for Insulin Resistance (HOMA-IR) [116]. This effect is probably due to the correction of an underlying latent $Mg^{2+}$ deficiency. Indeed, the supplementation with 360 mg of $Mg^{2+}$ for the same period does not improve insulin sensitivity in normomagnesemic T2D patients [117]. The administration of 250 mg of elemental $Mg^{2+}$ for 12 weeks improved the wound healing of diabetic foot ulcers, decreasing the lesion size, and ameliorating glucose metabolism [118]. Mg2+ could also affect glucose metabolism by modulating the concentration of inflammatory cytokines, such as IL-6. Although these data need to be confirmed, in prediabetic subjects, the supplementation with 380 mg of Mg2+ provides a trend of reduction in IL-6 plasmatic levels while there are no differences in the levels of C-reactive protein (CRP), Tumor Necrosis Factor-alpha (TNF-alpha), and Interleukin 10 (IL-10) [119]. It is noteworthy that, in apparently healthy runners fed a low Mg2+ diet, the administration of 500 mg of Mg2+ lowers IL-6 levels, reduces muscle soreness and increases post-exercise blood glucose [120].

$Mg^{2+}$ supplementation seems to improve blood pressure control and vascular resistance in patients with essential hypertension [121]. The administration of 300 mg of $Mg^{2+}$ for one month decreases systolic and diastolic pressures, systemic vascular resistance, and left cardiac work [122]. The oral $Mg^{2+}$ supplementation with 600 mg for 12 weeks is associated with moderate but consistent ambulatory blood pressure reduction in patients with mild hypertension [123]. This result can be explained by the evidence that $Mg^{2+}$ is a $Ca^{2+}$ antagonist, increases the synthesis of vasodilators such as prostacyclin and nitric oxide, and inhibits vascular calcifications through the modulation of TRPM7 [123,124]. An increase in the transcription of the $Mg^{2+}$ channel TRPM6 could explain the antihypertensive effects of $Mg^{2+}$ supplementation. The increase of TRPM6 mRNA expression is obtained with the administration of 360 mg of $Mg^{2+}$ for four months [124]. The positive effect of $Mg^{2+}$ supplementation on blood pressure is also reported in patients already undergoing drug treatment for hypertension. In thiazide-treated women, the administration of 600 mg of $Mg^{2+}$ improves endothelial function and subclinical atherosclerosis [125]. In hemodialysis patients, the administration of 440 mg of $Mg^{2+}$ for six months decreases carotid intimatemedia thickness, which is a marker of cardiovascular disease. This effect is not associated with an improvement of endothelial function measured by brachial artery flow-mediated dilatation and might be explained by the modulation of calcification through the regulation of calcium and phosphorus concentration in blood [126]. In disagreement with the aforementioned results, a randomized controlled trial on overweight and obese middle-aged and elderly adults did not report any improvement of endothelial function and cardiometabolic risk markers after supplementing 350 mg $Mg^{2+}$ daily for 24 weeks [126,127].

Since correcting $Mg^{2+}$ status lowers blood pressure, corrects lipid profile and ameliorates the control of glycemia, it is not surprising that $Mg^{2+}$ supplementation has positive effects in MetS. The supplementation of 380 mg of $Mg^{2+}$ for 16 weeks improves MetS by reducing blood pressure, hyperglycemia, and hypertriglyceridemia [128], because the correction of hypomagnesemia leads to changes in gene expression and proteomic profiling consistent with favorable effects on several metabolic pathways [85]. The effect of $Mg^{2+}$ on the lipid profile is still debated and appears to be mediated by the improvement of insulin resistance and appears to be present only if $Mg^{2+}$ supplementation corrects a previous deficiency [129]. The administration of 370 mg of $Mg^{2+}$ in healthy normomagnesemic young men with a family history of MetS does not show beneficial effects on blood pressure, vascular function, and glycolipid profile [130]. For MetS, as well as for T2D, the positive effect of $Mg^{2+}$ administration is only registered if the supplementation corrects a condition of hypomagnesemia.

Some critical issues emerge from the analysis of the literature and complicate the interpretation of the data (Table 1). First, there is no agreement on the dosages and timing of $Mg^{2+}$ supplementation in the treatment of MetS, T2D, and hypertension. Considering the literature just discussed [109–124], the dosage of $Mg^{2+}$ varies from 250 mg to 600 mg, with a median of 380 mg (95% confidence interval (CI) 300–500 mg). The time of $Mg^{2+}$ supplementation ranges between 7 days and six months, with a median of about three months (95% CI 4–24 weeks). Besides, there is no consensus on the type of $Mg^{2+}$ salt to use. The bioavailability of different Mg$^{2+}$ salts has been investigated in depth [131–134]. Magnesium sulfate, oxide, carbonate, chloride, citrate, malate, acetate, gluconate, lactate, aspartate, fumarate, acetyl taurate, bis-glycinate, and pidolate are all employed in Mg$^{2+}$ supplementation. In part, the differences in the bioavailability of Mg$^{2+}$ salt is due to their different solubility [135]. Although organic Mg$^{2+}$ salts were slightly more bioavailable than inorganic Mg$^{2+}$ salts, inorganic Mg$^{2+}$ salts have been administered to patients with interesting clinical outcomes. The choice of the type of Mg$^{2+}$ salt based on its bioavailability conditions its dosage and the possible side effects, especially as intestinal symptoms of osmotic dysentery [136] (Figure 3).

**Table 1.** A quick recap of the last and most relevant clinical trials describing the effects of Mg$^{2+}$ supplementation on obesity, metabolic syndrome (MetS), and type 2 diabetes (T2D).

| Author(s) | Year | Dosage of Mg$^{2+}$ Supplementation | Type of Salt | Timing of Mg$^{2+}$ Supplementation | Effects of Mg$^{2+}$ Supplementation | Ref. |
|---|---|---|---|---|---|---|
| Elderawi WA et al. | 2018 | 250 mg/day | Oxide, gluconate, lactate | 3 months | Improves glycemic control in T2D subjects with a reduction of glycated hemoglobin, insulin levels, C-peptide, and HOMA-IR. | [116] |
| Navarrete-Cortes A et al. | 2014 | 360 mg/day | Lactate | 3 months | No effects on insulin sensitivity. | [117] |
| Razzaghi R et al. | 2018 | 250 mg/day | Oxide | 12 weeks | Improves wound healing of diabetic foot ulcers, decreasing the lesion size, and ameliorating glucose metabolism. | [118] |
| Simental-Mendía LE et al. | 2012 | 380 mg/day | Chloride | 3 months | Reduces IL-6 plasmatic levels. | [119] |
| Steward CJ et al. | 2019 | 500 mg/day | Oxide, stearate | 7 days | Lowers IL-6 levels, reduces muscle soreness and increases post-exercise blood glucose. | [120] |
| Banjanin N et al. | 2018 | 300 mg/day | Oxide | 1 month | Decreases systolic and diastolic pressures, systemic vascular resistance, and left cardiac work. | [122] |
| Hatzistavri LS et al. | 2009 | 600 mg/day | Pidolate | 12 weeks | Reduces ambulatory blood pressure. | [123] |
| Rodríguez-Ramírez M et al. | 2017 | 360 mg/day | Lactate | 4 months | Increases TRPM6 mRNA relative expression. | [124] |
| Cunha AR et al. | 2017 | 600 mg twice a day | Chelate (not better specified) | 6 months | Improves endothelial function and subclinical atherosclerosis. | [125] |
| Mortazavi M et al. | 2013 | 440 mg 3 times per week | Oxide | 6 months | Decreases carotid intimate-media thickness, which is a marker of cardiovascular disease. | [126] |
| Joris PJ et al. | 2017 | 350 mg/day | Citrate | 24 weeks | No effect on endothelial function. | [127] |
| Rodríguez-Morán M et al. | 2018 | 380 mg/day | Chloride | 16 weeks | Improves MetS by reducing blood pressure, hyperglycemia, and hypertriglyceridemia. | [128] |
| Cosaro E et al. | 2014 | 370 mg twice a day | Pidolate | 8 weeks | Effects on blood pressure, vascular function, and glycolipid profile. | [130] |

HOMA-IR: Homeostasis Model Assessment-estimated for Insulin Resistance; IL-6: Interleukin-6; TRPM6: Transient Receptor Potential Melastatin 6.

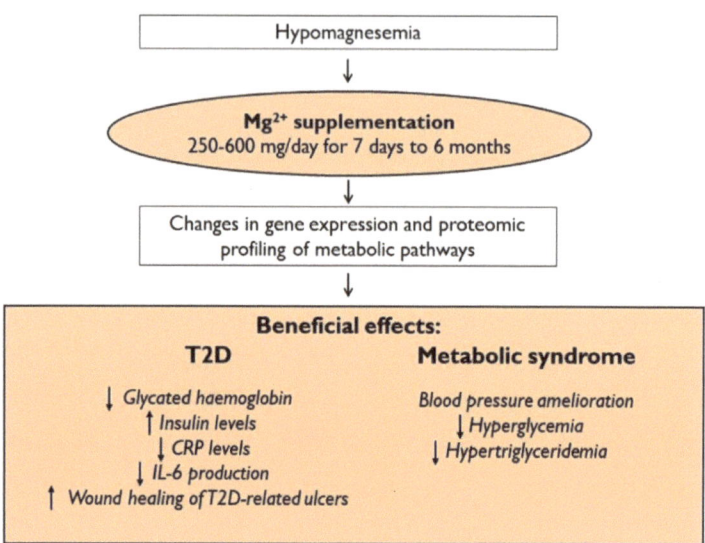

**Figure 3.** Beneficial effects of magnesium supplementation in hypomagnesemic patients with metabolic syndrome and type 2 diabetes.

## 8. Conclusions

Obesity, type 2 diabetes, and metabolic syndrome are intertwined conditions characterized by chronic low-grade inflammation partly attributable to $Mg^{2+}$ deficiency. In metabolic diseases, a low $Mg^{2+}$ status mainly due to unhealthy diets contributes to generate a pro-inflammatory environment that exacerbates metabolic derangement. $Mg^{2+}$ supplementation seems to foment the correction of this vicious loop, but at the moment it is hard to interpret whether $Mg^{2+}$ beneficial effects occur through a direct effect on metabolic pathways or an indirect action on inflammation, or both.

Several important points need to be clarified. At the clinical level, more studies are necessary to define which $Mg^{2+}$ salt and which dosage guarantee better outcomes. In addition, the investigation of microbiota in hypomagnesemic subjects might provide interesting hints and suggest targeted dietary approaches aimed at harmonizing the gut microbial ecosystem. In addition, biomarkers that grant the possibility of evaluating $Mg^{2+}$ homeostasis should be identified. At the cellular and molecular level, it is important to focus on the role of intracellular $Mg^{2+}$ in modulating cell function, from the regulation of metabolism to the release of inflammatory mediators.

Considering the worldwide prevalence of obesity, type 2 diabetes and metabolic syndrome, the correction of bad dietary habits and, eventually, the supplementation of $Mg^{2+}$ might represent an inexpensive but valuable tool to contain the occurrence and the progression of these conditions.

**Funding:** This research received no external funding. The APC was funded by University of Milan.

**Informed Consent Statement:** Not applicable.

**Data Availability Statement:** Data sharing is not applicable to this article.

**Acknowledgments:** The authors acknowledge support from the University of Milan through the APC initiative. Moreover, this work was developed as part of the PhD program in Nutrition Sciences, University of Milan.

**Conflicts of Interest:** The authors declare no conflict of interest.

## References

1. EFSA Panel on Dietetic Products, Nutrition and Allergies (NDA). Scientific Opinion on Dietary Reference Values for magnesium. *Efsa J.* **2015**, *13*, 4186. [CrossRef]
2. Jahnen-Dechent, W.; Ketteler, M. Magnesium basics. *CKJ Clin. Kidney J.* **2012**, *5*, 3–14. [CrossRef] [PubMed]
3. De Baaij, J.H.F.; Hoenderop, J.G.J.; Bindels, R.J.M. Magnesium in man: Implications for health and disease. *Physiol. Rev.* **2015**, *95*, 1–46. [CrossRef]
4. Saltiel, A.R.; Olefsky, J.M. Inflammatory mechanisms linking obesity and metabolic disease. *J. Clin. Investig.* **2017**, *127*, 1–4. [CrossRef] [PubMed]
5. Von Ehrlich, B.; Barbagallo, M.; Classen, H.G.; Guerrero-Romero, F.; Mooren, F.C.; Rodriguez-Moran, M.; Vierling, W.; Vormann, J.; Kisters, K. Significance of magnesium in insulin resistance, metabolic syndrome, and diabetes—Recommendations of the Association of Magnesium Research e.V. *Trace Elem. Electrolytes* **2017**, *34*, 124–129. [CrossRef]
6. Nielsen, F.H. Effects of magnesium depletion on inflammation in chronic disease. *Curr. Opin. Clin. Nutr. Metab. Care* **2014**, *17*, 525–530. [CrossRef]
7. Nielsen, F.H. Magnesium deficiency and increased inflammation: Current perspectives. *J. Inflamm. Res.* **2018**, *11*, 25–34. [CrossRef]
8. Mazidi, M.; Rezaie, P.; Banach, M. Effect of magnesium supplements on serum C-reactive protein: A systematic review and meta-analysis. *Arch. Med. Sci.* **2018**, *14*, 707–716. [CrossRef]
9. Lobionda, S.; Sittipo, P.; Kwon, H.Y.; Lee, Y.K. The role of gut microbiota in intestinal inflammation with respect to diet and extrinsic stressors. *Microorganisms* **2019**, *7*, 271. [CrossRef]
10. Oh, H.E.; Deeth, H.C. Magnesium in milk. *Int. Dairy J.* **2017**, *71*, 89–97. [CrossRef]
11. Gröber, U.; Schmidt, J.; Kisters, K. Magnesium in prevention and therapy. *Nutrients* **2015**, *7*, 8199–8226. [CrossRef] [PubMed]
12. Nielsen, F.H. The Problematic Use of Dietary Reference Intakes to Assess Magnesium Status and Clinical Importance. *Biol. Trace Elem. Res.* **2019**, *188*, 52–59. [CrossRef] [PubMed]
13. Costello, R.B.; Elin, R.J.; Rosanoff, A.; Wallace, T.C.; Guerrero-Romero, F.; Hruby, A.; Lutsey, P.L.; Nielsen, F.H.; Rodriguez-Moran, M.; Song, Y.; et al. Perspective: The Case for an Evidence-Based Reference Interval for Serum Magnesium: The Time Has Come. *Adv. Nutr. Int. Rev. J.* **2016**, *7*, 977–993. [CrossRef]
14. Lowenstein, F.W.; Stanton, M.F. Serum Magnesium Levels in The United States, 1971–1974. *J. Am. Coll. Nutr.* **1986**, *5*, 399–414. [CrossRef] [PubMed]
15. Nielsen, F.H. Guidance for the determination of status indicators and dietary requirements for magnesium. *Magnes. Res.* **2016**, *29*, 154–160. [CrossRef]
16. Razzaque, M.S. Magnesium: Are We Consuming Enough? *Nutrients* **2018**, *10*, 1863. [CrossRef]
17. Topf, J.M.; Murray, P.T. Hypomagnesemia and hypermagnesemia. *Rev. Endocr. Metab. Disord.* **2003**, *4*, 195–206. [CrossRef]
18. Al Alawi, A.M.; Majoni, S.W.; Falhammar, H. Magnesium and Human Health: Perspectives and Research Directions. *Int. J. Endocrinol.* **2018**, *2018*. [CrossRef]
19. Beaudart, C.; Locquet, M.; Touvier, M.; Reginster, J.Y.; Bruyère, O. Association between dietary nutrient intake and sarcopenia in the SarcoPhAge study. *Aging Clin. Exp. Res.* **2019**, *31*, 815–824. [CrossRef]
20. Van Dronkelaar, C.; Van Velzen, A.; Abdelrazek, M.; Van der Steen, A.; Weijs, P.J.M.; Tieland, M. Minerals and Sarcopenia; The Role of Calcium, Iron, Magnesium, Phosphorus, Potassium, Selenium, Sodium, and Zinc on Muscle Mass, Muscle Strength, and Physical Performance in Older Adults: A Systematic Review. *J. Am. Med. Dir. Assoc.* **2018**, *19*, 6–11.e3. [CrossRef]
21. Rude, R.K.; Gruber, H.E. Magnesium deficiency and osteoporosis: Animal and human observations. *J. Nutr. Biochem.* **2004**, *15*, 710–716. [CrossRef] [PubMed]
22. Ford, E.S.; Mokdad, A.H. Dietary magnesium intake in a national sample of US adults. *J. Nutr.* **2003**, *133*, 2879–2882. [CrossRef] [PubMed]
23. Olza, J.; Aranceta-Bartrina, J.; González-Gross, M.; Ortega, R.M.; Serra-Majem, L.; Varela-Moreiras, G.; Gil, Á. Reported dietary intake, disparity between the reported consumption and the level needed for adequacy and food sources of calcium, phosphorus, magnesium and vitamin D in the Spanish population: Findings from the ANIBES study. *Nutrients* **2017**, *9*, 168. [CrossRef] [PubMed]
24. Tarleton, E.K. Factors influencing magnesium consumption among adults in the United States. *Nutr. Rev.* **2018**, *76*, 526–538. [CrossRef] [PubMed]
25. Rondanelli, M.; Faliva, M.A.; Gasparri, C.; Peroni, G.; Naso, M.; Picciotto, G.; Riva, A.; Nichetti, M.; Infantino, V.; Alalwan, T.A.; et al. Micronutrients dietary supplementation advices for celiac patients on long-term gluten-free diet with good compliance: A review. *Medicine* **2019**, *55*, 337. [CrossRef] [PubMed]
26. Galland, L. Magnesium and inflammatory bowel disease. *Magnesium* **1988**, *7*, 78–83.
27. Kruis, W.; Phuong Nguyen, G. Iron Deficiency, Zinc, Magnesium, Vitamin Deficiencies in Crohn's Disease: Substitute or Not? *Dig. Dis.* **2016**, *34*, 105–111. [CrossRef]
28. Owczarek, D.; Rodacki, T.; Domagała-Rodacka, R.; Cibor, D.; Mach, T. Diet and nutritional factors in inflammatory bowel diseases. *World J. Gastroenterol.* **2016**, *22*, 895–905. [CrossRef]
29. Dinicolantonio, J.J.; O'keefe, J.H.; Wilson, W. Subclinical magnesium deficiency: A principal driver of cardiovascular disease and a public health crisis Coronary artery disease. *Open Hear.* **2018**, *5*, 668. [CrossRef]
30. Bateman, S.W. A Quick Reference on Magnesium. *Vet. Clin. N. Am. Small Anim. Pract.* **2017**, *47*, 235–239. [CrossRef]

31. Chrysant, S.G. Proton pump inhibitor-induced hypomagnesemia complicated with serious cardiac arrhythmias. *Expert Rev. Cardiovasc.* **2019**, *17*, 345–351. [CrossRef] [PubMed]
32. Grochowski, C.; Blicharska, E.; Baj, J.; Mierzwińska, A.; Brzozowska, K.; Forma, A.; MacIejewski, R. Serum iron, magnesium, copper, and manganese levels in alcoholism: A systematic review. *Molecules* **2019**, *24*, 1361. [CrossRef] [PubMed]
33. Maguire, D.; Ross, D.P.; Talwar, D.; Forrest, E.; Naz Abbasi, H.; Leach, J.P.; Woods, M.; Zhu, L.Y.; Dickson, S.; Kwok, T.; et al. Low serum magnesium and 1-year mortality in alcohol withdrawal syndrome. *Eur. J. Clin. Investig.* **2019**, *49*, e13152. [CrossRef] [PubMed]
34. Viering, D.H.H.M.; De Baaij, J.H.F.; Walsh, S.B.; Kleta, R.; Bockenhauer, D. Genetic causes of hypomagnesemia, a clinical overview. *Pediatr. Nephrol.* **2017**, *32*, 1123–1135. [CrossRef]
35. López-González, B.; Molina-López, J.; Florea, D.I.; Quintero-Osso, B.; Pérez De La Cruz, A.; Ma, E.; Del Pozo, P. Association between magnesium-deficient status and anthropometric and clinical-nutritional parameters in posmenopausal women. *Nutr Hosp.* **2014**, *29*, 658–664. [CrossRef]
36. Touitou, Y.; Godard, J.P.; Ferment, O.; Chastang, C.; Proust, J.; Bogdan, A.; Auzéby, A.; Touitou, C. Prevalence of magnesium and potassium deficiencies in the elderly. *Clin. Chem.* **1987**, *33*, 518–523. [CrossRef]
37. Nielsen, F.H. Magnesium, inflammation, and obesity in chronic disease. *Nutr. Rev.* **2010**, *68*, 333–340. [CrossRef]
38. Maier, J.A.; Castiglioni, S.; Locatelli, L.; Zocchi, M.; Mazur, A. Magnesium and inflammation: Advances and perspectives. *Semin. Cell Dev. Biol.* **2020**. [CrossRef]
39. Morais, J.B.S.; Severo, J.S.; Dos Santos, L.R.; De Sousa Melo, S.R.; De Oliveira Santos, R.; De Oliveira, A.R.S.; Cruz, K.J.C.; Do Nascimento Marreiro, D. Role of Magnesium in Oxidative Stress in Individuals with Obesity. *Biol. Trace Elem. Res.* **2017**, *176*, 20–26. [CrossRef]
40. Jiang, S.; Ma, X.; Li, M.; Yan, S.; Zhao, H.; Pan, Y.; Wang, C.; Yao, Y.; Jin, L.; Li, B. Association between dietary mineral nutrient intake, body mass index, and waist circumference in U.S. Adults using quantile regression analysis NHANES 2007–2014. *PeerJ* **2020**, *8*, e9127. [CrossRef]
41. Kelly, O.J.; Gilman, J.C.; Kim, Y.; Ilich, J.Z. Macronutrient Intake and Distribution in the Etiology, Prevention and Treatment of Osteosarcopenic Obesity. *Curr. Aging Sci.* **2016**, *10*, 83–105. [CrossRef]
42. Galan, P.; Preziosi, P.; Durlach, V.; Valeix, P.; Ribas, L.; Bouzid, D.; Favier, A.; Hercberg, S. Dietary magnesium intake in a French adult population. *Magnes. Res.* **1997**, *10*, 321–328. [CrossRef] [PubMed]
43. Lu, L.; Chen, C.; Yang, K.; Zhu, J.; Xun, P.; Shikany, J.M.; He, K. Magnesium intake is inversely associated with risk of obesity in a 30-year prospective follow-up study among American young adults. *Eur. J. Nutr.* **2020**, *59*, 3745–3753. [CrossRef] [PubMed]
44. Devaux, S.; Adrian, M.; Laurant, P.; Berthelot, A.; Quignard-Boulangé, A. Dietary magnesium intake alters age-related changes in rat adipose tissue cellularity. *Magnes. Res.* **2016**, *29*, 175–183. [CrossRef] [PubMed]
45. Castellanos-Gutiérrez, A.; Sánchez-Pimienta, T.G.; Carriquiry, A.; Da Costa, T.H.M.; Ariza, A.C. Higher dietary magnesium intake is associated with lower body mass index, waist circumference and serum glucose in Mexican adults. *Nutr. J.* **2018**, *17*, 114. [CrossRef]
46. He, K.; Liu, K.; Daviglus, M.L.; Morris, S.J.; Loria, C.M.; Van Horn, L.; Jacobs, D.R.; Savage, P.J. Magnesium intake and incidence of metabolic syndrome among young adults. *Circulation* **2006**, *113*, 1675–1682. [CrossRef]
47. Shamnani, G.; Rukadikar, C.; Gupta, V.; Singh, S.; Tiwari, S.; Bhartiy, S.; Sharma, P. Serum magnesium in relation with obesity. *Natl. J. Physiol. Pharm. Pharm.* **2018**, *8*, 1074–1077. [CrossRef]
48. Maguire, D.; Talwar, D.; Shiels, P.G.; McMillan, D. The role of thiamine dependent enzymes in obesity and obesity related chronic disease states: A systematic review. *Clin. Nutr. ESPEN* **2018**, *25*, 8–17. [CrossRef]
49. Mishra, S.; Padmanaban, P.; Deepti, G.N.; Sarkar, G.; Sumathi, S.; Toora, B.D. Serum magnesium and dyslipidemia in type-2 diabetes mellitus. *Biomed. Res.* **2012**, *23*, 295–300.
50. Ansari, M.R.; Maheshwari, N.; Shaikh, M.A.; Laghari, M.S.; Darshana; Lal, K.; Ahmed, K. Correlation of serum magnesium with dyslipidemia in patients on maintenance hemodialysis. *Saudi J. Kidney Dis. Transpl.* **2012**, *23*, 21–25. [CrossRef]
51. Deepti, R.; Nalini, G. Anbazhagan Relationship between hypomagnesemia and dyslipidemia in type 2 diabetes mellitus. *Asian J. Pharm. Res. Health Care* **2014**, *6*, 32–36.
52. Pereira-Santos, M.; Costa, P.R.F.; Assis, A.M.O.; Santos, C.A.S.T.; Santos, D.B. Obesity and vitamin D deficiency: A systematic review and meta-analysis. *Obes. Rev.* **2015**, *16*, 341–349. [CrossRef] [PubMed]
53. Rafiq, S.; Jeppesen, P.B. Body mass index, vitamin d, and type 2 diabetes: A systematic review and meta-analysis. *Nutrients* **2018**, *10*, 1182. [CrossRef] [PubMed]
54. Uwitonze, A.M.; Razzaque, M.S. Role of magnesium in vitamin d activation and function. *J. Am. Osteopath. Assoc.* **2018**, *118*, 181–189. [CrossRef] [PubMed]
55. Dai, Q.; Zhu, X.; Manson, J.A.E.; Song, Y.; Li, X.; Franke, A.A.; Costello, R.B.; Rosanoff, A.; Nian, H.; Fan, L.; et al. Magnesium status and supplementation influence Vitamin D status and metabolism: Results from a randomized trial. *Am. J. Clin. Nutr.* **2018**, *108*, 1249–1258. [CrossRef] [PubMed]
56. Al-Khalidi, B.; Kimball, S.M.; Rotondi, M.A.; Ardern, C.I. Standardized serum 25-hydroxyvitamin D concentrations are inversely associated with cardiometabolic disease in U.S. adults: A cross-sectional analysis of NHANES, 2001–2010. *Nutr. J.* **2017**, *16*, 16. [CrossRef] [PubMed]

57. Stokic, E.; Romani, A.; Ilincic, B.; Kupusinac, A.; Stosic, Z.; Isenovic, E.R. Chronic Latent Magnesium Deficiency in Obesity Decreases Positive Effects of Vitamin D on Cardiometabolic Risk Indicators. *Curr. Vasc. Pharm.* **2018**, *16*, 610–617. [CrossRef]
58. Farhanghi, M.A.; Mahboob, S.; Ostadrahimi, A. Obesity induced Magnesium deficiency can be treated by vitamin D supplementation. *J. Pak. Med. Assoc.* **2009**, *59*, 258–261.
59. Song, Y.; Ridker, P.M.; Manson, J.A.E.; Cook, N.R.; Buring, J.E.; Liu, S. Magnesium intake, C-reactive protein, and the prevalence of metabolic syndrome in middle-aged and older U.S. women. *Diabetes Care* **2005**, *28*, 1438–1444. [CrossRef]
60. McKeown, N.M.; Jacques, P.F.; Zhang, X.L.; Juan, W.; Sahyoun, N.R. Dietary magnesium intake is related to metabolic syndrome in older Americans. *Eur. J. Nutr.* **2008**, *47*, 210–216. [CrossRef]
61. Mirmiran, P.; Shab-Bidar, S.; Hosseini-Esfahani, F.; Asghari, G.; Hosseinpour-Niazi, S.; Azizi, F. Magnesium intake and prevalence of metabolic syndrome in adults: Tehran lipid and glucose study. *Public Health Nutr.* **2012**, *15*, 693–701. [CrossRef] [PubMed]
62. Choi, M.K.; Bae, Y.J. Relationship between dietary magnesium, manganese, and copper and metabolic syndrome risk in Korean Adults: The Korea national health and nutrition examination survey (2007-2008). *Biol. Trace Elem. Res.* **2013**, *156*, 56–66. [CrossRef] [PubMed]
63. Dibaba, D.T.; Xun, P.; Fly, A.D.; Yokota, K.; He, K. Dietary magnesium intake and risk of metabolic syndrome: A meta-analysis. *Diabet. Med.* **2014**, *31*, 1301–1309. [CrossRef]
64. Sarrafzadegan, N.; Khosravi-Boroujeni, H.; Lotfizadeh, M.; Pourmogaddas, A.; Salehi-Abargouei, A. Magnesium status and the metabolic syndrome: A systematic review and meta-analysis. *Nutrition* **2016**, *32*, 409–417. [CrossRef] [PubMed]
65. Yang, N.; He, L.; Li, Y.; Xu, L.; Ping, F.; Li, W.; Zhang, H. Reduced Insulin Resistance Partly Mediated the Association of High Dietary Magnesium Intake with Less Metabolic Syndrome in a Large Chinese Population. *Diabetes. Metab. Syndr. Obes.* **2020**, *13*, 2541–2550. [CrossRef]
66. Rasic-Milutinovic, Z.; Perunicic-Pekovic, G.; Jovanovic, D.; Gluvic, Z.; Cankovic-Kadijevic, M. Association of blood pressure and metabolic syndrome components with magnesium levels in drinking water in some Serbian municipalities. *J. Water Health* **2012**, *10*, 161–169. [CrossRef]
67. Fang, X.; Wang, K.; Han, D.; He, X.; Wei, J.; Zhao, L.; Imam, M.U.; Ping, Z.; Li, Y.; Xu, Y.; et al. Dietary magnesium intake and the risk of cardiovascular disease, type 2 diabetes, and all-cause mortality: A dose–response meta-analysis of prospective cohort studies. *BMC Med.* **2016**, *14*, 210. [CrossRef]
68. Zhang, W.; Iso, H.; Ohira, T.; Date, C.; Tamakoshi, A. Associations of dietary magnesium intake with mortality from cardiovascular disease: The JACC study. *Atherosclerosis* **2012**, *221*, 587–595. [CrossRef]
69. Veronese, N.; Watutantrige-Fernando, S.; Luchini, C.; Solmi, M.; Sartore, G.; Sergi, G.; Manzato, E.; Barbagallo, M.; Maggi, S.; Stubbs, B. Effect of magnesium supplementation on glucose metabolism in people with or at risk of diabetes: A systematic review and meta-analysis of double-blind randomized controlled trials. *Eur. J. Clin. Nutr.* **2016**, *70*, 1354–1359. [CrossRef]
70. Veronese, N.; Demurtas, J.; Pesolillo, G.; Celotto, S.; Barnini, T.; Calusi, G.; Caruso, M.G.; Notarnicola, M.; Reddavide, R.; Stubbs, B.; et al. Magnesium and health outcomes: An umbrella review of systematic reviews and meta-analyses of observational and intervention studies. *Eur. J. Nutr.* **2020**, *59*, 263–272. [CrossRef]
71. Rosique-Esteban, N.; Guasch-Ferré, M.; Hernández-Alonso, P.; Salas-Salvadó, J. Dietary magnesium and cardiovascular disease: A review with emphasis in epidemiological studies. *Nutrients* **2018**, *10*, 168. [CrossRef] [PubMed]
72. Maria De Lourdes, L.; Cruz, T.; Rodrigues, L.E.; Bomfim, O.; Melo, J.; Correia, R.; Porto, M.; Cedro, A.; Vicente, E. Serum and intracellular magnesium deficiency in patients with metabolic syndrome-Evidences for its relation to insulin resistance. *Diabetes Res. Clin. Pract.* **2009**, *83*, 257–262. [CrossRef]
73. Ghasemi, A.; Zahediasl, S.; Syedmoradi, L.; Azizi, F. Low serum magnesium levels in elderly subjects with metabolic syndrome. *Biol. Trace Elem. Res.* **2010**, *136*, 18–25. [CrossRef] [PubMed]
74. Wang, Y.; Wei, J.; Zeng, C.; Yang, T.; Li, H.; Cui, Y.; Xie, D.; Xu, B.; Liu, Z.; Li, J.; et al. Association between serum magnesium concentration and metabolic syndrome, diabetes, hypertension and hyperuricaemia in knee osteoarthritis: A cross-sectional study in Hunan Province, China. *BMJ Open* **2018**, *8*, e019159. [CrossRef] [PubMed]
75. Han, D.; Fang, X.; Su, D.; Huang, L.; He, M.; Zhao, D.; Zou, Y.; Zhang, R. Dietary Calcium Intake and the Risk of Metabolic Syndrome: A Systematic Review and Meta-Analysis. *Sci. Rep.* **2019**, *9*, 19046. [CrossRef] [PubMed]
76. Moore-Schiltz, L.; Albert, J.M.; Singer, M.E.; Swain, J.; Nock, N.L. Dietary intake of calcium and magnesium and the metabolic syndrome in the National Health and Nutrition Examination (NHANES) 2001-2010 data. *Br. J. Nutr.* **2015**, *114*, 924–935. [CrossRef]
77. Park, S.H.; Kim, S.K.; Bae, Y.J. Relationship between serum calcium and magnesium concentrations and metabolic syndrome diagnostic components in middle-aged Korean men. *Biol. Trace Elem. Res.* **2012**, *146*, 35–41. [CrossRef]
78. Dai, Q.; Shu, X.O.; Deng, X.; Xiang, Y.B.; Li, H.; Yang, G.; Shrubsole, M.J.; Ji, B.; Cai, H.; Chow, W.H.; et al. Modifying effect of calcium/magnesium intake ratio and mortality: A population based cohort study. *BMJ Open* **2013**, *3*, e002111. [CrossRef]
79. Dong, J.-Y.; Xun, P.; He, K.; Qin, L.-Q. Magnesium Intake and Risk of Type 2 Diabetes. *Diabetes Care* **2011**, *34*, 2116–2122. [CrossRef]
80. Bertinato, J.; Wang, K.C.; Hayward, S. Serum magnesium concentrations in the Canadian population and associations with diabetes, glycemic regulation, and insulin resistance. *Nutrients* **2017**, *9*, 296. [CrossRef]
81. Zhao, B.; Zeng, L.; Zhao, J.; Wu, Q.; Dong, Y.; Zou, F.; Gan, L.; Wei, Y.; Zhang, W. Association of magnesium intake with type 2 diabetes and total stroke: An updated systematic review and meta-analysis. *BMJ Open* **2020**, *10*, 32240. [CrossRef] [PubMed]
82. Barbagallo, M.; Dominguez, L.J. Magnesium metabolism in type 2 diabetes mellitus, metabolic syndrome and insulin resistance. *Arch. Biochem. Biophys.* **2007**, *458*, 40–47. [CrossRef] [PubMed]

83. Esmeralda, C.A.C.; Ibrahim, S.N.A.; David, P.E.; Maldonado, I.C.; David, A.S.; Escorza, M.A.Q.; Dealmy, D.G. Deranged fractional excretion of magnesium and serum magnesium levels in relation to retrograde glycaemic regulation in patients with type 2 diabetes mellitus. *Curr. Diabetes Rev.* **2020**, *17*, 91–100. [CrossRef] [PubMed]
84. Fang, X.; Han, H.; Li, M.; Liang, C.; Fan, Z.; Aaseth, J.; He, J.; Montgomery, S.; Cao, Y. Dose-Response Relationship between Dietary Magnesium Intake and Risk of Type 2 Diabetes Mellitus: A Systematic Review and Meta-Regression Analysis of Prospective Cohort Studies. *Nutrients* **2016**, *8*, 739. [CrossRef] [PubMed]
85. Chacko, S.A.; Sul, J.; Song, Y.; Li, X.; LeBlanc, J.; You, Y.; Butch, A.; Liu, S. Magnesium supplementation, metabolic and inflammatory markers, and global genomic and proteomic profiling: A randomized, double-blind, controlled, crossover trial in overweight individuals. *Am. J. Clin. Nutr.* **2011**, *93*, 463–473. [CrossRef]
86. Mooren, F.C.; Krüger, K.; Völker, K.; Golf, S.W.; Wadepuhl, M.; Kraus, A. Oral magnesium supplementation reduces insulin resistance in non-diabetic subjects—A double-blind, placebo-controlled, randomized trial. *Diabetes Obes. Metab.* **2011**, *13*, 281–284. [CrossRef]
87. Hruby, A.; Guasch-Ferré, M.; Bhupathiraju, S.N.; Manson, J.E.; Willett, W.C.; McKeown, N.M.; Hu, F.B. Magnesium Intake, Quality of Carbohydrates, and Risk of Type 2 Diabetes: Results From Three U.S. Cohorts. *Diabetes Care* **2017**, *40*, 1695–1702. [CrossRef]
88. Ashcroft, F.M.; Puljung, M.C.; Vedovato, N. Neonatal Diabetes and the KATP Channel: From Mutation to Therapy. *Trends Endocrinol. Metab.* **2017**, *28*, 377–387. [CrossRef]
89. Kostov, K. Effects of magnesium deficiency on mechanisms of insulin resistance in type 2 diabetes: Focusing on the processes of insulin secretion and signaling. *Int. J. Mol. Sci.* **2019**, *20*, 1351. [CrossRef]
90. Günther, T. The biochemical function of Mg2+ in insulin secretion, insulin signal transduction and insulin resistance. *Magnes. Res.* **2010**, *23*, 5–18. [CrossRef]
91. Gommers, L.M.M.; Hoenderop, J.G.J.; Bindels, R.J.M.; De Baaij, J.H.F. Hypomagnesemia in Type 2 Diabetes: A Vicious Circle? *Diabetes* **2016**, *65*, 3–13. [CrossRef] [PubMed]
92. Apell, H.J.; Hitzler, T.; Schreiber, G. Modulation of the Na,K-ATPase by Magnesium Ions. *Biochemistry* **2017**, *56*, 1005–1016. [CrossRef] [PubMed]
93. Grycova, L.; Sklenovsky, P.; Lansky, Z.; Janovska, M.; Otyepka, M.; Amler, E.; Teisinger, J.; Kubala, M. ATP and magnesium drive conformational changes of the Na+/K+-ATPase cytoplasmic headpiece. *Biochim. Biophys. Acta Biomembr.* **2009**, *1788*, 1081–1091. [CrossRef] [PubMed]
94. Nair, A.V.; Hocherb, B.; Verkaart, S.; Van Zeeland, F.; Pfab, T.; Slowinski, T.; Chen, Y.P.; Schlingmann, K.P.; Schaller, A.; Gallati, S.; et al. Loss of insulin-induced activation of TRPM6 magnesium channels results in impaired glucose tolerance during pregnancy. *Proc. Natl. Acad. Sci. USA* **2012**, *109*, 11324–11329. [CrossRef]
95. Hassan, S.A.U.; Ahmed, I.; Nasrullah, A.; Haq, S.; Ghazanfar, H.; Sheikh, A.B.; Zafar, R.; Askar, G.; Hamid, Z.; Khushdil, A.; et al. Comparison of Serum Magnesium Levels in Overweight and Obese Children and Normal Weight Children. *Cureus* **2017**, *9*, e1607. [CrossRef]
96. Zaakouk, A.M.; Hassan, M.A.; Tolba, O.A. Serum magnesium status among obese children and adolescents. *Egypt. Pediatr. Assoc. Gaz.* **2016**, *64*, 32–37. [CrossRef]
97. Huerta, M.G.; Roemmich, J.N.; Kington, M.L.; Bovbjerg, V.E.; Weltman, A.L.; Holmes, V.F.; Patrie, J.T.; Rogol, A.D.; Nadler, J.L. Magnesium deficiency is associated with insulin resistance in obese children. *Diabetes Care* **2005**, *28*, 1175–1181. [CrossRef]
98. Le Chatelier, E.; Nielsen, T.; Qin, J.; Prifti, E.; Hildebrand, F.; Falony, G.; Almeida, M.; Arumugam, M.; Batto, J.-M.; Kennedy, S.; et al. Richness of human gut microbiome correlates with metabolic markers. *Nature* **2013**, *500*, 541–546. [CrossRef]
99. Wang, M.; Monaco, M.H.; Donovan, S.M. Impact of early gut microbiota on immune and metabolic development and function. *Semin. Fetal Neonatal Med.* **2016**, *21*, 380–387. [CrossRef]
100. Parekh, P.J.; Balart, L.A.; Johnson, D.A. The influence of the gut microbiome on obesity, metabolic syndrome and gastrointestinal disease. *Clin. Transl. Gastroenterol.* **2015**, *6*, e91. [CrossRef]
101. Pyndt Jørgensen, B.; Winther, G.; Kihl, P.; Nielsen, D.S.; Wegener, G.; Hansen, A.K.; Sørensen, D.B. Dietary magnesium deficiency affects gut microbiota and anxiety-like behaviour in C57BL/6N mice. *Acta Neuropsychiatr.* **2015**, *27*, 307–311. [CrossRef] [PubMed]
102. Winther, G.; Pyndt Jørgensen, B.M.; Elfving, B.; Nielsen, D.S.; Kihl, P.; Lund, S.; Sørensen, D.B.; Wegener, G. Dietary magnesium deficiency alters gut microbiota and leads to depressive-like behaviour. *Acta Neuropsychiatr.* **2015**, *27*, 168–176. [CrossRef] [PubMed]
103. Luppino, F.S.; De Wit, L.M.; Bouvy, P.F.; Stijnen, T.; Cuijpers, P.; Penninx, B.W.J.H.; Zitman, F.G. Overweight, obesity, and depression: A systematic review and meta-analysis of longitudinal studies. *Arch. Gen. Psychiatry* **2010**, *67*, 220–229. [CrossRef] [PubMed]
104. Chang, P.; Friedenberg, F. Obesity and GERD. *Gastroenterol. Clin. N. Am.* **2014**, *43*, 161–173. [CrossRef]
105. Gommers, L.M.M.; Ederveen, T.H.A.; Van Der Wijst, J.; Overmars-Bos, C.; Kortman, G.A.M.; Boekhorst, J.; Bindels, R.J.M.; De Baaij, J.H.F.; Hoenderop, J.G.J. Low gut microbiota diversity and dietary magnesium intake are associated with the development of PPI-induced hypomagnesemia. *FASEB J.* **2019**, *33*, 11235–11246. [CrossRef]
106. Pachikian, B.D.; Neyrinck, A.M.; Deldicque, L.; De Backer, F.C.; Catry, E.; Dewulf, E.M.; Sohet, F.M.; Bindels, L.B.; Everard, A.; Francaux, M.; et al. Changes in intestinal bifidobacteria levels are associated with the inflammatory response in magnesium-deficient mice. *J. Nutr.* **2010**, *140*, 509–514. [CrossRef]
107. Cox, A.J.; West, N.P.; Cripps, A.W. Obesity, inflammation, and the gut microbiota. *Lancet Diabetes Endocrinol.* **2015**, *3*, 207–215. [CrossRef]

108. Saad, M.J.A.; Santos, A.; Prada, P.O. Linking gut microbiota and inflammation to obesity and insulin resistance. *Physiology* **2016**, *31*, 283–293. [CrossRef] [PubMed]
109. Schoeler, M.; Caesar, R. Dietary lipids, gut microbiota and lipid metabolism. *Rev. Endocr. Metab. Disord.* **2019**, *20*, 461–472. [CrossRef]
110. Semenkovich, C.F.; Danska, J.; Darsow, T.; Dunne, J.L.; Huttenhower, C.; Insel, R.A.; McElvaine, A.T.; Ratner, R.E.; Shuldiner, A.R.; Blaser, M.J. American Diabetes Association and JDRF Research Symposium: Diabetes and the Microbiome. *Diabetes* **2015**, *64*, 3967–3977. [CrossRef]
111. Caesar, R. Pharmacologic and Nonpharmacologic Therapies for the Gut Microbiota in Type 2 Diabetes. *Can. J. Diabetes* **2019**, *43*, 224–231. [CrossRef]
112. Nuli, R.; Cai, J.; Kadeer, A.; Zhang, Y.; Mohemaiti, P. Integrative Analysis Toward Different Glucose Tolerance-Related Gut Microbiota and Diet. *Front. Endocrinol. (Lausanne)* **2019**, *10*, 295. [CrossRef] [PubMed]
113. Thingholm, L.B.; Rühlemann, M.C.; Koch, M.; Fuqua, B.; Laucke, G.; Boehm, R.; Bang, C.; Franzosa, E.A.; Hübenthal, M.; Rahnavard, A.; et al. Obese Individuals with and without Type 2 Diabetes Show Different Gut Microbial Functional Capacity and Composition. *Cell Host Microbe* **2019**, *26*, 252–264.e10. [CrossRef]
114. Crowley, E.K.; Long-Smith, C.M.; Murphy, A.; Patterson, E.; Murphy, K.; O'Gorman, D.M.; Stanton, C.; Nolan, Y.M. Dietary supplementation with a magnesium-rich marine mineral blend enhances the diversity of gastrointestinal microbiota. *Mar. Drugs* **2018**, *16*, 216. [CrossRef] [PubMed]
115. Cazzola, R.; Della Porta, M.; Manoni, M.; Iotti, S.; Pinotti, L.; Maier, J.A. Going to the roots of reduced magnesium dietary intake: A tradeoff between climate changes and sources. *Helyon* **2020**, *6*, e05390. [CrossRef] [PubMed]
116. Elderawi, W.A.; Naser, I.A.; Taleb, M.H.; Abutair, A.S. The Effects of Oral Magnesium Supplementation on Glycemic Response among Type 2 Diabetes Patients. *Nutrients* **2018**, *11*, 44. [CrossRef] [PubMed]
117. Navarrete-Cortes, A.; Ble-Castillo, J.L.; Guerrero-Romero, F.; Cordova-Uscanga, R.; Juárez-Rojop, I.E.; Aguilar-Mariscal, H.; Tovilla-Zarate, C.A.; Del Rocio Lopez-Guevara, M. No effect of magnesium supplementation on metabolic control and insulin sensitivity in type 2 diabetic patients with normomagnesemia. *Magnes. Res.* **2014**, *27*, 48–56. [CrossRef]
118. Razzaghi, R.; Pidar, F.; Momen-Heravi, M.; Bahmani, F.; Akbari, H.; Asemi, Z. Magnesium Supplementation and the Effects on Wound Healing and Metabolic Status in Patients with Diabetic Foot Ulcer: A Randomized, Double-Blind, Placebo-Controlled Trial. *Biol. Trace Elem. Res.* **2018**, *181*, 207–215. [CrossRef]
119. Simental-Mendía, L.E.; Rodríguez-Morán, M.; Reyes-Romero, M.A.; Guerrero-Romero, F. No positive effect of oral magnesium supplementation in the decreases of inflammation in subjects with prediabetes: A pilot study. *Magnes. Res.* **2012**, *25*, 140–146. [CrossRef]
120. Steward, C.J.; Zhou, Y.; Keane, G.; Cook, M.D.; Liu, Y.; Cullen, T. One week of magnesium supplementation lowers IL-6, muscle soreness and increases post-exercise blood glucose in response to downhill running. *Eur. J. Appl. Physiol.* **2019**, *119*, 2617–2627. [CrossRef]
121. Dibaba, D.T.; Xun, P.; Song, Y.; Rosanoff, A.; Shechter, M.; He, K. The effect of magnesium supplementation on blood pressure in individuals with insulin resistance, prediabetes, or noncommunicable chronic diseases: A meta-analysis of randomized controlled trials. *Am. J. Clin. Nutr.* **2017**, *106*, 921–929. [CrossRef] [PubMed]
122. Banjanin, N.; Belojevic, G. Changes of blood pressure and hemodynamic parameters after oral magnesium supplementation in patients with essential hypertension—an intervention study. *Nutrients* **2018**, *10*, 581. [CrossRef] [PubMed]
123. Hatzistavri, L.S.; Sarafidis, P.A.; Georgianos, P.I.; Tziolas, I.M.; Aroditis, C.P.; Zebekakis, P.E.; Pikilidou, M.I.; Lasaridis, A.N. Oral magnesium supplementation reduces ambulatory blood pressure in patients with mild hypertension. *Am. J. Hypertens.* **2009**, *22*, 1070–1075. [CrossRef] [PubMed]
124. Rodríguez-Ramírez, M.; Rodríguez-Morán, M.; Reyes-Romero, M.A.; Guerrero-Romero, F. Effect of oral magnesium supplementation on the transcription of TRPM6, TRPM7, and SLC41A1 in individuals newly diagnosed of pre-hypertension. A randomized, double-blind, placebo-controlled trial. *Magnes. Res.* **2017**, *30*, 80–87. [CrossRef] [PubMed]
125. Cunha, A.R.; D'El-Rei, J.; Medeiros, F.; Umbelino, B.; Oigman, W.; Touyz, R.M.; Neves, M.F. Oral magnesium supplementation improves endothelial function and attenuates subclinical atherosclerosis in thiazide-treated hypertensive women. *J. Hypertens.* **2017**, *35*, 89–97. [CrossRef] [PubMed]
126. Mortazavi, M.; Moeinzadeh, F.; Saadatnia, M.; Shahidi, S.; McGee, J.C.; Minagar, A. Effect of magnesium supplementation on carotid intima-media thickness and flow-mediated dilatation among hemodialysis patients: A double-blind, randomized, placebo-controlled trial. *Eur. Neurol.* **2013**, *69*, 309–316. [CrossRef] [PubMed]
127. Joris, P.J.; Plat, J.; Bakker, S.J.L.; Mensink, R.P. Effects of long-term magnesium supplementation on endothelial function and cardiometabolic risk markers: A randomized controlled trial in overweight/obese adults. *Sci. Rep.* **2017**, *7*, 106. [CrossRef]
128. Rodríguez-Morán, M.; Simental-Mendía, L.E.; Gamboa-Gómez, C.I.; Guerrero-Romero, F. Oral Magnesium Supplementation and Metabolic Syndrome: A Randomized Double-Blind Placebo-Controlled Clinical Trial. *Adv. Chronic Kidney Dis.* **2018**, *25*, 261–266. [CrossRef]
129. Simental-Mendía, L.E.; Simental-Mendía, M.; Sahebkar, A.; Rodríguez-Morán, M.; Guerrero-Romero, F. Effect of magnesium supplementation on lipid profile: A systematic review and meta-analysis of randomized controlled trials. *Eur. J. Clin. Pharmacol.* **2017**, *73*, 525–536. [CrossRef]

130. Cosaro, E.; Bonafini, S.; Montagnana, M.; Danese, E.; Trettene, M.S.; Minuz, P.; Delva, P.; Fava, C. Effects of magnesium supplements on blood pressure, endothelial function and metabolic parameters in healthy young men with a family history of metabolic syndrome. *Nutr. Metab. Cardiovasc. Dis.* **2014**, *24*, 1213–1220. [CrossRef]
131. Firoz, M.; Graber, M. Bioavallability of US commercial magnesium preparations. *Magnes. Res.* **2001**, *14*, 257–262. [PubMed]
132. Verhas, M.; De, V.; Guéronnière, L.; Grognet, J.-M.; Paternot, J.; Hermanne, A.; Van Den Winkel, P.; Gheldof, R.; Martin, P.; Fantino, M.; et al. Magnesium bioavailability from mineral water. A study in adult men. *Eur. J. Clin. Nutr.* **2002**, *56*, 442–447. [CrossRef] [PubMed]
133. Coudray, C.; Rambeau, M.; Feillet-Coudray, C.; Gueux, E.; Tressol, J.C.; Mazur, A.; Rayssiguier, Y. Study of magnesium bioavailability from ten organic and inorganic Mg salts in Mg-depleted rats using a stable isotope approach. *Magnes. Res.* **2005**, *18*, 215–223. [PubMed]
134. Uysal, N.; Kizildag, S.; Yuce, Z.; Guvendi, G.; Kandis, S.; Koc, B.; Karakilic, A.; Camsari, U.M.; Ates, M. Timeline (Bioavailability) of Magnesium Compounds in Hours: Which Magnesium Compound Works Best? *Biol. Trace Elem. Res.* **2019**, *187*, 128–136. [CrossRef] [PubMed]
135. Lindberg, J.S.; Zobitz, M.M.; Poindexter, J.R.; Pak, C.Y.C. Magnesium bioavailability from magnesium citrate and magnesium oxide. *J. Am. Coll. Nutr.* **1990**, *9*, 48–55. [CrossRef]
136. Ates, M.; Kizildag, S.; Yuksel, O.; Hosgorler, F.; Yuce, Z.; Guvendi, G.; Kandis, S.; Karakilic, A.; Koc, B.; Uysal, N. Dose-Dependent Absorption Profile of Different Magnesium Compounds. *Biol. Trace Elem. Res.* **2019**, *192*, 244–251. [CrossRef]

*Review*

# Crosstalk of Magnesium and Serum Lipids in Dyslipidemia and Associated Disorders: A Systematic Review

Mihnea-Alexandru Găman [1,2,*], Elena-Codruța Dobrică [3,4], Matei-Alexandru Cozma [5], Ninel-Iacobus Antonie [6], Ana Maria Alexandra Stănescu [1], Amelia Maria Găman [3,7,*] and Camelia Cristina Diaconu [1,6,*]

1. Faculty of Medicine, "Carol Davila" University of Medicine and Pharmacy, 050474 Bucharest, Romania; alexandrazotta@yahoo.com
2. Department of Hematology, Center of Hematology and Bone Marrow Transplantation, Fundeni Clinical Institute, 022328 Bucharest, Romania
3. Department of Pathophysiology, University of Medicine and Pharmacy of Craiova, 200349 Craiova, Romania; codrutadobrica@yahoo.com
4. Department of Dermatology, "Elias" University Emergency Hospital, 011461 Bucharest, Romania
5. Department of Gastroenterology, Colentina Clinical Hospital, 20125 Bucharest, Romania; matei.cozma@gmail.com
6. Department of Internal Medicine, Clinical Emergency Hospital of Bucharest, 14461 Bucharest, Romania; antonieninel@yahoo.com
7. Clinic of Hematology, Filantropia City Hospital, 200143 Craiova, Romania
\* Correspondence: mihneagaman@yahoo.com (M.-A.G.); gamanamelia@yahoo.com (A.M.G.); drcameliadiaconu@gmail.com (C.C.D.)

**Citation:** Găman, M.-A.; Dobrică, E.-C.; Cozma, M.-A.; Antonie, N.-I.; Stănescu, A.M.A.; Găman, A.M.; Diaconu, C.C. Crosstalk of Magnesium and Serum Lipids in Dyslipidemia and Associated Disorders: A Systematic Review. *Nutrients* **2021**, *13*, 1411. https://doi.org/10.3390/nu13051411

Academic Editors: Sara Castiglioni, Giovanna Farruggia and Concettina Cappadone

Received: 18 March 2021
Accepted: 20 April 2021
Published: 22 April 2021

**Publisher's Note:** MDPI stays neutral with regard to jurisdictional claims in published maps and institutional affiliations.

**Copyright:** © 2021 by the authors. Licensee MDPI, Basel, Switzerland. This article is an open access article distributed under the terms and conditions of the Creative Commons Attribution (CC BY) license (https://creativecommons.org/licenses/by/4.0/).

**Abstract:** Dyslipidemia is a significant threat to public health worldwide and the identification of its pathogenic mechanisms, as well as novel lipid-lowering agents, are warranted. Magnesium (Mg) is a key element to human health and its deficiency has been linked to the development of lipid abnormalities and related disorders, such as the metabolic syndrome, type 2 diabetes mellitus, or cardiovascular disease. In this review, we explored the associations of Mg (dietary intake, Mg concentrations in the body) and the lipid profile, as well as the impact of Mg supplementation on serum lipids. A systematic search was computed in PubMed/MEDLINE and the Cochrane Library and 3649 potentially relevant papers were detected and screened (n = 3364 following the removal of duplicates). After the removal of irrelevant manuscripts based on the screening of their titles and abstracts (n = 3037), we examined the full-texts of 327 original papers. Finally, after we applied the exclusion and inclusion criteria, a number of 124 original articles were included in this review. Overall, the data analyzed in this review point out an association of Mg concentrations in the body with serum lipids in dyslipidemia and related disorders. However, further research is warranted to clarify whether a higher intake of Mg from the diet or via supplements can influence the lipid profile and exert lipid-lowering actions.

**Keywords:** magnesium; magnesemia; hypomagnesemia; lipids; cholesterol; triglycerides; dyslipidemia; hyperlipidemia; diabetes; metabolic syndrome

## 1. Introduction

Dyslipidemia has emerged as a significant threat to public health worldwide, with recent statistics revealing that its prevalence reaches 42.7% in China and 56.8% in the United States of America (USA). In addition, Lu et al. (2018) have pointed out that an alarming rate of the population suffering from this disorder is not aware of its lipid profile (26.7% in the USA versus 80.4% in China), leading to poor treatment and control rates of lipid abnormalities (13.2% and 4.6% in China versus 54.1 and 35.7% in the USA) [1]. Thus, identifying novel strategies to combat dyslipidemia are warranted, particularly due to its involvement in the development of and crosstalk with metabolic syndrome (MetS), type 2 diabetes mellitus (T2DM), cardiovascular disease, obesity, hypertension, chronic kidney disease (CKD), and others [1–3].

Magnesium (Mg) seems to play a key role in a myriad of disorders, e.g., MetS, T2DM, obesity, hypertension, and its deficiency has been regarded as highly prevalent, with Piuri et al. (2021) ranking it as the most common electrolyte imbalance in high-income countries [4–6]. Taking this information into consideration, we may hypothesize that there is a crosstalk between Mg and serum lipids which may impact on the pathogenesis of dyslipidemia and its associated comorbidities, as well as that Mg supplementation might provide health benefits in patients suffering from cardiometabolic disorders.

Therefore, the aim of this review is to explore the associations of Mg (dietary intake, Mg concentrations in the body) and the lipid profile, i.e., total cholesterol (TC), triglycerides (TG), high-density lipoprotein cholesterol (HDL-C), low-density lipoprotein cholesterol (LDL-C), other lipoproteins (Lp), or apolipoproteins (apo), as well as the impact of Mg supplementation on these variables based on data derived from high-quality evidence such as randomized clinical trials (RCTs).

## 2. Materials and Methods

The protocol employed in this systematic review was based on the Preferred Reporting Items for Systematic reviews and Meta-Analyses (PRISMA) checklist [7].

Four investigators (M.-A.G., E.-C.D., M.-A.C., and N.-I.A.) independently computed a literature search in PubMed/MEDLINE and Cochrane Library from the inception of these databases until 25 February 2021. The following keywords and combinations of words was employed: ("magnesium" OR "magnesemia*" OR "magnesaemia") AND ("Lp(a)" OR "Triglycerides" OR "Cholesterol" OR "Cholesterol" OR "Lipoproteins, HDL" OR "Cholesterol, HDL" OR "Cholesterol, LDL" OR "Lipoproteins, LDL" OR "Hyperlipidemias" OR "Dyslipidemias" OR "Hypercholesterolemia" OR "lipoprotein triglyceride" OR LDL OR HDL OR "Total cholesterol" OR TG OR Triglyceride OR Triacylglycerol OR TAG OR "lipid profile" OR "low density lipoprotein" OR "high density lipoprotein" OR "blood lipids" OR "lipids*" OR "triglycerid*" OR "trigly*" OR triacylglycerol OR cholesterol OR LDL-C OR HDL-C OR Hyperlipidemia OR Hyperlipidemic OR Dyslipidemia OR Dyslipidemic OR Hypercholesterolemia OR Hypercholesterolaemia OR Hypercholesterolemic OR hypercholesterolaemic) NOT (review OR mice OR rats OR rodents).

We decided for the following inclusion criteria: 1. Original articles/research letters evaluating the relationship between Mg and serum lipids in dyslipidemia and related disorders OR Original articles/research letters evaluating the effects of Mg intake/supplementation on serum lipids in dyslipidemia and related disorders in humans; 2. the subjects recruited in these original studies were adults (aged ≥18 years); 3. the papers were published in English, French, Italian, or Romanian (the languages spoken by the investigators); 4. the papers provided sufficient data regarding the relationship of Mg and serum lipids or the effects of Mg supplementation on serum lipids in dyslipidemia and related disorders; 5. the full-text of the papers could be downloaded/retrieved. We decided for the following exclusion criteria: 1. Reviews, letters to the editor, case reports, conference abstracts, grey literature; 2. the studies were conducted in vitro, on animals or in human subjects aged <18 years; 3. the papers were published in languages unknown to the authors (e.g., Chinese, Polish etc.); 4. the papers did not report sufficient data on the outcomes; 5. the full-text of the articles was unavailable to the investigators.

Relevant data were extracted independently by four investigators (M.-A.G., E.-C.D., M.-A.C., and N.-I.A.) and disagreements were resolved by consultation with the senior authors (C.C.D. and A.M.G.).

## 3. Results

A total of 3649 potentially relevant papers were detected and screened. A flowchart diagram of the detailed steps of the literature search process is illustrated in Figure 1. After we removed the duplicates and excluded the irrelevant manuscripts based on the screening of their titles and abstracts (n = 3322), we examined the full-texts of 327 original papers. Finally, after we applied the exclusion and inclusion criteria, a number of 124 original articles were included in this review.

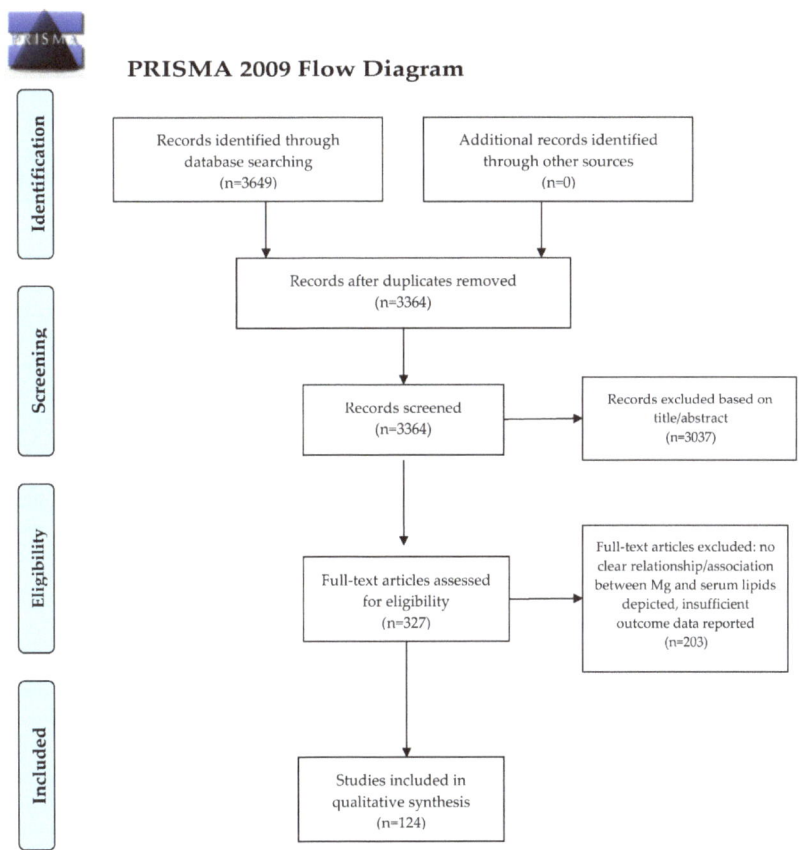

**Figure 1.** PRISMA 2009 Flow Diagram. From Moher, D.; Liberati, A.; Tetzlaff, J.; Altman, D.G. The Prisma Group. Preferred reporting items for systematic reviews and meta-analyses: The PRISMA statement. *PLoS Med.* **2009**, *6*, e1000097, doi:10.1371/journal.pmed.1000097. For more information, visit www.prisma-statement.org (accessed on 18 March 2021) [7].

### 3.1. Crosstalk of Magnesium, Serum Lipids, and Dyslipidemia

A total of 36 studies assessed the crosstalk of Mg, serum lipids, and dyslipidemia, of which the vast majority was focused on the relationship of serum Mg with the lipid profile and lipid abnormalities, including the impact of dietary Mg intake or Mg supplementation on these parameters (n = 25). Several studies tackled the Mg-serum lipids crosstalk in overweight/obesity, polycystic ovary syndrome, or nonalcoholic fatty liver disease (n = 11). The most relevant information of this subsection are summarized in Table 1.

#### 3.1.1. Magnesium, Serum Lipids, and Dyslipidemia

Barragán et al. (2020) evaluated the crosstalk between Mg concentrations, hypercholesterolemia, T2DM, and other cardiovascular risk factors in a Spanish cohort of 492 subjects. Hypomagnesemia was detected in nearly 19% of the study group, with no significant sex-differences ($p = 0.663$), including in terms of plasma Mg levels ($p = 0.106$). Females had higher TC ($p = 0.006$) and HDL-C ($p < 0.001$), lower TG ($p < 0.001$), similar LDL-C

($p$ = 0.781) versus men. Men were more likely to suffer from obesity ($p$ = 0.012), T2DM ($p$ = 0.039), hypertension ($p$ < 0.001), use lipid-lowering ($p$ = 0.040) or blood pressure-lowering drugs ($p$ < 0.001) versus females, but the prevalence of hypercholesterolemia was similar ($p$ = 0.186). Mg concentrations were increased in the plasma of hypercholesteremic subjects ($p$ = 0.001) and decreased in the plasma of diabetic subjects ($p$ = 0.009) as compared to individuals without hypercholesterolemia or T2DM, respectively. There was a significant association between T2DM but not hypercholesterolemia and hypomagnesemia (OR: 3.36, 95% CI: 1.26–8.96, $p$ = 0.016 and OR: 1.38, 95% CI: 0.81–2.35, $p$ = 0.233). However, there was an association between TC ($p$ = 0.01) and LDL-C ($p$ = 0.002), but HDL-C ($p$ = 0.933) or TG (0.959), and plasma Mg quartiles. Furthermore, patients in the fourth versus first quartile of Mg concentrations were more likely to be diagnosed with hypercholesterolemia (OR: 3.12; 95% CI: 1.66–5.85, $p$ < 0.001) [8]. Bersohn and Oelofse (1957) examined a number of forty seven healthy European and fifty three healthy African Bantu subjects. They revealed that the individuals with normal TC values had normal Mg values and the subjects who had a lower than normal TC had significantly higher Mg levels. Patients diagnosed with hypercholesterolemia presented significantly lower Mg values as compared to the healthy population [9]. Petersen et al. (1976) evaluated the 6-week intake of 3 g/day of Mg oxide in a group of seventeen patients suffering from hypercholesterolemia and/or hypertriglyceridemia in order to determine its effects on TC and TG. No significant relationship was registered between the initial values of serum Mg and TG ($r$ = 0.05, $p$ > 0.05) or TC ($r$ = −0.21, $p$ > 0.05). The authors discharged the hypothesis that 3 g/day of Mg has TC- or TG-lowering actions [10]. In a cross-sectional study, Liu et al. (2020) analyzed the relationship between Mg concentrations and dyslipidemia. Patients diagnosed with dyslipidemia had higher TC, TG, HDL-C, and LDL-C ($p$ < 0.001) and lower Mg levels ($p$ = 0.002) versus the subjects with normal serum lipids. Subjects in the fourth versus first quartile of serum Mg had an increased risk of dyslipidemia (OR = 1.4, 95% CI 1.0–1.9, $p$ = 0.023) in the unadjusted model, but this finding was not verified in the adjusted model [11].

3.1.2. Magnesium, Serum Lipids, and the Diet/dietary Interventions

In a retrospective analysis, Jin and Nicodemus-Johnson (2018) scrutinized 12,284 individuals from the United States' 2001–2013 National Health and Nutrition Examination Study (NHANES) and observed that, in females, Mg intake was positively associated with HDL-C levels, while it was negatively associated with the TC/HDL-C ratio. However, TG levels were negatively correlated with Mg intake in both genders [12]. Itoh et al. (1997) executed an RCT in which they administered Mg in 33 healthy Japanese subjects, discovering an increase in HDL-C and apoA1, as well as a significant decrease in LDL-C in the Mg group [13]. In Marken et al. (1989)'s RCT, the administration of 400 mg Mg oxide for 60 days to 50 healthy volunteers did not result in changes of TC, HDL-C, LDL-C, VLDL, or TG versus placebo [14]. According to Randell et al. (2008), a number of 1318 healthy individuals recruited in their study displayed significant positive correlations between Mg and TC, HDL-C, LDL-C, and TG [15]. In a cross-sectional study that administered Mg and potassium at variable doses in 529 healthy individuals, Guerrero-Romero et al. (2019) showed significant TG decreases ($p$ < 0.0005) and no HDL-C alterations in the subjects receiving recommended versus suboptimal Mg doses [16]. De Valk et al. investigated the link between serum Mg variations and lipolysis-induced TG generation, revealing that an elevation in serum Mg was parallel by an elevation in TG as well ($p$ < 0.001) [17]. Aslanabadi et al. (2014) explored the lipid-lowering effects of 1 daily liter of mineral-rich versus normal mineral water in an RCT conducted in 69 adults diagnosed with dyslipidemia (intervention group: 32 subjects, control group: 37 subjects). Despite the fact that the beverage which was enriched with Mg, calcium, sulfate, and bicarbonate lowered TC and LDL-C, the same outcome was reported in the control group as well and there was no statistical difference between the results. Both waters failed to exhibit an impact of TG or HDL-C [18]. According to Fu et al. (2012), the consumption of deep sea water (395 Mg mg/L) versus $MgCl_2$ fortified (386 Mg mg/L) or reverse osmotic water decreased TC, LDL-C, and HDL-C in a time-dependent manner [19]. Based on Nerbrand et al. (2003)'s research, there are no correlations between the content of Mg in the water and cardiovascular risk factors [20]. Luoma et al. (1973) studied the relationship of Mg concentrations in the drinking water with several cardiovascular risk factors (including TG and TC) in 300 men from four different Finnish rural districts, but found no association between these variables [21]. Low-energy dietary interventions which also involved Mg supplementation achieved reductions in TC, LDL-C, and TG, but also decreased HDL-C, according to a pre-post intervention evaluation of 49 subjects from the United States [22]. In de Los Rios (1963)'s research project, a number of twenty eight schizophrenic patients were prescribed a strictly controlled diet which included Mg in quantities above the recommended dietary values, but this nutrition experiment found no association between the alterations of Mg levels and TC [23]. A higher dietary consumption of Mg seems to be linked with oxidized LDL, a biomarker of oxidative stress, as concluded by Cocate et al. (2013) who assessed the intake of vegetables, fruit, and nutrients of 296 middle-aged males with a normal status of health [24]. Patients with multimorbidity (including hypercholesterolemia) are known to have smaller daily intakes of Mg, whereas an increased consumption of cereals (r = 0.60, $p < 0.0001$) and fruits and vegetables (r = 0.49, $p < 0.0001$) were associated with higher serum Mg levels and less multimorbidity [25]. There seems to be an inverse correlation between the intake of Mg and TC in both males ($p = 0.02$) and females ($p = 0.04$), according to Bain et al. (2015) who evaluated 4443 subjects aged 40–75 from the European Prospective Investigation into Cancer)-Norfolk cohort in order to assess the relationship between the risk of stroke and its most important risk factors and the dietary consumption of Mg [26]. In addition, Samavarchi Tehrani et al. (2020) recorded a significant association ($p = 0.012$) between the presence of dyslipidemia and Mg levels in 447 patients suffering a stroke of embolic origin [27]. Interestingly, Kim et al. (2014), based on best-fit models from stepwise linear regressions, discovered that the dietary intake of Mg predicts the concentrations of HDL-C and its subspecies. In their study group of 1566 individuals, the dietary consumption of Mg shared positive associations with HDL-C (coefficient ± SE: 4.79 ± 1.45; %HDL-C variation: 0.12%, $p = 0.001$) and its subfractions, HDL-2 (coefficient ± SE: 1.43 ± 0.61; %HDL-C variation: 0.028%, $p = 0.018$) and HDL-3 (coefficient ± SE: 2.98 ± 1.20; %HDL-C variation: 0.085%, $p = 0.013$), but not with apoA1 concentrations [28]. Kim and Choi (2013) investigated the dietary intake of Mg and its relationship with the lipid profile in 258 healthy Korean adults. Although men had a higher Mg daily consumption versus women, they registered higher TG ($p < 0.05$) and lower HDL-C ($p < 0.01$) values, yet similar TC and LDL-C levels. The atherogenic index of plasma was also elevated in males versus females ($p < 0.001$). However, the correlations between serum Mg and TC, HDL-C, LDL-C, TG or the atherogenic index of plasma did not reach statistical significance [29]. Cao et al. (2015) analyzed the relationships between serum and urinary Mg concentrations and the lipid profile in 2837 middle-aged/elderly Chinese, revealing a positive association between higher serum Mg and TC ($p < 0.001$), HDL-C ($p < 0.001$), LDL-C ($p = 0.001$), and TG ($p < 0.001$), and negative one with non-HDL-C/HDL-C ($p = 0.003$). There were also positive associations between the Mg/creatinine ratio in the urine and TC ($p = 0.004$), HDL-C ($p = 0.003$), and LDL-C ($p = 0.009$). However, there were some gender-based differences regarding these results: in males, the associations between serum Mg and LDL-C, TG or non-HDL-C/HDL-C, and those between urinary Mg/creatinine and HDL-C or LDL-C, were not statistically significant [30]. On the other hand, there are atomic absorption spectrophotometry studies conducted in postmenopausal females that show that Mg concentrations in red blood cells and not serum Mg concentrations correlate with serum TG (r = 0.287, $p = 0.011$) [31]. Data derived from the Cardiovascular Disease and Alimentary Comparison (CARDIAC) study also reinforced that hypercholesterolemia was more prevalent and TC concentrations were higher in individuals with lower versus higher 24-h urinary Mg/creatinine ratios ($p < 0.001$ for trend for both). Subjects in the lowest quintiles of 24-h Mg/creatinine urinary ratios were more likely to suffer from hypercholesterolemia (OR = 2.73; 95% CI 2.03 to 3.67; $p < 0.001$) versus the highest quintiles [32].

### 3.1.3. Magnesium, Serum Lipids, and Overweight and (or) Obesity

Guerrero-Romero and Rodriguez-Moran (2013) investigated the relationship between serum Mg and several metabolic phenotypes, namely healthy normal-weight (NW) versus metabolically obese normal weight (MONW) subjects, as well as obese versus metabolically healthy obese (MHO) patients. NW and MHO subjects exhibited higher serum Mg concentrations ($p = 0.04$ and $p = 0.01$, respectively) and lower TG ($p < 0.0005$ for both), yet similar HDL-C levels, versus their corresponding comparators. In MONW patients, there was a negative correlation between TG and serum Mg ($r = -0.530$, unreported $p$-value). Moreover, in both obese and in particular non-obese patients, low serum Mg concentrations were associated with the presence of hypertriglyceridemia (OR = 1.61, 95% CI: 1.5–2.46 and OR = 6.67, 95% CI 2.1–20.4, respectively) [33]. However, Mg levels did not correlate with TG concentrations in another study on obese individuals who were planning to undergo bariatric surgery [34]. In Rodriguez-Moran and Guerrero-Romero (2014)'s RCT, the 4-month daily administration of 30 mL of $MgCl_2$ 5% solution (equivalent to 382 mg of Mg) was compared to the administration of 30 mL of placebo solution in hypomagnesemic MONW subjects. Following the intervention, HDL-C increased ($p < 0.05$) and TG levels decreased ($p < 0.0001$) significantly [35]. In their RCT, Joris et al. (2017) examined the effect of long-term Mg supplementation on endothelial function and multiple cardiometabolic risk markers in subjects suffering from overweight or obesity. Fifty two subjects were assigned randomly to two different groups that received either a dose consisting of 350 mg Mg or placebo, yet no differences in TC, HDL-C, LDL-C, TG or non-esterified fatty acids were observed in the intervention versus control groups [36]. Guerrero-Romero et al. (2016) investigated the relationship between obesity and hypomagnesemia in six hundred and eighty one subjects. When separating the subjects based on their Mg levels, their findings reported a significant difference in the mean values of HDL-C ($1 \pm 0.3$ mmol/L in the low serum Mg group and $1.2 \pm 0.4$ mmol/L in the normal serum Mg group), as well as in TG ($1.9 \pm 1.4$ mmol/L in the low serum Mg group and $1.8 \pm 1.5$ mmol/L in the normal serum Mg group) [37]. Solati et al. (2019) executed a 6-month RCT in which they administered 300 mg Mg sulfate in the form of herbal supplements versus placebo in overweight subjects who did not suffer from T2DM. Mg administration increased HDL-C ($p < 0.001$) and HDL-C/TG ($p < 0.0001$) and lowered LDL-C ($p < 0.05$) and TG ($p < 0.05$) values [38].

**Table 1.** Magnesium and serum lipids interplay in patients with a normal health status and patients diagnosed with dyslipidemia or related disorders.

| Author and Year | Condition | Number of Patients | Method of Mg Determination | Main Results |
|---|---|---|---|---|
| Barragán et al. (2020) [8] | Cardiometabolic risk factors: T2DM, hypercholesterolemia, hypertension | 492 | Serum, urine (spectrometry) | Prevalence of hypoMg = 19% Hypercholesterolemia: Mg ↑; T2DM: Mg ↓ HypoMg-T2DM association (OR: 3.36, 95% CI: 1.26–8.96, $p = 0.016$) TC, LDL-C associated with Mg quartiles ↑ Hypercholesterolemia in 4th versus 1st quartile of Mg levels (OR: 3.12; 95% CI: 1.66–5.85, $p < 0.001$) |
| Bersohn and Oelofse (1957) [9] | Healthy | 100 | Serum (spectrophotometry) | Normal TC—normal Mg ↓ TC–↑ Mg Hypercholesterolemia: ↓ Mg |
| Petersen et al. (1976) [10] | Hypercholesterolemia Hypertriglyceridemia | 17 | Serum (spectrophotometry) | 3 g/day of MgO for 6 weeks: no effect on TC, TG |
| Liu et al. (2020) [11] | Dyslipidemia | 1466 | Serum (spectrophotometry) | ↑ TC, TG, HDL-C and LDL-C ($p < 0.001$) ↓ Mg ($p = 0.002$) ↑ Dyslipidemia in the 4th versus 1st Mg quartile (OR = 1.4, 95% CI 1.0–1.9, $p = 0.023$) |

Table 1. Cont.

| Author and Year | Condition | Number of Patients | Method of Mg Determination | Main Results |
|---|---|---|---|---|
| Jin and Nicodemus-Johnson (2018) [12] | Healthy | 12,284 | Serum (method unspecified) | (+) association of Mg intake and HDL-C in ♀<br>(−) association of Mg intake and TC/HDL-C ratio in ♀<br>(−) association of Mg intake and TG in ♀ and ♂ |
| Randell et al. (2008) [15] | Healthy | 1318 | Serum (spectroscopy) | (+) association of Mg and TC, HDL-C, LDL-C, TG |
| Guerrero-Romero et al. (2019) [16] | Healthy | 529 | Serum (method unspecified) | Mg and K supplementation ↓ TG |
| Fu et al. (2012) [19] | Healthy | 42 | Serum (method unspecified) | deep sea water (395 Mg mg/L) ↓ TC, LDL-C and HDL-C versus MgCl$_2$ fortified (386 Mg mg/L) or reverse osmotic water |
| Nerbrand et al. (2003) [20] | CV risk factors | 207 | Serum, whole blood, muscle, urine (method unspecified) | No correlation of Mg content in water and CV risk factors |
| Luoma et al. (1973) [21] | CV risk factors | 300 | Serum (method unspecified) | No correlation of Mg in drinking water and CV risk factors (TG, TC) |
| Balliett et al. (2013) [22] | Healthy | 49 | Serum (method unspecified) | Low-energy dietary interventions (+ Mg) ↓ TC, LDL-C, TG and HDL-C |
| de Los Rios (1963) [23] | Schizophrenia | 28 | Serum (method unspecified) | Controlled diet (Mg > RDV): no associations of ΔTC and ΔMg |
| Cocate et al. (2013) [24] | Healthy | 296 | Serum (method unspecified) | Association of ↑Mg intake and oxLDL |
| Ruel et al. (2014) [25] | Multimorbidity, Hypercholesterolemia | 1020 | Serum (method unspecified) | ↑ consumption of cereals (r = 0.60, $p < 0.0001$), fruits and vegetables (r = 0.49, $p < 0.0001$) associated with ↑ Mg and ↓ multimorbidity |
| Bain et al. (2015) [26] | Healthy | 4443 | Serum (method unspecified) | Inverse correlation of Mg intake and TC in ♂ ($p = 0.02$) and ♀ ($p = 0.04$) |
| Samavarchi Tehrani et al. (2020) [27] | Dyslipidemia | 447 | Serum (method unspecified) | Mg levels associated with dyslipidemia in embolic stroke ($p = 0.012$) |
| Kim et al. (2014) [28] | Healthy | 1566 | Serum (method unspecified) | (+) association of Mg intake and HDL-C ($p = 0.001$), HDL-2 ($p = 0.018$), HDL-3 ($p = 0.013$) |
| Kim and Choi (2013) [29] | Healthy | 258 | Serum (method unspecified) | ↑ Mg intake, ↑ TG ($p < 0.05$), ↓ HDL-C ($p < 0.01$), ↑ AIP ($p < 0.001$) in ♂ |
| Cao et al. (2015) [30] | Healthy | 2837 | Serum, urine (method unspecified) | (+) association of ↑ serum Mg and TC ($p < 0.001$), HDL-C ($p < 0.001$), LDL-C ($p = 0.001$), TG ($p < 0.001$)<br>(−) association of ↑ serum Mg and non-HDL-C/HDL-C ($p = 0.003$)<br>(+) association of Mg/creatinine ratio in urine and TC ($p = 0.004$), HDL-C ($p = 0.003$) and LDL-C ($p = 0.009$) |
| López-González et al. (2014) [31] | Post-menopause | 78 | Serum (method unspecified) | Mg in red blood cells, not serum Mg, correlates with TG (r = 0.287, $p = 0.011$) |
| Yamori et al. [32] | Hypercholesterolemia | 4211 | Serum, urine (method unspecified) | ↑ hypercholesterolemia, ↑ TC in lower versus higher 24-h urinary Mg/creatinine ratios ($p < 0.001$ for trend for both)<br>↑ hypercholesterolemia in the lowest versus the highest quintiles of 24-h Mg/creatinine urinary ratios (OR = 2.73; 95% CI 2.03 to 3.67; $p < 0.001$) |
| Guerrero-Romero and Rodriguez-Moran (2013) [33] | Overweight/obesity | 427 | Serum (method unspecified) | NW and MHO: ↑ Mg ($p = 0.04$ and $p = 0.01$, respectively), ↓ TG ($p < 0.0005$ for both)<br>MONW: (−) Mg-TG correlation<br>Obese and non-obese: ↓ Mg associated with hypertriglyceridemia (OR = 1.61, 95% CI: 1.5–2.46 and OR = 6.67, 95% CI 2.1–20.4, respectively) |

Table 1. Cont.

| Author and Year | Condition | Number of Patients | Method of Mg Determination | Main Results |
|---|---|---|---|---|
| Lefebvre et al. (2014) [34] | Obesity | 267 | Serum (method unspecified) | No Mg-TG correlation in candidates for bariatric surgery |
| Guerrero-Romero et al. (2016) [37] | Obesity | 681 | Serum (method unspecified) | ↓ HDL-C, ↑ TG in ↓ Mg versus normal Mg groups |
| Farsinejad-Marj et al. (2020) [39] | PCOS | 60 | Serum (method unspecified) | 250 mg/day Mg oxide for 8 weeks: no effect on TC, HDL-C, LDL-C, TG, TC/HDL-C, TG/HDL-C |
| Cutler et al. (2019) [40] | PCOS | 137 | Serum (method unspecified) | Mg-rich diet ↓ insulin resistance and ↑ HDL-C ($p = 0.02$ for both) |

Mg, magnesium. T2DM, type 2 diabetes mellitus. CV, cardiovascular. PCOS, polycystic ovary syndrome. hypoMg, hypomagnesemia. OR, odds ratio. CI, confidence interval. ↑, increased. ↓, decreased. (+), positive. (−), negative. ♂, male. ♀, female. MgO, Mg oxide. HDL-C, high-density lipoprotein cholesterol. LDL-C, low-density lipoprotein cholesterol. TC, total cholesterol. TG, triglycerides. VLDL, very low-density lipoprotein cholesterol. apoA1, apolipoprotein A1. oxLDL, oxidized LDL. RDV, recommended dietary value. Δ, variation. $MgCl_2$, Mg chloride. mg, milligrams. g, grams. mL, milliliter. K, potassium. MONW, metabolically obese normal-weight. NW, normal weight.

### 3.1.4. Magnesium, Serum Lipids, and Polycystic Ovary Syndrome (PCOS)

Mg supplementation (250 mg/day Mg oxide for 8 weeks) in females with polycystic ovary syndrome did not lead to significant changes in TC, HDL-C, LDL-C, TG, TC/HDL-C, or TG/HDL-C versus placebo [39]. Cutler et al. (2019) inspected the nutritional intakes of 87 women with PCOS and reported that females following a Mg-rich diet had less insulin resistance ($p = 0.02$) and higher HDL-C ($p = 0.02$) [40]. In Jamilian et al. (2019)'s RCT, 60 women with PCOS who received Mg+vitamin E displayed significant reductions in TG ($p = 0.001$) and VLDL ($p = 0.01$), but no alterations in LDL-C, HDL-C, or TC/HDL-C [41]. Moreover, in another RCT, Jamilian et al. (2017), TG ($p < 0.001$), VLDL ($p < 0.001$), and TC ($p = 0.04$) decreased significantly, but significant effect on LDL-C or HDL-C was seen in the 60 women with PCOS that received Mg, zinc, calcium, and vitamin D co-supplementation [42].

### 3.1.5. Magnesium, Serum Lipids, and Nonalcoholic Fatty Liver Disease (NAFLD)

Karandish et al. (2013) conducted an RCT in 34 NAFLD subjects versus 34 healthy controls in which they investigated the benefits of Mg supplementation in combination with a low-calorie diet and physical exercise on several biochemical parameters. NAFDL patients who received the intervention consisting in supplementation with Mg did not experience statistically significant alterations of TC, LDL-C, HDL-C, or TG concentrations, whereas LDL-C ($p = 0.000$) and TC ($p = 0.003$) concentrations dropped significantly in the control group who did not receive Mg and only practiced physical exercise and caloric restriction [43].

### 3.2. Crosstalk of Magnesium, Serum Lipids, and Metabolic Syndrome

A total of 16 studies assessed the crosstalk of Mg, serum lipids, and MetS, focusing on the relationship of serum Mg with the lipid profile in this metabolic disorder, including the impact of dietary Mg intake or Mg supplementation on serum lipids.

Ali et al. (2013) investigated the dietary intakes of 213 American Indians diagnosed with MetS who were enrolled into the Balance Study RCT. The authors detected that nearly 90% of the male subjects had Mg intakes below the daily recommended dose, mainly due to dietary patterns poor in whole grains and vegetables which are sources of Mg [44]. Similarly, Vajdi et al. (2020) reported that subjects who follow a nutrient pattern based on plant sources which is also rich in Mg depict lower odds of MetS ($p = 0.01$) and lower LDL-C ($p = 0.04$), but similar TC, TG, and HDL-C ($p > 0.05$ for all) when comparing the first to the fourth quartile of this diet [45]. However, Akbarzade et al. (2020) did not detect an association between a nutrient pattern which comprised a higher Mg intake and MetS components in Iranian adults [46]. Similarly, Mottaghian et al. (2020) did not discover any association between the Mg-rich dietary pattern and alterations of the serum lipids during a 3-year timeframe. Nevertheless, TG increased ($p < 0.05$ for trend) across the quartiles of the Mg-rich nutrient pattern [47]. Choi and Bae (2013) evaluated the intake of Mg and the risk of MetS in Korean adults, revealing that both males and females diagnosed with MetS had lower Mg intakes and percentages of the recommended nutrient intake of intake versus healthy controls. Moreover, nearly a half of the MetS subjects did not achieve the estimated average requirement of Mg. However, the authors did not detect a link between the risk of MetS and the intake of Mg after adjusting for potential confounders. Women with low HDL-C concentrations had a lower Mg consumption, yet the same finding was not verified in men. In both genders, higher TG levels could not be linked to Mg intakes [48]. Cano-Ibáñez et al. (2019) investigated the relationship between MetS and Mg concentrations in 6646 individuals and discovered that subjects with MetS have a lower dietary Mg intake. Low levels of education and male sex were linked with smaller dietary intakes of this nutrient [49]. Choi et al. (2014) detected no differences between MetS subjects and healthy controls in terms of serum Mg levels. However, when analyzing the hair mineral concentrations, they detected lower Mg concentrations ($p = 0.046$) and a higher sodium/magnesium ratio ($p = 0.013$) in MetS patients. Moreover, there was negative significant correlation between hair Mg and TG ($r = -0.125, p < 0.05$) and positive non-significant Mg-HDL-C correlation ($r = 0.093, p > 0.05$) [50]. Vanaelst et al. (2012) also analyzed hair Mg levels and depicted negative correlations between serum Mg and non-HDL-C ($r = -0.170, p = 0.030$) and metabolic score ($r = -0.257, p = 0.001$) in Belgian schoolgirls. Hair Mg concentrations were significantly lower ($p = 0.015$) in females with a metabolic score of more than 3 points [51]. However, Sun et al. (2013) detected lower Mg ($p < 0.001$) and HDL-C and higher TC, TG, and LDL-C levels in Chinese subjects with MetS versus controls [52]. Despite low serum and intramononuclear levels of Mg in non-diabetic females diagnosed with MetS, de Lourdes Lima de Souza e Silva (2014) did not register any changes in TG, TC, HDL-C, or LDL-C following supplementation with 400 mg of Mg chelate versus placebo in their 12-week RCT [53]. Rotter et al. (2015) evaluated the concentrations of heavy metals and bioelements in 313 Polish men aged 50–75 years and detected lower Mg levels in patients diagnosed with MetS ($p = 0.02$), T2DM ($p = 0.0001$), and hypertension ($p = 0.0001$). Overweight/obese and normal-weight individuals had similar Mg concentrations ($p = 0.41$). The authors depicted positive associations between Mg and TC ($r = 0.25; p < 0.001$) and LDL-C ($r = 0.26; p < 0.001$), however the Mg-HDL-C or Mg-TG correlations did not reach statistical significance [54]. Ghasemi et al. (2010) retrospectively analyzed 137 individuals aged >60 years and unmasked that patients with MetS, T2DM and hyperglycemia depict lower Mg levels [55]. Evangelopoulos et al. (2008) also exposed that Mg was positively correlated with HDL-C ($r = 0.18; p = 0.05$) in patients affected by MetS [56]. In a population-based research (192 MetS subjects versus 384 healthy controls), Guerrero-Romero and Rodríguez-Morán (2002) observed a strong connection of MetS dyslipidemia and Mg deficiency based on the Mg-HDL-C ($r = 0.36, p < 0.05$), Mg-TC ($r = -0.29, p < 0.05$) associations [57]. Yuan et al. (2016) found multiple statistically significant correlations between serum Mg, calcium, Ca/Mg levels and metabolic risk factors for MetS. Their study included two hundred and four MetS patients and two hundred and four healthy subjects as the control group. Multiple blood tests were performed and the values of serum Mg and blood lipids were determined. Correlation studies were performed and the following results were described: positive correlation between serum Mg and BMI ($r = 0.128, p < 0.05$), TC ($r = 0.254, p < 0.05$), and LDL-C ($r = 0.280, p < 0.05$) [58]. Rotter et al. (2016) observed statistically significant, positive correlations between Mg and TC ($r = 0.25, p < 0.0001$) and LDL-C ($r = 0.26, p < 0.0001$). Their results found no statistically significant correlations between Mg and HDL-C ($r = 0.009, p = 0.87$) or TG ($r = -0.06, p = 0.28$). Three hundred and thirteen men were involved in their research which had the objective of determining the relationship between serum Mg concentrations and the occurrence of metabolic and/or hormonal disorders [59].

The most relevant information of this subsection are summarized in Table 2.

Table 2. Magnesium and serum lipids interplay in MetS.

| Author and Year | Number of Patients | Method of Mg Determination | Main Results |
|---|---|---|---|
| Ali et al. (2013) [44] | 213 | Unspecified | 90% ♂Mg intake < daily recommended dose (↓ whole grains, vegetable intake) |
| Vajdi et al. (2020) [45] | 588 | Unspecified | ↓ odds of MetS ($p = 0.01$), ↓ LDL-C ($p = 0.04$), in the 1st versus 4th quartile of plant source-based diets (↑ Mg) |
| Akbarzade et al. (2020) [46] | 850 | Unspecified | No association of nutrient patterns with ↑ Mg intake and MetS components |
| Mottaghian et al. (2020) [47] | 1637 | Unspecified | No association of Mg-rich dietary pattern and lipid profile changes<br>↑ TG ($p < 0.05$ for trend) across the quartiles of the Mg-rich nutrient pattern |
| Choi and Bae (2013) [48] | 5136 | Unspecified | ♀and ♂: ↓ Mg intake, no link with MetS<br>♀with ↓ HDL-C: ↓ Mg intake<br>♀and ♂with ↑ TG: no link with Mg intake |
| Cano-Ibáñez et al. (2019) [49] | 6646 | Unspecified | MetS: ↓ Mg intake<br>↓ Mg intake in ♂and ↓ education |
| Choi et al. (2014) [50] | 456 | Serum (automatic analytical analyzer) | serum Mg similar in MetS versus controls<br>↓ hair Mg levels ($p = 0.046$) ↑ Na/Mg ratio ($p = 0.013$) in MetS<br>(−) correlation of hair Mg and TG ($r = -0.125$, $p < 0.05$) |
| Vanaelst et al. (2012) [51] | 166 | Unspecified | (−) correlations of serum Mg and non-HDL-C ($r = -0.170$, $p = 0.030$), metabolic score ($r = -0.257$, $p = 0.001$)<br>↓ hair Mg ($p = 0.015$) in ♀with metabolic score > 3 points |
| Sun et al. (2013) [52] | 7641 | Serum (biochemical analyzer) | ↓ Mg ($p < 0.001$), ↓ HDL-C, ↑ TC, ↑ TG, ↑ LDL-C |
| Rotter et al. (2015) [54] | 313 | Serum and whole blood (spectrometry) | ↓ Mg in MetS ($p = 0.02$), T2DM ($p = 0.0001$), HTN ($p = 0.0001$)<br>(+) associations of Mg and TC ($r = 0.25$; $p < 0.001$), LDL-C ($r = 0.26$; $p < 0.001$) |
| Ghasemi et al. (2010) [55] | 137 | Serum (spectrometry) | ↓ Mg in patients > 60 years with MetS, T2DM and hyperglycemia |
| Evangelopoulos et al. (2008) [56] | 117 | Serum (colorimetric reaction) | (+) Mg-HDL-C association ($r = 0.18$; $p = 0.05$) |
| Guerrero-Romero and Rodríguez-Morán (2002) [57] | 576 | Serum (colorimetric assay) | (+) Mg-HDL-C association ($r = 0.36$, $p < 0.05$)<br>(−) Mg-TC association ($r = -0.29$, $p < 0.05$) |
| Yuan et al. (2016) [58] | 408 | Serum (spectrometry) | (+) correlations: Mg-BMI ($r = 0.128$, $p < 0.05$), Mg-TC ($r = 0.254$, $p < 0.05$), Mg-LDL-C ($r = 0.280$, $p < 0.05$) |
| Rotter et al. (2016) [59] | 313 | Serum (spectrometry) | (+) correlations: Mg-TC ($r = 0.25$, $p < 0.0001$), Mg-LDL-C ($r = 0.26$, $p < 0.0001$) |

Mg, magnesium. MetS, metabolic syndrome. T2DM, type 2 diabetes mellitus. HTN, hypertension. hypoMg, hypomagnesemia. OR, odds ratio. CI, confidence interval. ↑, increased. ↓, decreased. (+), positive. (−), negative. ♂, male. ♀, female. MgO, Mg oxide. HDL-C, high-density lipoprotein cholesterol. LDL-C, low-density lipoprotein cholesterol. TC, total cholesterol. TG, triglycerides. VLDL, very low-density lipoprotein cholesterol. apoA1, apolipoprotein A1. oxLDL, oxidized LDL. RDV, recommended dietary value. Δ, variation. $MgCl_2$, Mg chloride. mg, milligrams. g, grams. mL, milliliter. K, potassium. MONW, metabolically obese normal-weight. NW, normal weight.

### 3.3. Magnesium, Serum Lipids, and Type 2 Diabetes Mellitus (T2DM)

A total of 40 studies assessed the crosstalk of Mg, serum lipids, and T2DM, focusing on the relationship of serum Mg with the lipid profile in this metabolic disorder, including the impact of dietary Mg intake or Mg supplementation on serum lipids. The associations of hypomagnesemia with the lipid profile was evaluated in eight papers, whereas the benefits of Mg supplementation (data mostly derived from RCTs) was scrutinized in 21 manuscripts.

Based on data from 5568 subjects enrolled in the Prevention of Renal and Vascular End-stage Disease (PREVEND) study, van Dijk et al. (2019) unveiled that Mg levels (assessed by both nuclear magnetic resonance spectroscopy and colorimetric assays) are lower in T2DM ($p < 0.001$ for both methods). In the entire study population, there was a negative correlation between Mg (measured by nuclear magnetic resonance spectroscopy but not colorimetric assays) and TG (r = $-0.073$, $p < 0.001$ and r ≤ $0.001$, $p = 0.99$, respectively). In T2DM subjects, there was a negative correlation of Mg (measured by nuclear magnetic resonance spectroscopy and colorimetric assays) and TG (r = $-0.184$, $p = 0.002$ and r = $-0.194$, $p = 0.001$, respectively). In the entire study population, according to the results of the multivariable linear regression, there was a correlation between Mg (measured by nuclear magnetic resonance spectroscopy) and low HDL-C concentrations (β = $-0.062$, $p < 0.001$), but not with high TG concentrations (β = $-0.011$, $p = 0.45$). There was no association between Mg levels measured by colorimetric assays and HDL-C or TG [60]. As compared to healthy controls, Rusu et al. (2013) demonstrated that T2DM patients, and particularly those suffering from peripheral arterial disease, have lower Mg ($p < 0.01$) and HDL-C and higher TG and TC levels [61]. Spiga et al. (2019) evaluated the associations of Mg concentrations and serum lipids in patients with impaired fasting glucose or T2DM. Mg and HDL-C levels decreased and TG increased as follows: patients with normal glucose tolerance → impaired fasting glucose → T2DM ($p < 0.02$, $p < 0.01$ and $p < 0.001$, respectively). Significant correlations of Mg with TC (r = $0.154$, $p < 0.001$), HDL-C (r = $0.113$, $p < 0.01$), LDL-C (r = $0.170$, $p < 0.001$), but not with TG (r = $0.01$, $p = 0.981$). Higher Mg concentrations were linked with a lower risk of T2DM (OR = $0.765$, 95% CI $0.629$–$0.932$, $p < 0.01$), including in the non-diabetic subjects who were followed-up for nearly 6 years (HR = $0.790$, 95% CI: $0.645$–$0.967$; $p = 0.022$) [62]. Esmeralda et al. (2021) investigated the link between TC, TG and serum/urinary Mg in T2DM versus healthy counterparts. T2DM subjects had higher TG ($p = 0.004$) and fractional excretion of Mg ($p = 0.01$), lower serum Mg ($p = 0.001$) and similar TC ($p = 0.31$) and urinary Mg ($p = 0.097$) versus controls. Nevertheless, no associations of the serum Mg or the fractional excretion of Mg with TC or TG were detected [63]. Gopal et al. (2019) concluded that serum Mg concentrations predict the development of proliferative retinopathy in patients with T2DM (optimum cut-off 1.7 mg/dL, sensitivity 92.86%, specificity 77.14%, AUC 0.837, SEM 0.06, 95% CI $0.70$–$0.92$). In T2DM, Mg levels decreased as following no retinopathy → non-proliferative retinopathy → proliferative retinopathy ($p < 0.01$). Surprisingly, patients with proliferative retinopathy had higher HDL-C values ($p < 0.05$). TC, TG, LDL-C, and VLDL concentrations did not differ significantly among the study groups [64]. Hruby et al. (2017) studied the link between the intake of Mg and the risk of T2DM in three cohorts from the United States of America, detecting a 15% T2DM-risk reduction in individuals with a higher dietary intake of Mg. Hypercholesterolemia was more frequent in the fifth (12.7% for 427–498 mg/day) versus first (8.7% for 242–275 mg/day) quartile of Mg intake, whereas in women the data were conflicting: in one cohort, hypercholesterolemia was more prevalent in the fifth (7.8% for 357–418 mg/day) versus first (6.2% for 187–218 mg/day) quartile of Mg intake, whereas in the other cohort the results were opposite (15.4% for 213–245 mg/day versus 14.4% for 385–448 mg/day) [65]. In Anetor et al. (2002)'s research on 40 T2DM patients from Nigeria, only TC displayed a significant positive correlation with Mg levels (r = $0.6$; $p < 0.001$) [66]. Corica et al. (2006) analyzed 290 T2DM and detected that serum Mg was significantly lower in individuals with low HDL-C ($p < 0.001$) and high TG ($p < 0.001$) [67]. Romero and Moran (2000) evaluated 180 subjects with impaired glucose regulation (un/controlled T2DM and IFG) versus 190 healthy controls and demonstrated that decreased Mg levels are associated with decreased HDL-C, regardless of blood glucose values ($p = 0.01$ for the T2DM groups, $p = 0.05$ for the IFG group and $p = 0.03$ for the control group) [68]. Yu et al. (2018) reported that a group of 8163 Chinese T2DM adults, classified based on Mg levels, exhibited significant elevations in serum lipids, except for HDL-C, across progressive concentrations of serum Mg (from low Mg levels of ≤$0.65$ mmol/L normal levels $0.65$–$0.95$ mmol/L, high levels ≥$0.95$ mmol/L) ($p < 0.05$), regardless of whether they suffered from central obesity or not. The generalized linear model showed that after the full adjustment for demographic characteristics, lifestyle, dietary, and clinical factors, TG, TC, HDL-C, and LDL-C were significantly higher in subjects with Mg levels ≥0.95 mmol/L versus those with lower Mg levels ($p < 0.05$) [69]. Kurstjens et al. (2016) investigated the determinants of serum Mg levels in a T2DM cohort (n = 395). Multiple blood parameters were investigated including serum Mg levels, HDL-C, LDL-C, TC, TG, and others, and correlation studies were performed. Their results showed a statistically significant negative correlation between TG and serum Mg (r = −0.273, $p < 0.001$), as well as a positive correlation between HDL-C and Mg (r = 0.156, $p = 0.002$) [70].

The most relevant information of this subsection are summarized in Table 3.

**Table 3.** Magnesium and serum lipids interplay in T2DM.

| Author and Year | Number of Patients | Method of Mg Determination | Main Results |
|---|---|---|---|
| van Dijk et al. (2019) [60] | 5568 | Serum (xylidyl blue test) | (−) correlation of Mg (measured by nuclear magnetic resonance spectroscopy and colorimetric assays) and TG (r = −0.184, $p = 0.002$ and r = −0.194, $p = 0.001$, respectively) |
| Rusu et al. (2013) [61] | 154 | Serum (automated multianalyzer) | T2DM + peripheral arterial disease: ↓ Mg ($p < 0.01$), ↓ HDL-C, ↑ TG, ↑ TC |
| Spiga et al. (2019) [62] | 589 | Serum (colorimetric assay) | ↓ Mg, ↓ HDL-C, ↑ TG in normal glucose tolerance → impaired fasting glucose → T2DM ($p < 0.02$, $p < 0.01$ and $p < 0.001$, respectively) (+) correlations: Mg and TC (r = 0.154, $p < 0.001$), HDL-C (r = 0.113, $p < 0.01$), LDL-C (r = 0.170, $p < 0.001$) ↑ Mg: ↓ risk of T2DM (OR = 0.765, 95% CI 0.629–0.932, $p < 0.01$) |
| Esmeralda et al. (2021) [63] | 62 | Serum and urine analysis (methods unspecified) | T2DM: ↑ TG ($p = 0.004$), ↑ fractional excretion of Mg ($p = 0.01$), ↓ serum Mg ($p = 0.001$) versus controls |
| Gopal et al. (2019) [64] | 90 | Serum (calmagite colorimetric test) | Mg concentrations predict proliferative retinopathy development in T2DM (optimum cut-off 1.7 mg/dL, sensitivity 92.86%, specificity 77.14%, AUC 0.837, SEM 0.06, 95% CI 0.70–0.92) Mg ↓: no retinopathy → non-proliferative retinopathy → proliferative retinopathy ($p < 0.01$) proliferative retinopathy: ↑ HDL-C ($p < 0.05$), similar TC, TG, LDL-C, VLDL |
| Hruby et al. (2017) [65] | 202,743 | Unspecified | 15% T2DM-risk ↓ in individuals with ↑ dietary intake of Mg ↑ Hypercholesterolemia in the 5th (12.7% for 427–498 mg/day) versus 1st first (8.7% for 242–275 mg/day) quartile of Mg intake |
| Anetor et al. (2002) [66] | 60 | Serum (spectrophotometry) | (+) association of Mg and TC (r = 0.6; $p < 0.001$) |
| Corica et al. (2006) [67] | 290 | Serum (ion selective analyzer) | ↓ Mg: ↓ HDL-C ($p < 0.001$) and ↑ TG ($p < 0.001$) |
| Romero and Moran (2000) [68] | 390 | Serum (chemical autoanalyzer) | ↓ Mg associated with ↓ HDL-C ($p = 0.01$ in T2DM; $p = 0.05$ in IFG; $p = 0.03$ for controls) |
| Yu et al. (2018) [69] | 8163 | Serum (xylidyl blue test) | ↑ serum lipids, except for HDL-C, across progressive Mg concentrations (from ↓ Mg of ≤0.65 mmol/L, normal 0.65–0.95 mmol/L, ↑ levels ≥0.95 mmol/L) ($p < 0.05$) ↑ TG, TC, HDL-C and LDL-C (Mg ≥0.95 mmol/L versus ↓ Mg, $p < 0.05$) |
| Kurstjens et al. (2016) [70] | 395 | Serum (spectrophotometry) | (−) negative correlation of Mg and TG (r = −0.273, $p < 0.001$) (+) positive correlation of Mg and HDL-C (r = 0.156, $p = 0.002$) |

Mg, magnesium. MetS, metabolic syndrome. T2DM, type 2 diabetes mellitus. HTN, hypertension. hypoMg, hypomagnesemia. OR, odds ratio. CI, confidence interval. ↑, increased. ↓, decreased. (+), positive. (−), negative. ♂, male. ♀, female. MgO, Mg oxide. HDL-C, high-density lipoprotein cholesterol. LDL-C, low-density lipoprotein cholesterol. TC, total cholesterol. TG, triglycerides. VLDL, very low-density lipoprotein cholesterol. apoA1, apolipoprotein A1. oxLDL, oxidized LDL. RDV, recommended dietary value. Δ, variation. MgCl$_2$, Mg chloride. mg, milligrams. g, grams. mL, milliliter. K, potassium. MONW, metabolically obese normal-weight. NW, normal weight.

### 3.3.1. Hypomagnesemia

Rasheed et al. (2012) compared laboratory variables in 219 T2DM patients and 100 healthy controls, depicting a higher prevalence of hypomagnesemia, higher TG and LDL-C and lower HDL-C concentrations in the study group [71]. Srinivasan el al. (2012) also unmasked that, in T2DM, hypomagnesemia not normomagnesemia positively correlated with TG concentrations ($p < 0.05$) [72]. Pokharel et al. (2017) researched the association of serum Mg and different cardiovascular risk factors in 150 Nepalese T2DM subjects versus 50 controls. Hypomagnesemia was discovered in 50% of T2DM individuals versus 0% in the healthy counterparts. An inverse correlation between serum Mg and TC ($r = -0.219$; $p < 0.01$) and LDL-C ($r = -0.168$; $p < 0.05$) was detected. Mg concentrations were similar ($p > 0.05$) between overweight, hypertensive, and dyslipidemic T2DM patients and the control group. The authors concluded that based on their study a nutritional supplement of Mg was warranted for the prevention and minimization of chronic T2DM systemic complications such as insulin resistance and dyslipidemia [73]. Hyassat et al. (2014) analyzed 1105 overweight/obese subjects diagnosed with T2DM and detected a prevalence of hypomagnesemia of 19% (95% CI: 16.8–21.4%) in the study group. Patients who were dyslipidemic or were prescribed statins were more likely to associate low serum Mg concentrations ($p = 0.022$ and $p < 0.001$, respectively). Low serum Mg levels remained associated with the administration of statins in the multivariate logistic regression as well (OR = 1.56, 95% CI: 1.1–2.2) [74]. Waanders et al. (2019) investigated the crosstalk between T2DM and hypomagnesemia in 929 Dutch individuals, demonstrating that Mg concentrations correlated with TC ($r = 0.142$, $p < 0.001$), TC/HDL-C ratio ($r = 0.114$, $p < 0.001$), and LDL-C ($r = 0.166$, $p < 0.001$), but not with HDL-C ($r = -0.003$, $p = 0.923$) or TG ($r = 0.002$, $p = 0.941$). The results of the stepwise multivariable regression revealed a significant association of serum Mg with LDL-C ($\beta = 0.141$, $p = 0.001$) [75]. In Huang et al. (2012)'s cross-sectional observational study on 210 T2DM subjects aged >65 years, the appraisal of nutritional habits unmasked an insufficient Mg intake in >88% and hypomagnesemia in 37% of the individuals. Mg intake was positively correlated with HDL-C ($r = 0.192$; $p = 0.005$) [76]. Corsonello et al. (2000) conducted a study on 90 T2DM patients (30 patients without albuminuria, 30 with microalbuminuria and 30 with proteinuria) and observed a decrease in serum Mg and an increase in TG as the urinary protein loss increased ($p < 0.001$). However, the presence of hypomagnesemia was not associated with hypercholesterolemia, despite higher TC being detected in the patients with hypomagnesemia [77]. Shardha et al. (2014) investigated the relationship between serum Mg levels and serum lipids in T2DM individuals suffering from hypokalemia. TG (234.50 mg/dl versus 169 mg/dL; $p = 0.001$) and LDL-C (123 mg/dL versus 105 mg/dL; $p < 0.001$) were higher and HDL-C (39 mg/dL versus 46 mg/dL; $p < 0.001$) was lower in hypokalemic T2DM individuals who associated hypomagnesemia [78].

The most relevant information of this subsection are summarized in Table 4.

Table 4. Interplay of hypomagnesemia and T2DM.

| Author and Year | Number of Patients | Method of Mg Determination | Main Results |
|---|---|---|---|
| Rasheed et al. (2012) [71] | 319 | Serum (spectrophotometry) | ↑ prevalence of hypoMg in T2DM hypoMg: ↑ TG, ↑ LDL-C, ↓ HDL-C |
| Srinivasan el al. (2012) [72] | 30 | Serum (calmagite colorimetric test) | association of hypoMg and TG ($p < 0.05$) |
| Pokharel et al. (2017) [73] | 300 | Serum (xylidyl blue test) | ↑ prevalence of hypoMg in T2DM: 50% (−) correlation of Mg and TC (r = −0.219; $p < 0.01$), LDL-C (r = −0.168; $p < 0.05$) |
| Hyassat et al. (2014) [74] | 1105 | Serum (colorimetric assay) | ↑ prevalence of hypoMg in overweight/obese T2DM: 19% dyslipidemia, statin use associated with hypoMg ($p = 0.022$ and $p < 0.001$, respectively) hypoMg-statin use association (OR = 1.56, 95% CI: 1.1–2.2) |
| Waanders et al. (202p) [75] | 929 | Serum (colorimetric assay) | (+) associations of Mg and TC (r = 0.142, $p < 0.001$), TC/HDL-C ratio (r = 0.114, $p < 0.001$), LDL-C (r = 0.166, $p < 0.001$) stepwise multivariable regression: Mg associated with LDL-C (β = 0.141, $p = 0.001$) |
| Huang et al. (2012) [76] | 210 | Serum (methylthymol blue method) | T2DM, >65 years: insufficient Mg intake (>88%), hypoMg (37%) (+) association of Mg intake and HDL-C (r = 0.192; $p = 0.005$) |
| Corsonello et al. (2000) [77] | 110 | Serum (ion selective analyzer) | ↓ Mg, ↑ TG with the ↑ urinary protein loss ($p < 0.001$) hypoMg: ↑ TC |
| Shardha et al. (2014) [78] | 358 | Serum (method unspecified) | ↑ TG ($p = 0.001$), ↑ LDL-C ($p < 0.001$), ↓ HDL-C ($p < 0.001$) in T2DM + hypoK + hypoMg |

Mg, magnesium. MetS, metabolic syndrome. T2DM, type 2 diabetes mellitus. HTN, hypertension. hypoMg, hypomagnesemia. OR, odds ratio. CI, confidence interval. ↑, increased. ↓, decreased. (+), positive. (−), negative. ♂, male. ♀, female. MgO, Mg oxide. HDL-C, high-density lipoprotein cholesterol. LDL-C, low-density lipoprotein cholesterol. TC, total cholesterol. TG, triglycerides. VLDL, very low-density lipoprotein cholesterol. apoA1, apolipoprotein A1. oxLDL, oxidized LDL. RDV, recommended dietary value. Δ, variation. $MgCl_2$, Mg chloride. mg, milligrams. g, grams. mL, milliliter. K, potassium. MONW, metabolically obese normal-weight. NW, normal weight. hypoK, hypokalemia.

### 3.3.2. Magnesium Supplementation in RCTs and Interventional Studies

Song et al. (2006)'s meta-analysis of RCTs unmasked that supplementation with Mg elevated HDL-C in T2DM subjects (+0.08 mmol/L, 95% CI 0.03–0.14, $p = 0.36$ for heterogeneity), but failed to produce effects on TG, TC or LDL-C [79]. Hamedifard et al. (2020) conducted a 12-week RCT in which they compared the administration of 250 mg Mg oxide plus 150 mg zinc sulfate versus placebo in 60 T2DM subjects suffering from coronary heart disease. The intervention did not yield any significant changes in TC, TC/HDL-C ratio, LDL-C, VLDL or TG, however it resulted in a notable increase in HDL-C (β = 2.09 mg/dL, 95% CI, 0.05, 4.13; $p = 0.04$) versus placebo [80]. Al-Daghri et al. (2014) supplemented a cohort of 126 Saudi T2DM subjects with vitamin D and detected a significant positive correlation between serum Mg and TG (r = 0.32, $p = 0.04$) in males. In females, there was a positive correlation (r = 0.36, $p = 0.006$) between HDL-C and the variation in Mg concentrations following 6 months of vitamin D supplementation [81]. Rashvand et al. (2019) explored the impact of 500 mg Mg oxide versus 1000 mg choline bitartrate versus Mg + choline co-supplementation versus placebo on the lipid profile of 96 T2DM subjects in a 2-month RCT. Mg supplementation alone raised serum Mg concentrations ($p = 0.02$), but failed to alter the values of TG ($p = 0.24$), TC ($p = 0.48$), LDL-C ($p = 0.89$), HDL-C ($p = 0.09$), or LDL-C/HDL-C ($p = 0.62$). Choline supplementation alone did not have any impact on the lipid profile as well. However, the Mg+choline intervention elevated Mg levels ($p < 0.001$ versus $p = 0.02$ for Mg supplementation alone), decreased TG ($p = 0.04$), increased HDL-C ($p = 0.01$), but did not alter TC ($p = 0.92$), LDL-C

($p$ = 0.44) or LDL-C/HDL-C ($p$ = 0.08) [82]. Dietary interventions, such as honey enriched with Mg, cinnamon, and chromium, was also able to decrease LDL-C (−0.29 mmol/L; 95%CI −0.57 to −0.23; $p$ = 0.039) and TC (−0.37 mmol/L; 95% CI −0.073 to −0.008; $p$ = 0.046) values versus standard honey in an RCT involving subjects diagnosed with T2DM, yet TG and HDL-C values remained unaltered [83]. In their recent meta-analysis of 4 RCTs, Dehbalaei et al. (2021) analyzed the effects of combined 250 mg/day of Mg and 400 IU/day of vitamin E supplementation for 6–12 weeks on the lipid profile in 119 participants suffering from T2DM-induced foot ulcers, gestational diabetes or polycystic ovary syndrome versus 118 controls. The intervention yielded a reduction in TC (WMD: −15.89 mg/dL, 95% CI: 24.39, 7.39, $p$ < 0.001), LDL-C (WMD: −11.37 mg/dL, 95% CI: 19.32, 3.41, $p$ = 0.005), and TG (WMD: −26.97 mg/dL, 95% CI: 46.03, 7.90, $p$ = 0.006), but did not influence HDL-C (WMD: 1.59 mg/dL, 95% CI: 0.17, 3.35, $p$ = 0.076) [84]. According to Yokota et al. (2004), a 30-day supplementation with 300 mg/day of Mg did not modify TC, HDL-C, or TG in T2DM subjects [85]. Djurhuus et al. (2001) monitored the effects of 24 weeks of oral supplementation with Mg in patients with type 1 diabetes and hypomagnesemia. Although TG levels increased in the first week, at 24 weeks the authors noted a reduction in LDL-C, TC, and apoB. Intravenous administration of Mg resulted in a more pronounced decrease of the same parameters [86]. Ham and Shon (2020) conducted an RCT in which they evaluated the 8-week administration of deep-sea water enriched with Mg versus placebo in 74 individuals with prediabetes. Following the intervention, the authors recorded a significant reduction in LDL-C ($p$ = 0.003) and TC ($p$ = 0.006), but HDL-C and TG remained unchanged [87]. However, in their RCT, Cosaro et al. (2014) detected no benefits of Mg supplementation in terms of TC, LDL-C, or TG reduction and HDL-C elevation in a small study group of 7 males aged 23–33 years who had a positive family history for T2DM/MetS versus 7 healthy controls [88]. Asemi et al. (2015) conducted an RCT to investigate the effects of Mg supplementation in females suffering from gestational diabetes and concluded that the placebo group had higher values of TC ($p$ = 0.01), VLDL ($p$ = 0.005), and TG ($p$ = 0.005) at the end of the intervention versus the baseline. In both the placebo and the Mg-deficient women's group, HDL-C, LDL-C and TC/HDL-C ratio did not change significantly throughout the RCT [89]. In Navarrete-Cortes et al. (2014)'s RCT, a 3-month administration of 360 mg Mg daily failed to lead to any changes in TC, TG, LDL-C, or HDL-C in T2DM subjects with normomagnesemia versus placebo [90]. Another RCT, conducted by Guerrero-Romero et al. (2015), explored the benefits of 30 mL of $MgCl_2$ 5% solution (equivalent to 382 mg Mg) versus placebo in 59 individuals diagnosed with hypomagnesemia and prediabetes versus controls. At the end of the intervention which lasted for 4 months, the study group displayed an elevation in HDL-C (+4.7 ± 10.5 mg/dL versus −3.9 ± 11.3 mg/dL, $p$ = 0.04) and a reduction in TG (−57.1 ± 80.7 mg/dL versus −30.9 ± 87.7, $p$ = 0.009) as compared to placebo [91]. In the RCT performed by Solati et al. (2014), 25 T2DM patients who received 300 mg/day of Mg sulfate for a period of 3 months were compared to 22 T2DM patients who received placebo. Following Mg prescription, the intervention group displayed lower LDL-C (93.63 ± 24.58 mg/dL versus 120.4 ± 34.86 mg/dL; $p$ < 0.01) and non-HDL-C (125.30 ± 23.19 mg/dL versus 152.16 ± 37.05 mg/dL; $p$ < 0.001) levels at the end of the RCT, but TG, TC, and HDL-C concentrations were similar to the placebo group [92]. De Valk et al. (1998) supplemented 34 patients with controlled T2DM with Mg aspartate for 3 months, but failed to detect significant changes in serum lipids in the intervention versus control group [93]. Similarly, the 90-day administration of 600 mg/day of Mg in 56 T2DM patients did not alter TC, HDL-C, or LDL-C [94]. Talari et al. executed an RCT in which 54 Iranian T2DM patients on hemodialysis were prescribed 250 mg/day of Mg oxide for 24 weeks. There was a significant decrease in TC ($p$ = 0.02) and LDL-C ($p$ = 0.01) in the group receiving Mg [95]. In another 12-week RCT, Afzali et al. (2019) studied the effects of 250 mg/day Mg oxide + 400 IU vitamin E in 57 patients with diabetic foot. The intervention group exhibited a decrease in TG ($p$ = 0.04) and LDL-C ($p$ = 0.03), as well as an increase in HDL-C ($p$ = 0.01) [96]. In a cross-sectional study, Brandao-Lima et al. (2019) researched the impact of zinc, potassium, calcium, and Mg administration at different concentrations in 95 T2DM patients. Higher levels of TG ($p = 0.01$), TC ($p = 0.097$), LDL-C ($p = 0.0867$), and HDL-C ($p = 0.0247$) were observed in the group that received the aforementioned elements in lower concentrations [97]. In a 6-week RCT performed by Karamali M. et al. (2018), 60 patients with gestational diabetes mellitus were given either Mg-zinc-calcium-vitamin D co-supplements or placebo (n = 30 each group). The combined supplementation significantly decreased TG ($-0.27 \pm 0.89$ versus $+0.36 \pm 0.39$ mmol/L, $p = 0.001$) and VLDL ($-0.13 \pm 0.40$ versus $+0.16 \pm 0.18$ mmol/L, $p = 0.001$) as compared to placebo [98]. Sedeghian et al. (2020) conducted a 12-week RCT to assess the effects of Mg sulfate supplementation in 80 patients with early diabetic nephropathy, but did not detect significant changes in serum Mg or TC, LDL-C, HDL-C, TC/HDL-C, or TG [99].

### 3.4. Crosstalk of Magnesium, Serum Lipids and Cardiovascular Disorders

A total of 16 studies assessed the crosstalk of Mg, serum lipids, and cardiovascular disorders, focusing on the relationship of serum Mg with the lipid profile in atherosclerosis, angina pectoris, acute myocardial infarction, coronary heart disease, coronary artery calcifications, as well as hypertension (n = 7).

#### 3.4.1. Atherosclerosis, Angina Pectoris and Acute Myocardial Infarction

The prevalence of hypercholesterolemia was equally balanced among serum Mg concentrations in 414 patients younger than 50 years old who were subjected to drug-eluding stent implantation following an acute coronary syndrome [100]. Qazmooz et al. (2020) analyzed the crosstalk between several trace elements (including Mg) and subjects with atherosclerosis versus unstable angina versus healthy controls. Patients with atherosclerosis had higher Mg levels versus controls and unstable angina patients. In the subjects diagnosed with atherosclerosis or unstable angina, elevated TC, Castelli index 1 (zTC–zHDL-C) and lower HDL-C levels were recorded as compared to controls. In addition, there were significant differences in terms of LDL-C, TG, atherogenic index of plasma (zTG–zHDL-C), Castelli index 2 (zLDL-C–zHDL-C) between study groups. TG and the atherogenic index of plasma (zTG–zHDL-C) increased as follows: controls → atherosclerosis → unstable angina. LDL-C and the Castelli index 2 (zLDL-C–zHDL-C) increased as follows: controls → unstable angina → atherosclerosis. Based on the results of the multiple regression analysis, Mg was one of the explanatory variables accounting for the variance in the atherogenic index of plasma ($\beta = -0.205$, $t = -3.036$, $p = 0.003$), Castelli index 1 ($\beta = -0.179$, $t = -2.633$, $p = 0.009$), Castelli index 2 ($\beta = -0.143$, $t = -1.983$, $p = 0.049$) and HDL-C ($\beta = 0.157$, $t = 2.106$, $p = 0.037$) [101]. Brown et al. (1958) had also sought to examine the interplay of Mg, serum lipids and myocardial infarction in 186 adults who attended the Cardiovascular Health Center at Albany Medical College annually, but did not discover statistically significant associations of Mg and serum cholesterol, total lipids or alpha/beta lipoproteins [102]. The findings of Mahalle el al. (2012) regarding the crosstalk of Mg and serum lipids in 300 patients with known cardiovascular disease also delineate that TC, LDL-C, VLDL and TG are higher and HDL-C is lower in subjects with low serum Mg [103].

#### 3.4.2. Coronary Heart Disease

In the ARIC study, Liao et al. (1998) looked into the correlation between serum Mg levels and risk factors for coronary heart disease, including TC, TG, and HDL-C. The nearly 13,000 patients enrolled exhibited an increase in TC and HDL-C, as well as a decrease in LDL-C, with increasing serum Mg values [104]. Farshidi et al. (2020) conducted a 6-month RCT in which they prescribed 300 mg/day of Mg sulfate to 32 patients with coronary heart disease versus 32 subjects who received placebo. At the 3-month evaluation, the authors demonstrated a decrease in oxLDL and TC/HDL-C ratio versus placebo. Patients with one atherosclerotic vessel benefited from a reduction in LDL-C, whereas patients with more affected vessels registered an elevation in HDL-C. At 6 months, there were significant alterations of oxLDL, LDL-C, and HDL-C in subjects with one interested vessel, as well as

LDL-C alterations in patients with at least two atherosclerotic vessels. Notable changes in the concentrations of oxLDL receptors was seen. During the 3 months between evaluations, oxLDL decreased in subjects with only one vessel interested by atherosclerosis [105]. Petersen et al. (1977) examined a cohort of seventy three men and women and evaluated the relationship between serum and erythrocyte Mg and several indicators of coronary heart disease, detecting a significant inverse correlation of serum Mg and systolic blood pressure ($r = -0.31$, $p < 0.01$), and a significant positive correlation of erythrocyte Mg and TC ($r = 0.25$, $p < 0.05$) [106].

### 3.4.3. Coronary Artery Calcifications

Posadas-Sanchez et al. (2016) sought to examine the association between serum Mg levels and coronary artery calcification. The authors included a total of one thousand two hundred and seventy six subjects in their study. Blood serum values of Mg were compared to the values of LDL-C, HDL-C, TG, TC, apolipoproteins A and B, and using regression models, correlation studies were performed. After dividing the study participants according to their serum Mg quartiles, four groups were formed. None of the lipid profile components showed significant differences based on Mg quartiles, with the mean values of LDL-C, HDL-C, TG, apolipoprotein A and B being similar patients with hypomagnesemia and patients with normal serum Mg [107]. Lee et al. (2015) examined the relationship between low serum Mg and coronary artery calcification in a cross-sectional study which included 34,553 subjects who underwent coronary multi-detector computer tomography and serum Mg measurement as part of a health program in Korea. After differentiating the cohort based on their serum Mg levels, and dividing them into three groups (low < 1.9 mg/dL, normal 1.9–2.3 mg/dL, high > 2.3 mg/dL), the mean values of TC, LDL-C and HDL-C were compared, but no differences in serum lipids were seen based on the Mg subgroups [108].

### 3.4.4. Hypertension

Zemel et al. (1990) explored the benefits of Mg supplementation during a 3-month RCT, but hypertensive patients displayed no significant changes in serum lipids [109]. On the other hand, Motoyama et al. (1989)'s prescription of 600 mg of Mg oxide for 4 weeks to 21 hypertensive men resulted in significant decreases in TG and free fatty acids [110]. The 4-month supplementation with 2.5 $MgCl_2$ (450 mg Mg) employed by Guerrero-Romero and Rodriguez-Moran (2008) in their RCT led to significant improvements in HDL-C levels versus placebo in subjects diagnosed with hypertension [111]. In the prospective, 3-year observational Esfandiari et al. (2017)'s study tracking the effects of the adherence to the DASH diet, it was found that a higher questionnaire-assessed DASH score was associated with a higher Mg consumption and lower TC ($p < 0.05$) [112]. Cunha et al. (2016) performed an RCT to investigate oral Mg supplementation and its effects on the improvement of endothelial function and subclinical atherosclerosis in thiazide-treated hypertensive women. There was a significant difference in HDL-C and LDL-C between the placebo group (n = 18) and the Mg supplemented group (n = 17) at the start of the trial compared to the mean values determined at the end of the clinical trial, but Mg administration failed to alter the HDL-C ($p = 0.720$) or LDL-C ($p = 0.058$) concentrations [113]. As a strategy to combat hypertension, Karppanen et al. (1984) examined the benefits of KCl–NaCl–$MgCl_2$ (low in sodium) salts as a replacement for the common table salt (NaCl) used for food preparation. Multiple measurements were made before, after, and during the time that the patients received the salt mixture, including TC and TG. Although serum Mg significantly increased, no statistically significant changes in TC or TG were seen [114]. Delva et al. (1998) discovered that low intra-lymphocytic Mg levels correlated with an increase in TG, regardless of the patients' hypertensive or normotensive status [115].

The most relevant information of this subsection are summarized in Table 5.

Table 5. Interplay of magnesium and serum lipids in cardiovascular disorders.

| Author and Year | Condition | Number of Patients | Method of Mg Determination | Main Results |
|---|---|---|---|---|
| Qazmooz et al. (2020) [101] | atherosclerosis versus unstable angina | 178 | Serum (spectrophotometry) | ↑ Mg in atherosclerosis versus controls and unstable angina Mg explained the variance in AIP ($\beta = -0.205$, $t = -3.036$, $p = 0.003$), Castelli index 1 ($\beta = -0.179$, $t = -2.633$, $p = 0.009$), Castelli index 2 ($\beta = -0.143$, $t = -1.983$, $p = 0.049$), HDL-C ($\beta = 0.157$, $t = 2.106$, $p = 0.037$) |
| Brown et al. (1958) [102] | myocardial infarction | 1225 | Serum (spectrophotometry) | Mg not associated with serum cholesterol, total lipids, $\alpha/\beta$ lipoproteins |
| Mahalle el al. (2012) [103] | cardiovascular disease | 300 | Serum (xylidyl blue test) | ↓ Mg: ↑ TC, ↑ LDL-C, ↑ VLDL, ↑ TG and ↓ HDL-C |
| Liao et al. (1998) [104] | CHD | 13,922 | Serum (calmagite colorimetric test) | ↑ Mg = ↑ TC, ↑ HDL-C, ↓ LDL-C |
| Petersen et al. (1977) [106] | CHD | 73 | Serum, erythrocytes (spectrophotometry) | (+) correlation of erythrocyte Mg and TC ($r = 0.25$, $p < 0.05$) |
| Posadas-Sanchez et al. (2016) [107] | CAC | 1276 | Serum (xylidyl blue test) | hypoMg and normoMg: similar LDL-C, HDL-C, TG, apolipoprotein A/B |
| Lee et al. (2015) [108] | CAC | 34,553 | Serum (colorimetric assay) | no TC, LDL-C, HDL-C difference across Mg subgroups: ↓ < 1.9 mg/dL, normal 1.9–2.3 mg/dL, ↑ > 2.3 mg/dL |
| Esfandiari et al. (2017) [112] | HTN | 927 | Unspecified | ↑ questionnaire-assessed DASH score = ↑ Mg and ↓ TC ($p < 0.05$) |
| Karppanen et al. (1984) [114] | HTN | 126 | Serum (method unspecified) | KCl–NaCl–MgCl$_2$ salts versus common table salt (NaCl) ↑ Mg but no effect on TC, TG |
| Delva et al. (1998) [115] | HTN | 52 | Intralymphocyte (fluorimetric test) | ↓ intra-lymphocytic Mg associated with ↑ TG |

Mg, magnesium. MetS, metabolic syndrome. T2DM, type 2 diabetes mellitus. HTN, hypertension. hypoMg, hypomagnesemia. OR, odds ratio. CI, confidence interval. ↑, increased. ↓, decreased. (+), positive. (−), negative. MgO, Mg oxide. HDL-C, high-density lipoprotein cholesterol. LDL-C, low-density lipoprotein cholesterol. TC, total cholesterol. TG, triglycerides. VLDL, very low-density lipoprotein cholesterol. apoA1, apolipoprotein A1. oxLDL, oxidized LDL. RDV, recommended dietary value. Δ, variation. MgCl$_2$, Mg chloride. mg, milligrams. g, grams. mL, milliliter. K, potassium. MONW, metabolically obese normal-weight. NW, normal weight.

### 3.5. Crosstalk of Magnesium, Chronic Kidney Disease, and Hemodialysis

A total of 16 studies assessed the crosstalk of Mg, serum lipids, and kidney disorders, focusing on the relationship of serum Mg with the lipid profile in chronic kidney disease (CKD) or in patients undergoing hemo-/peritoneal dialysis. The impact of dietary Mg intake or Mg supplementation on serum lipids was also scrutinized.

#### 3.5.1. Chronic Kidney Disease

Toprak et al. (2017) evaluated the impact of hypomagnesemia on erectile dysfunction in 372 elderly, non-T2DM, stage 3 and 4 CKD patients and argued that the subjects with hypomagnesemia were more likely to suffer from obesity ($p = 0.003$), MetS ($p = 0.026$), have increased waist circumference ($p = 0.043$) and low HDL-C ($p = 0.009$) [116]. Khatami et al. (2013) evaluated the relationship between serum Mg and the lipid profile in 103 patients diagnosed with end-stage renal disease who were receiving hemodialysis. Serum Mg was similar between patients who had a history of dyslipidemia or had received statins, and there were no differences in terms of HDL-C, LDL-C, or apoprotein(a) levels between subjects with low versus high serum Mg concentrations. However, TC ($p = 0.03$) and TG ($p = 0.04$) were elevated in individuals with high serum Mg levels, though no correlations were detected between Mg and TC, TG, LDL-C, HDL-C or apoprotein(a) concentrations [117]. Dey et al. (2015) also evaluated the links between serum Mg concentrations and the lipid profile in 90 patients diagnosed with CKD. CKD subjects had lower serum and urinary Mg and higher TC, LDL-C, and non-HDL-C ($p < 0.001$ for all) versus controls, yet no differences were recorded in terms of HDL-C, TG, or VLDL concentrations. In patients suffering from CKD, there was a positive correlation of serum Mg with HDL-C

($r = 0.326$, $p = 0.002$) and negative correlations with TC ($r = -0.247$, $p = 0.019$), LDL-C ($r = -0.303$, $p = 0.004$), and non-HDL-C ($r = -0.289$, $p = 0.006$). Serum Mg also correlated with the Framingham risk score ($r = -0.939$, $p < 0.001$), the presence of MetS ($r = -0.830$, $p < 0.001$) and CKD severity ($r = -0.245$, $p = 0.02$) [118]. Cambray et al. (2020) analyzed the Mg-lipids-atherosclerosis crosstalk in 1754 CKD patients, revealing that Mg and TG concentrations displayed a tendency to increase ($p < 0.001$ for trend, both groups) and TC, LDL-C, and HDL-C displayed a tendency to decrease ($p < 0.001$ for trend for all) as CKD severity advanced toward dialysis. Mg levels were correlated with the presence of T2DM ($r = -0.070$, $p = 0.003$), hypertension ($r = 0.053$, $p = 0.028$), BMI ($r = -0.053$, $p = 0.027$), however no associations of Mg with the presence of dyslipidemia or TC, TG, HDL-C, or LDL-c were detected. In the multivariate linear effects model for carotid intima-media thickness, the authors reported associations between the aforementioned marker of atherosclerosis and TC ($\beta = -0.006$, SE = 0.003 $p = 0.02$), HDL-C ($\beta = 0.006$, SE = 0.003, $p = 0.03$), LDL-C ($\beta = 0.006$, SE = 0.003, $p = 0.04$), and TG ($\beta = 0.001$, SE = 0.0005, $p = 0.014$), but not with Mg ($\beta = -0.12$, SE = 0.099, $p = 0.23$). Despite these findings, when looking closely at the interactions of Mg with serum lipids, namely Mg-TC ($\beta = 0.008$, SE = 0.003, $p = 0.011$), Mg-HDL-C ($\beta = -0.007$, SE = 0.003, $p = 0.016$), Mg-LDL-C ($\beta = -0.007$, SE = 0.003, $p = 0.03$), and Mg-TG ($\beta = -0.0014$, SE = 0.0005, $p = 0.01$), Mg concentrations and carotid intima-media thickness were associated [119]. In a pilot study conducted on kidney transplant patients suffering from hypomagnesemia, the administration of Mg oxide significantly decreased TC and LDL-C, but did not alter TC/HDL-C, TG or apolipoprotein fractions [120].

### 3.5.2. Hemodialysis

Liu et al. (2013) also evaluated the associations between serum Mg and the lipid profile in hemodialysis patients, revealing that subjects suffering from hypo- versus hypermagnesemia had higher HDL-C ($p < 0.05$) but similar LDL-C, TC, TG, and lipoprotein-a concentrations. Among the aforementioned components of the lipid profile, Mg levels only correlated with HDL-C ($r = -0.028$, $p = 0.024$) [121]. Hemodialysis reduced serum Mg levels ($1.11 \pm 0.14$ mmol/L versus $0.97 \pm 0.10$ mmol/L, $p < 0.05$) in 148, with Han et al. (2020) reporting that pre-hemodialysis Mg concentrations are correlated with TC [$\beta = 0.03$ (0.006, 0.05), $p = 0.016$; $\beta = -0.003$ ($-0.004$, $-0.0009$), $p = 0.003$; $\beta = 0.03$ (0.006, 0.05), $p = 0.02$] according to different un/adjusted models [122]. According to Mortazavi et al. (2013)'s RCT, a 6-month Mg supplementation reduced LDL-C ($p = 0.04$) but did not affect HDL-C, TC, or TG in 54 Iranian subjects undergoing hemodialysis [123]. Shimohata et al. (2019) investigated 83 patients without T2DM who were undergoing hemodialysis in order to assess the link between mortality and serum Mg concentrations and revealed an univariate association between Mg levels and HDL-C ($r = 0.284$, $p = 0.009$) but not LDL-C ($r = 0.075$, $p = 0.499$). However, when analyzed by multiple regression, the Mg-HDL-C association did not reach statistical significance (coefficient: 0.004; $\beta = 0.196$; $p = 0.105$). Nevertheless, the mortality was higher (log rank = 4.951; $p = 0.026$) in patients with Mg < 2.5 mg/dL versus $\geq 2.5$ mg/dL [124]. Robles et al. (1997) have also shown that in patients receiving hemodialysis there are significant correlations of low serum Mg levels with increased LDL-C, VLDL-C, and apoB levels ($p < 0.001$) [125]. Tamura et al. (2019) scrutinized 392 patients undergoing hemodialysis for 4 years and concluded that low Mg levels are associated with higher mortality rates. In addition, in their paper, there were positive Mg–TC ($p = 0.257$), and Mg–TG ($p = 0.0279$) associations. However, HDL-C remained unchanged regardless of the Mg concentrations ($p = 0.097$) [126]. Ansari et al. (2012) observed significant positive correlations between serum Mg and lipoprotein-a ($r = 0.40$, $p < 0.007$), HDL-C ($r = 0.31$, $p < 0.01$) and TG ($r = 0.35$, $p < 0.005$), but not with LDL-C or TC in 50 patients receiving hemodialysis [127]. Similarly, Baradaran and Nasri (2004) studied 36 hemodialysis subjects and unmasked significant positive correlations of serum Mg with lipoprotein (a) ($r = 0.65$, $p < 0.05$) and TG ($r = 0.32$, $p < 0.05$), but not with TC, HDL-C, or LDL-C [128]. HDL-C ($r = 0.315$, $p = 0.003$) was positively associated with Mg concentrations in Ikee et al. (2016)'s cross-sectional study which included eighty-six patients undergoing hemodialysis [129].

Mitwalli et al. (2016) studied the significance of lower Mg levels in the serum of Saudi dialysis patients. One hundred and fifteen patients partook in this retrospective study: seventy patients were on hemodialysis and forty-five were on peritoneal dialysis. The subjects' serum values of Mg, TC, and TG were determined, showing that patients that underwent peritoneal dialysis had lower Mg levels compared to hemodialysis patients. The correlation studies did not find statistically significant associations between Mg and TC or TG variations [130]. While studying the association of hypomagnesemia with increased mortality among patients that underwent peritoneal dialysis in a research which included two hundred and fifty three subjects, Cai et al. (2016) reported a positive association between low serum Mg and TG (r = 0.160, $p$ = 0.011), but the correlation with TC did not reach statistical significance ($p$ = 0.929) [131].

The most relevant information of this subsection are summarized in Table 6.

Table 6. Interplay of magnesium, serum lipids, and the kidney.

| Author and Year | Condition | Number of Patients | Method of Mg Determination | Main Results |
|---|---|---|---|---|
| Toprak et al. (2017) [116] | stage 3 and 4 CKD, erectile dysfunction | 372 | Unspecified | hypoMg: ↑ obesity ($p$ = 0.003), ↑ MetS ($p$ = 0.026), ↓ HDL-C ($p$ = 0.009) |
| Khatami et al. (2013) [117] | end-stage renal disease + HD | 103 | Serum (spectrophotometry) | ↑ Mg = ↑ TC ($p$ = 0.03), ↑ TG ($p$ = 0.04) |
| Dey et al. (2015) [118] | CKD | 180 | Serum (chemical autoanalyzer) | ↓ serum and urinary Mg ↑ TC, LDL-C, non-HDL-C ($p$ < 0.001 for all) (+) correlation of serum Mg and HDL-C (r = 0.326, $p$ = 0.002) (−) correlations of serum Mg and TC (r = −0.247, $p$ = 0.019), LDL-C (r = −0.303, $p$ = 0.004), non-HDL-C (r = −0.289, $p$ = 0.006), Framingham risk score (r = −0.939, $p$ < 0.001), the presence of MetS (r = −0.830, $p$ < 0.001), CKD severity (r = −0.245, $p$ = 0.02) |
| Cambray et al. (2020) [119] | CKD | 1754 | Serum (Mg reagent) | ↑ CKD severity = ↑ Mg and ↑ TG ($p$ < 0.001 for trend) ↑ CKD severity = ↓ TC, LDL-C, HDL-C ($p$ < 0.001 for) associations of Mg with T2DM (r = −0.070, $p$ = 0.003), HTN (r = 0.053, $p$ = 0.028), BMI (r = −0.053, $p$ = 0.027) association of CIMT and Mg-TC (β = 0.008, SE = 0.003, $p$ = 0.011), Mg-HDL-C (β = −0.007, SE = 0.003, $p$ = 0.016), Mg-LDL-C (β = −0.007, SE = 0.003, $p$ = 0.03), Mg-TG (β = −0.0014, SE = 0.0005, $p$ = 0.01) interactions |
| Gupta et al. (1999) [120] | kidney transplant | 14 | Serum (method unspecified) | MgO in hypoMg: ↓ TC, ↓ LDL-C |
| Liu et al. (2013) [121] | HD | 98 | Serum (colorimetric assay) | ↑ HDL-C ($p$ < 0.05) in hypo- versus hyperMg similar LDL-C, TC, TG, lipoprotein-a (−) Mg-HDL-C correlation (r = −0.028, $p$ = 0.024) |
| Han et al. (2020) [122] | HD | 148 | Serum (toluidine blue assay) | HD ↓ Mg ($p$ < 0.05) pre-HD Mg correlated with TC [β = 0.03 (0.006, 0.05), $p$ = 0.016; β = −0.003 (−0.004, −0.0009), $p$ = 0.003; β = 0.03 (0.006, 0.05), $p$ = 0.02] |
| Shimohata et al. (2019) [124] | HD | 83 | Serum (chemical autoanalyzer) | Mg-HDL-C association (r = 0.284, $p$ = 0.009) multiple regression: no Mg-HDL-C (coefficient: 0.004; β = 0.196; $p$ = 0.105) ↑ mortality for Mg < 2.5 versus ≥2.5 mg/dL (log rank = 4.951; $p$ = 0.026) |
| Robles et al. (1997) [125] | HD | 25 | Serum (method unspecified) | ↓ Mg correlated with ↑ LDL-C, ↑ VLDL-C, ↑ apoB ($p$ < 0.001) |
| Tamura et al. (2019) [126] | HD | 392 | Serum (method unspecified) | ↓ Mg associated with ↑ mortality (+) Mg–TC ($p$ = 0.257) and Mg–TG ($p$ = 0.0279) associations |

Table 6. Cont.

| Author and Year | Condition | Number of Patients | Method of Mg Determination | Main Results |
| --- | --- | --- | --- | --- |
| Ansari et al. (2012) [127] | HD | 50 | Serum (standard method) | (+) correlations of Mg and lipoprotein-a (r = 0.40, $p$ < 0.007), HDL-C (r = 0.31, $p$ < 0.01), TG (r = 0.35, $p$ < 0.005) |
| Baradaran and Nasri (2004) [128] | HD | 36 | Dialysis fluid (method unspecified) | (+) correlations of Mg and lipoprotein (a) (r = 0.65, $p$ < 0.05), TG (r = 0.32, $p$ < 0.05) |
| Ikee et al. (2016) [129] | HD | 86 | Serum (xylidyl blue test) | (+) association of Mg and. HDL-C (r = 0.315, $p$ = 0.003) |
| Mitwalli et al. (2016) [130] | HD, PD | 115 | Dialysis fluid (Mg reagent) | Mg in PD compared to HD no associations of Mg and TC or TG variations |
| Cai et al. (2016) [131] | PD | 253 | Dialysis fluid (chemical autoanalyzer) | (+) association of hypoMg and TG (r = 0.160, $p$ = 0.011) |

Mg, magnesium. MetS, metabolic syndrome. T2DM, type 2 diabetes mellitus. HTN, hypertension. hypoMg, hypomagnesemia. OR, odds ratio. CI, confidence interval. ↑, increased. ↓, decreased. (+), positive. (−), negative. ♂, male. ♀, female. MgO, Mg oxide. HDL-C, high-density lipoprotein cholesterol. LDL-C, low-density lipoprotein cholesterol. TC, total cholesterol. TG, triglycerides. VLDL, very low-density lipoprotein cholesterol. apoA1, apolipoprotein A1. oxLDL, oxidized LDL. RDV, recommended dietary value. Δ, variation. $MgCl_2$, Mg chloride. mg, milligrams. g, grams. mL, milliliter. K, potassium. MONW, metabolically obese normal-weight. NW, normal weight. CKD, chronic kidney disease. HD, hemodialysis. PD, peritoneal dialysis. CIMT, carotid intima-media thickness.

Finally, the effects of Mg supplementation (alone, in combination or as part of dietary interventions) as depicted in RCTs are reported in Table 7.

Table 7. Effects of Mg supplementation in RCTs.

| Authors and Year | Country | Intervention | Duration | Condition | No. Subjects | Effects on the Lipid Profile |
| --- | --- | --- | --- | --- | --- | --- |
| Itoh et al. (1997) [13] | Japan | $Mg(OH)_2$ ~411–548 mg Mg/day | 4 weeks | Healthy | 33 | HDL-C, apoA1: ↑ LDL-C, TC/HDL-C: ↓ |
| Marken et al. (1989) [14] | USA | 800 mg/day MgO | 60 days | Healthy | 50 | TC, HDL-C, LDL-C, VLDL, TG: no effect |
| Aslanabadi et al. (2014) [18] | Iran | 1L/day Mg-rich miner water | 1 month | Dyslipidemia | 69 | TC, HDL-C, LDL-C, TG: no effect |
| Rodriguez-Moran and Guerrero-Romero (2014) [35] | Mexico | 30 mL/day $MgCl_2$ 5% solution ~382 mg Mg | 4 months | MONW + Hypomagne-semia | 47 | HDL-C: ↑ TG: ↓ |
| Joris et al. (2017) [36] | The Netherlands | 350 mg/day Mg | 24 weeks | Overweight/Obesity | 52 | TC, HDL-C, LDL-C, TG, non-esterified fatty acids: no effect |
| Solati et al. (2019) [38] | Iran | herbal supplement ~300 mg/day $MgSO_4$ | 6 months | Overweight | 70 | HDL-C, HDL-C/TG: ↑ LDL-C, TG: ↓ |
| Jamilian et al. (2019) [41] | Iran | 250 mg/day Mg + 400 mg/day vitamin E | 12 weeks | PCOS | 60 | TG, VLDL: ↓ HDL-C, LDL-C, TC/HDL-C: no effect |
| Jamilian et al. (2017) [42] | Iran | 200 mg/day Mg + 800 mg/day Ca + 8 mg/day Zn + 400 IU vitamin D | 12 weeks | PCOS | 60 | TC, TG, VLDL: ↓ HDL-C, LDL-C: no effect |
| Karandish et al. (2013) [43] | Iran | 350 mg/day Mg + low-calorie diet + physical exercise | 90 days | NAFLD | 68 | TC, HDL-C, LDL-C, TG: no effect |
| de Lourdes Lima de Souza e Silva (2014) [53] | Brazil | 400 mg/day Mg chelate | 12 weeks | MetS | 72 | TC, HDL-C, LDL-C, TG: no effect |
| Hamedifard et al. (2020) [80] | Iran | 250 mg/day MgO + 150 mg/day $ZnSO_4$ | 12 weeks | T2DM + CHD | 60 | HDL-C: ↑ TC, TC/HDL-C, LDL-C, VLDL, TG: no effect |

Table 7. Cont.

| Authors and Year | Country | Intervention | Duration | Condition | No. Subjects | Effects on the Lipid Profile |
|---|---|---|---|---|---|---|
| Rashvand et al. (2019) [82] | Iran | 500 mg/day MgO + 1000 mg/day choline bitartrate | 2 months | T2DM | 96 | HDL-C: ↑<br>TG: ↓<br>TC, LDL-C/HDL-C, LDL-C: no effect |
| Whitfield et al. (2016) [83] | New Zealand | 53.5 g Mg, Cr, cinnamon enriched honey | 40 days | T2DM | 12 | TC, LDL-C: ↓<br>HDL-C, TG: no effect |
| Ham and Shon (2020) [87] | Korea | 440 mL/day Mg-enriched deep sea water | 8 weeks | Prediabetes | 74 | TC, LDL-C: ↓<br>HDL-C, TG: no effect |
| Cosaro et al. (2014) [88] | Italy | 16.2 mmol/day Mg pidolate | 8 weeks | Healthy men with a positive family history for T2DM/MetS | 14 | TC, HDL-C, LDL-C, TG: no effect |
| Asemi et al. (2015) [89] | Iran | 250 mg/day MgO | 6 weeks | Gestational diabetes + Mg deficiency | 70 | HDL-C, LDL-C, TC/HDL-C: no effect |
| Navarrete-Cortes et al. (2014) [90] | Mexico | 360 mg/day Mg lactate | 3 months | T2DM + normomagnesemia | 98 | TC, HDL-C, LDL-C, TG: no effect |
| Guerrero-Romero et al. (2015) [91] | Mexico | 30 mL/day MgCl$_2$ 5% solution ~382 mg of Mg | 4 months | Prediabetes + hypomagnesemia | 116 | HDL-C: ↑<br>TG: ↓ |
| Solati et al. (2014) [92] | Iran | 300 mg/day MgSO$_4$ | 3 months | T2DM | 54 | LDL-C, non-HDL-C: ↓<br>TC, HDL-C, TG: no effect |
| De Valk et al. (1998) [93] | The Netherlands | 15 mmol/day Mg aspartate | 3 months | Controlled T2DM | 50 | TC, HDL-C, TG: no effect |
| Talari et al. (2019) [95] | Iran | 250 mg/day MgO | 24 weeks | T2DM + HD | 54 | TC, LDL-C ↓ |
| Afzali et al. (2019) [96] | Iran | 250 mg/day MgO + 400 IU/day vitamin E | 12 weeks | T2DM + diabetic foot | 57 | HDL-C: ↑<br>LDL-C, TG: ↓ |
| Karamali M. et al. (2018) [98] | Iran | 200 mg/day Mg + 800 mg/day Ca + 8 mg/day Zn + 400 IU vitamin D | 6 weeks | Gestational diabetes | 60 | VLDL, TG: ↓ |
| Sadeghian et al. (2020) [99] | Iran | 250 mg/day MgO | 12 weeks | T2DM nephropathy | 80 | TC, LDL-C, HDL-C, TC/HDL-C, TG: no effect |
| Farshidi et al. (2020) [105] | Iran | 300 mg/day MgSO$_4$ | 6 months | CHD | 64 | HDL-C: ↑<br>oxLDL, TC/HDL-C, LDL-C: ↓ |
| Zemel et al. (1990) [109] | USA | 40 mmol/day Mg aspartate | 3 months | Hypertension | 13 | TC, HDL-C, LDL-C, TG: no effect |
| Guerrero-Romero and Rodriguez-Moran (2008) [111] | Mexico | 2.5 g/day MgCl$_2$ (450 mg Mg) | 4 months | Hypertension | 82 | HDL-C: ↑<br>TG: no effect |
| Cunha et al. (2016) [113] | Brazil | 1200 mg/day Mg chelate | 6 months | Hypertension in women prescribed thiazides | 35 | HDL-C, LDL-C: no effect |
| Mortazavi et al. (2013) [123] | Iran | 440 mg MgO x3/week | 6 months | HD | 54 | LDL-C: ↓<br>TC, HDL-C, TG: no effect |

Mg, magnesium. RCTs, randomized clinical trials. ↑, increased. ↓, decreased. mg, milligrams. g, grams. USA, United States of America. mL, milliliter. Mg(OH)$_2$, Mg hydroxide. MgO, Mg oxide. MgCl$_2$, Mg chloride. Ca, calcium. Zn, Zinc. Cr, chromium. IU, international units. MONW, metabolically obese normal-weight. PCOS, polycystic ovary syndrome. NAFLD, nonalcoholic fatty liver disease. MetS, metabolic syndrome. T2DM, type 2 diabetes mellitus. CHD, coronary heart disease. HD, hemodialysis. HDL-C, high-density lipoprotein cholesterol. LDL-C, low-density lipoprotein cholesterol. TC, total cholesterol. TG, triglycerides. VLDL, very low-density lipoprotein cholesterol. apoA1, apolipoprotein A1. oxLDL, oxidized LDL.

## 4. Discussion

In this systematic review, we focused on depicting the relationship between Mg levels and serum lipids in dyslipidemia and associated disorders. We analyzed a total of 124 studies conducted on patients diagnosed with dyslipidemia, MetS, T2DM, cardiovascular or kidney disorders, yet due to the heterogeneity of the analyzed papers it is difficult to conclude to which extent Mg levels are linked to serum lipids concentrations. Although we aimed to investigate both "classical" (TC, TG, HDL-C and LDL-C) and "non-classical" components (apolipoproteins, Lp(a), oxLDL etc.) of the lipid profile, most studies were focused on examining the crosstalk of Mg levels and TC, TG, HDL-C, and LDL-C.

Mg deficits may arise both from primary (insufficient intake, decreased absorption or elevated excretion) and secondary causes, e.g., disorders that accompany the advancement in age, several comorbidities (T2DM, MetS) or it can occur due to the use of certain medications such as loop diuretics [132]. In addition, particular attention should be given to the methods employed in the measurement of Mg concentrations. Barbagallo et al. (2014) demonstrated that, in elderly patients diagnosed with T2DM, serum ionized rather than total serum Mg may emerge as a superior predictor of the subclinical deficit of this micronutrient. In addition, they also detected that TG may be a confounding factor in the crosstalk between Mg levels and markers of glucose metabolism. For example, after multiple adjustments for TG, BMI, and glomerular filtration rate, the associations of serum total Mg with FPG and HbA1c, respectively, failed to reach statistical significance. However, serum-ionized Mg remained associated with these variables despite multiple adjustments [133]. Moreover, in their recent umbrella review of systematic reviews and meta-analyses of observational and intervention studies focused on the crosstalk between Mg concentrations and health outcomes, Veronese et al. (2019) evidenced that an elevated intake of this micronutrient can result in a reduction of the risk of both stroke and T2DM. However, Mg intake was not linked to any other cardiovascular endpoints based on their results [134].

The potential lipid-lowering effects of Mg warrant further investigation, with a myriad of studies linking the serum concentrations of this micronutrient to cardiometabolic disorders, e.g., obesity, T2DM, MetS, cardiovascular disorders, neurological ailments, and even cancer, all of which are worldwide public health threats [5,135,136]. Mg supplements stand out as one of the most popular supplements in Europe and the United States [137]. In particular, Mg orotate supplementation, due to the Mg-fixing capacity of this salt, has exerted health benefits [137,138]. For example, in an RCT, patients with heart failure who were prescribed Mg orotate had better 1-year survival versus subjects receiving placebo [138]. Similarly, patients diagnosed with concomitant heart failure and hypertension who were administered Mg orotate registered a decrease in both blood pressure and N-terminal (NT)-pro hormone BNP (NT-proBNP) [139].

In terms of Mg supplementation, the most reliable data included in our paper were obtained from 29 RCTs with a total number of 1724 subjects who received different forms of Mg supplementation for a period of time ranging from 4 to 24 weeks [13,14,18,35,36,38, 41–43,53,80,82,83,87–93,95,96,98,99,105,109,111,113,123]. Overall, the vast majority of the analyzed RCTs reported no variations in HDL-C (n = 19), LDL-C (n = 16), TG (n = 17), TC (n = 14), TC/HDL-C (n = 4), VLDL (n = 2), or LDL-C/HDL-C (n = 1). However, some RCTs reported elevations of the HDL-C (n = 8), apoA1 (n = 1), or HDL-C/TG (n = 1), as well as reductions in LDL-C (n = 7), TC (n = 4), TG (n = 7), TC/HDL-C (n = 2), or VLDL (n = 3) in the participants exposed to the Mg intervention [13,14,18,35,36,38,41–43,53,80,82,83,87–93,95,96,98,99,105,109,111,113,123].

Our systematic review has several strengths and limitations. A major strength of our manuscripts is that we systematically analyzed a considerable number of studies that focused on the assessment of the relationship between Mg levels and serum lipids concentrations. Although lengthy, we strived to summarize all the available evidence regarding this research question and clarify the crosstalk between Mg and serum lipids in dyslipidemia and associated disorders. We also reported data derived from RCTs conducted with the purpose of evaluating the impact of Mg supplementation on serum lipids values. However, due to multiple reasons (the data analyzed was extremely heterogeneous, most studies included small sample sizes and no control group, some studies assessed serum Mg whereas other assessed hair or urinary Mg concentrations, the methods employed to measure Mg levels were heterogeneous, the crosstalk between Mg and serum lipids was evaluated in a myriad of disorders), we did not perform a meta-analysis of the gathered evidence. Thus, further research is needed to clarify the relationship between Mg and serum lipids levels, as well as the effects of Mg supplementation on the lipid metabolism.

## 5. Conclusions

Mg remains an important micronutrient for human health, and its putative role in the pathogenesis of dyslipidemia and associated disorders, i.e., MetS, T2DM and cardiovascular disease, has been documented in a myriad of studies. However, the mechanisms activated by Mg in its interplay with serum lipids have been elucidated, and further research is warranted to explore the lipid-lowering effects of Mg supplementation or whether a higher dietary intake of this element might emerge as a spearhead in the therapeutic *armamentarium* of dyslipidemia and closely linked diseases.

**Author Contributions:** Conceptualization, M.-A.G. and C.C.D.; methodology, M.-A.G.; software, M.-A.G., E.-C.D., M.-A.C., and N.-I.A.; validation, M.-A.G., E.-C.D., M.-A.C., and N.-I.A.; formal analysis, M.-A.G., E.-C.D., M.-A.C., and N.-I.A.; investigation, M.-A.G., E.-C.D., M.-A.C., and N.-I.A.; resources, M.-A.G., E.-C.D., M.-A.C., and N.-I.A.; data curation, M.-A.G., E.-C.D., M.-A.C., and N.-I.A.; writing—original draft preparation, M.-A.G., E.-C.D., M.-A.C., N.-I.A., and A.M.A.S.; writing—review and editing, M.-A.G., A.M.G., and C.C.D.; visualization, M.-A.G. and C.C.D.; supervision, M.-A.G. and C.C.D.; project administration, M.-A.G. and C.C.D.; funding acquisition, A.M.A.S., C.C.D. All authors have read and agreed to the published version of the manuscript.

**Funding:** This research received no external funding.

**Conflicts of Interest:** The authors declare no conflict of interest.

## References

1. Lu, Y.; Wang, P.; Zhou, T.; Lu, J.; Spatz, E.S.; Nasir, K.; Jiang, L.; Krumholz, H.M. Comparison of Prevalence, Awareness, Treatment, and Control of Cardiovascular Risk Factors in China and the United States. *J. Am. Heart Assoc.* **2018**, *7*, e007462. [CrossRef] [PubMed]
2. Epingeac, M.E.; Gaman, M.A.; Diaconu, C.C.; Gaman, A.M. Crosstalk between Oxidative Stress and Inflammation in Obesity. *Rev. Chim.* **2020**, *71*, 228–232. [CrossRef]
3. Munteanu, M.A.; Gheorghe, G.; Stanescu, A.M.A.; Bratu, O.G.; Diaconu, C.C. What Is New Regarding the Treatment of Dyslipidemia in the 2019 European Society of Cardiology Guidelines? *Arch. Balk. Med. Union* **2019**, *54*, 749–752. [CrossRef]
4. Orlova, S.; Dikke, G.; Pickering, G.; Konchits, S.; Starostin, K.; Bevz, A. Magnesium Deficiency Questionnaire: A New Non-Invasive Magnesium Deficiency Screening Tool Developed Using Real-World Data from Four Observational Studies. *Nutrients* **2020**, *12*, 2062. [CrossRef] [PubMed]
5. Piuri, G.; Zocchi, M.; Della Porta, M.; Ficara, V.; Manoni, M.; Zuccotti, G.V.; Pinotti, L.; Maier, J.A.; Cazzola, R. Magnesium in Obesity, Metabolic Syndrome, and Type 2 Diabetes. *Nutrients* **2021**, *13*, 320. [CrossRef]
6. Rosanoff, A.; Costello, R.B.; Johnson, G.H. Effectively Prescribing Oral Magnesium Therapy for Hypertension: A Categorized Systematic Review of 49 Clinical Trials. *Nutrients* **2021**, *13*, 195. [CrossRef] [PubMed]
7. Moher, D.; Liberati, A.; Tetzlaff, J.; Altman, D.G. The PRISMA Group. Preferred reporting items for systematic reviews and meta-analyses: The PRISMA statement. *PLoS Med.* **2009**, *6*, e1000097. [CrossRef]
8. Barragán, R.; Llopis, J.; Portolés, O.; Sorlí, J.V.; Coltell, O.; Rivas-García, L.; Asensio, E.M.; Ortega-Azorín, C.; Corella, D.; Sánchez-González, C. Influence of Demographic and Lifestyle Variables on Plasma Magnesium Concentrations and Their Associations with Cardiovascular Risk Factors in a Mediterranean Population. *Nutrients* **2020**, *12*, 1018. [CrossRef]
9. Bersohn, I.; Oelofse, P.J. Correlation of Serum-Magnesium and Serum-Cholesterol Levels in South African Bantu and European Subjects. *Lancet* **1957**, *269*, 1020–1021. [CrossRef]
10. Petersen, B.; Christiansen, C.; Hansen, P.F. Treatment of Hypercholesterolaemia and Hypertriglyceridaemia with Magnesium. *Acta Med. Scand.* **1976**, *200*, 59–61. [CrossRef]
11. Liu, A.; Xu, P.; Gong, C.; Zhu, Y.; Zhang, H.; Nie, W.; Zhou, X.; Liang, X.; Xu, Y.; Huang, C.; et al. High Serum Concentration of Selenium, but Not Calcium, Cobalt, Copper, Iron, and Magnesium, Increased the Risk of Both Hyperglycemia and Dyslipidemia in Adults: A Health Examination Center Based Cross-Sectional Study. *J. Trace Elem. Med. Biol.* **2020**, *59*, 126470. [CrossRef]

12. Jin, H.; Nicodemus-Johnson, J. Gender and Age Stratified Analyses of Nutrient and Dietary Pattern Associations with Circulating Lipid Levels Identify Novel Gender and Age-Specific Correlations. *Nutrients* **2018**, *10*, 1760. [CrossRef] [PubMed]
13. Itoh, K.; Kawasaka, T.; Nakamura, M. The Effects of High Oral Magnesium Supplementation on Blood Pressure, Serum Lipids and Related Variables in Apparently Healthy Japanese Subjects. *Br. J. Nutr.* **1997**, *78*, 737–750. [CrossRef]
14. Marken, P.A.; Weart, C.W.; Carson, D.S.; Gums, J.G.; Lopes-Virella, M.F. Effects of Magnesium Oxide on the Lipid Profile of Healthy Volunteers. *Atherosclerosis* **1989**, *77*, 37–42. [CrossRef]
15. Randell, E.W.; Mathews, M.; Gadag, V.; Zhang, H.; Sun, G. Relationship between Serum Magnesium Values, Lipids and Anthropometric Risk Factors. *Atherosclerosis* **2008**, *196*, 413–419. [CrossRef] [PubMed]
16. Guerrero-Romero, F.; Rodríguez-Morán, M. The Ratio Potassium-to-Magnesium Intake and High Blood Pressure. *Eur. J. Clin. Invest.* **2019**, *49*, e13093. [CrossRef]
17. de Valk, H.W.; Bianchi, R.; van Rijn, H.J.; Erkelens, D.W. Acute Exogenous Elevation of Plasma Free Fatty Acids Does Not Influence the Plasma Magnesium Concentration. *Clin. Chem. Lab. Med.* **1998**, *36*, 115–117. [CrossRef]
18. Aslanabadi, N.; Habibi Asl, B.; Bakhshalizadeh, B.; Ghaderi, F.; Nemati, M. Hypolipidemic Activity of a Natural Mineral Water Rich in Calcium, Magnesium, and Bicarbonate in Hyperlipidemic Adults. *Adv. Pharm. Bull.* **2014**, *4*, 303–307.
19. Fu, Z.-Y.; Yang, F.L.; Hsu, H.-W.; Lu, Y.-F. Drinking Deep Seawater Decreases Serum Total and Low-Density Lipoprotein-Cholesterol in Hypercholesterolemic Subjects. *J. Med. Food* **2012**, *15*, 535–541. [CrossRef] [PubMed]
20. Nerbrand, C.; Agréus, L.; Lenner, R.A.; Nyberg, P.; Svärdsudd, K. The Influence of Calcium and Magnesium in Drinking Water and Diet on Cardiovascular Risk Factors in Individuals Living in Hard and Soft Water Areas with Differences in Cardiovascular Mortality. *BMC Public Health* **2003**, *3*, 21. [CrossRef]
21. Luoma, H.; Helminen, S.K.J.; Ranta, H.; Rytömaa, I.; Meurman, J.H. Relationships between the Fluoride and Magnesium Concentrations in Drinking Water and Some Components in Serum Related to Cardiovascular Diseases in Men from Four Rural Districts in Finland. *Scand. J. Clin. Lab. Investig.* **1973**, *32*, 217–224. [CrossRef] [PubMed]
22. Balliett, M.; Burke, J.R. Changes in Anthropometric Measurements, Body Composition, Blood Pressure, Lipid Profile, and Testosterone in Patients Participating in a Low-Energy Dietary Intervention. *J. Chiropr. Med.* **2013**, *12*, 3–14. [CrossRef] [PubMed]
23. de Los Rios, M.G. Serum Magnesium and Serum Cholesterol Changes in Man. *Am. J. Clin. Nutr.* **1961**, *9*, 315–319. [CrossRef]
24. Cocate, P.G.; Natali, A.J.; de Oliveira, A.; Longo, G.Z.; Rita de Cássia, G.A.; Maria do Carmo, G.P.; dos Santos, E.C.; Buthers, J.M.; de Oliveira, L.L.; Hermsdorff, H.H.M. Fruit and Vegetable Intake and Related Nutrients Are Associated with Oxidative Stress Markers in Middle-Aged Men. *Nutrition* **2014**, *30*, 660–665. [CrossRef]
25. Ruel, G.; Shi, Z.; Zhen, S.; Zuo, H.; Kröger, E.; Sirois, C.; Lévesque, J.-F.; Taylor, A.W. Association between Nutrition and the Evolution of Multimorbidity: The Importance of Fruits and Vegetables and Whole Grain Products. *Clin. Nutr.* **2014**, *33*, 513–520. [CrossRef] [PubMed]
26. Bain, L.K.M.; Myint, P.K.; Jennings, A.; Lentjes, M.A.H.; Luben, R.N.; Khaw, K.-T.; Wareham, N.J.; Welch, A.A. The Relationship between Dietary Magnesium Intake, Stroke and Its Major Risk Factors, Blood Pressure and Cholesterol, in the EPIC-Norfolk Cohort. *Int. J. Cardiol.* **2015**, *196*, 108–114. [CrossRef]
27. Samavarchi Tehrani, S.; Khatami, S.H.; Saadat, P.; Sarfi, M.; Ahmadi Ahangar, A.; Daroie, R.; Firouzjahi, A.; Maniati, M. Association of Serum Magnesium Levels with Risk Factors, Severity and Prognosis in Ischemic and Hemorrhagic Stroke Patients. *Caspian J. Intern. Med.* **2020**, *11*, 83–91. [PubMed]
28. Kim, D.S.; Burt, A.A.; Ranchalis, J.E.; Jarvik, L.E.; Eintracht, J.F.; Furlong, C.E.; Jarvik, G.P. Effects of Dietary Components on High-Density Lipoprotein Measures in a Cohort of 1,566 Participants. *Nutr. Metab.* **2014**, *11*, 44. [CrossRef] [PubMed]
29. Kim, M.-H.; Choi, M.-K. Seven Dietary Minerals (Ca, P, Mg, Fe, Zn, Cu, and Mn) and Their Relationship with Blood Pressure and Blood Lipids in Healthy Adults with Self-Selected Diet. *Biol. Trace Elem. Res.* **2013**, *153*, 69–75. [CrossRef]
30. Cao, Y.; Wang, C.; Guan, K.; Xu, Y.; Su, Y.-X.; Chen, Y.-M. Association of Magnesium in Serum and Urine with Carotid Intima-Media Thickness and Serum Lipids in Middle-Aged and Elderly Chinese: A Community-Based Cross-Sectional Study. *Eur. J. Nutr.* **2016**, *55*, 219–226. [CrossRef]
31. López-González, B.; Molina-López, J.; Florea, D.I.; Quintero-Osso, B.; Pérez de la Cruz, A.; Planells del Pozo, E.M. Association between Magnesium-Deficient Status and Anthropometric and Clinical-Nutritional Parameters in Posmenopausal Women. *Nutr. Hosp.* **2014**, *29*, 658–664. [PubMed]
32. Yamori, Y.; Sagara, M.; Mizushima, S.; Liu, L.; Ikeda, K.; Nara, Y.; CARDIAC Study Group. An Inverse Association between Magnesium in 24-h Urine and Cardiovascular Risk Factors in Middle-Aged Subjects in 50 CARDIAC Study Populations. *Hypertens. Res.* **2015**, *38*, 219–225. [CrossRef] [PubMed]
33. Guerrero-Romero, F.; Rodriguez-Moran, M. Serum Magnesium in the Metabolically-Obese Normal-Weight and Healthy-Obese Subjects. *Eur. J. Intern. Med.* **2013**, *24*, 639–643. [CrossRef] [PubMed]
34. Lefebvre, P.; Letois, F.; Sultan, A.; Nocca, D.; Mura, T.; Galtier, F. Nutrient Deficiencies in Patients with Obesity Considering Bariatric Surgery: A Cross-Sectional Study. *Surg. Obes. Relat. Dis.* **2014**, *10*, 540–546. [CrossRef] [PubMed]
35. Rodríguez-Moran, M.; Guerrero-Romero, F. Oral Magnesium Supplementation Improves the Metabolic Profile of Metabolically Obese, Normal-Weight Individuals: A Randomized Double-Blind Placebo-Controlled Trial. *Arch. Med. Res.* **2014**, *45*, 388–393. [CrossRef]
36. Joris, P.J.; Plat, J.; Bakker, S.J.L.; Mensink, R.P. Effects of Long-Term Magnesium Supplementation on Endothelial Function and Cardiometabolic Risk Markers: A Randomized Controlled Trial in Overweight/Obese Adults. *Sci. Rep.* **2017**, *7*, 106. [CrossRef]

37. Guerrero-Romero, F.; Flores-García, A.; Saldaña-Guerrero, S.; Simental-Mendía, L.E.; Rodríguez-Morán, M. Obesity and Hypomagnesemia. *Eur. J. Intern. Med.* **2016**, *34*, 29–33. [CrossRef]
38. Solati, M.; Kazemi, L.; Shahabi Majd, N.; Keshavarz, M.; Pouladian, N.; Soltani, N. Oral Herbal Supplement Containing Magnesium Sulfate Improve Metabolic Control and Insulin Resistance in Non-Diabetic Overweight Patients: A Randomized Double Blind Clinical Trial. *Med. J. Islam. Repub. Iran* **2019**, *33*, 2. [CrossRef]
39. Farsinejad-Marj, M.; Azadbakht, L.; Mardanian, F.; Saneei, P.; Esmaillzadeh, A. Clinical and Metabolic Responses to Magnesium Supplementation in Women with Polycystic Ovary Syndrome. *Biol. Trace Elem. Res.* **2020**, *196*, 349–358. [CrossRef]
40. Cutler, D.A.; Pride, S.M.; Cheung, A.P. Low Intakes of Dietary Fiber and Magnesium Are Associated with Insulin Resistance and Hyperandrogenism in Polycystic Ovary Syndrome: A Cohort Study. *Food Sci. Nutr.* **2019**, *7*, 1426–1437. [CrossRef]
41. Jamilian, M.; Sabzevar, N.K.; Asemi, Z. The Effect of Magnesium and Vitamin E Co-Supplementation on Glycemic Control and Markers of Cardio-Metabolic Risk in Women with Polycystic Ovary Syndrome: A Randomized, Double-Blind, Placebo-Controlled Trial. *Horm. Metab. Res.* **2019**, *51*, 100–105. [CrossRef] [PubMed]
42. Jamilian, M.; Maktabi, M.; Asemi, Z. A Trial on the Effects of Magnesium-Zinc-Calcium-Vitamin D Co-Supplementation on Glycemic Control and Markers of Cardio-Metabolic Risk in Women with Polycystic Ovary Syndrome. *Arch. Iran. Med.* **2017**, *20*, 640–645. [PubMed]
43. Karandish, M.; Tamimi, M.; Shayesteh, A.A.; Haghighizadeh, M.H.; Jalali, M.T. The Effect of Magnesium Supplementation and Weight Loss on Liver Enzymes in Patients with Nonalcoholic Fatty Liver Disease. *J. Res. Med. Sci.* **2013**, *18*, 573–579. [PubMed]
44. Ali, R.; Lee, E.T.; Knehans, A.W.; Zhang, Y.; Yeh, J.; Rhoades, E.R.; Jobe, J.B.; Ali, T.; Johnson, M.R. Dietary Intake among American Indians with Metabolic Syndrome—Comparison to Dietary Recommendations: The Balance Study. *Int. J. Health Nutr.* **2013**, *4*, 33–45.
45. Vajdi, M.; Farhangi, M.A.; Nikniaz, L. Diet-Derived Nutrient Patterns and Components of Metabolic Syndrome: A Cross-Sectional Community-Based Study. *BMC Endocr. Disord.* **2020**, *20*, 69. [CrossRef]
46. Akbarzade, Z.; Amini, M.R.; Djafari, F.; Yarizadeh, H.; Mohtashaminia, F.; Majdi, M.; Bazshahi, E.; Djafarian, K.; Clark, C.C.T.; Shab-Bidar, S. Association of Nutrient Patterns with Metabolic Syndrome and Its Components in Iranian Adults. *Clin. Nutr. Res.* **2020**, *9*, 318–331. [CrossRef]
47. Mottaghian, M.; Salehi, P.; Teymoori, F.; Mirmiran, P.; Hosseini-Esfahani, F.; Azizi, F. Nutrient Patterns and Cardiometabolic Risk Factors among Iranian Adults: Tehran Lipid and Glucose Study. *BMC Public Health* **2020**, *20*, 653. [CrossRef]
48. Choi, M.-K.; Bae, Y.-J. Relationship between Dietary Magnesium, Manganese, and Copper and Metabolic Syndrome Risk in Korean Adults: The Korea National Health and Nutrition Examination Survey (2007–2008). *Biol. Trace Elem. Res.* **2013**, *156*, 56–66. [CrossRef] [PubMed]
49. Cano-Ibáñez, N.; Gea, A.; Ruiz-Canela, M.; Corella, D.; Salas-Salvadó, J.; Schröder, H.; Navarrete-Muñoz, E.M.; Romaguera, D.; Martínez, J.A.; Barón-López, F.J.; et al. Diet Quality and Nutrient Density in Subjects with Metabolic Syndrome: Influence of Socioeconomic Status and Lifestyle Factors. A Cross-Sectional Assessment in the PREDIMED-Plus Study. *Clin. Nutr.* **2020**, *39*, 1161–1173. [CrossRef]
50. Choi, W.-S.; Kim, S.-H.; Chung, J.-H. Relationships of Hair Mineral Concentrations with Insulin Resistance in Metabolic Syndrome. *Biol. Trace Elem. Res.* **2014**, *158*, 323–329. [CrossRef] [PubMed]
51. Vanaelst, B.; Huybrechts, I.; Michels, N.; Flórez, M.R.; Aramendía, M.; Balcaen, L.; Resano, M.; Vanhaecke, F.; Bammann, K.; Bel-Serrat, S.; et al. Hair Minerals and Metabolic Health in Belgian Elementary School Girls. *Biol. Trace Elem. Res.* **2013**, *151*, 335–343. [CrossRef]
52. Sun, X.; Du, T.; Huo, R.; Yu, X.; Xu, L. Impact of HbA1c Criterion on the Definition of Glycemic Component of the Metabolic Syndrome: The China Health and Nutrition Survey 2009. *BMC Public Health* **2013**, *13*, 1045. [CrossRef] [PubMed]
53. Lima de Souza, E.; Silva, M.d.L.; Cruz, T.; Rodrigues, L.E.; Ladeia, A.M.; Bomfim, O.; Olivieri, L.; Melo, J.; Correia, R.; Porto, M.; et al. Magnesium Replacement Does Not Improve Insulin Resistance in Patients with Metabolic Syndrome: A 12-Week Randomized Double-Blind Study. *J. Clin. Med. Res.* **2014**, *6*, 456–462. [CrossRef] [PubMed]
54. Rotter, I.; Kosik-Bogacka, D.; Dołęgowska, B.; Safranow, K.; Lubkowska, A.; Laszczyńska, M. Relationship between the Concentrations of Heavy Metals and Bioelements in Aging Men with Metabolic Syndrome. *Int. J. Environ. Res. Public Health* **2015**, *12*, 3944–3961. [CrossRef]
55. Ghasemi, A.; Zahediasl, S.; Syedmoradi, L.; Azizi, F. Low Serum Magnesium Levels in Elderly Subjects with Metabolic Syndrome. *Biol. Trace Elem. Res.* **2010**, *136*, 18–25. [CrossRef]
56. Evangelopoulos, A.A.; Vallianou, N.G.; Panagiotakos, D.B.; Georgiou, A.; Zacharias, G.A.; Alevra, A.N.; Zalokosta, G.J.; Vogiatzakis, E.D.; Avgerinos, P.C. An Inverse Relationship between Cumulating Components of the Metabolic Syndrome and Serum Magnesium Levels. *Nutr. Res.* **2008**, *28*, 659–663. [CrossRef]
57. Guerrero-Romero, F.; Rodríguez-Morán, M. Low Serum Magnesium Levels and Metabolic Syndrome. *Acta Diabetol.* **2002**, *39*, 209–213. [CrossRef]
58. Yuan, Z.; Liu, C.; Tian, Y.; Zhang, X.; Ye, H.; Jin, L.; Ruan, L.; Sun, Z.; Zhu, Y. Higher Levels of Magnesium and Lower Levels of Calcium in Whole Blood Are Positively Correlated with the Metabolic Syndrome in a Chinese Population: A Case-Control Study. *Ann. Nutr. Metab.* **2016**, *69*, 125–134. [CrossRef]

59. Rotter, I.; Kosik-Bogacka, D.; Dołęgowska, B.; Safranow, K.; Karakiewicz, B.; Laszczyńska, M. Relationship between Serum Magnesium Concentration and Metabolic and Hormonal Disorders in Middle-Aged and Older Men. *Magnes. Res.* **2015**, *28*, 99–107. [CrossRef]
60. van Dijk, P.R.; Schutten, J.C.; Jeyarajah, E.J.; Kootstra-Ros, J.E.; Connelly, M.A.; Bakker, S.J.L.; Dullaart, R.P.F. Blood Mg2+ Is More Closely Associated with Hyperglycaemia than with Hypertriacylglycerolaemia: The PREVEND Study. *Diabetologia* **2019**, *62*, 1732–1734. [CrossRef]
61. Rusu, M.; Cristea, V.; Frențiu, T.; Măruțoiu, C.; Rusu, L.D. Magnesium and Selenium in Diabetics with Peripheral Artery Disease of the Lower Limbs. *Clujul Med.* **2013**, *86*, 235–239.
62. Spiga, R.; Mannino, G.C.; Mancuso, E.; Averta, C.; Paone, C.; Rubino, M.; Sciacqua, A.; Succurro, E.; Perticone, F.; Andreozzi, F.; et al. Are Circulating Mg$^{2+}$ Levels Associated with Glucose Tolerance Profiles and Incident Type 2 Diabetes? *Nutrients* **2019**, *11*, 2460. [CrossRef]
63. Esmeralda, C.A.C.; David, P.E.; Maldonado, I.C.; Ibrahim, S.N.A.; David, A.S.; Escorza, M.A.Q.; Dealmy, D.G. Deranged Fractional Excretion of Magnesium and Serum Magnesium Levels in Relation to Retrograde Glycaemic Regulation in Patients with Type 2 Diabetes Mellitus. *Curr. Diabetes Rev.* **2021**, *17*, 91–100. [CrossRef] [PubMed]
64. Niranjan, G.; Srinivasan, A.R.; Srikanth, K.; Pruthu, G.; Reeta, R.; Ramesh, R.; Anitha, R.; Valli, V.M. Evaluation of Circulating Plasma VEGF-A, ET-1 and Magnesium Levels as the Predictive Markers for Proliferative Diabetic Retinopathy. *Indian J. Clin. Biochem.* **2019**, *34*, 352–356. [CrossRef]
65. Hruby, A.; Guasch-Ferré, M.; Bhupathiraju, S.N.; Manson, J.E.; Willett, W.C.; McKeown, N.M.; Hu, F.B. Magnesium Intake, Quality of Carbohydrates, and Risk of Type 2 Diabetes: Results from Three U.s. Cohorts. *Diabetes Care* **2017**, *40*, 1695–1702. [CrossRef]
66. Anetor, J.I.; Senjobi, A.; Ajose, O.A.; Agbedana, E.O. Decreased Serum Magnesium and Zinc Levels: Atherogenic Implications in Type-2 Diabetes Mellitus in Nigerians. *Nutr. Health* **2002**, *16*, 291–300. [CrossRef]
67. Corica, F.; Corsonello, A.; Ientile, R.; Cucinotta, D. Serum ionized magnesium levels in relation to metabolic syndrome in type 2 diabetic patients. *J. Am. Col. Nutr.* **2006**, *25*, 210–215. [CrossRef]
68. Guerrero-Romero, F.; Rodríguez-Morán, M. Hypomagnesemia Is Linked to Low Serum HDL-Cholesterol Irrespective of Serum Glucose Values. *J. Diabetes Complicat.* **2000**, *14*, 272–276. [CrossRef]
69. Yu, L.; Zhang, J.; Wang, L.; Li, S.; Zhang, Q.; Xiao, P.; Wang, K.; Zhuang, M.; Jiang, Y. Association between Serum Magnesium and Blood Lipids: Influence of Type 2 Diabetes and Central Obesity. *Br. J. Nutr.* **2018**, *120*, 250–258. [CrossRef]
70. Kurstjens, S.; de Baaij, J.H.F.; Bouras, H.; Bindels, R.J.M.; Tack, C.J.J.; Hoenderop, J.G.J. Determinants of Hypomagnesemia in Patients with Type 2 Diabetes Mellitus. *Eur. J. Endocrinol.* **2017**, *176*, 11–19. [CrossRef]
71. Rasheed, H.; Elahi, S.; Ajaz, H. Serum Magnesium and Atherogenic Lipid Fractions in Type II Diabetic Patients of Lahore, Pakistan. *Biol. Trace Elem. Res.* **2012**, *148*, 165–169. [CrossRef]
72. Srinivasan, A.R.; Niranjan, G.; Kuzhandai Velu, V.; Parmar, P.; Anish, A. Status of Serum Magnesium in Type 2 Diabetes Mellitus with Particular Reference to Serum Triacylglycerol Levels. *Diabetes Metab. Syndr.* **2012**, *6*, 187–189. [CrossRef]
73. Pokharel, D.R.; Khadka, D.; Sigdel, M.; Yadav, N.K.; Kafle, R.; Sapkota, R.M.; Jha, S.K. Association of Serum Magnesium Level with Poor Glycemic Control and Renal Functions in Nepalese Patients with Type 2 Diabetes Mellitus. *Diabetes Metab. Syndr.* **2017**, *11*, S417–S423. [CrossRef]
74. Hyassat, D.; Al Sitri, E.; Batieha, A.; El-Khateeb, M.; Ajlouni, K. Prevalence of Hypomagnesaemia among Obese Type 2 Diabetic Patients Attending the National Center for Diabetes, Endocrinology and Genetics (NCDEG). *Int. J. Endocrinol. Metab.* **2014**, *12*, e17796. [CrossRef]
75. Waanders, F.; Dullaart, R.P.F.; Vos, M.J.; Hendriks, S.H.; van Goor, H.; Bilo, H.J.G.; van Dijk, P.R. Hypomagnesaemia and Its Determinants in a Contemporary Primary Care Cohort of Persons with Type 2 Diabetes. *Endocrine* **2020**, *67*, 80–86. [CrossRef]
76. Huang, J.-H.; Lu, Y.-F.; Cheng, F.-C.; Lee, J.N.-Y.; Tsai, L.-C. Correlation of Magnesium Intake with Metabolic Parameters, Depression and Physical Activity in Elderly Type 2 Diabetes Patients: A Cross-Sectional Study. *Nutr. J.* **2012**, *11*, 41. [CrossRef]
77. Corsonello, A.; Ientile, R.; Buemi, M.; Cucinotta, D.; Mauro, V.N.; Macaione, S.; Corica, F. Serum Ionized Magnesium Levels in Type 2 Diabetic Patients with Microalbuminuria or Clinical Proteinuria. *Am. J. Nephrol.* **2000**, *20*, 187–192. [CrossRef]
78. Shardha, A.K.; Vaswani, A.S.; Faraz, A.; Alam, M.T.; Kumar, P. Frequency and Risk Factors Associated with Hypomagnesaemia in Hypokalemic Type-2 Diabetic Patients. *J. Coll. Physicians Surg. Pak.* **2014**, *24*, 830–835.
79. Song, Y.; He, K.; Levitan, E.B.; Manson, J.E.; Liu, S. Effects of Oral Magnesium Supplementation on Glycaemic Control in Type 2 Diabetes: A Meta-Analysis of Randomized Double-Blind Controlled Trials: Review Article. *Diabet. Med.* **2006**, *23*, 1050–1056. [CrossRef]
80. Hamedifard, Z.; Farrokhian, A.; Reiner, Ž.; Bahmani, F.; Asemi, Z.; Ghotbi, M.; Taghizadeh, M. The Effects of Combined Magnesium and Zinc Supplementation on Metabolic Status in Patients with Type 2 Diabetes Mellitus and Coronary Heart Disease. *Lipids Health Dis.* **2020**, *19*, 112. [CrossRef]
81. Al-Daghri, N.M.; Alkharfy, K.M.; Khan, N.; Alfawaz, H.A.; Al-Ajlan, A.S.; Yakout, S.M.; Alokail, M.S. Vitamin D Supplementation and Serum Levels of Magnesium and Selenium in Type 2 Diabetes Mellitus Patients: Gender Dimorphic Changes. *Int. J. Vitam. Nutr. Res.* **2014**, *84*, 27–34. [CrossRef]

82. Rashvand, S.; Mobasseri, M.; Tarighat-Esfanjani, A. Effects of Choline and Magnesium Concurrent Supplementation on Coagulation and Lipid Profile in Patients with Type 2 Diabetes Mellitus: A Pilot Clinical Trial. *Biol. Trace Elem. Res.* **2020**, *194*, 328–335. [CrossRef]
83. Whitfield, P.; Parry-Strong, A.; Walsh, E.; Weatherall, M.; Krebs, J.D. The Effect of a Cinnamon-, Chromium- and Magnesium-Formulated Honey on Glycaemic Control, Weight Loss and Lipid Parameters in Type 2 Diabetes: An Open-Label Cross-over Randomised Controlled Trial. *Eur. J. Nutr.* **2016**, *55*, 1123–1131. [CrossRef]
84. Dehbalaei, M.G.; Ashtary-Larky, D.; Amarpoor Mesrkanlou, H.; Talebi, S.; Asbaghi, O. The Effects of Magnesium and Vitamin E Co-Supplementation on Some Cardiovascular Risk Factors: A Meta-Analysis. *Clin. Nutr. Espen* **2021**, *41*, 110–117. [CrossRef]
85. Yokota, K.; Kato, M.; Lister, F.; Ii, H.; Hayakawa, T.; Kikuta, T.; Kageyama, S.; Tajima, N. Clinical Efficacy of Magnesium Supplementation in Patients with Type 2 Diabetes. *J. Am. Coll. Nutr.* **2004**, *23*, 506S–509S. [CrossRef] [PubMed]
86. Djurhuus, M.S.; Klitgaard, N.A.; Pedersen, K.K.; Blaabjerg, O.; Altura, B.M.; Altura, B.T.; Henriksen, J.E. Magnesium Reduces Insulin-Stimulated Glucose Uptake and Serum Lipid Concentrations in Type 1 Diabetes. *Metabolism* **2001**, *50*, 1409–1417. [CrossRef]
87. Ham, J.Y.; Shon, Y.H. Natural Magnesium-Enriched Deep-Sea Water Improves Insulin Resistance and the Lipid Profile of Prediabetic Adults: A Randomized, Double-Blinded Crossover Trial. *Nutrients* **2020**, *12*, 515. [CrossRef]
88. Cosaro, E.; Bonafini, S.; Montagnana, M.; Danese, E.; Trettene, M.S.; Minuz, P.; Delva, P.; Fava, C. Effects of Magnesium Supplements on Blood Pressure, Endothelial Function and Metabolic Parameters in Healthy Young Men with a Family History of Metabolic Syndrome. *Nutr. Metab. Cardiovasc. Dis.* **2014**, *24*, 1213–1220. [CrossRef]
89. Asemi, Z.; Karamali, M.; Jamilian, M.; Foroozanfard, F.; Bahmani, F.; Heidarzadeh, Z.; Benisi-Kohansal, S.; Surkan, P.J.; Esmaillzadeh, A. Magnesium Supplementation Affects Metabolic Status and Pregnancy Outcomes in Gestational Diabetes: A Randomized, Double-Blind, Placebo-Controlled Trial. *Am. J. Clin. Nutr.* **2015**, *102*, 222–229. [CrossRef]
90. Navarrete-Cortes, A.; Ble-Castillo, J.L.; Guerrero-Romero, F.; Cordova-Uscanga, R.; Juárez-Rojop, I.E.; Aguilar-Mariscal, H.; Tovilla-Zarate, C.A.; Lopez-Guevara, M.D.R. No Effect of Magnesium Supplementation on Metabolic Control and Insulin Sensitivity in Type 2 Diabetic Patients with Normomagnesemia. *Magnes. Res.* **2014**, *27*, 48–56. [CrossRef]
91. Guerrero-Romero, F.; Simental-Mendía, L.E.; Hernández-Ronquillo, G.; Rodriguez-Morán, M. Oral Magnesium Supplementation Improves Glycaemic Status in Subjects with Prediabetes and Hypomagnesaemia: A Double-Blind Placebo-Controlled Randomized Trial. *Diabetes Metab.* **2015**, *41*, 202–207. [CrossRef] [PubMed]
92. Solati, M.; Ouspid, E.; Hosseini, S.; Soltani, N.; Keshavarz, M.; Dehghani, M. Oral Magnesium Supplementation in Type II Diabetic Patients. *Med. J. Islam. Repub. Iran* **2014**, *28*, 67.
93. de Valk, H.W.; Verkaaik, R.; van Rijn, H.J.; Geerdink, R.A.; Struyvenberg, A. Oral Magnesium Supplementation in Insulin-Requiring Type 2 Diabetic Patients. *Diabet. Med.* **1998**, *15*, 503–507. [CrossRef]
94. Eriksson, J.; Kohvakka, A. Magnesium and Ascorbic Acid Supplementation in Diabetes Mellitus. *Ann. Nutr. Metab.* **1995**, *39*, 217–223. [CrossRef]
95. Talari, H.R.; Zakizade, M.; Soleimani, A.; Bahmani, F.; Ghaderi, A.; Mirhosseini, N.; Eslahi, M.; Babadi, M.; Mansournia, M.A.; Asemi, Z. Effects of Magnesium Supplementation on Carotid Intima-Media Thickness and Metabolic Profiles in Diabetic Haemodialysis Patients: A Randomised, Double-Blind, Placebo-Controlled Trial. *Br. J. Nutr.* **2019**, *121*, 809–817. [CrossRef] [PubMed]
96. Afzali, H.; Jafari Kashi, A.H.; Momen-Heravi, M.; Razzaghi, R.; Amirani, E.; Bahmani, F.; Gilasi, H.R.; Asemi, Z. The Effects of Magnesium and Vitamin E Co-Supplementation on Wound Healing and Metabolic Status in Patients with Diabetic Foot Ulcer: A Randomized, Double-Blind, Placebo-Controlled Trial: Supplementation and Diabetic Foot. *Wound Repair Regen.* **2019**, *27*, 277–284. [CrossRef]
97. Brandão-Lima, P.N.; Carvalho, G.B.d.; Santos, R.K.F.; Santos, B.D.C.; Dias-Vasconcelos, N.L.; Rocha, V.D.S.; Barbosa, K.B.F.; Pires, L.V. Intakes of Zinc, Potassium, Calcium, and Magnesium of Individuals with Type 2 Diabetes Mellitus and the Relationship with Glycemic Control. *Nutrients* **2018**, *10*, 1948. [CrossRef]
98. Karamali, M.; Bahramimoghadam, S.; Sharifzadeh, F.; Asemi, Z. Magnesium–Zinc–Calcium–Vitamin D Co-Supplementation Improves Glycemic Control and Markers of Cardiometabolic Risk in Gestational Diabetes: A Randomized, Double-Blind, Placebo-Controlled Trial. *Appl. Physiol. Nutr. Metab.* **2018**, *43*, 565–570. [CrossRef]
99. Sadeghian, M.; Azadbakht, L.; Khalili, N.; Mortazavi, M.; Esmaillzadeh, A. Oral Magnesium Supplementation Improved Lipid Profile but Increased Insulin Resistance in Patients with Diabetic Nephropathy: A Double-Blind Randomized Controlled Clinical Trial. *Biol. Trace Elem. Res.* **2020**, *193*, 23–35. [CrossRef]
100. An, G.; Du, Z.; Meng, X.; Guo, T.; Shang, R.; Li, J.; An, F.; Li, W.; Zhang, C. Association between Low Serum Magnesium Level and Major Adverse Cardiac Events in Patients Treated with Drug-Eluting Stents for Acute Myocardial Infarction. *PLoS ONE* **2014**, *9*, e98971. [CrossRef]
101. Qazmooz, H.A.; Smesam, H.N.; Mousa, R.F.; Al-Hakeim, H.K.; Maes, M. Trace Element, Immune and Opioid Biomarkers of Unstable Angina, Increased Atherogenicity and Insulin Resistance: Results of Machine Learning. *J. Trace Elem. Med. Biol.* **2021**, *64*, 126703. [CrossRef]
102. Brown, D. Magnesium-Lipid Relations in Health and in Patients with Myocardial Infarction. *Lancet* **1958**, *272*, 933–935. [CrossRef]
103. Mahalle, N.; Kulkarni, M.V.; Naik, S.S. Is Hypomagnesaemia a Coronary Risk Factor among Indians with Coronary Artery Disease? *J. Cardiovasc. Dis. Res.* **2012**, *3*, 280–286. [CrossRef]

104. Liao, F.; Folsom, A.R.; Brancati, F.L. Is Low Magnesium Concentration a Risk Factor for Coronary Heart Disease? The Atherosclerosis Risk in Communities (ARIC) Study. *Am. Heart J.* **1998**, *136*, 480–490. [CrossRef]
105. Farshidi, H.; Sobhani, A.R.; Eslami, M.; Azarkish, F.; Eftekhar, E.; Keshavarz, M.; Soltani, N. Magnesium Sulfate Administration in Moderate Coronary Artery Disease Patients Improves Atherosclerotic Risk Factors: A Double-Blind Clinical Trial Study. *J. Cardiovasc. Pharmacol.* **2020**, *76*, 321–328. [CrossRef] [PubMed]
106. Petersen, B.; Schroll, M.; Christiansen, C.; Transbol, I. Serum and Erythrocyte Magnesium in Normal Elderly Danish People. Relationship to Blood Pressure and Serum Lipids. *Acta Med. Scand.* **1977**, *201*, 31–34. [CrossRef] [PubMed]
107. Posadas-Sánchez, R.; Posadas-Romero, C.; Cardoso-Saldaña, G.; Vargas-Alarcón, G.; Villarreal-Molina, M.T.; Pérez-Hernández, N.; Rodríguez-Pérez, J.M.; Medina-Urrutia, A.; Jorge-Galarza, E.; Juárez-Rojas, J.G.; et al. Serum Magnesium Is Inversely Associated with Coronary Artery Calcification in the Genetics of Atherosclerotic Disease (GEA) Study. *Nutr. J.* **2015**, *15*, 22. [CrossRef] [PubMed]
108. Lee, S.Y.; Hyun, Y.Y.; Lee, K.B.; Kim, H. Low Serum Magnesium Is Associated with Coronary Artery Calcification in a Korean Population at Low Risk for Cardiovascular Disease. *Nutr. Metab. Cardiovasc. Dis.* **2015**, *25*, 1056–1061. [CrossRef]
109. Zemel, P.C.; Zemel, M.B.; Urberg, M.; Douglas, F.L.; Geiser, R.; Sowers, J.R. Metabolic and Hemodynamic Effects of Magnesium Supplementation in Patients with Essential Hypertension. *Am. J. Clin. Nutr.* **1990**, *51*, 665–669. [CrossRef]
110. Motoyama, T.; Sano, H.; Fukuzaki, H. Oral Magnesium Supplementation in Patients with Essential Hypertension. *Hypertension* **1989**, *13*, 227–232. [CrossRef]
111. Guerrero-Romero, F.; Rodríguez-Morán, M. The Effect of Lowering Blood Pressure by Magnesium Supplementation in Diabetic Hypertensive Adults with Low Serum Magnesium Levels: A Randomized, Double-Blind, Placebo-Controlled Clinical Trial. *J. Hum. Hypertens.* **2009**, *23*, 245–251. [CrossRef]
112. Esfandiari, S.; Bahadoran, Z.; Mirmiran, P.; Tohidi, M.; Azizi, F. Adherence to the Dietary Approaches to Stop Hypertension Trial (DASH) Diet Is Inversely Associated with Incidence of Insulin Resistance in Adults: The Tehran Lipid and Glucose Study. *J. Clin. Biochem. Nutr.* **2017**, *61*, 123–129. [CrossRef] [PubMed]
113. Cunha, A.R.; D'El-Rei, J.; Medeiros, F.; Umbelino, B.; Oigman, W.; Touyz, R.M.; Neves, M.F. Oral Magnesium Supplementation Improves Endothelial Function and Attenuates Subclinical Atherosclerosis in Thiazide-Treated Hypertensive Women. *J. Hypertens.* **2017**, *35*, 89–97. [CrossRef]
114. Karppanen, H.; Tanskanen, A.; Tuomilehto, J.; Puska, P.; Vuori, J.; Jäntti, V.; Seppänen, M.-L. Safety and Effects of Potassium- and Magnesium-Containing Low Sodium Salt Mixtures. *J. Cardiovasc. Pharmacol.* **1984**, *6*, S236. [CrossRef]
115. Delva, P.; Pastori, C.; Degan, M.; Montesi, G.; Lechi, A. Intralymphocyte Free Magnesium and Plasma Triglycerides. *Life Sci.* **1998**, *62*, 2231–2240. [CrossRef]
116. Toprak, O.; Sarı, Y.; Koç, A.; Sarı, E.; Kırık, A. The Impact of Hypomagnesemia on Erectile Dysfunction in Elderly, Non-Diabetic, Stage 3 and 4 Chronic Kidney Disease Patients: A Prospective Cross-Sectional Study. *Clin. Interv. Aging* **2017**, *12*, 437–444. [CrossRef]
117. Khatami, M.R.; Mirchi, E.; Khazaeipour, Z.; Abdollahi, A.; Jahanmardi, A. Association between Serum Magnesium and Risk Factors of Cardiovascular Disease in Hemodialysis Patients. *Iran. J. Kidney Dis.* **2013**, *7*, 47–52.
118. Dey, R.; Rajappa, M.; Parameswaran, S.; Revathy, G. Hypomagnesemia and Atherogenic Dyslipidemia in Chronic Kidney Disease: Surrogate Markers for Increased Cardiovascular Risk. *Clin. Exp. Nephrol.* **2015**, *19*, 1054–1061. [CrossRef]
119. Cambray, S.; Ibarz, M.; Bermudez-Lopez, M.; Marti-Antonio, M.; Bozic, M.; Fernandez, E.; Valdivielso, J.M. Magnesium Levels Modify the Effect of Lipid Parameters on Carotid Intima Media Thickness. *Nutrients* **2020**, *12*, 2631. [CrossRef]
120. Gupta, B.K.; Glicklich, D.; Tellis, V.A. Magnesium Repletion Therapy Improves Lipid Metabolism in Hypomagnesemic Renal Transplant Recipients: A Pilot Study. *Transplantation* **1999**, *67*, 1485–1487. [CrossRef] [PubMed]
121. Liu, F.; Zhang, X.; Qi, H.; Wang, J.; Wang, M.; Zhang, Y.; Yan, H.; Zhuang, S. Correlation of Serum Magnesium with Cardiovascular Risk Factors in Maintenance Hemodialysis Patients—a Cross-Sectional Study. *Magnes. Res.* **2013**, *26*, 100–108. [CrossRef] [PubMed]
122. Han, Z.; Zhou, L.; Liu, R.; Feng, L. The Effect of Hemodialysis on Serum Magnesium Concentration in Hemodialysis Patients. *Ann. Palliat. Med.* **2020**, *9*, 1134–1143. [CrossRef] [PubMed]
123. Mortazavi, M.; Moeinzadeh, F.; Saadatnia, M.; Shahidi, S.; McGee, J.C.; Minagar, A. Effect of Magnesium Supplementation on Carotid Intima-Media Thickness and Flow-Mediated Dilatation among Hemodialysis Patients: A Double-Blind, Randomized, Placebo-Controlled Trial. *Eur. Neurol.* **2013**, *69*, 309–316. [CrossRef]
124. Shimohata, H.; Yamashita, M.; Ohgi, K.; Tsujimoto, R.; Maruyama, H.; Takayasu, M.; Hirayama, K.; Kobayashi, M. The Relationship between Serum Magnesium Levels and Mortality in Non-Diabetic Hemodialysis Patients: A 10-Year Follow-up Study. *Hemodial. Int.* **2019**, *23*, 369–374. [CrossRef]
125. Robles, N.R.; Escola, J.M.; Albarran, L.; Espada, R. Correlation of Serum Magnesium and Serum Lipid Levels in Hemodialysis Patients. *Nephron* **1998**, *78*, 118–119. [CrossRef] [PubMed]
126. Tamura, T.; Unagami, K.; Okazaki, M.; Komatsu, M.; Nitta, K. Serum Magnesium Levels and Mortality in Japanese Maintenance Hemodialysis Patients. *Blood Purif.* **2019**, *47* (Suppl. 2), 88–94. [CrossRef]
127. Ansari, M.R.; Maheshwari, N.; Shaikh, M.A.; Laghari, M.S.; Lal, K.; Ahmed, K. Correlation of Serum Magnesium with Dyslipidemia in Patients on Maintenance Hemodialysis. *Saudi J. Kidney Dis. Transpl.* **2012**, *23*, 21–25.
128. Nasri, H.; Baradaran, A. Correlation of Serum Magnesium with Dyslipidemia in Maintenance Hemodialysis Patients. *Acta Med. (Hradec Kral.)* **2004**, *47*, 263–265. [CrossRef]

129. Ikee, R.; Toyoyama, T.; Endo, T.; Tsunoda, M.; Hashimoto, N. Impact of Sevelamer Hydrochloride on Serum Magnesium Concentrations in Hemodialysis Patients. *Magnes. Res.* **2016**, *29*, 184–190. [CrossRef]
130. Mitwalli, A.H. Why Are Serum Magnesium Levels Lower in Saudi Dialysis Patients? *J. Taibah Univ. Med. Sci.* **2017**, *12*, 41–46. [CrossRef]
131. Cai, K.; Luo, Q.; Dai, Z.; Zhu, B.; Fei, J.; Xue, C.; Wu, D. Hypomagnesemia Is Associated with Increased Mortality among Peritoneal Dialysis Patients. *PLoS ONE* **2016**, *11*, e0152488. [CrossRef]
132. Barbagallo, M.; Veronese, N.; Dominguez, L.J. Magnesium in Aging, Health and Diseases. *Nutrients* **2021**, *13*, 463. [CrossRef]
133. Barbagallo, M.; Di Bella, G.; Brucato, V.; D'Angelo, D.; Damiani, P.; Monteverde, A.; Belvedere, M.; Dominguez, L.J. Serum Ionized Magnesium in Diabetic Older Persons. *Metabolism* **2014**, *63*, 502–509. [CrossRef]
134. Veronese, N.; Demurtas, J.; Pesolillo, G.; Celotto, S.; Barnini, T.; Calusi, G.; Caruso, M.G.; Notarnicola, M.; Reddavide, R.; Stubbs, B.; et al. Magnesium and Health Outcomes: An Umbrella Review of Systematic Reviews and Meta-Analyses of Observational and Intervention Studies. *Eur. J. Nutr.* **2020**, *59*, 263–272. [CrossRef] [PubMed]
135. Fiorentini, D.; Cappadone, C.; Farruggia, G.; Prata, C. Magnesium: Biochemistry, Nutrition, Detection, and Social Impact of Diseases Linked to Its Deficiency. *Nutrients* **2021**, *13*, 1136. [CrossRef] [PubMed]
136. Găman, M.-A.; Cozma, M.-A.; Dobrică, E.-C.; Bacalbașa, N.; Bratu, O.G.; Diaconu, C.C. Dyslipidemia: A Trigger for Coronary Heart Disease in Romanian Patients with Diabetes. *Metabolites* **2020**, *10*, 195. [CrossRef] [PubMed]
137. Frank, J.; Kisters, K.; Stirban, O.A.; Obeid, R.; Lorkowski, S.; Wallert, M.; Egert, S.; Podszun, M.C.; Eckert, G.P.; Pettersen, J.A.; et al. The Role of Biofactors in the Prevention and Treatment of Age-Related Diseases. *Biofactors* **2021**. [CrossRef] [PubMed]
138. Stepura, O.B.; Martynow, A.I. Magnesium Orotate in Severe Congestive Heart Failure (MACH). *Int. J. Cardiol.* **2009**, *131*, 293–295. [CrossRef] [PubMed]
139. Kisters, K.; Gremmler, B.; Gröber, U. Magnesium deficiency in hypertensive heart disease. *J. Hypertens.* **2015**, *33*, e273. [CrossRef]

Article

# Magnesium Levels Modify the Effect of Lipid Parameters on Carotid Intima Media Thickness

Serafi Cambray [1,*], Merce Ibarz [2], Marcelino Bermudez-Lopez [1], Manuel Marti-Antonio [1], Milica Bozic [1], Elvira Fernandez [1], Jose M. Valdivielso [1,*] and on behalf of the NEFRONA Investigators

[1] Vascular and Renal Translational Research Group, Institute for Biomedical Research Pifarré Foundation, IRBLleida Av. Rovira Roure 80, 25198 Lleida, Spain; mbermudez@irblleida.cat (M.B.-L.); mmartia@irblleida.cat (M.M.-A.); milica.bozic@irblleida.udl.cat (M.B.); elvirafgiraldez@gmail.com (E.F.)

[2] Indicators and Specifications of the Quality in the Clinical Laboratory Group, Institute for Biomedical Research Pifarré Foundation, IRBLleida, 25198 Lleida, Spain; mibarz.lleida.ics@gencat.cat

* Correspondence: scambray@irblleida.cat (S.C.); valdivielso@irblleida.cat (J.M.V.)

Received: 15 July 2020; Accepted: 25 August 2020; Published: 28 August 2020

**Abstract:** Classical risk factors of atherosclerosis in the general population show paradoxical effects in chronic kidney disease (CKD) patients. Thus, low low-density lipoprotein (LDL) cholesterol levels have been associated with worse cardiovascular outcomes. Magnesium (Mg) is a divalent cation whose homeostasis is altered in CKD. Furthermore, Mg levels have been associated with cardiovascular health. The present study aims to understand the relationships of Mg and lipid parameters with atherosclerosis in CKD. In this analysis, 1754 participants from the Observatorio Nacional de Atherosclerosis en Nefrologia (NEFRONA) cohort were included. Carotid intima media thickness (cIMT) was determined in six arterial territories, and associated factors were investigated by linear regression. cIMT correlated positively with being male, Caucasian, a smoker, diabetic, hypertensive, dyslipidemic and with increased age, BMI, and triglyceride levels, and negatively with levels of HDL cholesterol. First-order interactions in linear regression analysis showed that Mg was an effect modifier on the influence of lipidic parameters. Thus, cIMT predicted values were higher when triglycerides or LDL levels were high and Mg levels were low. On the contrary, when Mg levels were high, this effect disappeared. In conclusion, Mg acts as an effect modifier between lipidic parameters and atherosclerotic cardiovascular disease. Therefore, Mg levels, together with lipidic parameters, should be taken into account when assessing atherosclerotic risk.

**Keywords:** magnesium; cholesterol; atherosclerosis; first-order interaction; cardiovascular risk

## 1. Introduction

According to the Cardiovascular Disease Statistics from the European Society of Cardiology, in 2015 there were 11 million new cases of cardiovascular disease (CVD) reported in Europe, and the majority of countries showed an increase in cases from 1990 [1]. Its prevalence was reported to be 83.5 million people, of which 35.7 million people showed peripheral vascular disease [1]. Moreover, CVD also led to 64 million disability-adjusted life years in the European population [1].

Among risk factors associated with CVD, plasma lipids play a key role in the initiation and progression of atheromatous disease [2]. Epidemiological studies show that increased concentrations of low-density lipoprotein cholesterol (LDL-C) are associated with an increased risk of cardiovascular events [3,4], while the contrary effect has been demonstrated for high-density lipoprotein cholesterol [5] (HDL-C). Triglyceride levels are inversely associated with HDL-C levels [6] and, despite not showing atherogenic properties per se, are considered to be an important biomarker of CVD, due to their association with atherogenic remnant lipoproteins containing apo CIII [7]. Triglycerides have important implications for chronic kidney disease (CKD) patients, a population that presents a high incidence of atherosclerotic events [8]. Additionally, traditional lipid profiles are not associated with increased cardiovascular risk in this population. CKD patients present low levels of HDL-C and hypertriglyceridemia [9], but levels of LDL-C or total cholesterol are not usually modified, or are even low in advanced stages [10]. Some works point to the importance of LDL-C particle size [11] or the levels of oxidized LDL-C [12]. Nevertheless, the link between altered lipid metabolism and the higher presence of atheromasias in CKD patients is not clear.

Currently, to prevent cardiovascular events in high risk populations, most efforts have been directed towards lowering plasma LDL-C and triglycerides, by means of pharmacological treatment [13], dietary intervention [14,15], or diet supplements [16–18]. Among diet supplements, magnesium (Mg) was one of the first to be used for lowering serum lipids. During the 1980s, it was demonstrated that Mg deficiency was associated with hypertension and vascular calcifications, while increasing its dietary intake prevented atheroma in experimental animals with normal renal function [19], and in experimental uremia [20]. Later on, during the 1990s, different studies in patients showed that lower Mg plasma concentrations were associated with atherosclerosis [21], and that Mg supplementation lowered plasma cholesterol, LDL, and triglyceride concentrations [22,23]. Recent studies also found an inverse correlation between Mg levels, carotid intima–media thickness (cIMT) [24,25] and peripheral artery disease [26,27], while many epidemiological studies and clinical trials reported Mg as a key player in cardiovascular health [28–31], even in hemodialysis patients [32]. However, randomized clinical trials with Mg supplementation have yielded conflicting results [33,34]. CKD patients often present an altered Mg balance. Thus, the decrease in glomerular filtration rate (GFR) can induce Mg retention, whereas in other cases tubular dysfunction and diuretic use can induce hypomagnesemia. Therefore, the altered Mg levels, alongside the specific dyslipidemia found in those patients, make them a very interesting group to investigate the possible interactions of both variables in atherosclerosis.

Despite the fact that some data showed that Mg levels can affect cIMT and that, according to some studies, it could also affect plasma lipid concentrations, no studies have analyzed the possible interactions of both variables in atherosclerosis. In the present study, we aim to study the impact of Mg on cIMT, and how Mg interacts with plasma lipids and other known CVD risk factors in a CKD population form the Observatorio Nacional de Atherosclerosis en Nefrologia (NEFRONA) cohort.

## 2. Materials and Methods

### 2.1. Study Design

The NEFRONA study was designed to assess the utility of noninvasive vascular imaging techniques and plasma biomarkers to predict cardiovascular events and mortality in CKD patients [35,36]. Briefly, CKD and non-CKD volunteers aged 18 to 75 years were recruited throughout Spanish primary care centers and renal units from 2009 to 2012. Patients with a history of CVD, remarkable carotid stenosis, active infections (tuberculosis and human immunodeficiency virus), pregnancy, less than twelve month of life expectancy, and with any organ transplantation or carotid artery surgery were excluded from the study. Of the NEFRONA study, 1754 subjects had available serum samples to measure Mg levels. Out of those, 40 presented with missing data on CKD status or cIMT values. Thus, 1754 volunteers were used in the present study. In total, 1542 presented with CKD (629 Stage 3; 528 Stages 4–5; 385 dialysis), and 212 were non-CKD controls (glomerular filtration rate > 60 mL/min/1.73 $m^2$). The ethics committee of each hospital approved the protocol of the study, and all volunteers were included after signing an informed consent. The research followed the principles of the Declaration of Helsinki. Non-CKD controls were included as per the protocol of the NEFRONA study and as a reference with normal Mg and lower cIMT values.

## 2.2. Clinical Data and Mg Determination

A nurse and two specifically trained technicians collected the following data: gender, age, body mass index, systolic and diastolic blood pressure (SBP and DBP), pulse pressure, and smoking status. Information about presence of diabetes, hypertension, and dyslipidemia was obtained from clinical records. Fasting blood samples were also collected by the same team and were stored at −80 °C in the Biobank of the RedInRen in the University of Alcala de Henares (Madrid). Biochemical analysis was performed as described previously [37]. The determination of Mg in serum samples was performed using the Mg reagent from Beckman Coulter (Brea, CA USA; Ref. OSR6189), following the manufacturer's instructions.

## 2.3. Atherosclerosis Assessment

Atherosclerosis assessment was performed as previously described [38]. Briefly, carotid ultrasound measurements were performed in three territories of both carotid arteries (bifurcation, internal, and common carotid arteries). Plaques were defined according to the Mannheim Carotid Intima–Media Thickness (cIMT) Consensus and the American Society of Echocardiography as a cIMT lumen protrusion ≥ 1.5 mm [39,40]. More extended protocols and data about plaque prevalence in the entire NEFRONA cohort have been published previously [37]. In the present analysis, we used the average cIMT of the territories that did not show atheroma plaque.

## 2.4. Statistical Analysis

Absolute frequencies (and percentage) or mean (and standard deviation) were used to describe qualitative and quantitative variables, respectively. Pearson correlation coefficients were calculated to analyze the relationships between cIMT and Mg levels with other clinical variables. Multivariate regression linear models were used to assess the association of clinical variables with cIMT. To specifically assess the joint association of Mg with the rest of the explanatory variables on cIMT, all possible first-order interactions were considered in the model, and a backwards stepwise algorithm was used to select the significant ones. A graph showing the predicted values of cIMT when the rest of the variables of the model are set to zero, and showing the interactions of continuous values of lipid parameters and Mg levels was drawn. All analyses were performed using R, setting the threshold of significance at 0.05.

## 3. Results

### 3.1. Clinical Characteristics

The study comprised 1754 volunteers with a mean age of 58.6 ± 12.7 years. Significant differences were found between CKD stages in all the parameters of the study, except in the percentage of smokers, which was not significantly different between CKD stages. Interestingly, Mg levels increased as renal function decreased, whereas cIMT levels showed the opposite tendency. LDL levels also showed a significant tendency to decrease as renal function impairment worsened (Table 1).

**Table 1.** Clinical characteristics of the cohort.

| Variable | All n = 1754 (100%) | Control n = 212 (12.1%) | CKD 2–3 n = 629 (35.9%) | CKD 4–5 n = 528 (30.1%) | Dialysis n = 385 (21.9%) | p-Value (CKD Groups) | p-Value (Trend CKD Groups) |
|---|---|---|---|---|---|---|---|
| Sex (Male) | 1040 (59.3%) | 100 (47.2%) | 425 (67.6%) | 299 (56.6%) | 216 (56.1%) | <0.001 | 0.512 |
| Race (Caucasian) | 1704 (97.1%) | 209 (98.6%) | 620 (98.6%) | 514 (97.3%) | 361 (93.8%) | <0.001 | <0.001 |
| Age, years | 58.6 (12.7) | 51.7 (12.4) | 62.15 (11.2) | 58.8 (12.5) | 56.4 (13.6) | <0.001 | 0.001 |
| Smoker (Yes) | 977 (55.7%) | 123 (58%) | 363 (57.7%) | 279 (52.8%) | 212 (55.1%) | 0.347 | 0.211 |
| Diabetes (Yes) | 373 (21.3%) | 0 (0%) | 170 (27%) | 127 (24.1%) | 76 (19.7%) | <0.001 | 0.003 |
| Hypertension (Yes) | 1498 (85.4%) | 67 (31.6) | 569 (90.5%) | 507 (96%) | 355 (92.8%) | <0.001 | <0.001 |
| Dyslipidemia (Yes) | 1170 (66.7%) | 52 (24.5) | 467 (74.2%) | 399 (75.6%) | 252 (65.5%) | <0.001 | <0.001 |
| BMI (kg/m$^2$) | 28.3 (5.21) | 27.7 (4.5) | 29.2 (4.7) | 28.5 (5.5) | 26.9 (5.6) | <0.001 | 0.026 |
| SBP (mmHg) | 138 (20.4) | 131 (17.9) | 138 (18.7) | 141 (18.9) | 138 (24.8) | 0.004 | <0.001 |
| DBP (mmHg) | 79.8 (10.9) | 79 (10.2) | 79.6 (9.7) | 81 (10.2) | 79 (13.5) | <0.001 | 0.792 |
| Pulse pressure (mmHg) | 58.7 (16.7) | 52.1 (12.6) | 58.8 (15.4) | 60.3 (16.8) | 59.9 (19.4) | <0.001 | <0.001 |
| Total Cholesterol (mg/dL) | 180 (39.9) | 203 (32.3) | 184 (37.5) | 177 (38.5) | 164 (42.2) | <0.001 | <0.001 |
| HDL Cholesterol (mg/dL) | 50.1 (14.9) | 52.9 (13.8) | 50.8 (14.8) | 50.1 (15.1) | 46.9 (15.1) | <0.001 | <0.001 |
| LDL Cholesterol (mg/dL) | 104 (34.0) | 127 (30.2) | 107 (31.6) | 100 (33.0) | 91.9 (34.5) | <0.001 | <0.001 |
| Triglycerides (mg/dL) | 141 (84.6) | 116 (71.7) | 145 (83.5) | 143 (88.6) | 146 (85.5) | <0.001 | <0.001 |
| Glucose (mg/dL) | 107 (42.1) | 96 (12.4) | 110 (38.4) | 107 (45.1) | 108 (52.5) | <0.001 | 0.003 |
| Calcium (mmol/L) | 2.33 (0.152) | 2.34 (0.094) | 2.37 (0.119) | 2.34 (0.144) | 2.24 (0.197) | <0.001 | <0.001 |
| Phosphorus (mmol/L) | 1.259 (0.332) | 1.09 (0.158) | 1.06 (0.18) | 1.292 (0.245) | 1.55 (0.442) | <0.001 | <0.001 |
| Sodium (mmol/L) | 141 (2.93) | 141 (2.25) | 141 (2.65) | 141 (2.80) | 139 (3,22) | <0.001 | <0.001 |
| Potassium (mmol/L) | 4.79 (0.60) | 4.4 (0.39) | 4.7 (0.50) | 4.9 (0.54) | 4.9 (0.80) | <0.001 | <0.001 |
| Magnesium (mmol/L) | 0.83 (0.15) | 0.82 (0.08) | 0.80 (0.12) | 0.84 (0.14) | 0.88 (0.22) | <0.001 | <0.001 |
| cIMT | 0.73 (0.14) | 0.70 (0.13) | 0.75 (0.14) | 0.71 (0.14) | 0.72 (0.14) | <0.001 | 0.569 |

Absolute frequency (percentage) and mean (standard deviation) are shown for qualitative and quantitative variables, respectively. Abbreviations: Body Mass Index (BMI), Systolic Blood Pressure (SBP), Diastolic Blood Pressure (DBP), Carotid Intima–Media Thickness (cIMT), Chronic Kidney Disease (CKD).

### 3.2. Correlations of Clinical Variables with cIMT and Mg Levels

Results for the correlation matrix of cIMT and Mg levels with the clinical variables of the study are shown in Table 2. cIMT correlated positively with being male, Caucasian, smoker, diabetic, hypertensive, dyslipidemic and with increased age, BMI, and triglyceride levels. Furthermore, it correlated negatively with levels of HDL cholesterol. Interestingly, no significant correlation was found with total or LDL cholesterol levels. Mg levels positively correlated with the presence of hypertension, and negatively with being Caucasian, diabetic, and with level of BMI. The correlation matrix of the variables with cIMT was calculated for the controls (Supplementary Table S1) as a sensitivity analysis. Some of the significant correlations were lost (due to the lower sample size), but most of the coefficients were similar.

**Table 2.** Bivariate correlation coefficients between cIMT and Magnesium levels with other variables.

| | cIMT | | Magnesium | |
|---|---|---|---|---|
| Variable | r | p-Value | r | p-Value |
| Sex (Male) | 0.223 | <0.001 | −0.003 | 0.906 |
| Race (Caucasian) | 0.083 | 0.001 | −0.079 | 0.001 |
| Age, years | 0.528 | <0.001 | −0.035 | 0.141 |
| Smoker | 0.105 | <0.001 | −0.018 | 0.445 |
| Diabetes | 0.157 | <0.001 | −0.070 | 0.003 |
| Hypertension | 0.137 | <0.001 | 0.053 | 0.028 |
| Dyslipidemia | 0.094 | <0.001 | 0.015 | 0.536 |
| BMI | 0.158 | <0.001 | −0.053 | 0.027 |

Table 2. *Cont.*

|  | cIMT | | Magnesium | |
|---|---|---|---|---|
| Variable | r | p-Value | r | p-Value |
| Total Cholesterol | 0.021 | 0.392 | −0.043 | 0.075 |
| HDL Cholesterol | −0.071 | 0.007 | 0.001 | 0.967 |
| LDL Cholesterol | 0.007 | 0.781 | −0.009 | 0.724 |
| Triglycerides | 0.082 | 0.001 | 0.009 | 0.709 |
| Magnesium | −0.003 | 0.914 | - | - |
| cIMT | - | - | −0.003 | 0.914 |

Pearson's correlation coefficients (r) and *p*-value are shown for each variable. Abbreviations: Body Mass Index (BMI), Carotid Intima–Media Thickness (cIMT).

### 3.3. Association of Clinical Variables with cIMT

The results from the multivariate main effects model, which included all the variables studied, showed that age, being male or a smoker, pulse pressure, potassium, and CKD stage were factors that were significantly associated with cIMT. Mg levels did not reach statistical significance in this model, underlying the fact that Mg had no association with cIMT when considering its effect in the multivariate model independently of the rest of variables (Table 3). However, interestingly, the results from the model including the interactions revealed that Mg levels were significantly associated with cIMT, when interactions with cholesterol levels (total, LDL-C, and HDL-C) and triglycerides are taken into account. Sex, age, tobacco, pulse pressure, CKD stage, and potassium levels were significantly associated with cIMT in this multivariate model (Table 4).

Table 3. Multivariate linear main effects model for cIMT.

|  | All Cohort | | |
|---|---|---|---|
|  | Beta | SE | p-Value |
| Intercept | 0.14 | 0.18 | 0.43 |
| Magnesium | 0.002 | 0.02 | 0.92 |
| Sex, male | 0.034 | 0.007 | <0.00001 |
| Race, Caucasian | 0.006 | 0.02 | 0.74 |
| Age | 0.005 | 0.0003 | <0.00001 |
| Current smoker | 0.02 | 0.007 | 0.01 |
| Diabetes | −0.0002 | 0.01 | 0.98 |
| Hypertension | 0.01 | 0.01 | 0.35 |
| Dyslipidemia | −0.005 | 0.007 | 0.49 |
| BMI | 0.0003 | 0.0007 | 0.65 |
| Pulse pressure | 0.0007 | 0.0002 | 0.0004 |
| CKD Stage, 2–3 | −0.03 | 0.013 | 0.015 |
| CKD Stage, 4–5 | −0.06 | 0.014 | 0.00006 |
| CKD Stage, dialysis | −0.016 | 0.015 | 0.31 |
| Total Cholesterol | 0.0005 | 0.0004 | 0.25 |
| HDL Cholesterol | −0.0006 | 0.0005 | 0.24 |
| LDL Cholesterol | −0.0002 | 0.0004 | 0.57 |
| Triglycerides | −0.00002 | 0.00009 | 0.82 |
| Glucose | 0.0001 | 0.0001 | 0.25 |
| Calcium | −0.001 | 0.006 | 0.82 |
| Phosphorus | −0.002 | 0.004 | 0.59 |
| Sodium | 0.0007 | 0.001 | 0.53 |
| Potassium | 0.014 | 0.006 | 0.016 |

Estimated parameters (beta), standard error (SE) and *p*-value shown for each variable. Abbreviations: Body Mass Index (BMI), Chronic Kidney Disease (CKD).

**Table 4.** Multivariate linear effects model for cIMT with first-order interactions.

|  | All Cohort | | |
|---|---|---|---|
|  | **Beta** | **SE** | ***p*-Value** |
| Intercept | 0.35 | 0.089 | 0.00007 |
| Sex, male | 0.04 | 0.007 | <0.00001 |
| Age (years) | 0.006 | 0.0003 | <0.00001 |
| Current smoker | 0.018 | 0.007 | 0.006 |
| Pulse pressure | 0.0007 | 0.0002 | 0.0002 |
| CKD stage, 2–3 | −0.027 | 0.01 | 0.01 |
| CKD stage, 4–5 | −0.05 | 0.01 | 0.00001 |
| CKD stage, dialysis | −0.014 | 0.01 | 0.26 |
| Magnesium | −0.12 | 0.099 | 0.23 |
| Total Cholesterol | −0.006 | 0.003 | 0.02 |
| HDL Cholesterol | 0.006 | 0.003 | 0.03 |
| LDL Cholesterol | 0.006 | 0.003 | 0.04 |
| Triglycerides | 0.001 | 0.0005 | 0.014 |
| Potassium | 0.013 | 0.006 | 0.016 |
| Interactions |  |  |  |
| Magnesium→Total Cholesterol | 0.008 | 0.003 | 0.011 |
| Magnesium→HDL Cholesterol | −0.007 | 0.003 | 0.016 |
| Magnesium→LDL Cholesterol | −0.007 | 0.003 | 0.03 |
| Magnesium→Triglycerides | −0.0014 | 0.0005 | 0.01 |

Estimated parameters (beta), standard error (SE) and *p*-value shown for each significant variable considering interactions.

### 3.4. Visualization of the Association of the Interaction of Mg and Plasma Lipids with cIMT

To better visualize the interactions between plasma lipids and Mg, we performed an interaction graph for continuous data. As seen in Figure 1a, cIMT predicted values were higher when triglycerides levels were high and Mg levels were low. On the contrary, when Mg levels were higher than 1 mmol/L, the relationship between high triglyceride levels and increased cIMT disappeared. A similar tendency was seen when LDL-C levels, Mg, and cIMT were plotted (Figure 1b). The interaction between HDL-C and Mg showed that the protective effect of high HDL-C levels disappears when Mg levels are low (Figure 1c). Finally, a paradoxical effect of total cholesterol levels is depicted, predicting higher cIMT when both total cholesterol and Mg are high (Figure 1d).

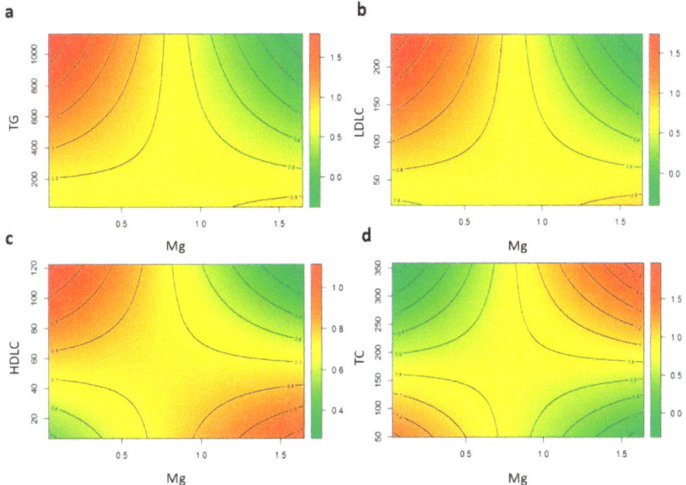

**Figure 1.** Interaction Graphs for Continuous Data. (**a**) Predicted cIMT with different levels of triglycerides (TG) and magnesium (Mg). (**b**) Predicted cIMT with different levels of LDL cholesterol (LDLC) and magnesium. (**c**) Predicted cIMT with different levels of HDL cholesterol (HDLC) and magnesium. (**d**) Predicted cIMT with different levels of total cholesterol (TC) and magnesium.

## 4. Discussion

The present study identifies, for the first time, the role of Mg as an effect modifier on the association of lipid parameters with cIMT in CKD patients. Thus, high LDL and triglyceride levels affect cIMT only when Mg levels are low. In the same model, high HDL levels associate with lower cIMT only when Mg levels are high. This result could be behind some of the paradoxical effects of lipid parameters on atherosclerosis in renal patients, in which Mg levels are often modified by the disease.

Hypertriglyceridemia and high cholesterol levels are related to atherosclerosis [2,41], and current mathematical models estimate that lipid-lowering therapies could avoid a substantial number of cardiovascular events [42]. Among the different dietary supplements proposed for lipid-lowering treatment, Mg has been used not only to prevent cardiovascular events in patients suffering from chronic kidney disease [43], but also for metabolic syndrome treatment [44] as well as to improve cardiovascular health in overweight and obese people [45]. In renal patients, however, the relationship between lipid parameters and atherosclerosis is not so clear. Thus, high LDL cholesterol levels have sometimes been found to be unrelated to cardiovascular disease in CKD patients. In other cases, the phenomenon called reverse causality has also been shown in this population, finding an association of lower LDL levels with higher cardiovascular risk [46]. Although malnutrition has been shown to partially explain this paradoxical effect, hypomagnesemia, which is associated with malnutrition, could be part of the exact mechanism [47]. The same different effect on cardiovascular risk has also been observed with HDL, which seems to lose its association with cardiovascular risk as renal function declines [48]. This report also shows that the interaction between Mg and triglycerides has an impact on cIMT. Pioneering work from the ARIC study found a relationship between serum Mg levels and cIMT that was maintained after adjusting for age, but that disappeared after adjusting for other risk factors [49]. Indeed, and in agreement with previous literature reports, Mg has been recently identified as a possible independent risk factor for carotid atherosclerosis [50], and a meta-analysis showed an inverse relationship between circulating and dietary Mg and cardiovascular risk [30]. As far as we know, none of the works that identified a relationship between serum Mg levels and cIMT have considered the impact of Mg–lipid interaction on cIMT.

One of the possible explanations for our results is that Mg is an important antioxidant, and its deficiency has been related to an increase in oxidative stress biomarkers [51], and to an increase in lipid peroxidation [52,53]. High levels of these molecules have been associated with cardiovascular events [54]. A proposed mechanism of its atherogenic potential includes the recruitment and retention of macrophages, the secretion of cytokines by macrophages and endothelial cells, the proliferation of smooth muscle cells, and lymphocyte chemotaxis [55]. Therefore, according to our findings, high Mg levels could mitigate lipid peroxidation even with high plasma lipid profiles, thereby reducing its impact on cIMT. Another possible mechanism to explain our results could be the prevention of the oxidation of LDL by Mg. There is a clear relationship between the levels of oxidized LDL and atheromatous disease [56], and CKD patients showed increased levels of oxidized LDL [57]. In addition, it has been reported that oxidized LDL concentrations are higher in subjects with low Mg levels [58,59]. Thus, future studies should aim to correlate Mg levels with oxidized LDL levels and atheromatous disease in CKD patients. Finally, it is worth mentioning that many studies have proposed an active role of Mg in retarding vascular calcification of the medial arterial layer. However, this process is not related to lipid peroxidation, but to the binding of phosphate, decreasing calcium phosphate deposition [60], and regulating smooth muscle cells transdifferentiation [61].

This study has some limitations. First, its cross-sectional nature impeded us from making predictions—only associations. Furthermore, and in order to obtain a relatively wide degree of variation in Mg levels, a population with controls, different stages of CKD, and hemodialysis were included. This last group is especially sensible for the current analysis due to its susceptibility to present atherosclerotic disease and a plethora of additional pro-atherogenic factors (such as dialysis vintage, type of dialysis, etc.) that cannot be included in the whole analysis as possible confounders. However, CKD stage was included in the logistic model and did not reach statistical significance, allowing for the extrapolation of our results to our whole population. The second limitation is that we only measured plasma Mg, whose levels could differ from intracellular Mg [62,63] (which is responsible for the majority of reactions involving it as coenzyme [31]), and from levels obtained by the twenty-four hour excretion of magnesium in urine, which could more accurately reflect Mg balance. However, on the one hand, plasma lipid peroxidation is the possible mechanism by which Mg could be involved in the modification of cardiovascular events [54], and therefore serum Mg levels could better correlate with cIMT than tissue levels. On the other hand, Mg excretion in urine should be carefully taken in CKD patients [64]; consequently, for the current study, plasma Mg is a valid quantification method. As for the strengths of the current work, we would like to hallmark the relatively large cohort, which allowed us to adjust the regression models by ten different variables.

In summary, the current study shows an interaction of Mg with lipid parameters that modifies its effect on cITM. This result could shed some light on the effect of lipid levels on atherosclerosis in conditions in which Mg levels are modified and highlight that the important effect of this mineral on cardiovascular physiology could be more complex than initially thought. Although this analysis suggests around 1 nmol/L as the cutoff point of serum Mg that could influence the deleterious effect of lipids on atherosclerosis, further specifically designed studies are needed. In conclusion, Mg levels should be measured together with lipid parameters in order to assess the risk of atherosclerosis, especially in situations in which alterations in serum Mg are expected.

**Supplementary Materials:** The following are available online at http://www.mdpi.com/2072-6643/12/9/2631/s1, Table S1: Bivariate correlation coefficients between cIMT and Magnesium levels with other variables in control volunteers.

**Author Contributions:** Conceptualization, S.C., E.F., and J.M.V.; methodology, M.I. and M.M.-A.; formal analysis, S.C., M.B.-L., and M.M.-A.; data curation, M.B.; writing—original draft preparation, S.C.; writing—review and editing, S.C. and J.M.V.; funding acquisition, E.F. and J.M.V. All authors have read and agreed to the published version of the manuscript.

**Funding:** This research was funded by Instituto de salud Carlos III; grants PI18/00610 and ISCIII-RETIC REDinREN RD016/009.

**Conflicts of Interest:** The authors declare no conflict of interest.

## References

1. Timmis, A.; Townsend, N.; Gale, C.; Grobbee, R.; Maniadakis, N.; Flather, M.; Wilkins, E.; Wright, L.; Vos, R.; Bax, J.; et al. European Society of Cardiology: Cardiovascular Disease Statistics 2017. *Eur. Heart J.* **2018**, *39*, 508–579. [CrossRef]
2. Ference, B.A.; Graham, I.; Tokgozoglu, L.; Catapano, A.L. Impact of Lipids on Cardiovascular Health: JACC Health Promotion Series. *J. Am. Coll. Cardiol.* **2018**, *72*, 1141–1156. [CrossRef] [PubMed]
3. Lewington, S.; Whitlock, G.; Clarke, R.; Sherliker, P.; Emberson, J.; Halsey, J.; Qizilbash, N.; Peto, R.; Collins, R. Blood cholesterol and vascular mortality by age, sex, and blood pressure: A meta-analysis of individual data from 61 prospective studies with 55,000 vascular deaths. *Lancet* **2007**, *370*, 1829–1839.
4. Ference, B.A.; Ginsberg, H.N.; Graham, I.; Ray, K.K.; Packard, C.J.; Bruckert, E.; Hegele, R.A.; Krauss, R.M.; Raal, F.J.; Schunkert, H.; et al. Low-density lipoproteins cause atherosclerotic cardiovascular disease. 1. Evidence from genetic, epidemiologic, and clinical studies. A consensus statement from the European Atherosclerosis Society Consensus Panel. *Eur. Heart J.* **2017**, *38*, 2459–2472. [CrossRef]
5. Toth, P.P.; Barter, P.J.; Rosenson, R.S.; Boden, W.E.; Chapman, M.J.; Cuchel, M.; D'Agostino, R.B., Sr.; Davidson, M.H.; Davidson, W.S.; Heinecke, J.W.; et al. High-density lipoproteins: A consensus statement from the National Lipid Association. *J. Clin. Lipidol.* **2013**, *7*, 484–525. [CrossRef]
6. Varbo, A.; Benn, M.; Tybjærg-Hansen, A.; Jørgensen, A.B.; Frikke-Schmidt, R.; Nordestgaard, B.G. Remnant cholesterol as a causal risk factor for ischemic heart disease. *J. Am. Coll. Cardiol.* **2013**, *61*, 427–436. [CrossRef]
7. Miller, M.; Stone, N.J.; Ballantyne, C.; Bittner, V.; Criqui, M.H.; Ginsberg, H.N.; Goldberg, A.C.; Howard, W.J.; Jacobson, M.S.; Kris-Etherton, P.M.; et al. Triglycerides and cardiovascular disease: A scientific statement from the American Heart Association. *Circulation* **2011**, *123*, 2292–2333. [CrossRef]
8. Sarnak, M.J.; Levey, A.S.; Schoolwerth, A.C.; Coresh, J.; Culleton, B.; Hamm, L.L.; McCullough, P.A.; Kasiske, B.L.; Kelepouris, E.; Klag, M.J.; et al. Kidney disease as a risk factor for development of cardiovascular disease: A statement from the American Heart Association Councils on Kidney in Cardiovascular Disease, High Blood Pressure Research, Clinical Cardiology, and Epidemiology and Prevention. *Circulation* **2003**, *108*, 2154–2169. [CrossRef]
9. Ferro, C.J.; Mark, P.B.; Kanbay, M.; Sarafidis, P.; Heine, G.H.; Rossignol, P.; Massy, Z.A.; Mallamaci, F.; Valdivielso, J.M.; Malyszko, J.; et al. Lipid management in patients with chronic kidney disease. *Nat. Rev. Nephrol.* **2018**, *14*, 727–749. [CrossRef]
10. Arroyo, D.; Betriu, A.; Martinez-Alonso, M.; Vidal, T.; Valdivielso, J.M.; Fernández, E. Observational multicenter study to evaluate the prevalence and prognosis of subclinical atheromatosis in a Spanish chronic kidney disease cohort: Baseline data from the NEFRONA study. *BMC Nephrol.* **2014**, *15*, 168. [CrossRef]
11. Bermudez-Lopez, M.; Forne, C.; Amigo, N.; Bozic, M.; Arroyo, D.; Bretones, T.; Alonso, N.; Cambray, S.; Del Pino, M.D.; Mauricio, D.; et al. An in-depth analysis shows a hidden atherogenic lipoprotein profile in non-diabetic chronic kidney disease patients. *Expert Opin. Ther. Targets* **2019**, *23*, 619–630. [CrossRef] [PubMed]
12. Florens, N.; Calzada, C.; Lyasko, E.; Juillard, L.; Soulage, C.O. Modified Lipids and Lipoproteins in Chronic Kidney Disease: A New Class of Uremic Toxins. *Toxins* **2016**, *8*, 376. [CrossRef] [PubMed]
13. Hegele, R.A.; Tsimikas, S. Lipid-Lowering Agents. *Circ. Res.* **2019**, *124*, 386–404. [CrossRef]
14. Yokoyama, Y.; Levin, S.M.; Barnard, N.D. Association between plant-based diets and plasma lipids: A systematic review and meta-analysis. *Nutr. Rev.* **2017**, *75*, 683–698. [CrossRef]
15. Schwingshackl, L.; Hoffmann, G. Comparison of effects of long-term low-fat vs high-fat diets on blood lipid levels in overweight or obese patients: A systematic review and meta-analysis. *J. Acad. Nutr. Diet.* **2013**, *113*, 1640–1661. [CrossRef]
16. Jorat, M.V.; Tabrizi, R.; Mirhosseini, N.; Lankarani, K.B.; Akbari, M.; Heydari, S.T.; Mottaghi, R.; Asemi, Z. The effects of coenzyme Q10 supplementation on lipid profiles among patients with coronary artery disease: A systematic review and meta-analysis of randomized controlled trials. *Lipids Health Dis.* **2018**, *17*, 230. [CrossRef]
17. Bianconi, V.; Mannarino, M.R.; Sahebkar, A.; Cosentino, T.; Pirro, M. Cholesterol-Lowering Nutraceuticals Affecting Vascular Function and Cardiovascular Disease Risk. *Curr. Cardiol. Rep.* **2018**, *20*, 53. [CrossRef]

18. Martini, D.; Chiavaroli, L.; González-Sarrías, A.; Bresciani, L.; Palma-Duran, S.A.; Dall'Asta, M.; Deligiannidou, G.E.; Massaro, M.; Scoditti, E.; Combet, E.; et al. Impact of Foods and Dietary Supplements Containing Hydroxycinnamic Acids on Cardiometabolic Biomarkers: A Systematic Review to Explore Inter-Individual Variability. *Nutrients* **2019**, *11*, 1805. [CrossRef]
19. Rayssiguier, Y. Magnesium, lipids and vascular diseases. Experimental evidence in animal models. *Magnesium* **1986**, *5*, 182–190.
20. Kaesler, N.; Goettsch, C.; Weis, D.; Schurgers, L.; Hellmann, B.; Floege, J.; Kramann, R. Magnesium but not nicotinamide prevents vascular calcification in experimental uraemia. *Nephrol. Dial. Transplant.* **2020**, *35*, 65–73. [CrossRef]
21. Iskra, M.; Patelski, J.; Majewski, W. Concentrations of calcium, magnesium, zinc and copper in relation to free fatty acids and cholesterol in serum of atherosclerotic men. *J. Trace Elem. Electrolytes Health Dis.* **1993**, *7*, 185–188.
22. Kirsten, R.; Heintz, B.; Nelson, K.; Sieberth, H.G.; Oremek, G.; Hasford, J.; Speck, U. Magnesium pyridoxal 5-phosphate glutamate reduces hyperlipidaemia in patients with chronic renal insufficiency. *Eur. J. Clin. Pharmacol.* **1988**, *34*, 133–137. [CrossRef]
23. Singh, R.B.; Rastogi, S.S.; Mani, U.V.; Seth, J.; Devi, L. Does dietary magnesium modulate blood lipids? *Biol. Trace Elem. Res.* **1991**, *30*, 59–64. [CrossRef]
24. De Oliveira Otto, M.C.; Alonso, A.; Lee, D.H.; Delclos, G.L.; Jenny, N.S.; Jiang, R.; Lima, J.A.; Symanski, E.; Jacobs, D.R., Jr.; Nettleton, J.A. Dietary micronutrient intakes are associated with markers of inflammation but not with markers of subclinical atherosclerosis. *J. Nutr.* **2011**, *141*, 1508–1515. [CrossRef]
25. Hashimoto, T.; Hara, A.; Ohkubo, T.; Kikuya, M.; Shintani, Y.; Metoki, H.; Inoue, R.; Asayama, K.; Kanno, A.; Nakashita, M.; et al. Serum magnesium, ambulatory blood pressure, and carotid artery alteration: The Ohasama study. *Am. J. Hypertens.* **2010**, *23*, 1292–1298. [CrossRef]
26. Sun, X.; Zhuang, X.; Huo, M.; Feng, P.; Zhang, S.; Zhong, X.; Zhou, H.; Guo, Y.; Hu, X.; Du, Z.; et al. Serum magnesium and the prevalence of peripheral artery disease: The Atherosclerosis Risk in Communities (ARIC) study. *Atherosclerosis* **2019**, *282*, 196–201. [CrossRef]
27. Menez, S.; Ding, N.; Grams, M.E.; Lutsey, P.L.; Heiss, G.; Folsom, A.R.; Selvin, E.; Coresh, J.; Jaar, B.G.; Matsushita, K. Serum magnesium, bone-mineral metabolism markers and their interactions with kidney function on subsequent risk of peripheral artery disease: The Atherosclerosis Risk in Communities Study. *Nephrol. Dial. Transplant.* **2020**. [CrossRef]
28. Rosique-Esteban, N.; Guasch-Ferré, M.; Hernández-Alonso, P.; Salas-Salvadó, J. Dietary Magnesium and Cardiovascular Disease: A Review with Emphasis in Epidemiological Studies. *Nutrients* **2018**, *10*, 168. [CrossRef]
29. DiNicolantonio, J.J.; Liu, J.; O'Keefe, J.H. Magnesium for the prevention and treatment of cardiovascular disease. *Open Heart* **2018**, *5*, e000775. [CrossRef]
30. Del Gobbo, L.C.; Imamura, F.; Wu, J.H.; de Oliveira Otto, M.C.; Chiuve, S.E.; Mozaffarian, D. Circulating and dietary magnesium and risk of cardiovascular disease: A systematic review and meta-analysis of prospective studies. *Am. J. Clin. Nutr.* **2013**, *98*, 160–173. [CrossRef]
31. Severino, P.; Netti, L.; Mariani, M.V.; Maraone, A.; D'Amato, A.; Scarpati, R.; Infusino, F.; Pucci, M.; Lavalle, C.; Maestrini, V.; et al. Prevention of Cardiovascular Disease: Screening for Magnesium Deficiency. *Cardiol. Res. Pract.* **2019**, *2019*, 4874921. [CrossRef]
32. Tzanakis, I.P.; Stamataki, E.E.; Papadaki, A.N.; Giannakis, N.; Damianakis, N.E.; Oreopoulos, D.G. Magnesium retards the progress of the arterial calcifications in hemodialysis patients: A pilot study. *Int. Urol. Nephrol.* **2014**, *46*, 2199–2205. [CrossRef]
33. Woods, K.L.; Fletcher, S.; Roffe, C.; Haider, Y. Intravenous magnesium sulphate in suspected acute myocardial infarction: Results of the second Leicester Intravenous Magnesium Intervention Trial (LIMIT-2). *Lancet* **1992**, *339*, 1553–1558. [CrossRef]
34. ISIS-4. A randomised factorial trial assessing early oral captopril, oral mononitrate, and intravenous magnesium sulphate in 58,050 patients with suspected acute myocardial infarction. ISIS-4 (Fourth International Study of Infarct Survival) Collaborative Group. *Lancet* **1995**, *345*, 669–685. [CrossRef]

35. Junyent, M.; Martínez Alonso, M.; Borràs, M.; Betriu i Bars, M.; Coll, B.; Marco Mayayo, M.P.; Sarró, F.; Valdivielso Revilla, J.M.; Fernández i Giráldez, E. Usefulness of imaging techniques and novel biomarkers in the prediction of cardiovascular risk in patients with chronic kidney disease in Spain: The NEFRONA project. *Nefrologia* **2010**, *30*, 119–126.
36. Junyent, M.; Martínez, M.; Borràs, M.; Coll, B.; Valdivielso, J.M.; Vidal, T.; Sarró, F.; Roig, J.; Craver, L.; Fernández, E. Predicting cardiovascular disease morbidity and mortality in chronic kidney disease in Spain. The rationale and design of NEFRONA: A prospective, multicenter, observational cohort study. *BMC Nephrol.* **2010**, *11*, 14. [CrossRef]
37. Betriu, A.; Martinez-Alonso, M.; Arcidiacono, M.V.; Cannata-Andia, J.; Pascual, J.; Valdivielso, J.M.; Fernández, E. Prevalence of subclinical atheromatosis and associated risk factors in chronic kidney disease: The NEFRONA study. *Nephrol. Dial. Transplant.* **2014**, *29*, 1415–1422. [CrossRef]
38. Gracia, M.; Betriu, À.; Martínez-Alonso, M.; Arroyo, D.; Abajo, M.; Fernández, E.; Valdivielso, J.M. Predictors of Subclinical Atheromatosis Progression over 2 Years in Patients with Different Stages of CKD. *Clin. J. Am. Soc. Nephrol.* **2016**, *11*, 287–296. [CrossRef]
39. Stein, J.H.; Korcarz, C.E.; Hurst, R.T.; Lonn, E.; Kendall, C.B.; Mohler, E.R.; Najjar, S.S.; Rembold, C.M.; Post, W.S. Use of carotid ultrasound to identify subclinical vascular disease and evaluate cardiovascular disease risk: A consensus statement from the American Society of Echocardiography Carotid Intima-Media Thickness Task Force. Endorsed by the Society for Vascular Medicine. *J. Am. Soc. Echocardiogr.* **2008**, *21*, 93–111.
40. Touboul, P.J.; Hennerici, M.G.; Meairs, S.; Adams, H.; Amarenco, P.; Desvarieux, M.; Ebrahim, S.; Fatar, M.; Hernandez, R.H.; Kownator, S.; et al. Mannheim intima-media thickness consensus. *Cerebrovasc. Dis.* **2004**, *18*, 346–349. [CrossRef]
41. Peng, J.; Luo, F.; Ruan, G.; Peng, R.; Li, X. Hypertriglyceridemia and atherosclerosis. *Lipids Health Dis.* **2017**, *16*, 233. [CrossRef] [PubMed]
42. Cannon, C.P.; Khan, I.; Klimchak, A.C.; Sanchez, R.J.; Sasiela, W.J.; Massaro, J.M.; D'Agostino, R.B.; Reynolds, M.R. Simulation of impact on cardiovascular events due to lipid-lowering therapy intensification in a population with atherosclerotic cardiovascular disease. *Am. Heart J.* **2019**, *216*, 30–41. [CrossRef]
43. Leenders, N.H.J.; Vervloet, M.G. Magnesium: A Magic Bullet for Cardiovascular Disease in Chronic Kidney Disease? *Nutrients* **2019**, *11*, 455. [CrossRef]
44. Guerrero-Romero, F.; Jaquez-Chairez, F.O.; Rodríguez-Morán, M. Magnesium in metabolic syndrome: A review based on randomized, double-blind clinical trials. *Magnes. Res.* **2016**, *29*, 146–153. [CrossRef] [PubMed]
45. Joris, P.J.; Plat, J.; Bakker, S.J.; Mensink, R.P. Long-term magnesium supplementation improves arterial stiffness in overweight and obese adults: Results of a randomized, double-blind, placebo-controlled intervention trial. *Am. J. Clin. Nutr.* **2016**, *103*, 1260–1266. [CrossRef]
46. Massy, Z.A.; de Zeeuw, D. LDL cholesterol in CKD–to treat or not to treat? *Kidney Int.* **2013**, *84*, 451–456. [CrossRef]
47. Jahnen-Dechent, W.; Ketteler, M. Magnesium basics. *Clin. Kidney J.* **2012**, *5* (Suppl. 1), i3–i14. [CrossRef]
48. Zewinger, S.; Speer, T.; Kleber, M.E.; Scharnagl, H.; Woitas, R.; Lepper, P.M.; Pfahler, K.; Seiler, S.; Heine, G.H.; März, W.; et al. HDL cholesterol is not associated with lower mortality in patients with kidney dysfunction. *J. Am. Soc. Nephrol.* **2014**, *25*, 1073–1082. [CrossRef]
49. Ma, J.; Folsom, A.R.; Melnick, S.L.; Eckfeldt, J.H.; Sharrett, A.R.; Nabulsi, A.A.; Hutchinson, R.G.; Metcalf, P.A. Associations of serum and dietary magnesium with cardiovascular disease, hypertension, diabetes, insulin, and carotid arterial wall thickness: The ARIC study. Atherosclerosis Risk in Communities Study. *J. Clin. Epidemiol.* **1995**, *48*, 927–940. [CrossRef]
50. Rodríguez-Ortiz, M.E.; Gómez-Delgado, F.; de Larriva, A.P.A.; Canalejo, A.; Gómez-Luna, P.; Herencia, C.; López-Moreno, J.; Rodríguez, M.; López-Miranda, J.; Almadén, Y. Serum Magnesium is associated with Carotid Atherosclerosis in patients with high cardiovascular risk (CORDIOPREV Study). *Sci. Rep.* **2019**, *9*, 8013. [CrossRef]
51. Zheltova, A.A.; Kharitonova, M.V.; Iezhitsa, I.N.; Spasov, A.A. Magnesium deficiency and oxidative stress: An update. *Biomedicine* **2016**, *6*, 20. [CrossRef] [PubMed]
52. Rayssiguier, Y.; Gueux, E.; Bussière, L.; Durlach, J.; Mazur, A. Dietary magnesium affects susceptibility of lipoproteins and tissues to peroxidation in rats. *J. Am. Coll. Nutr.* **1993**, *12*, 133–137. [CrossRef] [PubMed]

53. Scibior, A.; Gołębiowska, D.; Niedźwiecka, I. Magnesium can protect against vanadium-induced lipid peroxidation in the hepatic tissue. *Oxid. Med. Cell. Longev.* **2013**, *2013*, 802734. [CrossRef] [PubMed]
54. Walter, M.F.; Jacob, R.F.; Bjork, R.E.; Jeffers, B.; Buch, J.; Mizuno, Y.; Mason, R.P. Circulating lipid hydroperoxides predict cardiovascular events in patients with stable coronary artery disease: The PREVENT study. *J. Am. Coll. Cardiol.* **2008**, *51*, 1196–1202. [CrossRef]
55. Feingold, K.R.; Grunfeld, C. Introduction to Lipids and Lipoproteins. In *Endotext*; Feingold, K.R., Anawalt, B., Boyce, A., Eds.; MDText.com, Inc.: South Dartmouth, MA, USA, 2 February 2018.
56. Gao, S.; Liu, J. Association between circulating oxidized low-density lipoprotein and atherosclerotic cardiovascular disease. *Chronic Dis. Transl. Med.* **2017**, *3*, 89–94. [CrossRef]
57. Vaziri, N.D. Role of dyslipidemia in impairment of energy metabolism, oxidative stress, inflammation and cardiovascular disease in chronic kidney disease. *Clin. Exp. Nephrol.* **2014**, *18*, 265–268. [CrossRef]
58. Cocate, P.G.; Natali, A.J.; de Oliveira, A.; Longo, G.Z.; Rita de Cássia, G.A.; Maria do Carmo, G.P.; dos Santos, E.C.; Buthers, J.M.; de Oliveira, L.L.; Hermsdorff, H.H.M. Fruit and vegetable intake and related nutrients are associated with oxidative stress markers in middle-aged men. *Nutrition* **2014**, *30*, 660–665. [CrossRef]
59. Wegner, M.; Araszkiewicz, A.; Zozulińska-Ziółkiewicz, D.; Wierusz-Wysocka, B.; Pioruńska-Mikołajczak, A.; Pioruńska-Stolzmann, M. The relationship between concentrations of magnesium and oxidized low density lipoprotein and the activity of platelet activating factor acetylhydrolase in the serum of patients with type 1 diabetes. *Magnes. Res.* **2010**, *23*, 97–104.
60. Louvet, L.; Büchel, J.; Steppan, S.; Passlick-Deetjen, J.; Massy, Z.A. Magnesium prevents phosphate-induced calcification in human aortic vascular smooth muscle cells. *Nephrol. Dial. Transplant.* **2013**, *28*, 869–878. [CrossRef]
61. Braake, A.D.; Shanahan, C.M.; Baaij, J.F.H. Magnesium Counteracts Vascular Calcification Passive Interference or Active Modulation? *Arterioscler. Thromb. Vasc. Biol.* **2017**, *37*, 1431–1445. [CrossRef]
62. De Lourdes Lima, M.; Cruz, T.; Rodrigues, L.E.; Bomfim, O.; Melo, J.; Correia, R.; Porto, M.; Cedro, A.; Vicente, E. Serum and intracellular magnesium deficiency in patients with metabolic syndrome—Evidences for its relation to insulin resistance. *Diabetes Res. Clin. Pract.* **2009**, *83*, 257–262. [CrossRef] [PubMed]
63. Shah, S.A.; Clyne, C.A.; Henyan, N.; Migeed, M.; Yarlagadda, R.; Silver, B.B.; Kluger, J.; White, C.M. Impact of magnesium sulfate on serum magnesium concentrations and intracellular electrolyte concentrations among patients undergoing radio frequency catheter ablation. *Conn. Med.* **2008**, *72*, 261–265. [PubMed]
64. Elin, J.R. Assessment of magnessium status for diagnosis and therapy. *Magnes. Res.* **2010**, *23*, 194–198.

© 2020 by the authors. Licensee MDPI, Basel, Switzerland. This article is an open access article distributed under the terms and conditions of the Creative Commons Attribution (CC BY) license (http://creativecommons.org/licenses/by/4.0/).

Article

# Magnesium Metabolism in Chronic Alcohol-Use Disorder: Meta-Analysis and Systematic Review

Flora O. Vanoni [1], Gregorio P. Milani [2,3,4,*], Carlo Agostoni [2,3], Giorgio Treglia [5,6], Pietro B. Faré [7], Pietro Camozzi [8], Sebastiano A. G. Lava [9], Mario G. Bianchetti [1] and Simone Janett [8]

- [1] Family Medicine Institute, Faculty of Biomedical Science, Università della Svizzera Italiana, 6900 Lugano, Switzerland; flora.vanoni@hotmail.com (F.O.V.); mario.bianchetti@usi.ch (M.G.B.)
- [2] Pediatric Unit, Fondazione IRCCS Ca' Granda Ospedale Maggiore Policlinico, 20122 Milan, Italy; carlo.agostoni@unimi.it
- [3] Department of Clinical Sciences and Community Health, Università degli Studi di Milano, 20122 Milan, Italy
- [4] Pediatric Institute of Southern Switzerland, Ospedale San Giovanni, 6500 Bellinzona, Switzerland
- [5] Academic Education, Research and Innovation Area, General Directorate, Ente Ospedaliero Cantonale, 6500 Bellinzona, Switzerland; giorgio.treglia@eoc.ch
- [6] Faculty of Biomedical Science, Università della Svizzera Italiana, 6900 Lugano, Switzerland
- [7] Department of Internal Medicine, Ente Ospedaliero Cantonale, 6600 Locarno, Switzerland; pietroBenedetto.Fare@eoc.ch
- [8] Department of Internal Medicine, Ente Ospedaliero Cantonale, 6500 Bellinzona, Switzerland; pietro.camozzi@eoc.ch (P.C.); Simone.Janett@eoc.ch (S.J.)
- [9] Pediatric Cardiology Unit, Department of Pediatrics, Centre Hospitalier Universitaire Vaudois, University of Lausanne, 1011 Lausanne, Switzerland; webmaster@sebastianolava.ch
- * Correspondence: milani.gregoriop@gmail.com; Tel.: +39-(0)2550-38727; Fax: +39-(0)2550-32918

**Abstract:** Chronic alcohol-use disorder has been imputed as a possible cause of dietary magnesium depletion. The purpose of this study was to assess the prevalence of hypomagnesemia in chronic alcohol-use disorder, and to provide information on intracellular magnesium and on its renal handling. We carried out a structured literature search up to November 2020, which returned 2719 potentially relevant records. After excluding non-significant records, 25 were retained for the final analysis. The meta-analysis disclosed that both total and ionized circulating magnesium are markedly reduced in chronic alcohol-use disorder. The funnel plot and the Egger's test did not disclose significant publication bias. The $I^2$-test demonstrated significant statistical heterogeneity between studies. We also found that the skeletal muscle magnesium content is reduced and the kidney's normal response to hypomagnesemia is blunted. In conclusion, magnesium depletion is common in chronic alcohol-use disorder. Furthermore, the kidney plays a crucial role in the development of magnesium depletion.

**Keywords:** magnesium; hypomagnesemia; depletion; kidney; diet; alcohol-use; electrolytes

## 1. Introduction

Chronic alcohol-use disorder is a frequent, disabling condition [1], which is often associated with electrolyte derangements such as metabolic acidosis or alkalosis, hypokalemia, hyponatremia, hypocalcemia, and hypophosphatemia [2]. Magnesium depletion has also been reported [2].

Magnesium balance is a function of intake, distribution between the extra- and the intracellular compartments, and excretion. Approximately one-third is absorbed in the small bowel. In the extracellular fluid, magnesium is ionized and bound to either small anions or proteins, primarily albumin. However, most of the body's magnesium is inside cells, bound to adenosine triphosphate and other intracellular nucleotides and enzymes, and an integral component of bone mineral. The renal handling of magnesium differs from that of other ions because the major sites of transport are the loop of Henle and the distal convoluted tubule [3,4].

There is a need to summarize the knowledge about the metabolism of magnesium in chronic alcohol-use disorder. Hence, we reviewed the literature addressing the magnesium balance in this condition. The purposes of our study were to assess the intake, the intestinal metabolism, the extra- and intracellular levels, and the kidney handling of magnesium in this condition.

## 2. Materials and Methods

### 2.1. Data Sources and Searches

This review was accomplished following the Preferred Reporting Items for Systematic Reviews and Meta-Analyses recommendations [5]. We carried out a structured [6] literature search up to November 2020 with no date and language limits in the databases of the United States National Library of Medicine, Excerpta Medica, and Web of Science. We also reviewed reference lists of retained articles and our personal files. The search strategy incorporated following terms: ("magnesium" OR "ionized magnesium" OR "intracellular magnesium" OR "hypomagnesemia" OR "magnesium depletion" OR "magnesium deficiency" OR "dietary magnesium" OR "intestinal magnesium absorption") AND ("alcohol" OR "alcohol-use disorder" OR "alcoholism" OR "withdrawal").

### 2.2. Selection Criteria

We retained full-length, observational reports addressing the prevalence of hypomagnesemia, either total or ionized, in subjects affected by chronic alcohol-use disorder. Of interest were reports including 5 or more chronic alcohol-use disorder subjects who were investigated with respect to their blood magnesium concentration. Studies addressing the intracellular magnesium concentration and the kidney magnesium handling were also included. Chronic alcohol-use disorder subjects in treatment with drugs with a potential to cause hypomagnesemia such as aminoglycosides, calcineurin inhibitors, diuretics, or proton-pump inhibitors were excluded [4,7,8].

### 2.3. Definitions

The diagnosis of chronic alcohol-use disorder made in the original reports was retained. The kidney normally reduces the magnesium excretion to very low levels following depletion of this ion [4,9]. Consequently, in subjects with hypomagnesemia and normal blood creatinine, a 24-h magnesium excretion greater than 500 µmol or a fractional magnesium clearance above $2.0 \times 10^{-2}$ indicates renal magnesium wasting [4,9].

### 2.4. Data Extraction

All data were extracted using a pilot-tested sheet. We extracted the following data from the retrieved studies that were used for the pooled analyses: number of patients with chronic alcohol-use disorder and healthy subjects; mean and standard deviation of total and ionized circulating magnesium level in patients with chronic alcohol-use disorder and healthy subjects; number of cases of total hypomagnesemia in patients with chronic alcohol-use disorder; laboratory technique.

### 2.5. Analysis

Results are expressed as mean ± SD or as frequency. The two-tailed t-test was used for non-pooled analysis. Significance was set at $p < 0.05$.

We conducted three different pooled analyses: first, an analysis of observational studies investigating mean difference of total magnesium level in patients with chronic alcohol-use disorder compared to healthy subjects; second, an analysis of observational studies investigating mean difference of ionized circulating magnesium level in patients with chronic alcohol-use disorder compared to healthy subjects; third, an analysis of the prevalence of total hypomagnesemia in chronic alcohol-use disorder. The comparison of mean total and ionized circulating magnesium level in patients with chronic alcohol-use disorder and healthy subjects was expressed as mean difference (MD) from the retrieved articles and a pooled MD was calculated for these parameters. The prevalence of total hypomagnesemia in chronic alcohol-use disorder was calculated from the retrieved studies and a pooled prevalence was calculated through a meta-analysis. A random-effects model was used for statistical pooling of data. Pooled data represented weighted mean or proportion, which were related to the sample size of the individual studies. Pooled results were presented with 95% confidence intervals (95%-CI) and displayed using forest plots.

An $I^2$-test was also performed to test for heterogeneity between studies; this test describes the percentage of variation across studies that is due to heterogeneity rather than chance. The presence of significant heterogeneity was defined as an I-square value of more than 50%. For publication bias, evaluation funnel plots and Egger's test were used when a sufficient number of studies was available. Statistical analyses were performed using StatsDirect statistical software (StatsDirect Ltd., Birkenhead, UK).

## 3. Results
*3.1. Search Results*

The literature search returned 2719 potentially relevant records (Figure 1). After excluding non-significant records, 25 potentially eligible reports were considered [10–34]: 13 from the United States of America, 2 from Greece, 2 from Italy, and one each from Croatia, Denmark, Finland, Israel, Poland, Singapore, Sweden, and the United Kingdom. All reports were written in English. No reports about dietary intake or intestinal metabolism of magnesium were identified.

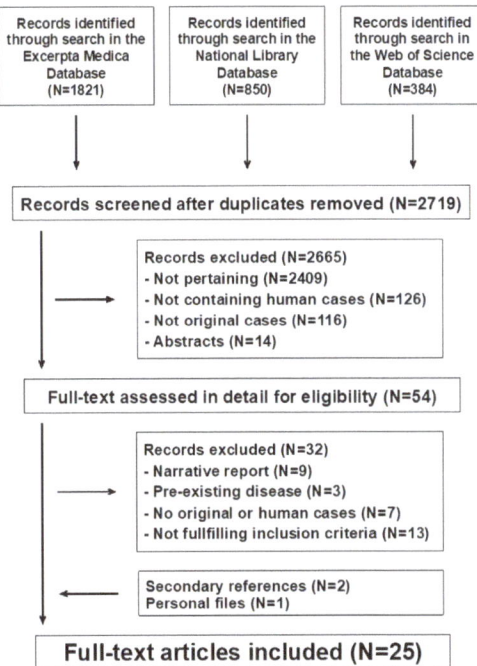

Figure 1. Magnesium metabolism in chronic alcohol-use disorder. Flowchart of the literature search process.

*3.2. Extracellular Magnesium Concentration*

3.2.1. Total Magnesium

Thirteen reports compared the total circulating magnesium level in 546 subjects affected by chronic alcohol-use disorder and 660 healthy controls [11,12,15,21–23,25,28–30,32–34].

Magnesium was determined by colorimetry in eight and by atomic absorption spectrometry in the remaining five reports. Circulating magnesium was found to be significantly reduced in nine of the mentioned studies [11,12,15,21–23,25,29,32].

On the other hand, 12 reports [10,13,14,16–20,24,26,27,31] addressed the prevalence of hypomagnesemia in 538 subjects with chronic alcohol-use disorder. Colorimetry was employed in nine and atomic absorption spectrometry in three reports. Hypomagnesemia was observed in 199 (27%) of the patients.

### 3.2.2. Ionized Magnesium

Three reports [29,32,34] determined by direct potentiometry the ionized magnesium concentration in 171 subjects with alcohol-use disorder and 165 healthy controls. A tendency to ionized hypomagnesemia was noted in all the mentioned reports.

### 3.2.3. Pooled Analyses

Pooled MD of total magnesium level in patients with chronic alcohol-use disorder compared to healthy subjects, taking into account 13 studies (1206 patients, 546 with chronic alcohol-use disorder and 660 healthy subjects), was −0.86 (95%-CI: −1.27 to −0.45) mmol/L (Figure 2).

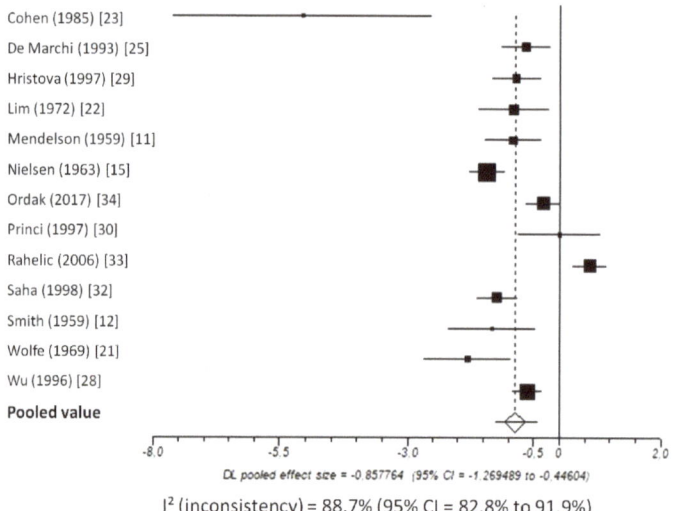

**Figure 2.** Forest plot of individual studies and pooled mean difference of magnesemia between alcohol group and controls, including 95% confidence intervals (95%-CI). The size of the squares is related to the weight of each study. Number of patients and control subjects in each study: Cohen [23]: controls $n$ = 5, patients $n$ = 5; De Marchi [25]: controls $n$ = 42, patients $n$ = 30; Hristova [29]: controls $n$ = 40, patients $n$ = 31; Lim [22]: controls $n$ = 87, patients $n$ = 9; Mendelson [11]: controls $n$ = 18, patients $n$ = 50; Nielsen [15]: controls $n$ = 157, patients $n$ = 48; Ordak [34]: controls $n$ = 50, patients $n$ = 100; Princi [30]: controls $n$ = 14, patients $n$ = 10; Rahelic [33]: controls $n$ = 50, patients $n$ = 105; Saha [32]: controls $n$ = 75, patients $n$ = 40; Smith [12]: controls $n$ = 13, patients $n$ = 12; Wolfe [21]: controls $n$ = 12, patients $n$ = 18; Wu [28]: controls $n$ = 97, patients $n$ = 88.

$I^2$-test was 88.7% (95%-CI: 82.8–91.9) demonstrating significant statistical heterogeneity between studies. No significant publication bias was found through the visual analysis of funnel plot (Figure 3) and Egger's test ($p$ = 0.15).

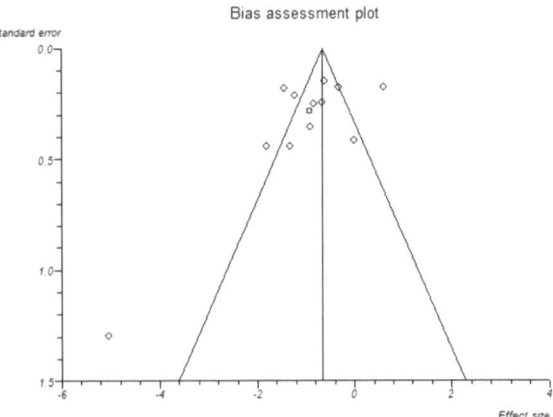

**Figure 3.** Bias assessment plot (funnel plot) about the pooled mean difference of magnesemia between alcohol group and controls. The visual analysis does not show a significant asymmetry and the presence of a publication bias is not demonstrated.

Pooled mean difference of total magnesium level in patients with alcohol-use disorder compared to controls based on the different assays was as follows: atomic absorption spectrometry: −1.60 (95%-CI = −2.67 to −0.05); $I^2$: 93%; colorimetry: −0.79 (95%-CI = −1.10 to −0.48); $I^2$: 76%

Pooled MD of ionized magnesium level in patients with chronic alcohol-use disorder compared to healthy subjects, taking into account 3 studies (336 patients, 171 with chronic alcohol-use disorder and 165 healthy subjects), was −1.03 (95%-CI: −1.49 to −0.57) mmol/L (Figure 4).

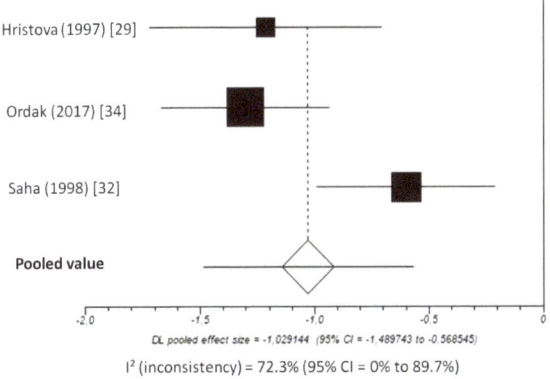

**Figure 4.** Forest plot of individual studies and pooled mean difference of ionized magnesium between alcohol group and controls, including 95% confidence intervals (95%-CI). The size of the squares is related to the weight of each study. Number of patients and control subjects in each study: Hristova [29]: controls $n = 40$, patients $n = 31$; Ordak [34]: controls $n = 50$, patients $n = 100$; Saha [32]: controls $n = 75$, patients $n = 40$.

The $I^2$-test was 72.3% (95%-CI: 0–89.7%), demonstrating significant statistical heterogeneity between studies. No analysis of publication bias was performed due to the limited number of studies available for this analysis.

Pooled prevalence of total hypomagnesemia in patients with chronic alcohol-use disorder, taking into account 12 studies (538 patients), was 44.4% (95%-CI: 31.7 to 57.4) (Figure 5).

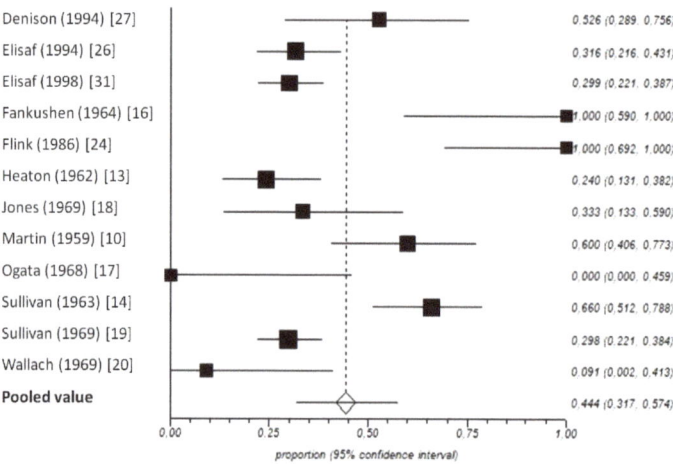

**Figure 5.** Forest plot of individual studies and pooled proportion of hypomagnesemia in patients with alcoholism, including 95% confidence intervals (95%-CI). The size of the squares is related to the weight of each study. Number of subjects in each study: Denison [27]: $n = 19$; Elisaf [26]: $n = 79$; Elisaf [31]: $n = 127$; Fankushen [16]: $n = 7$; Flink [24]: $n = 10$; Heaton [13]: $n = 50$; Jones [18]: $n = 18$; Martin [10]: $n = 30$; Ogata [17]: $n = 6$; Sullivan [14]: $n = 50$; Sullivan [19]: $n = 131$; Wallach [20]: $n = 11$.

The $I^2$-test was 87.7% (95%-CI: 80.5–91.4), demonstrating significant statistical heterogeneity between studies. No significant publication bias was found through the visual analysis of funnel plot (Figure 6) and Egger's test ($p = 0.32$).

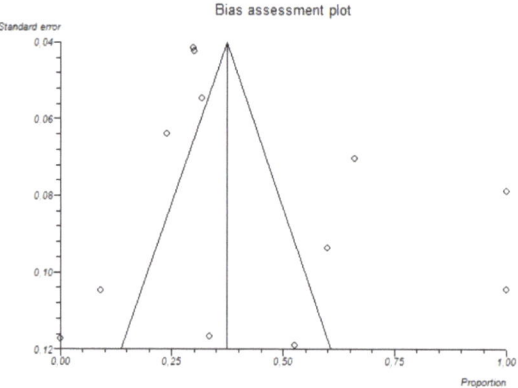

**Figure 6.** Bias assessment plot (funnel plot) showing the pooled proportion of hypomagnesemia in patients with alcoholism. The visual analysis does not show significant asymmetry and the presence of a publication bias is not demonstrated.

## 3.3. Intracellular Magnesium
### 3.3.1. Intra-Erythrocytic Magnesium

The total red blood cell magnesium concentration was investigated in four reports. Two reports analyzed [13,33] the prevalence of low total intra-erythrocytic magnesium level in 69 patients and found a reduced value in just three (4.3%) cases. Two reports [12,34] compared total red blood cell magnesium in chronic alcohol-use disorder subjects and healthy controls. Smith et al. [12] found quite meaningfully reduced content in patients compared to controls ($1.60 \pm 0.38$ mmol/L, $n = 12$, vs. $2.65 \pm 0.17$ mmol/L, $n = 13$, $p < 0.0001$).

More recently, Ordak et al. [34] assessed both total and ionized red blood cell magnesium. The total magnesium level was found to be similar in patients and controls ($1.64 \pm 0.80$ mmol/L, $n = 100$, vs. $1.60 \pm 0.70$ mmol/L, $n = 50$, $p = 0.764$). By contrast, the ionized magnesium level was significantly lower in chronic alcohol-use disorder patients than in healthy controls ($0.36 \pm 0.10$ mmol/L; $n = 100$, vs. $0.71 \pm 0.14$ mmol/L, $n = 50$, $p < 0.0001$).

### 3.3.2. Intra-Lymphocytic Magnesium

The total intra-lymphocytic magnesium concentration was investigated in two small reports. Princi et al. [30] found almost identical ($p = 0.799$) intra-lymphocytic magnesium levels in chronic alcohol-use disorder patients ($56.9 \pm 22.4$ nmol/mg protein; $n = 10$) and healthy controls ($60.2 \pm 35.8$ nmol/mg protein; $n = 14$). Also Cohen et al. [23] found similar ($p = 0.077$) intralymphocytic levels in both groups: $36.6 \pm 2.3$ mmol/kg lymphocyte-dry weight ($n = 5$) versus $42.1 \pm 5.6$ mmol/kg lymphocyte-dry weight ($n = 5$).

### 3.3.3. Bone Magnesium Content

Two small studies addressed the bone magnesium concentration [22,23]. Cohen et al. [23] found a slightly (by 17%) but significantly lower ($p < 0.03$) magnesium trabecular bone content in patients ($175 \pm 23$ mmol/kg; $n = 5$) than in controls ($209 \pm 18$ mmol/kg; $n = 5$). The cortical bone magnesium content, however, was found to be similar in the two groups (no quantitative data provided in the report). Lim et al. [22] included exclusively hypomagnesemic patients and found a slightly (by 9.1%) but significantly lower ($p < 0.03$) content in patients ($101 \pm 5$ mmol/kg; $n = 9$) than in healthy controls ($111 \pm 12$ mmol/kg; $n = 47$).

### 3.3.4. Skeletal Muscle Magnesium Content

The magnesium concentration in skeletal muscle [18] was shown to be reduced ($p < 0.001$) by 16% in subjects with chronic alcohol-use disorder ($30.6 \pm 3.4$ mmol/kg; $n = 20$) as compared with controls ($36.3 \pm 2.3$ mmol/kg; $n = 12$). Similarly, Lim et al. [22] found a reduced skeletal muscle magnesium concentration in 9 out of 10 subjects with chronic alcohol-use disorder.

## 3.4. Kidney Magnesium Handling

Five reports investigated the kidneys' magnesium handling in chronic alcohol-use disorder. Sullivan et al. [14] found a similar urinary magnesium excretion in chronic alcohol-use disorder patients with normal ($n = 11$) and reduced ($n = 24$) total circulating magnesium. De Marchi et al. [25] assessed the fractional renal magnesium clearance in a group of 61 chronic alcohol-use disorders patients with a tendency towards hypomagnesemia and 42 healthy controls. This value was significantly higher ($p < 0.0001$) in patients ($3.3 \pm 1.8 \times 10^{-2}$) than in controls ($1.9 \pm 1.2 \times 10^{-2}$). In the 18 patients with hypomagnesemia, the latter authors found a fractional magnesium clearance above $2.0 \times 10^{-2}$, i.e., features consistent with renal magnesium wasting. Three further studies [16,20,26] investigated a total of 42 chronic alcohol-use disorder patients with hypomagnesemia and found in all cases a 24-h urinary magnesium excretion of more than 0.5 mmol, i.e., features consistent with renal magnesium wasting.

## 4. Discussion

Magnesium plays an essential role in the physiology of the brain, heart, and skeletal muscles; has anti-inflammatory properties; and acts as a calcium-antagonist [3,4]. The results of this meta-analysis and systematic review on magnesium metabolism in chronic alcohol-use disorder may be summarized in three points: (1) both total and ionized circulating magnesium are markedly reduced; (2) skeletal muscle magnesium content is also reduced; (3) the normal response to hypomagnesemia of the kidney is blunted.

Circulating magnesium exists in the ionized state, the most interesting with respect to physiological properties, and in the undissociated form [3,4,35]. Since, in chronic alcohol-use disorder, hypoalbuminemia is common [36], it has been speculated that in this condition low total circulating magnesium is spurious and brought about by coexisting hypoalbuminemia. Our study demonstrates that, in chronic alcohol-use disorder, hypomagnesemia is not spurious.

The total body magnesium content is approximately 1000 mmol. Approximately 1% is in the extracellular fluid compartment, 60–65% is found in bone, 20% is localized in the muscle, and the remainder is found in other tissues [3,4]. Our analysis indicates that hypomagnesemia is habitually linked with skeletal muscle magnesium depletion (in these cells magnesium is vital for glycolysis, formation of cyclic adenosine monophosphate, and reactions that use or produce energy). On the other hand, the results addressing the intra-erythrocytic, intra-lymphocytic, and bone magnesium values appear to be rather inconsistent.

The kidney plays a crucial role in magnesium homeostasis and one can distinguish hypomagnesemia due to poor dietary intake or impaired intestinal metabolism from depletion caused by kidney wasting by measuring the 24-h urinary excretion or the fractional clearance of this ion [4,9]. The present review documents the presence of inappropriately high magnesium excretion in hypomagnesemic patients with chronic alcohol-use disorder. It is therefore concluded that an impaired kidney magnesium handling plays a critical role in the development of magnesium deficiency.

The most relevant strength of the study relates to the comprehensive and exhaustive analysis, which aimed at surveying the entirety of literature on this issue. Furthermore, we did not find a significant publication bias in the pooled analyses. The limitations of the available literature on magnesium metabolism in chronic alcohol-use disorder are the failure to clearly define what is meant by this condition, the inexistent documentation of nutritional status [37,38], the absence of studies addressing the dietary intake and the intestinal metabolism of magnesium [37,38], the paucity of data on common comorbidities such as liver disease or pancreatitis [39,40], and the likely inaccurate evaluation of the effect of drug medication [7].

Bearing in mind the mentioned limitations, future research should take into consideration the following seven points:

1. Definition of alcohol-use disorder: The DSM-5 criteria for alcohol-use disorders, its severity and its various subtypes should be used to enroll patients in future investigations.
2. Nutritional status: Nutritional status is often impaired in alcohol-use disorder [37,38]. In fact, affected subjects often ingest too little of essential nutrients. Furthermore, alcohol may prevent the body from properly absorbing and using nutrients [37,38]. The determination of mid-arm muscle circumference, triceps skinfold thickness, mid-arm circumference, and subscapular skinfold thickness is imperative for the clinical evaluation of nutritional status. The gold standard for the diagnosis of poor intestinal absorption is the determination for fat on 72-h stool collection. Near-infrared analysis is superior because it allows for simultaneous measurement of fecal fat, nitrogen, and carbohydrates in a single sample. Finally, stable isotopes represent a useful technique for determining intestinal absorption of magnesium [41].
3. Dietary magnesium intake: The strategy to collect information on dietary magnesium intake is nowadays an image-assisted dietary food record [42].

4. Liver disease: Alcohol-associated liver disease might play a role in modulating the severity of magnesium depletion [39]. Magnesium metabolism should be investigated separately in patients without and with liver diseases.
5. Pancreatitis: Attacks of pancreatitis are common in chronic alcohol-use disorder. Since hypomagnesemia can occur in this condition, we advise the determination of lipase in future studies [40].
6. Drug-induced magnesium depletion: The reports selected for this review did not include subjects with drug-induced hypomagnesemia. Since an association between hypomagnesemia and proton pump inhibitors was first suspected in 2006 [7], we speculate that these agents were sometimes not specifically imputed as a cause of hypomagnesemia. As a consequence, the possible role of drug-induced hypomagnesemia should be carefully addressed in future studies.
7. Magnesium status: Ideally, total circulating magnesium level should be measured by atomic absorption spectrometry. It is true, however, that many colorimetric methods have been validated. Potentiometric sensors are employed for the measurement of blood-ionized magnesium. Since the selectivity of these sensors for magnesium over calcium is unsatisfactory, the results are adjusted for calcium interference. It has been suggested that magnesium content in red blood cells, peripheral lymphocytes, bones, and skeletal muscles might be an index of magnesium status. However, the available evidence does not support the use of intracellular magnesium levels as an indicator of overall bodily magnesium status [3,4,43].

## 5. Management

No study has so far investigated the management of magnesium depletion in chronic alcohol-use disorder. Hypomagnesemia is mostly mild and presents without or minimal symptoms [2]. In this setting, avoidance of drugs with a potential to induce hypomagnesemia (e.g., the substitution of proton-pump inhibitors with histamine type-2 receptor antagonists respectively of thiazides with potassium-sparing diuretics) is advised. Owing to the effects of alcohol on kidney tubular function, replacement is not expected to be very effective, as in inherited disorders of renal magnesium handling [4]. The restoration of tubular function and magnesium status occurs approximately 4 weeks after alcohol abstinence [25]. After intravenous administration, only a small portion of magnesium is retained because most is excreted in the urine [4,25]. For this reason, repeated oral dosing may be preferred to repair the deficit. Intravenous replacement is required for symptomatic cardiac arrhythmias such as torsade de pointes or neuromuscular irritability [2]. The prescription of magnesium for alcohol withdrawal including delirium tremens has been debated. However, there is so far insufficient evidence to prescribe magnesium in this setting [44].

Future studies are warranted to explore the risk stratification of magnesium depletion and the intracellular magnesium metabolism in chronic alcohol-use disorder.

Magnesium is deemed to be the fifth but forgotten electrolyte [3]. In conclusion, this review demonstrates that magnesium depletion is common in chronic alcohol-use disorder. Contrary to what common sense would suggest, the kidney plays a crucial role in the development of magnesium depletion. We urge that patients with chronic alcohol-use disorder have frequently blood magnesium measured. This suggestion is supported by recent data that implicate magnesium depletion in high mortality in this condition [45,46].

**Author Contributions:** Conceptualization, F.O.V., G.P.M., S.A.G.L., M.G.B. and S.J.; methodology, C.A., G.T., P.B.F., P.C. and M.G.B.; software G.T.; formal analysis, F.O.V., G.T. and M.G.B.; data curation, G.P.M., C.A., G.T., P.B.F., P.C., S.A.G.L. and S.J.; writing—original draft preparation, F.O.V., G.P.M., G.T., S.A.G.L., M.G.B. and S.J., writing—review and editing, C.A., G.T., P.B.F. and P.C.; supervision, M.G.B. All authors have read and agreed to the published version of the manuscript.

**Funding:** This research received no external funding.

**Institutional Review Board Statement:** Not applicable.

**Informed Consent Statement:** Not applicable.

**Data Availability Statement:** Data are available upon the corresponding author after reasonable request.

**Conflicts of Interest:** The authors declare no conflict of interest.

## References

1. Spanagel, R. Alcoholism: A systems approach from molecular physiology to addictive behavior. *Physiol. Rev.* **2009**, *89*, 649–705. [CrossRef] [PubMed]
2. Palmer, B.F.; Clegg, D.J. Electrolyte disturbances in patients with chronic alcohol-use disorder. *N. Engl. J. Med.* **2017**, *377*, 1368–1377. [CrossRef]
3. Elin, R.J. Magnesium metabolism in health and disease. *Dis. Mon.* **1988**, *34*, 161–218. [CrossRef]
4. Agus, Z.S. Mechanisms and causes of hypomagnesemia. *Curr. Opin. Nephrol. Hypertens.* **2016**, *25*, 301–307. [CrossRef]
5. Liberati, A.; Altman, D.G.; Tetzlaff, J.; Mulrow, C.; Gøtzsche, P.C.; Ioannidis, J.P.; Clarke, M.; Devereaux, P.J.; Kleijnen, J.; Moher, D. The PRISMA statement for reporting systematic reviews and meta-analyses of studies that evaluate health care interventions: Explanation and elaboration. *Ann. Intern. Med.* **2009**, *151*, W65–W94. [CrossRef] [PubMed]
6. Krupski, T.L.; Dahm, P.; Fesperman, S.F.; Schardt, C.M. How to perform a literature search. *J. Urol.* **2008**, *179*, 1264–1270. [CrossRef] [PubMed]
7. Janett, S.; Camozzi, P.; Peeters, G.G.; Lava, S.A.; Simonetti, G.D.; Goeggel Simonetti, B.; Bianchetti, M.G.; Milani, G.P. Hypomagnesemia induced by long-term treatment with proton-pump inhibitors. *Gastroenterol. Res. Pract.* **2015**, *2015*, 951768. [CrossRef] [PubMed]
8. Santi, M.; Milani, G.P.; Simonetti, G.D.; Fossali, E.F.; Bianchetti, M.G.; Lava, S.A.G. Magnesium in cystic fibrosis—Systematic review of the literature. *Pediatr. Pulmonol.* **2016**, *51*, 196–202. [CrossRef] [PubMed]
9. Scoble, J.E.; Sweny, P.; Varghese, Z.; Moorhead, J. Investigating hypomagnesaemia. *Lancet* **1987**, *329*, 276. [CrossRef]
10. Martin, H.E.; McCuskey, C., Jr.; Tupikova, N. Electrolyte disturbance in acute alcoholism: With particular reference to magnesium. *Am. J. Clin. Nutr.* **1959**, *7*, 191–196. [CrossRef]
11. Mendelson, J.; Wexler, D.; Kubzansky, P.; Leiderman, H.; Solomon, P. Serum magnesium in delirium tremens and alcoholic hallucinosis. *J. Nerv. Ment. Dis.* **1959**, *128*, 352–357. [CrossRef] [PubMed]
12. Smith, W.O.; Hammarsten, J.F. Intracellular magnesium in delirium tremens and uremia. *Am. J. Med. Sci.* **1959**, *237*, 413–417. [CrossRef] [PubMed]
13. Heaton, F.W.; Pyrah, L.N.; Beresford, C.C.; Bryson, R.W.; Martin, D.F. Hypomagnesæmia in chronic alcoholism. *Lancet* **1962**, *280*, 802–805. [CrossRef]
14. Sullivan, J.F.; Lankford, H.G.; Swartz, M.J.; Farrell, C. Magnesium metabolism in alcoholism. *Am. J. Clin. Nutr.* **1963**, *13*, 297–303. [CrossRef]
15. Nielsen, J. Magnesium metabolism in acute alcoholics. *Dan. Med. Bull.* **1963**, *10*, 225–233. [CrossRef]
16. Fankushen, D.; Raskin, D.; Dimich, A.; Wallach, S. The significance of hypomagnesemia in alcoholic patients. *Am. J. Med.* **1964**, *37*, 802–812. [CrossRef]
17. Ogata, M.; Mendelson, J.H.; Mello, N.K. Electrolyte and osmolality in alcoholics during experimentally induced intoxication. *Psychosom. Med.* **1968**, *30*, 463–488. [CrossRef]
18. Jones, J.E.; Shane, S.R.; Jacobs, W.H.; Flink, E.B. Magnesium balance studies in chronic alcoholism. *Ann. N. Y. Acad. Sci.* **1969**, *162*, 934–946. [CrossRef]
19. Sullivan, J.F.; Wolpert, P.W.; Williams, R.; Egan, J.D. Serum magnesium in chronic alcoholism. *Ann. N. Y. Acad. Sci.* **1969**, *162*, 947–962. [CrossRef]
20. Wallach, S.; Dimich, A. Radiomagnesium turnover studies in hypomagnesemic states. *Ann. N. Y. Acad. Sci.* **1969**, *162*, 963–972. [CrossRef]
21. Wolfe, S.M.; Victor, M. The relationship of hypomagnesemia and alkalosis to alcohol withdrawal symptoms. *Ann. N. Y. Acad. Sci.* **1969**, *162*, 973–984. [CrossRef]
22. Lim, P.; Jacob, E. Magnesium status of alcoholic patients. *Metabolism* **1972**, *21*, 1045–1051. [CrossRef]
23. Cohen, L.; Laor, A.; Kitzes, R. Lymphocyte and bone magnesium in alcohol-associated osteoporosis. *Magnesium* **1985**, *4*, 148–152.
24. Flink, E.B. Magnesium deficiency in alcoholism. *Alcohol. Clin. Exp. Res.* **1986**, *10*, 590–594. [CrossRef] [PubMed]
25. De Marchi, S.; Cecchin, E.; Basile, A.; Bertotti, A.; Nardini, R.; Bartoli, E. Renal tubular dysfunction in chronic alcohol abuse—Effects of abstinence. *N. Engl. J. Med.* **1993**, *329*, 1927–1934. [CrossRef] [PubMed]
26. Elisaf, M.; Merkouropoulos, M.; Tsianos, E.V.; Siamopoulos, K.C. Acid-base and electrolyte abnormalities in alcoholic patients. *Miner. Electrolyte Metab.* **1994**, *20*, 274–281.
27. Denison, H.; Jern, S.; Jagenburg, R.; Wendestam, C.; Wallerstedt, S. Influence of increased adrenergic activity and magnesium depletion on cardiac rhythm in alcohol withdrawal. *Br. Heart J.* **1994**, *72*, 554–560. [CrossRef]
28. Wu, C.; Kenny, M.A. Circulating total and ionized magnesium after ethanol ingestion. *Clin. Chem.* **1996**, *42*, 625–629. [CrossRef] [PubMed]
29. Hristova, E.N.; Rehak, N.N.; Cecco, S.; Ruddel, M.; Herion, D.; Eckardt, M.; Linnoila, M.; Elin, R.J. Serum ionized magnesium in chronic alcoholism: Is it really decreased? *Clin. Chem.* **1997**, *43*, 394–399. [CrossRef] [PubMed]

30. Princi, T.; Artero, M.; Malusà, N.; Uxa, L.; Livia, V.; Reina, G. Serum and intracellular magnesium concentrations in intoxicated chronic alcoholic and control subjects. *Drug Alcohol. Depend.* **1997**, *46*, 119–122. [CrossRef]
31. Elisaf, M.; Bairaktari, E.; Kalaitzidis, R.; Siamopoulos, K.C. Hypomagnesemia in alcoholic patients. *Alcohol. Clin. Exp. Res.* **1998**, *22*, 134. [CrossRef]
32. Saha, H.; Harmoinen, A.; Karvonen, A.L.; Mustonen, J.; Pasternack, A. Serum ionized versus total magnesium in patients with intestinal or liver disease. *Clin. Chem. Lab. Med.* **1998**, *36*, 715–718. [CrossRef]
33. Rahelić, D.; Kujundzić, M.; Romić, Z.; Brkić, K.; Petrovecki, M. Serum concentration of zinc, copper, manganese and magnesium in patients with liver cirrhosis. *Coll. Antropol.* **2006**, *30*, 523–528.
34. Ordak, M.; Maj-Zurawska, M.; Matsumoto, H.; Bujalska-Zadrozny, M.; Kieres-Salomonski, I.; Nasierowski, T.; Muszynska, E.; Wojnar, M. Ionized magnesium in plasma and erythrocytes for the assessment of low magnesium status in alcohol dependent patients. *Drug Alcohol. Depend.* **2017**, *178*, 271–276. [CrossRef] [PubMed]
35. Rooney, M.R.; Rudser, K.D.; Alonso, A.; Harnack, L.; Saenger, A.K.; Lutsey, P.L. Circulating ionized magnesium: Comparisons with circulating total magnesium and the response to magnesium supplementation in a randomized controlled trial. *Nutrients* **2020**, *12*, 263. [CrossRef]
36. Khaderi, S.A. Alcohol and alcoholism. *Clin. Liver Dis.* **2019**, *23*, 1–10. [CrossRef]
37. Green, P.H. Alcohol, nutrition and malabsorption. *Clin. Gastroenterol.* **1983**, *12*, 563–574. [CrossRef]
38. Glória, L.; Cravo, M.; Camilo, M.E.; Resende, M.; Cardoso, J.N.; Oliveira, A.G.; Leitão, C.N.; Mira, F.C. Nutritional deficiencies in chronic alcoholics: Relation to dietary intake and alcohol consumption. *Am. J. Gastroenterol.* **1997**, *92*, 485–489.
39. Kourkoumpetis, T.; Sood, G. Pathogenesis of alcoholic liver disease: An update. *Clin. Liver Dis.* **2019**, *23*, 71–80. [CrossRef]
40. Haber, P.S.; Kortt, N.C. Alcohol use disorder and the gut. *Addiction* **2021**, *116*, 658–667. [CrossRef]
41. Nikaki, K.; Gupte, G.L. Assessment of intestinal malabsorption. *Best Pract. Res. Clin. Gastroenterol.* **2016**, *30*, 225–325. [CrossRef] [PubMed]
42. EFSA Panel on Dietetic Products, Nutrition and Allergies (NDA). Scientific opinion on dietary reference values for magnesium. *EFSA J.* **2015**, *13*, 4186. [CrossRef]
43. Reddy, S.T.; Soman, S.S.; Yee, J. Magnesium balance and measurement. *Adv. Chronic Kidney Dis.* **2018**, *25*, 224–229. [CrossRef] [PubMed]
44. Airagnes, G.; Ducoutumany, G.; Laffy-Beaufils, B.; Le Faou, A.L.; Limosin, F. Alcohol withdrawal syndrome management: Is there anything new? *Rev. Med. Interne* **2019**, *40*, 373–379. [CrossRef]
45. Maguire, D.; Talwar, D.; Burns, A.; Catchpole, A.; Stefanowicz, F.; Robson, G.; Ross, D.P.; Young, D.; Ireland, A.; Forrest, E.; et al. A prospective evaluation of thiamine and magnesium status in relation to clinicopathological characteristics and 1-year mortality in patients with alcohol withdrawal syndrome. *J. Transl. Med.* **2019**, *17*, 384. [CrossRef]
46. Maguire, D.; Ross, D.P.; Talwar, D.; Forrest, E.; Naz Abbasi, H.; Leach, J.P.; Woods, M.; Zhu, L.Y.; Dickson, S.; Kwok, T.; et al. Low serum magnesium and 1-year mortality in alcohol withdrawal syndrome. *Eur. J. Clin. Investig.* **2019**, *49*, e13152. [CrossRef]

Article

# Association of Magnesium Intake with Liver Fibrosis among Adults in the United States

Meng-Hua Tao [1,*] and Kimberly G. Fulda [2]

[1] Department of Biostatistics and Epidemiology, University of North Texas Health Science Center, Fort Worth, TX 76107, USA
[2] Department of Family Medicine and Osteopathic Manipulative Medicine, NorTex, University of North Texas Health Science Center, Fort Worth, TX 76107, USA; kimberly.fulda@unthsc.edu
* Correspondence: menghua.tao@unthsc.edu; Tel.: +817-735-0520; Fax: +817-735-0446

**Abstract:** Liver fibrosis represents the consequences of chronic liver injury. Individuals with alcoholic or nonalcoholic liver diseases are at high risk of magnesium deficiency. This study aimed to evaluate the association between magnesium and calcium intakes and significant liver fibrosis, and whether the associations differ by alcohol drinking status. Based on the National Health and Nutrition Examination Survey (NHANES) 2017–2018, the study included 4166 participants aged >18 years who completed the transient elastography examination and had data available on magnesium intake. The median liver stiffness of 8.2 kPa was used to identify subjects with significant fibrosis ($\geq$F2). The age-adjusted prevalence of significant fibrosis was 12.81%. Overall total magnesium intake was marginally associated with reduced odds of significant fibrosis ($p$ trend = 0.14). The inverse association of total magnesium intake with significant fibrosis was primarily presented among those who had daily calcium intake <1200 mg. There were no clear associations for significant fibrosis with calcium intake. Findings suggest that high total magnesium alone may reduce risk of significant fibrosis. Further studies are needed to confirm these findings.

**Keywords:** magnesium; calcium; significant liver fibrosis; epidemiology

**Citation:** Tao, M.-H.; Fulda, K.G. Association of Magnesium Intake with Liver Fibrosis among Adults in the United States. *Nutrients* **2021**, *13*, 142. https://doi.org/10.3390/nu130 10142

Received: 4 December 2020
Accepted: 28 December 2020
Published: 2 January 2021

**Publisher's Note:** MDPI stays neutral with regard to jurisdictional claims in published maps and institutional affiliations.

**Copyright:** © 2021 by the authors. Licensee MDPI, Basel, Switzerland. This article is an open access article distributed under the terms and conditions of the Creative Commons Attribution (CC BY) license (https://creativecommons.org/licenses/by/4.0/).

## 1. Introduction

Chronic liver disease is a substantial worldwide problem [1]. Its major consequence is accumulation of extracellular matrix within the liver, leading to the development of cirrhosis, liver failure, or even liver cancer [2]. Nonalcoholic steatohepatitis (NASH), chronic hepatitis C infection, and alcohol abuse are the main causes of liver fibrosis in Western countries [2]. NASH is a subtype of nonalcoholic fatty liver disease (NAFLD) [3]. As the most common liver disease in the world [4], NAFLD is considered as the hepatic manifestation of metabolic syndrome [3] and is associated with obesity and type 2 diabetes [5]. Previous studies have also shown that liver fibrosis stage is associated with long-term outcomes of patients with NAFLD [6]. Inflammation, oxidant stress, and insulin resistance not only play critical roles in the progression of hepatic fibrosis in NAFLD patients [2], but they can occur and stimulate liver fibrosis among patients with hepatitis C or B infection [2,7] or alcoholic liver diseases [2,8].

Magnesium status is closely linked with liver function and may be related to the etiology of chronic liver disease. In the liver, mast cells contribute to liver fibrosis [9]; animal studies have shown that low-magnesium diet increases the levels of mRNA known to be expressed by mast cells in the liver and induce the emergence of mast cells around portal triads of the liver in Sprague–Dawley rats [9]. An in vivo study showed that extracellular magnesium deficiency modulates the expression levels of molecules related to oxidative stress/antioxidant response in HepG2 human hepatoma cells [10]. Serum magnesium concentration is also significantly low in patients with alcoholic steatosis, nonalcoholic steatosis, or NASH [11,12]. Previous studies found that chronic alcohol consumption leads to a decrease in liver magnesium content, while deficient magnesium levels in alcoholic liver disease patients, in turn, disrupt normal metabolism and lead to greater lipid deposition in the liver [13]. Recent studies suggest that high magnesium intake may be associated with reduced risk of fatty liver disease [14] and mortality due to liver disease, particularly among those with fatty liver disease or alcoholic drinkers [15]. Animal studies have shown that magnesium deficiency induces inflammatory response [16]; randomized controlled trials further report that magnesium treatments significantly decrease concentrations of C-reactive protein (CRP) among patients with metabolic syndrome [17] or high risk of inflammation [18]. Randomized controlled trials also show that magnesium supplementation improves insulin sensitivity in patients with type 2 diabetes or non-diabetic individuals with insulin resistance [19–21].

In this study, we investigated whether magnesium intake is associated with the prevalence of liver fibrosis in adults. Previous studies have examined the association of calcium intake with type 2 diabetes [22]; however, few studies have investigated the role of calcium intake in relation to liver fibrosis [23]. Calcium intake may interact with intake of magnesium in the development of many chronic diseases including fatty liver disease and prediabetes [14]. Therefore, we hypothesized that intake of calcium may also be associated with risk of liver fibrosis. To test this hypothesis, we examined the association between calcium intake and liver fibrosis and investigated whether the associations varied by alcohol drinking status among US adults utilizing data from the National Health and Nutrition Examination Survey (NHANES) conducted in 2017 and 2018.

## 2. Materials and Methods

### 2.1. Study Population

This study utilized data from one cycle of the National Health and Nutrition Examination Survey (NHANES) conducted between 2017 and 2018 in which liver ultrasound transient elastography examination was performed. The NHANES is an ongoing survey program designed to assess health and nutrition in a nationally representative sample of the non-institutionalized US population. A detailed description of the study design has been published elsewhere [24]. The survey is maintained and administered by the National Center for Health Statistics (NCHS) of the Centers for Disease Control and Prevention (CDC) [24]. In 2017–2018, 9254 persons completed the interview. In our study, participants who were aged less than 19 years at time of the survey and did not complete the liver ultrasound transient elastography examination were excluded (N = 4262). Pregnant or lactating females, participants with unreliable dietary data or autoimmune liver disease, and participants with missing data for magnesium, calcium intake, or confounders (age, sex, race/ethnicity, education, body mass index (BMI), high-density lipoprotein (HDL) cholesterol) were further excluded from the analyses (N = 826). As a result, 4166 participants were included in the final analysis. All participants provided written informed consent and the NCHS Research Ethics Review Board approved the survey protocol (Protocol #2011-17, 2018-01).

### 2.2. Ascertainment of Outcomes

In the 2017–2018 cycle of NHANES, transient elastography measurements were conducted in the NHANES Mobile Examination Center (MEC) by trained health technicians, using FibroScan®model 502 V2 Touch equipped with a medium or extra-large probe. Briefly, transient elastography is a widely used and validated technique to quantitatively assess tissue stiffness. It is considered as a reliable, non-invasive method to assess liver fibrosis [25,26]. All participants aged 12 years and over were eligible with exclusion for participants who were unable to lie down on the exam table, pregnant at the time of their exam, had an implanted electronic medical device, or were wearing a bandage or had lesions on the site where measurements would be taken. Only participant with complete exams (i.e., fasting time of at least 3 h, 10 or more complete stiffness measures, and a liver stiffness interquartile range/median <30%) were included in the current analysis. Several meta-analyses [25,27] and a recent prospective study [28] analyzed, assessed, and reported optimal stiffness cutoff values to define different stages of fibrosis among subjects with NAFLD. The transient elastography cutoff values of 8.2, 9.7, and 13.6 (Kpa) were used to define METAVIR (Meta-analysis of Histological Data in Viral Hepatitis) fibrosis stage F2, F3, and F4 fibrosis, respectively [25,28]. A median liver stiffness of 8.2 (kPa) was used to define cases of significant fibrosis ($\geq$F2).

### 2.3. Assessments of Nutrient Intake

Details of the protocol and dietary information collection methods have been described previously [29]. Briefly, daily dietary intake information was obtained through 24-h recall interviews and a 30-day dietary supplement questionnaire. Two 24-h recalls were conducted for each participant in NHANES 2017–2018. The first dietary recall was collected in person by trained interviewers in the NHANES MEC and the second dietary recall was completed by trained interviewers via telephone 3–10 days after the MEC interview [29]. To keep intake information consistent, only the in-person dietary recall for all participants was used in the current analysis. Total intakes of magnesium, calcium, and other nutrients were calculated by summing intake from foods and dietary supplements.

### 2.4. Assessments of Covariates

In the NHANES, race/ethnicity was categorized by using survey questions on race and Hispanic origin: non-Hispanic White referring to whites who are not of Hispanic origin; non-Hispanic Black referring to blacks without Hispanic origin; Hispanic referring to all Hispanics regardless of race; non-Hispanic Asian including Asians without Hispanic origin; and Other Race including American Indians or Alaska Natives, Native Hawaiians or other Pacific Islanders, and multiracial persons. The amount of daily alcohol beverage consumption was also collected in the 24-h recalls. Daily alcohol consumption was categorized into non-drinkers (0 g), low (<31.32 g), and high ($\geq$31.32 g) based upon the median daily alcohol intake among non-cases. Body mass index (BMI, measured weight/height$^2$) was classified into three categories: <25.0, 25.0–29.9, and $\geq$30.0, categories of under/normal weight, overweight, and obesity, respectively. Based on Physical Activity Guidelines recommendation of at least 75 min of vigorous or 150 min of moderate physical activity in a typical week [30], participants were classified into physically inactive (no), less active (<the recommendation), and active ($\geq$the recommendation). Participants with type 2 diabetes were identified as having any of the following: (1) hemoglobin A1C concentration $\geq$6.5% [31] or (2) self-report of diabetes diagnosis. Participants with hepatitis B or C virus (HBV or HCV) infection were identified by positive testing results [32,33] or self-report of hepatitis B or C infection. Participants were asked whether they had ever smoked $\geq$100 cigarettes in their lifetime and whether they smoked currently to identify current and former smokers. Participants were defined as former smokers if they did not smoke currently but had ever smoked $\geq$100 cigarettes in the past. Blood high-density lipoprotein (HDL) cholesterol was measured based on standard laboratory methods.

### 2.5. Statistical Analysis

All statistical analyses were conducted in SAS 9.4 software (SAS Institute, Cary, NC, USA) using the "Survey" procedure to estimate variance after incorporating the complex, multistage, clustered probability sampling design of the NHANES [34]. Characteristics and covariates were compared between those with and without significant fibrosis using Rao–Scott chi-square test for categorical variables and Student's *t*-test for continuous variables. The age-adjusted prevalence of significant fibrosis was estimated, stratified by race/ethnicity and age groups using the 2000 US Census as the standard population. Logistic regressions were used to estimate odds ratios (ORs) and 95% confidence intervals (95% CIs) for associations between magnesium intake and significant fibrosis. Dietary and total magnesium and calcium intake were categorized into quartiles based on intakes among non-cases, using the lowest category as the reference. Potential confounders included in analyses were gender, race/ethnicity, education, BMI status, alcohol intake status, physical activity status, HCV infection status, and medical history of diabetes. Age and HDL, as well as daily intakes of total energy, calcium, magnesium, fiber, and phosphate, were included as continuous variables for adjustment in models. We also assessed potential confounding by smoking and HBV status, and these factors did not alter point estimates by $\geq 10\%$ and were excluded from the final models. Tests for dose–response relationship were estimated by fitting models with exposure variables included as continuous variables.

Stratified analyses by gender, calcium:magnesium intake ratio (<2.62, $\geq$2.62, according to physiological range of the ratio and previous reports on the ratio in the US population [14,35]; the second quartile of the ratio in our population was used as the cut-point), daily calcium intake (<1200 mg or $\geq$1200 mg), daily alcohol drinking status (no, yes), and race/ethnicity were further conducted. Possible interaction between magnesium or calcium intake and potential effect modifiers were examined in the logistic regression model by evaluation of multiplicative interactions. All reported $p$-values are two-sided with statistical significance evaluated at 0.05.

## 3. Results

Table 1 summarizes selected characteristics by significant liver fibrosis status. A total of 628 participants presented $\geq$F2 liver fibrosis, with age-adjusted prevalence of 12.81%. Compared to participants without significant fibrosis, those with $\geq$F2 fibrosis were older, more likely to be male, obese, and physically inactive, and had a history of HCV infection or diabetes, lower HDL level, and higher total energy intake.

Associations of intakes of magnesium and calcium with the odds of significant fibrosis are presented in Table 2. After adjustment for intakes of energy, fiber, phosphates and total calcium, age, race/ethnicity, gender, and other potential confounders, total magnesium intake was marginally associated with lower odds of significant fibrosis (OR = 0.53, 95% CI, 0.35–1.10; $p$ trend = 0.14). When examining dietary magnesium intake, no significant association was observed. Neither dietary nor total intake of calcium was associated with odds of significant fibrosis.

We further conducted stratified analyses for total magnesium intake by the ratio of calcium:magnesium intake (Table 3). Among those with a high calcium:magnesium intake ratio ($\geq$2.62), participants who consumed high total magnesium tended to have lower odds of significant fibrosis; however, the association was not statistically significant. Among those with a low calcium:magnesium intake ratio (<2.62), the pattern suggested a positive association between high magnesium intake and odds of significant fibrosis. Again, the association was not statistically significant, possibly due to a smaller sample size in this strata. In addition, there was no significant interaction between calcium:magnesium ratio and intake of magnesium. On the other hand, among those who had a daily total calcium intake <1200 mg, compared to those with the lowest quartile intake, participants with total magnesium intake at the highest quartile had reduced odds of significant fibrosis (OR = 0.35, 95% CI, 0.16–0.77; $p$ trend = 0.02). The test for interaction between magnesium intake and calcium intake was not significant ($p$ interaction = 0.44). No significant interactions were found between magnesium intake and other potential effect modifiers including daily alcohol drinking status ($p$ interaction = 0.38) and gender ($p$ interaction = 0.75). However, the reduced odds of significant fibrosis in relation to high total magnesium intake tended to be stronger in non-drinkers (OR = 0.45, 95% CI, 0.18–1.09; $p$ trend = 0.09) and males (OR = 0.47, 95% CI, 0.23–0.99; $p$ trend = 0.12). Similarly, we found the inverse association of total magnesium intake with odds of significant fibrosis only among non-Hispanic whites (OR = 0.34, 95% CI, 0.12–0.98 the highest vs. the lowest quartile; $p$ trend = 0.10) (data not shown). No clear associations were observed among other racial/ethnic groups, possibly due to smaller sample sizes for these racial/ethnic groups. Again, no significant interaction was found for total magnesium intake and race/ethnicity.

Table 1. Participant characteristics by significant liver fibrosis status, National Health and Nutrition Examination Survey (NHANES) 2017–2018.

| Characters | Yes (n = 628) | No (n = 3538) | p-Value |
|---|---|---|---|
| Age (years) ‡ | 52.4 (1.36) | 47.7 (0.68) | <0.01 |
| Sex, n (%) * | | | <0.01 |
| Male | 371 (58.8) | 1698 (48.1) | |
| Female | 257 (41.2) | 1840 (51.9) | |
| Race/ethnicity, n (%) * | | | 0.44 |
| Non-Hispanic White | 230 (65.3) | 1501 (64.5) | |
| Non-Hispanic Black | 161 (11.8) | 950 (10.2) | |
| Hispanic | 158 (15.4) | 956 (15.1) | |
| Non-Hispanic Asian | 53 (3.4) | 549 (5.4) | |
| Other races [1] | 26 (4.1) | 210 (4.8) | |
| Education, n (%) * | | | 0.03 |
| Less than high school | 131 (11.2) | 637 (10.0) | |
| High school | 156 (31.9) | 852 (27.3) | |
| Some college | 229 (35.4) | 1147 (30.2) | |
| College graduate | 112 (21.5) | 902 (32.5) | |
| BMI group, n (%) * | | | <0.01 |
| <25 | 89 (13.3) | 975 (28.4) | |
| 25–30 | 130 (18.0) | 1200 (32.8) | |
| ≥30 | 409 (68.7) | 1363 (38.8) | |
| Physical activity level, n (%) * | | | <0.01 |
| Active | 142 (22.8) | 961 (30.5) | |
| Less active | 107 (22.2) | 792 (26.5) | |
| Inactive | 379 (54.9) | 1785 (43.0) | |
| Smoking status, n (%) * | | | 0.14 |
| Never | 327 (51.7) | 2050 (57.9) | |
| Former | 186 (30.8) | 842 (24.9) | |
| Current | 115 (17.4) | 646 (17.2) | |
| Daily alcohol drinking status, n (%) | | | 0.73 |
| Non-drinkers | 499 (73.2) | 2785 (74.2) | |
| Low drinkers (<31.32 g) | 63 (12.1) | 399 (13.2) | |
| High drinkers (≥31.32 g) | 66 (14.7) | 354 (12.6) | |
| History of diabetes, n (%) | | | <0.01 |
| Yes | 239 (32.3) | 622 (11.8) | |
| Having HBV infection, n (%) | | | 0.09 |
| Yes | 43 (3.9) | 231 (3.8) | |
| Having HCV infection, n (%) | | | <0.01 |
| Yes | 48 (7.7) | 64 (1.9) | |
| Laboratory features ‡ | | | |
| HDL (mmol/L) | 1.3 (0.03) | 1.4 (0.01) | <0.01 |
| Median CAP (dB/m) | 306.8 (4.36) | 258.9 (2.07) | <0.01 |
| Median stiffness (kPa) | 13.8 (0.54) | 4.7 (0.04) | <0.01 |
| Daily intake of nutrients ‡ | | | |
| Total energy (kcal) | 2325.6 (45.30) | 2176.3 (18.39) | 0.01 |
| Dietary calcium (mg) | 1005.9 (36.95) | 968.5 (14.34) | 0.33 |
| Dietary magnesium (mg) | 310.7 (11.8) | 304.9 (4.28) | 0.57 |
| Total calcium (mg) | 1101.1 (37.81) | 1061.4 (17.27) | 0.33 |
| Total magnesium (mg) | 329.3 (16.91) | 329.8 (6.00) | 0.97 |

‡ Values shown are mean and (standard error); * unweighted frequency counts and weighted percentages shown.
[1] Other races include American Indian or Alaska Native, Native Hawaiian or other Pacific Islander, and multiracial persons.

Table 2. Associations between intakes of magnesium and calcium and odds of significant liver fibrosis, NHANES 2017–2018.

| Daily Intake of Nutrients (mg) | Liver Fibrosis Status | | Model 1 | Model 2 |
|---|---|---|---|---|
| | Yes | No | OR (95% CI) [1] | OR (95% CI) [2] |
| Dietary magnesium | | | | |
| $Q_1$ < 205.93 | 190 | 996 | Referent | Referent |
| $Q_2$ 205.93–279.12 | 154 | 866 | 0.74 (0.51–1.06) | 0.73 (0.50–1.06) |
| $Q_3$ 279.13–375.20 | 143 | 857 | 0.96 (0.60–1.54) | 0.87 (0.49–1.54) |
| $Q_4$ ≥ 375.21 | 141 | 819 | 0.99 (0.67–1.45) | 0.70 (0.35–1.38) |
| $p_{trend}$ | | | 0.77 | 0.47 |
| Total magnesium intake | | | | |
| $Q_1$ < 212.99 | 189 | 1009 | 1.00 | 1.00 |
| $Q_2$ 212.99–294.38 | 162 | 861 | 0.70 (0.52–0.95) | 0.70 (0.51–0.97) |
| $Q_3$ 294.39–400.26 | 134 | 853 | 0.91 (0.56–1.47) | 0.70 (0.38–1.28) |
| $Q_4$ ≥ 400.27 | 143 | 815 | 0.85 (0.54–1.34) | 0.53 (0.25–1.10) |
| $p_{trend}$ | | | 0.74 | 0.14 |
| Dietary calcium | | | | |
| $Q_1$ < 574.06 | 183 | 1047 | Referent | Referent |
| $Q_2$ 574.06–855.22 | 153 | 893 | 1.41 (0.86–2.32) | 1.54 (0.97–2.45) |
| $Q_3$ 855.23–1238.68 | 160 | 822 | 1.22 (0.84–1.77) | 1.14 (0.77–1.67) |
| $Q_4$ ≥ 1238.69 | 132 | 776 | 1.29 (0.85–1.96) | 1.03 (0.59–1.79) |
| $p_{trend}$ | | | 0.31 | 0.77 |
| Total calcium intake | | | | |
| $Q_1$ < 628.54 | 177 | 1051 | 1.00 | 1.00 |
| $Q_2$ 628.54–945.25 | 170 | 883 | 1.41 (0.86–2.32) | 1.50 (0.97–2.33) |
| $Q_3$ 945.25–1356.52 | 146 | 823 | 1.19 (0.76–1.84) | 1.10 (0.72–1.68) |
| $Q_4$ ≥ 1356.53 | 135 | 781 | 1.23 (0.78–1.94) | 1.14 (0.71–1.84) |
| $p_{trend}$ | | | 0.53 | 0.99 |

[1] Adjusted for age; [2] further adjusted for gender, race/ethnicity, education, BMI, physical activity status, HCV infection status, status of diabetes, HDL level, and intakes of total energy, alcohol, fiber, and phosphate. Intakes of calcium and magnesium are mutually adjusted.

Table 3. Associations between intake of total magnesium and odds of significant fibrosis by gender, calcium:magnesium ratio, intake of calcium, and daily alcohol drinking status, NHANES 2017–2018.

| Total Magnesium Intake (mg/Day) | Significant Liver Fibrosis | | | |
|---|---|---|---|---|
| | Yes | No | OR (95% CI) [1] | p for Trend |
| | Males | | | |
| $Q_1$ < 212.99 | 97 | 357 | 1.00 | |
| $Q_2$ 212.99–294.38 | 88 | 368 | 0.58 (0.36–0.95) | |
| $Q_3$ 294.39–400.26 | 84 | 449 | 0.67 (0.32–1.43) | |
| $Q_4$ ≥ 400.27 | 102 | 524 | 0.47 (0.23–0.99) | 0.12 |
| | Females | | | |
| $Q_1$ < 212.99 | 92 | 652 | 1.00 | |
| $Q_2$ 212.99–294.38 | 74 | 493 | 0.77 (0.52–1.15) | |
| $Q_3$ 294.39–400.26 | 50 | 404 | 0.65 (0.35–1.20) | |
| $Q_4$ ≥ 400.27 | 41 | 291 | 0.63 (0.18–2.19) | 0.36 |
| p for interaction | | | | 0.75 |
| | Calcium:Magnesium ratio < 2.62 | | | |
| $Q_1$ < 212.99 | 46 | 320 | 1.00 | |
| $Q_2$ 212.99–294.38 | 61 | 274 | 1.83 (1.05–3.21) | |
| $Q_3$ 294.39–400.26 | 56 | 333 | 1.33 (0.53–3.31) | |
| $Q_4$ ≥ 400.27 | 70 | 436 | 0.70 (0.26–1.91) | 0.44 |

Table 3. Cont.

| Total Magnesium Intake (mg/Day) | Significant Liver Fibrosis | | | |
|---|---|---|---|---|
| | Yes | No | OR (95% CI) [1] | p for Trend |
| | Calcium:Magnesium ratio ≥ 2.62 | | | |
| $Q_1$ < 212.99 | 143 | 689 | 1.00 | |
| $Q_2$ 212.99–294.38 | 101 | 587 | 0.52 (0.35–0.78) | |
| $Q_3$ 294.39–400.26 | 78 | 520 | 0.59 (0.29–1.21) | |
| $Q_4$ ≥ 400.27 | 73 | 379 | 0.59 (0.23–1.49) | 0.29 |
| p for interaction | | | | 0.30 |
| | Calcium < 1200 mg/day | | | |
| $Q_1$ < 212.99 | 171 | 941 | 1.00 | |
| $Q_2$ 212.99–294.38 | 128 | 681 | 0.66 (0.46–0.96) | |
| $Q_3$ 294.39–400.26 | 92 | 555 | 0.56 (0.29–1.10) | |
| $Q_4$ ≥ 400.27 | 53 | 329 | 0.35 (0.16–0.77) | 0.02 |
| | Calcium ≥ 1200 mg/day | | | |
| $Q_1$ <212.99 | 18 | 68 | 1.00 | |
| $Q_2$ 212.99–294.38 | 34 | 180 | 0.71 (0.20–1.06) | |
| $Q_3$ 294.39–400.26 | 42 | 298 | 0.53 (0.20–1.42) | |
| $Q_4$ ≥400.27 | 90 | 486 | 0.38 (0.16–1.43) | 0.43 |
| p for interaction | | | | 0.44 |
| | Daily alcohol drinking: No | | | |
| $Q_1$ < 212.99 | 161 | 858 | 1.00 | |
| $Q_2$ 212.99–294.38 | 134 | 683 | 0.72 (0.49–1.06) | |
| $Q_3$ 294.39–400.26 | 106 | 650 | 0.69 (0.39–1.24) | |
| $Q_4$ ≥ 400.27 | 98 | 594 | 0.45 (0.18–1.09) | 0.09 |
| | Daily alcohol drinking: Yes | | | |
| $Q_1$ < 212.99 | 28 | 151 | 1.00 | |
| $Q_2$ 212.99–294.38 | 28 | 178 | 0.60 (0.33–1.09) | |
| $Q_3$ 294.39–400.26 | 28 | 203 | 0.57 (0.23–1.40) | |
| $Q_4$ ≥ 400.27 | 45 | 221 | 0.68 (0.30–1.51) | 0.51 |
| p for interaction | | | | 0.51 |

[1] Adjusted for age, gender, race/ethnicity, education, BMI status, physical activity status, alcohol intake status, HCV infection status, HBV infection status, status of diabetes, HDL level, and intakes of total energy, fiber, phosphate, and total calcium.

Intake of total calcium was not significantly associated with significant fibrosis in both males and females (Table 4). No significant association or interaction was found for total calcium intake in different groups of calcium:magnesium intake or alcohol drinking status. The association between intake of calcium and significant fibrosis did not differ by race/ethnicity.

Table 4. Associations between intake of total calcium and odds of significant fibrosis by gender, calcium:magnesium ratio, and daily alcohol drinking status, NHANES 2017–2018.

| Total Calcium Intake (mg/Day) | Significant Liver Fibrosis | | | |
|---|---|---|---|---|
| | Yes | No | OR (95% CI) [1] | p for Trend |
| | | Males | | |
| $Q_1$ < 628.54 | 95 | 450 | 1.00 | |
| $Q_2$ 628.54–945.25 | 95 | 414 | 1.56 (0.96–2.54) | |
| $Q_3$ 945.25–1356.52 | 91 | 393 | 1.07 (0.70–1.63) | |
| $Q_4$ ≥ 1356.53 | 90 | 441 | 1.10 (0.58–2.07) | 0.77 |

**Table 4.** Cont.

| Total Calcium Intake (mg/Day) | Significant Liver Fibrosis | | OR (95% CI) [1] | p for Trend |
|---|---|---|---|---|
| | Yes | No | | |
| | Females | | | |
| $Q_1$ < 628.54 | 82 | 601 | 1.00 | |
| $Q_2$ 628.54–945.25 | 75 | 469 | 1.51 (0.77–3.00) | |
| $Q_3$ 945.25–1356.52 | 55 | 430 | 1.23 (0.57–2.63) | |
| $Q_4$ ≥ 1356.53 | 45 | 340 | 1.37 (0.50–3.75) | 0.64 |
| p for interaction | | | | 0.99 |
| | Calcium:Magnesium ratio < 2.62 | | | |
| $Q_1$ < 628.54 | 125 | 737 | 1.00 | |
| $Q_2$ 628.54–945.25 | 64 | 370 | 1.89 (0.93–3.83) | |
| $Q_3$ 945.25–1356.52 | 32 | 181 | 2.40 (1.18–4.85) | |
| $Q_4$ ≥ 1356.53 | 12 | 75 | 1.72 (0.32–9.20) | 0.12 |
| | Calcium:Magnesium ratio ≥ 2.62 | | | |
| $Q_1$ < 628.54 | 52 | 314 | 1.00 | |
| $Q_2$ 628.54–945.25 | 106 | 513 | 1.99 (0.69–2.09) | |
| $Q_3$ 945.25–1356.52 | 114 | 642 | 0.82 (0.47–1.42) | |
| $Q_4$ ≥ 1356.53 | 123 | 706 | 0.91 (0.46–1.81) | 0.45 |
| p for interaction | | | | 0.17 |
| | Daily alcohol drinking: No | | | |
| $Q_1$ < 628.54 | 151 | 837 | 1.00 | |
| $Q_2$ 628.54–945.25 | 139 | 695 | 1.29 (0.85–1.96) | |
| $Q_3$ 945.25–1356.52 | 104 | 636 | 0.99 (0.69–1.42) | |
| $Q_4$ ≥ 1356.53 | 105 | 617 | 0.92 (0.53–1.61) | 0.51 |
| | Daily alcohol drinking: Yes | | | |
| $Q_1$ < 628.54 | 26 | 214 | 1.00 | |
| $Q_2$ 628.54–945.25 | 31 | 188 | 2.17 (0.66–7.12) | |
| $Q_3$ 945.25–1356.52 | 42 | 187 | 1.50 (0.37–6.11) | |
| $Q_4$ ≥ 1356.53 | 30 | 164 | 2.40 (0.37–15.77) | 0.51 |
| p for interaction | | | | 0.59 |

[1] Adjusted for age, gender, race/ethnicity, education, BMI status, physical activity status, alcohol intake status, HCV infection status, HBV infection status, status of diabetes, HDL level, and intakes of total energy, fiber, phosphate, and total magnesium.

## 4. Discussion

Utilizing data from the recent NHANES cycle, a nationally representative sample of the US general population, results suggested that there is an association between high total magnesium intake and lower odds of significant liver fibrosis. Moreover, the inverse association between magnesium intake and significant fibrosis appeared stronger among males, non-Hispanic whites, subjects who had calcium intake <1200 mg per day, and subjects who did not drink alcohol, although interactions were not statistically significant. On the other hand, there was no association between calcium intake and significant fibrosis.

Our finding of the inverse association between total magnesium intake and significant fibrosis is consistent with previous studies. Rodríguez-Hernández et al. [36] reported a positive association between low serum magnesium and risk of NASH in a study of obese subjects. A recent study using data from NHANES III found a significant association of total magnesium intake with reduced odds of fatty liver disease [14]. In particular, our results suggest an inverse association between total magnesium intake and odds of significant fibrosis among subjects who did not drink alcoholic beverages. Our novel findings are consistent with previous results that serum magnesium concentrations were significantly lower in individuals with NASH [12,36]. This finding is important because the prevalence of NAFLD has been increasing steadily in the last three decades in the US with an estimated prevalence of 32% in adults [37]. Meanwhile, there are steady increases in the rates of obesity and diabetes in the US [37], both of which are major risk factors of NAFLD [5]. It has been reported that 60% of American adults do not meet the Estimated Average Requirement (EAR) for magnesium in NHANES 2001–2008 [38], and overweight/obese adults had higher prevalence of inadequate intake of magnesium than normal-weight adults [38,39]. Further large studies are needed to confirm the results.

Previous studies have shown the importance of the balance between magnesium and calcium in relation to their physiological functions. Calcium can directly or indirectly compete with magnesium for (re)absorption in the intestine and kidney [40]. Clinical trials consistently show that a high calcium and low or insufficient magnesium diet can interfere with magnesium absorption, resulting in depressed absorption of magnesium [40,41]. In agreement with prior studies on fatty liver disease [14], our study found that total magnesium intake was associated with reduced odds of significant liver fibrosis only among those with daily calcium intake less than 1200 mg, suggesting that the beneficial effects of magnesium could be suppressed when calcium intake is more than the Dietary Reference Intakes (DRIs).

The inverse association between total magnesium intake and significant fibrosis was limited among non-Hispanic whites in our study. Previous studies found that non-Hispanic whites had higher magnesium intake than non-Hispanic blacks and Hispanics [39], which may be in part due to disparities in socioeconomic status and educational attainment between racial/ethnic groups. The relatively small number of minority participants still limited our ability to detect weak associations in each minority group. Further studies are needed to understand specific associations in minority populations.

The strengths of this study include using NHANES data with nationally representative samples and a relatively large number of adults with transient elastography examination, providing the power to detect weak associations. However, several limitations common to observational studies should be mentioned. Due to the nature of cross-sectional studies, the temporal sequences may not be clear. However, the suggested inverse association between magnesium intake and significant fibrosis is consistent with previous findings on the associations of magnesium intake with fatty liver disease [14], metabolic syndrome, and insulin resistance [17,20,21]. Although the transient elastography measurement is a widely used non-invasive method to assess liver fibrosis [25,26], it can be limited by patient obesity, the presence of perihepatic ascites, and limited selection of an appropriate sampling area [42]. However, the transient elastography has been shown as a validated and reliable technique and has been recommended to discriminate significant ($F \geq 2$) from non-significant fibrosis (F0–F1) by the World Federation for Ultrasound in Medicine and Biology [42]. In addition, misclassification may have occurred in the analyses since there is no well-defined cutoff for significant fibrosis utilizing the transient elastography measurement. However, this misclassification is likely to be nondifferential. Alcohol intake is an important risk factor for liver fibrosis; we adjusted for the amount of alcohol intake based on the 24-h dietary recall due to the unavailability of data from the alcohol use questionnaire (ALQ) in the NHANES 2017–18 cycle. Previous studies showed that the frequency of participants consuming some amount of alcoholic beverages estimated through the 24-h dietary recall was lower than the frequency of drinking some alcoholic beverages at least once in the past one-year period obtained from the ALQ [43]. The residual confounding by alcohol intake in this study may dilute or even mask the associations of magnesium intake with liver fibrosis. The 24-h dietary recall used in NHANES has been extensively evaluated [29]; however, the self-reported dietary recall is likely to have both random and systematic errors, particularly in energy intake [44]. Moreover, recall bias from self-reported diet may also exist [45]. Although multiple 24-h dietary recalls are used as the gold standard measure in large-scale nutrition epidemiological studies, a one-time, 24-h dietary recall may not capture long-term dietary exposures [29]. We adjusted for many potential confounding factors including physical activity, daily alcohol drinking status, and several medical conditions associated with liver fibrosis, which enabled us to capture the association more accurately.

In conclusion, our findings suggest that participants who had high intake of magnesium may have reduced odds of having significant liver fibrosis, whereas high intake of calcium was not associated with change in risk. In particular, the inverse association may be limited among those who had daily calcium intake less than 1200 mg and those who did not drink alcohol. Further studies, such as prospective cohort studies, are warranted to confirm the present results.

**Author Contributions:** M.-H.T.: study design, data analyses and interpretation, drafting of the manuscript, review of the manuscript. K.G.F.: data interpretation and review of the manuscript. All authors have read and agreed to the published version of the manuscript.

**Funding:** Tao's effort was partially supported by the National Institute on Minority Health and Health Disparities of the National Institute of Health under Award U54MD006882.

**Institutional Review Board Statement:** The study was conducted according to the guidelines of the Declaration of Helsinki, and approved by the NCHS Research Ethics Review Board (Protocol #2011-17, 2018-01)

**Informed Consent Statement:** Informed consent was obtained from all subjects involved in the study.

**Data Availability Statement:** Publicly available datasets were analyzed in this study. This data can be found here: https://wwwn.cdc.gov/nchs/nhanes/.

**Acknowledgments:** We thank the investigators, the staff, and the participants of NHANES for their valuable contribution.

**Conflicts of Interest:** No potential conflicts of interests were disclosed.

## References

1. Asrani, S.K.; Devarbhavi, H.; Eaton, J.; Kamath, P.S. Burden of liver diseases in the world. *J. Hepatol.* **2019**, *70*, 151–171. [CrossRef] [PubMed]
2. Friedman, S. Liver fibrosis—From bench to bedside. *J. Hepatol.* **2003**, *38* (Suppl. 1), 38–52. [CrossRef]
3. Matteoni, C.A.; Younossi, Z.M.; Gramlich, T.; Boparai, N.; Liu, Y.C.; McCullough, A.J. Nonalcoholic fatty liver disease: A spectrum of clinical and pathological severity. *Gastroenterology* **1999**, *116*, 1413–1419. [CrossRef]
4. Younossi, Z.M.; Koenig, A.B.; Abdelatif, D.; Fazel, Y.; Henry, L.; Wymer, M. Global epidemiology of nonalcoholic fatty liver disease-meta-analytic assessment of prevalence, incidence, and outcomes. *Hepatology* **1999**, *64*, 73–84. [CrossRef] [PubMed]
5. Younossi, Z.M. Nonalcoholic fatty liver disease-a global public health perspective. *J. Hepatol.* **2019**, *70*, 531–544. [CrossRef]
6. Hagström, H.; Nasr, P.; Ekstedt, M.; Hammar, U.; Stål, P.; Hultcrantz, R.; Kechagias, S. Fibrosis stage but not nash predicts mortality and time to development of severe liver disease in biopsy-proven nafld. *J. Hepatol.* **2017**, *67*, 1265–1273. [CrossRef]
7. Patel, S.; Jinjuvadia, R.; Patel, R.; Liangpunsakul, S. Insulin resistance is associated with significant liver fibrosis in chronic hepatitis c patients: A systemic review and meta-analysis. *J. Clin. Gastroenterol.* **2016**, *50*, 80–84. [CrossRef]
8. Carr, R.M.; Correnti, J.C. Insulin resistance in clinical and experimental alcoholic liver disease. *Ann. N. Y. Acad. Sci.* **2015**, *1353*, 1–20. [CrossRef]
9. Takemoto, S.; Yamamoto, A.; Tomonaga, S.; Funaba, M.; Matsui, T. Magnesium deficiency induces the emergence of mast cells in the liver of rats. *J. Nutr. Sci. Vitaminol.* **2013**, *59*, 560–563. [CrossRef]
10. Shigematsu, M.; Tomonaga, S.; Shimokawa, F.; Murakami, M.; Imamura, T.; Matsui, T.; Funaba, M. Regulatory responses of hepatocytes, macrophages and vascular endothelial cells to magnesium deficiency. *J. Nutr. Biochem.* **2018**, *56*, 35–47. [CrossRef]
11. Turecky, L.; Kupcova, V.; Szantova, M.; Uhlikova, E.; Viktorinova, A.; Czirfusz, A. Serum magnesium levels in patients with alcoholic and non-alcoholic fatty liver. *Bratisl. Lek. Listy* **2006**, *107*, 58–61. [PubMed]
12. Eshraghian, A.; Nikeghbalian, S.; Geramizadeh, B.; Malek-Hosseini, S.A. Serum magnesium concentration is independently associated with non-alcoholic fatty liver and non-alcoholic steatohepatitis. *United Eur. Gastroenterol. J.* **2018**, *6*, 97–103. [CrossRef] [PubMed]
13. Liu, M.; Yang, H.; Mao, Y. Magnesium and liver disease. *Ann. Transl. Med.* **2019**, *7*, 578. [CrossRef] [PubMed]
14. Li, W.; Zhu, X.; Song, Y.; Fan, L.; Wu, L.; Kabagambe, E.K.; Hou, L.; Shrubsole, M.J.; Liu, J.; Dai, Q. Intakes of magnesium, calcium and risk of fatty liver disease and prediabetes. *Public Health Nutr.* **2018**, *21*, 2088–2095. [CrossRef]
15. Wu, L.; Zhu, X.; Fan, L.; Kabagambe, E.K.; Song, Y.; Tao, M.; Zhong, X.; Hou, L.; Shrubsole, M.J.; Liu, J.; et al. Magnesium intake and mortality due to liver diseases: Results from the third national health and nutrition examination survey cohort. *Sci. Rep.* **2017**, *7*, 17913. [CrossRef]
16. Nielsen, F.H. Magnesium deficiency and increased inflammation: Current perspectives. *J. Inflamm. Res.* **2018**, *11*, 25–34. [CrossRef]

17. Kim, H.N.; Kim, S.H.; Eun, Y.M.; Song, S.W. Effects of zinc, magnesium, and chromium supplementation on cardiometabolic risk in adults with metabolic syndrome: A double-blind, placebo-controlled randomised trial. *J. Trace Elem. Med. Biol.* **2018**, *48*, 166–171. [CrossRef]
18. Nielsen, F.H.; Johnson, L.K.; Zeng, H. Magnesium supplementation improves indicators of low magnesium status and inflammatory stress in adults older than 51 years with poor quality sleep. *Magnes. Res.* **2010**, *23*, 158–168.
19. Mooren, F.C.; Krüger, K.; Völker, K.; Golf, S.W.; Wadepuhl, M.; Kraus, A. Oral magnesium supplementation reduces insulin resistance in non-diabetic subjects - a double-blind, placebo-controlled, randomized trial. *Diabetes Obes. Metab.* **2011**, *13*, 281–284. [CrossRef]
20. Veronese, N.; Watutantrige-Fernando, S.; Luchini, C.; Solmi, M.; Sartore, G.; Sergi, G.; Manzato, E.; Barbagallo, M.; Maggi, S.; Stubbs, B. Effect of magnesium supplementation on glucose metabolism in people with or at risk of diabetes: A systematic review and meta-analysis of double-blind randomized controlled trials. *Eur. J. Clin. Nutr.* **2016**, *70*, 1354–1359. [CrossRef]
21. Song, Y.; He, K.; Levitan, E.B.; Manson, J.E.; Liu, S. Effects of oral magnesium supplementation on glycaemic control in type 2 diabetes: A meta-analysis of randomized double-blind controlled trials. *Diabetes Metab.* **2006**, *23*, 1050–1056. [CrossRef] [PubMed]
22. Pittas, A.G.; Lau, J.; Hu, F.B.; Dawson-Hughes, B. The role of vitamin d and calcium in type 2 diabetes. A systematic review and meta-analysis. *J. Clin. Endocrinol. Metab.* **2007**, *92*, 2017–2029. [CrossRef] [PubMed]
23. Kim, M.J.; Lee, K.J. Analysis of the dietary factors associated with suspected pediatric nonalcoholic fatty liver disease and potential liver fibrosis: Korean national health and nutrition examination survey 2014–2017. *BMC Pediatr.* **2020**, *20*, 121. [CrossRef] [PubMed]
24. Chen, T.C.; Clark, J.; Riddles, M.K.; Mohadjer, L.K.; Fakhouri, T.H.I. National health and nutrition examination survey, 2015–2018: Sample design and estimation procedures. National center for health statistics. *Vital Health Stat 2* **2020**, *184*, 1–35.
25. Xiao, G.; Zhu, S.; Xiao, X.; Yan, L.; Yang, J.; Wu, G. Comparison of laboratory tests, ultrasound, or magnetic resonance elastography to detect fibrosis in patients with nonalcoholic fatty liver disease: A meta-analysis. *Hepatology* **2017**, *66*, 1486–1501. [CrossRef]
26. Jiang, W.; Huan, S.; Teng, H.; Wang, P.; Wu, M.; Zhou, X.; Ran, H. Diagnostic accuracy of point shear wave elastography and transient elastography for staging hepatic fibrosis in patients with non-alcoholic fatty liver disease: A meta-analysis. *BMJ Open* **2018**, *8*, e021787. [CrossRef]
27. Tsochatzis, E.A.; Gurusamy, K.S.; Ntaoula, S.; Cholongitas, E.; Davidson, B.R.; Burroughs, A.K. Elastography for the diagnosis of severity of fibrosis in chronic liver disease: A meta-analysis of diagnostic accuracy. *J. Hepatol.* **2011**, *54*, 650–659. [CrossRef]
28. Eddowes, P.J.; Sasso, M.; Allison, M.; Tsochatzis, E.; Anstee, Q.M.; Sheridan, D.; Guha, I.N.; Cobbold, J.F.; Deeks, J.J.; Paradis, V.; et al. Accuracy of fibroscan controlled attenuation parameter and liver stiffness measurement in assessing steatosis and fibrosis in patients with nonalcoholic fatty liver disease. *Gastroenterology* **2019**, *156*, 1717–1730. [CrossRef]
29. Ahluwalia, N.; Dwyer, J.; Terry, A.; Moshfegh, A.; Johnson, C. Update on nhanes dietary data: Focus on collection, release, analytical considerations, and uses to inform public policy. *Adv. Nutr.* **2016**, *7*, 121–134. [CrossRef]
30. *Physical Activity Guidelines for Americans*, 2nd ed.; Department of Health and Human Services: Washington, DC, USA, 2018.
31. Forouhi, N.G.; Wareham, N.J. Epidemiology of diabetes. *Medicine (Abingdon)* **2014**, *42*, 698–702. [CrossRef]
32. Centers for Disease Control and Prevention. Testing for hcv infection: An update of guidance for clinicians and laboratorians. *MMWR* **2013**, *62*, 1–4.
33. Krajden, M.; McNabb, G.; Petric, M. The laboratory diagnosis of hepatitis b virus. *Can. J. Infect. Dis. Med. Microbiol.* **2005**, *16*, 65–72. [CrossRef] [PubMed]
34. Division of the National Health and Nutrition Examination Surveys. *The National Health and Nutrition Examination Survey (Nhanes) Analytic and Reporting Guidelines*; Government Publishing Office: Washington, DC, USA, 2020.
35. Tao, M.H.; Dai, Q.; Millen, A.E.; Nie, J.; Edge, S.B.; Trevisan, M.; Shields, P.G.; Freudenheim, J.L. Associations of intakes of magnesium and calcium and survival among women with breast cancer: Results from western new york exposures and breast cancer (web) study. *Am. J. Cancer Res.* **2015**, *6*, 105–113. [PubMed]
36. Rodríguez-Hernández, H.; Gonzalez, J.L.; Rodríguez-Morán, M.; Guerrero-Romero, F. Hypomagnesemia, insulin resistance, and non-alcoholic steatohepatitis in obese subjects. *Arch. Med. Res.* **2005**, *36*, 362–366. [CrossRef]
37. Younossi, Z.M.; Stepanova, M.; Younossi, Y.; Golabi, P.; Mishra, A.; Rafiq, N.; Henry, L. Epidemiology of chronic liver disease in the USA in the past three decades. *Gut* **2020**, *69*, 564–568. [CrossRef]
38. Agarwal, S.; Reider, C.; Brooks, J.R.; Fulgoni, V.L., 3rd. Comparison of prevalence of inadequate nutrient intake based on body weight status of adults in the united states: An analysis of nhanes 2001–2008. *J. Am. Coll. Nutr.* **2015**, *34*, 126–134. [CrossRef]
39. Liu, J.; Zhu, X.; Fulda, K.G.; Chen, S.; Tao, M.H. Comparison of dietary micronutrient intakes by body weight status among mexican-american and non-hispanic black women aged 19–39 years: An analysis of nhanes 2003–2014. *Nutrients* **2019**, *11*, 2846. [CrossRef]
40. Hoenderop, J.G.; Bindels, R.J. Epithelial ca2+ and mg2+ channels in health and disease. *J. Am. Soc. Nephrol.* **2005**, *16*, 15–26. [CrossRef]
41. Nielsen, F.H.; Milne, D.B.; Gallagher, S.; Johnson, L.; Hoverson, B. Moderate magnesium deprivation results in calcium retention and altered potassium and phosphorus excretion by postmenopausal women. *Magnes. Res.* **2007**, *20*, 19–31.
42. Sigrist, R.M.S.; Liau, J.; Kaffas, A.E.; Chammas, M.C.; Willmann, J.K. Ultrasound elastography: Review of techniques and clinical applications. *Theranostics* **2017**, *7*, 1303–1329. [CrossRef]

43. Guenther, P.M.; Bowman, S.A.; Goldman, J.D. Alcoholic Beverage Consumption by Adults 21 Years and Over in the United States: Results from the National Health and Nutrition Examinationsurvey, 2003–2006. Technical Report. Center for Nutrition Policy and Promotion and Agriculturalresearch Service, U.S. Department of Agriculture. Available online: www.Cnpp.Usda.Gov/publications/dietaryguidelines/2010/meeting5/alcoholicbeveragesconsumption.Pdf (accessed on 3 November 2020).
44. Murakami, K.; Livingstone, M.B. Prevalence and characteristics of misreporting of energy intake in us adults: Nhanes 2003–2012. *Br. J. Nutr.* **2015**, *114*, 1294–1303. [CrossRef] [PubMed]
45. Roark, R.A.; Niederhauser, V.P. Fruit and vegetable intake: Issues with definition and measurement. *Public Health Nutr.* **2013**, *16*, 2–7. [CrossRef] [PubMed]

Article

# Hypomagnesemia Is a Risk Factor for Infections after Kidney Transplantation: A Retrospective Cohort Analysis

Balazs Odler [1,2], Andras T. Deak [1,2], Gudrun Pregartner [3], Regina Riedl [3], Jasmin Bozic [1], Christian Trummer [4], Anna Prenner [1,2], Lukas Söllinger [1], Marcell Krall [1], Lukas Höflechner [1], Carina Hebesberger [1,2], Matias S. Boxler [1], Andrea Berghold [3], Peter Schemmer [2,5], Stefan Pilz [4] and Alexander R. Rosenkranz [1,2,*]

[1] Division of Nephrology, Department of Internal Medicine, Medical University of Graz, A-8036 Graz, Austria; balazs.odler@medunigraz.at (B.O.); andras.deak@medunigraz.at (A.T.D.); jasmin.bozic@stud.medunigraz.at (J.B.); anna.prenner@medunigraz.at (A.P.); lukas.soellinger@stud.medunigraz.at (L.S.); marcell.krall@medunigraz.at (M.K.); lukas.hoeflechner@klinikum-graz.at (L.H.); carina.hebesberger@medunigraz.at (C.H.); matias.boxle@medunigraz.at (M.S.B.)

[2] Transplant Center Graz, Medical University of Graz, A-8036 Graz, Austria; peter.schemmer@medunigraz.at

[3] Institute of Medical Informatics, Statistics and Documentation, Medical University of Graz, A-8036 Graz, Austria; gudrun.pregartner@medunigraz.at (G.P.); regina.riedl@medunigraz.at (R.R.); andrea.berghold@medunigraz.at (A.B.)

[4] Division of Endocrinology and Diabetology, Department of Internal Medicine, Medical University of Graz, A-8036 Graz, Austria; christian.trummer@medunigraz.at (C.T.); stefan.pilz@medunigraz.at (S.P.)

[5] General, Visceral and Transplant Surgery, Department of Surgery, Medical University of Graz, A-8036 Graz, Austria

* Correspondence: alexander.rosenkranz@medunigraz.at; Tel.: +43-316-38512170; Fax: +43-316-38514426

Citation: Odler, B.; Deak, A.T.; Pregartner, G.; Riedl, R.; Bozic, J.; Trummer, C.; Prenner, A.; Söllinger, L.; Krall, M.; Höflechner, L.; et al. Hypomagnesemia Is a Risk Factor for Infections after Kidney Transplantation: A Retrospective Cohort Analysis. Nutrients 2021, 13, 1296. https://doi.org/10.3390/nu13041296

Academic Editors: Sara Castiglioni, Giovanna Farruggia and Concettina Cappadone

Received: 22 March 2021
Accepted: 7 April 2021
Published: 14 April 2021

**Publisher's Note:** MDPI stays neutral with regard to jurisdictional claims in published maps and institutional affiliations.

Copyright: © 2021 by the authors. Licensee MDPI, Basel, Switzerland. This article is an open access article distributed under the terms and conditions of the Creative Commons Attribution (CC BY) license (https://creativecommons.org/licenses/by/4.0/).

**Abstract:** Introduction: Magnesium ($Mg^{2+}$) deficiency is a common finding in the early phase after kidney transplantation (KT) and has been linked to immune dysfunction and infections. Data on the association of hypomagnesemia and the rate of infections in kidney transplant recipients (KTRs) are sparse. Methods: We conducted a single-center retrospective cohort study of KTRs transplanted between 2005 and 2015. Laboratory data, including serum $Mg^{2+}$ (median time of the $Mg^{2+}$ measurement from KT: 29 days), rate of infections including mainly urinary tract infections (UTI), and common transplant-related viral infections (CMV, polyoma, EBV) in the early phase after KT were recorded. The primary outcome was the incidence of infections within one year after KT, while secondary outcomes were hospitalization due to infection, incidence rates of long-term (up to two years) infections, and all-cause mortality. Results: We enrolled 376 KTRs of whom 229 patients (60.9%) suffered from $Mg^{2+}$ deficiency defined as a serum $Mg^{2+}$ < 0.7 mmol/L. A significantly higher incidence rate of UTIs and viral infections was observed in patients with versus without $Mg^{2+}$ deficiency during the first year after KT (58.5% vs. 47.6%, $p = 0.039$ and 69.9% vs. 51.7%, $p < 0.001$). After adjustment for potential confounders, serum $Mg^{2+}$ deficiency remained an independent predictor of both UTIs and viral infections (odds ratio (OR): 1.73, 95% CI: 1.04–2.86, $p = 0.035$ and OR: 2.05, 95% CI: 1.23–3.41, $p = 0.006$). No group differences according to $Mg^{2+}$ status in hospitalizations due to infections and infection incidence rates in the 12–24 months post-transplant were observed. In the Cox regression analysis, $Mg^{2+}$ deficiency was not significantly associated with all-cause mortality (HR: 1.15, 95% CI: 0.70–1.89, $p = 0.577$). Conclusions: KTRs suffering from $Mg^{2+}$ deficiency are at increased risk of UTIs and viral infections in the first year after KT. Interventional studies investigating the effect of $Mg^{2+}$ supplementation on $Mg^{2+}$ deficiency and viral infections in KTRs are needed.

**Keywords:** magnesium; kidney transplantation; infection; urinary tract infection; viral; incidence; immunosuppression; immunity

## 1. Introduction

Magnesium ($Mg^{2+}$) is an essential trace element and the second most abundant intracellular mineral in the body [1]. It exerts several crucial functions in the human body, including actions as a cofactor in cell proliferation and cellular energy metabolism, and it serves as a cofactor of many enzymatic processes [2–4].

Accumulating evidence suggests a possible link between $Mg^{2+}$ and the immune system. $Mg^{2+}$ plays an essential role in controlling immune function by exerting an extended effect on several immune processes, such as immunoglobulin synthesis, immune cell adherence, antibody-dependent cytolysis, and regulation of Th1/Th2 responses [5,6]. Recent studies on different $Mg^{2+}$ permeable ion channels and transporters provided new insights into the role of $Mg^{2+}$ in immune responses [7]. These findings led to the discovery that mutations in $Mg^{2+}$ transport systems are the underlying cause of a mild form of combined immune deficiency (CID) named X-linked immunodeficiency with magnesium defect (XMEN) [8], thus supporting the notion that $Mg^{2+}$ signaling is critical for natural killer (NK) and $CD8^+$ T-cell function.

Among kidney transplant recipients (KTRs), a high incidence of hypomagnesemia is observed, which seems to be related to the widespread use of calcineurin inhibitors (CNIs), and especially tacrolimus [9]. This phenomenon can mainly be explained by CNI-induced downregulation of transport proteins in the renal tubules leading to renal $Mg^{2+}$ excretion and wasting [10–13]. The net state of immunosuppression, including induction, maintenance, and anti-rejection therapies, is mainly recognized to increase susceptibility to infections and malignancies, some of which are infection triggered [14].

Opportunistic infections are the most critical complications and a significant cause of graft loss and mortality in KTRs [15,16]. In dialysis patients, hypomagnesemia was associated with an increased risk of death due to infection [17]. In XMEN, $Mg^{2+}$ supplementation increases intracellular $Mg^{2+}$ and normalizes Epstein–Barr virus (EBV) cellular immune response, thus leading to effective viral load suppression [18]. In KTRs, a lower serum $Mg^{2+}$ concentration was associated with an increased risk of severe infections [19]. Nevertheless, data on the association between mild transplant-related infections and $Mg^{2+}$ levels in KTRs in the daily routine practice in the transplant out- and inpatient setting are partly lacking. Knowledge gained from such data may potentially guide our clinical practice on how to deal with diagnosing and treating $Mg^{2+}$ deficiency in KTRs.

Considering the limited data on the relationship between infections after kidney transplantation (KT) and hypomagnesemia, we aimed to examine this association in a cohort of KTRs using retrospective data from a single-center transplant database.

## 2. Materials and Methods

### 2.1. Study Design and Patients' Characteristics

We conducted a single-center retrospective cohort study of 376 consecutive KTRs transplanted at the Transplant Center Graz (Medical University of Graz) from 1 January 2005 to 31 December 2015. Patients with concomitant serum $Mg^{2+}$ and 25(OH)-vitamin D measurement within three months after KT were identified using our database as published previously (Figure 1) [20].

Male and female patients aged above 18 years were eligible for inclusion in the study if: (1) they received at least one kidney allograft, (2) had a serum $Mg^{2+}$ and 25(OH)-vitamin D measurement within 3 months after KT. Patients were excluded if age was <18 years, they had combined organ transplantation (e.g., pancreas–kidney), or KT follow-up took place at another center. Data were collected after last KT if the patient had more than one KT during the observation period. Data on baseline patient characteristics, primary renal disease, comorbidities, renal replacement therapy (RRT), cytomegalovirus (CMV)-status, transplantation-related data, as well as data on delayed graft function and laboratory findings were collected. Induction (IL2-receptor antagonists (IL-2Ra) or anti-thymocyte globulin (ATG)), maintenance immunosuppression (CNIs, antiproliferative agents (including mycophenolate mofetil (MMF), mycophenolic acid (MPA) and azathioprine (AZA)) or mTOR inhibitors (mTORi—sirolimus, everolimus)) therapies were also recorded. All patients received a standardized corticosteroid therapy with an initial dose of 500 mg prednisolone at the day of transplantation and subsequent tapering to 5 mg per day at month three after KT.

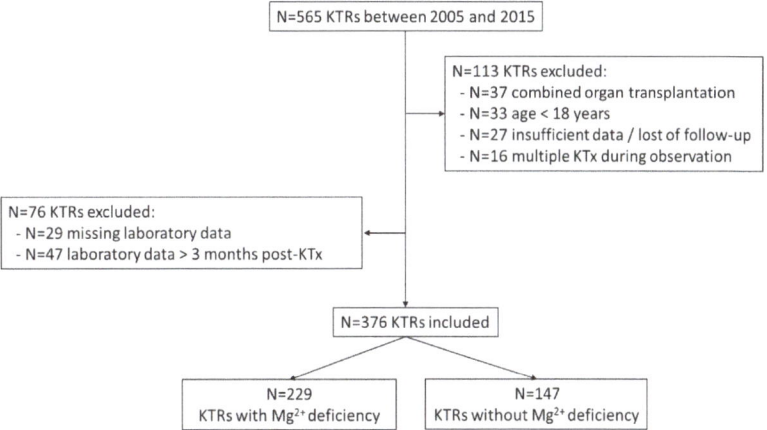

**Figure 1.** Selection of study population with concomitant serum $Mg^{2+}$ and 25(OH)-vitamin D measurements. (KTR: kidney transplant recipient; KTx: kidney transplantation; $Mg^{2+}$: magnesium).

All data were derived retrospectively from the KT and electronic medical records of our center and the Austria Dialysis and Transplantation Registry (OEDTR), as stated previously [20]. The date of KT (index date) was registered for the calculation of time to outcome event. As stated previously, in more than 98% of cases, the ethnicity of the patients was Caucasian, representing the Austrian ethnical background, and was not further specified [20].

### 2.2. Type of Infections, Laboratory, and Clinical Definitions

Data on the most common transplant-related infections, namely urinary tract infections (UTIs) and viral infections, including CMV, polyomavirus (polyoma), and Epstein–Barr virus (EBV), were recorded from 0–12 and 12–24 months after KT. Severe infections requiring hospitalization, defined by the principal diagnosis in the physician's letter after discharge from the hospital, were also recorded. The diagnostic criteria of UTI in the study were the presence of a positive urine culture ($\geq 10^5$ cfu/mL). CMV, polyoma, and EBV quantitation were done using a real-time polymerase chain reaction (rtPCR) procedure validated by the Diagnostic and Research Institute of Hygiene, Microbiology and Environmental Medicine at the Medical University of Graz. Additionally, if PCR was not available, CMV infections identified by the pp65 antigenemia assay method were also collected.

Serum $Mg^{2+}$ concentrations were measured and validated in the Laboratory Medicine Institute of the Medical University of Graz (normal range of 0.7 to 1.10 mmol/L). $Mg^{2+}$ deficiency was defined as $Mg^{2+} < 0.7$ mmol/L, whereas $Mg^{2+}$ levels $\geq 0.7$ mmol/L were considered as sufficient using a single $Mg^{2+}$ measurement. The estimated glomerular filtration rate (eGFR) was calculated using the CKD-EPI (Chronic Kidney Disease Epidemiology Collaboration) Creatinine Equation [21]. Serum levels of tacrolimus and cyclosporine A were collected within one week of $Mg^{2+}$ measurements.

Pre-transplant diabetes mellitus (DM) was defined according to the American Diabetes Association (ADA) Guidelines [22] or as an intake of glucose-lowering drugs according to the patient records.

Delayed graft function was defined as acute kidney injury (AKI) that occurred in the first week of KT, which necessitated dialysis intervention [23].

*2.3. Outcomes*

The primary outcome of this study was the incidence of UTIs and viral infections (CMV, polyoma, and/or EBV) within one year after KT. Secondary outcomes were hospitalization due to infection, incidence rates of long-term (up to two years) infections, as well as all-cause mortality.

All-cause mortality data were requested from the database of the Main Association of Austrian Social Insurance Institutions (Hauptverband der österreichischen Sozialversicherungsträger, last accessed on 15 August 2019). Data on infection-related mortality during the first year after KT, including those with death due to sepsis, were also collected from OEDTR.

*2.4. Statistical Analysis*

Continuous parameters are summarized as the median and interquartile range (IQR), whereas categorical parameters are presented as absolute and relative frequencies. Differences between $Mg^{2+}$ deficient and non-deficient patients were assessed either with the Mann–Whitney U or $\chi^2$ test. To identify factors associated with infections, a logistic regression analysis was performed using clinical and known risk factors and possible confounders (age at time of KT, sex, body mass index (BMI), dialysis modality (hemodialysis), dialysis vintage (<1 vs. $\geq$1 year), living kidney donation, donor and recipient CMV seropositivity, glomerulonephritis (GN) as primary renal disease, maintenance immunosuppression (defined as highest tertile of tacrolimus and cyclosporine A serum level at the time of the $Mg^{2+}$ measurement), nicotine abuse, diabetes mellitus (DM), delayed graft function, eGFR, albumin, and previous KT). To further analyze the influence of $Mg^{2+}$ deficiency on the incidence of infections, the analyses for $Mg^{2+}$ deficiency were adjusted for all variables listed above. Additionally, logistic regression analyses were performed to identify possible risk factors leading to an $Mg^{2+}$ deficiency and to test the influence of the induction therapy on infection risk. Kaplan–Meier and Cox proportional hazard regression analyses were performed to assess the influence of $Mg^{2+}$ deficiency on overall mortality. Results are presented as either odds ratios (ORs, logistic regression) or hazard ratios (HRs, Cox regression) with the respective 95% confidence intervals (CIs). A $p$-value < 0.05 was considered statistically significant, and all analyses were performed using SAS version 9.4 (SAS Institute, Cary, NC, USA).

## 3. Results

*3.1. Patients' Characteristics and Laboratory Findings*

The study population consisted of 376 KTRs. The majority of the patients (258, 68.6%) were male, and the median age was 52.0 years (IQR 41.0–62.0). Detailed patient characteristics and laboratory findings of the study population are shown in Table 1. The median $Mg^{2+}$ concentration of the whole study population was 0.67 (IQR 0.61–0.74) mmol/L, measured after a median time of 29 days (IQR: 20–45) after KT. A total of 229 patients (60.9%) had magnesium deficiency with a median serum $Mg^{2+}$ concentration of 0.63 (IQR 0.58–0.66) mmol/L. KTRs with a $Mg^{2+}$ deficiency had significantly higher rates of donor CMV seropositivity than patients without a deficiency (median 51.8% vs. 39.0%, $p$ = 0.017). Other than that, the study groups were similar regarding clinical characteristics, underlying kidney diseases, comorbidities, and medication (Table 1).

Table 1. Baseline patient characteristics and laboratory results in the whole study population and stratified according to the presence and absence of $Mg^{2+}$ deficiency.

| | Whole Study Population (N = 376) | Patients without $Mg^{2+}$ Deficiency (N = 147) | Patients with $Mg^{2+}$ Deficiency (N = 229) | p-Value |
|---|---|---|---|---|
| Age at time of KT (years) | 52.0 (41.0–62.0) | 50.0 (40.0–61.0) | 53.0 (42.0–62.0) | 0.133 |
| Gender (female) | 118 (31.4) | 41 (27.9) | 77 (33.6) | 0.242 |
| BMI (kg/m$^2$) | 24.9 (22.1–28.1) | 24.4 (22.1–28.1) | 25.2 (22.3–28.0) | 0.372 |
| Nicotine abuse | 169 (44.9) | 71 (48.3) | 98 (42.8) | 0.295 |
| **Dialysis-related data** | | | | |
| Hemodialysis | 293 (77.9) | 114 (77.6) | 179 (78.2) | 0.888 |
| Peritoneal dialysis | 62 (16.5) | 25 (17.0) | 37 (16.2) | 0.829 |
| Preemptive KT | 21 (5.6) | 8 (5.4) | 13 (5.7) | 0.923 |
| Dialysis vintage (months) | 40.5 (22.0–71.0) | 35.0 (18.0–58.0) | 42.0 (22.0–75.0) | 0.208 |
| **Comorbidities** | | | | |
| Diabetes mellitus | 58 (15.4) | 28 (19.0) | 30 (13.1) | 0.119 |
| Dyslipidemia | 197 (52.4) | 75 (51.0) | 122 (53.3) | 0.669 |
| Hypertension | 357 (94.9) | 138 (93.9) | 219 (95.6) | 0.448 |
| Coronary heart disease | 37 (9.8) | 16 (10.9) | 21 (9.2) | 0.586 |
| **Transplantation-related data** | | | | |
| Living kidney donation | 47 (12.5) | 19 (12.9) | 28 (12.2) | 0.842 |
| Previous KT | 80 (21.3) | 27 (18.4) | 53 (23.1) | 0.269 |
| Donor CMV seropositivity | 170 (46.8) | 55 (39.0) | 115 (51.8) | **0.017** |
| Recipient CMV seropositivity | 225 (61.8) | 85 (59.9) | 140 (63.1) | 0.539 |
| GN as primary renal disease | 130 (34.6) | 50 (34.0) | 80 (34.9) | 0.855 |
| Delayed graft function | 133 (35.4) | 56 (38.1) | 77 (33.6) | 0.376 |
| **Immunosuppression** | | | | |
| CNI | 374 (99.5) | 145 (98.6) | 229 (100) | 0.077 |
| mTOR inhibitor | 3 (0.8) | 2 (1.4) | 1 (0.4) | 0.326 |
| Antiproliferative agents | 372 (98.9) | 144 (98.0) | 228 (99.6) | 0.139 |
| **Laboratory data *** | | | | |
| Leukocytes (10 × 9/L) | 8.3 (6.3–10.7) | 8.7 (6.4–10.6) | 8.1 (6.3–10.7) | 0.471 |
| C-reactive protein (mg/dL) | 2.7 (1.0–5.9) | 2.8 (1.0–7.2) | 2.4 (1.0–4.9) | 0.151 |
| Parathyroid hormone (pg/mL) | 155.2 (105.8–234.8) | 155.7 (94.6–237.1) | 153.6 (108.5–228.7) | 0.644 |
| Calcium, total (mmol/L) | 2.4 (2.3–2.5) | 2.4 (2.3–2.5) | 2.4 (2.3–2.5) | 0.239 |
| Phosphate (mg/dL) | 2.3 (1.9–2.9) | 2.5 (2.0–3.3) | 2.2 (1.7–2.7) | **<0.001** |
| Bicarbonate (mmol/L) | 22.1 (20.1–24.5) | 22.2 (19.9–24.9) | 22.0 (20.3–24.2) | 0.594 |
| Creatinine (mg/dL) | 1.5 (1.3–1.9) | 1.7 (1.3–2.3) | 1.5 (1.3–1.8) | **0.001** |
| eGFR (ml/min/1.73 m$^2$) | 47.0 (35.5–59.8) | 41.7 (31.0–59.6) | 49.2 (38.4–60.4) | **0.009** |
| Albumin (g/dL) | 3.9 (3.5–4.3) | 3.8 (3.4–4.4) | 3.9 (3.5–4.3) | 0.950 |
| 25(OH)-vitamin D (ng/mL) | 23.7 (15.2–31.4) | 21.1 (14.0–29.2) | 25.2 (17.3–32.8) | **0.007** |
| **Magnesium supplementation** | 57 (15.2) | 25 (17.0) | 32 (14.0) | 0.424 |

Statistically significant p-values appear in boldface type (p < 0.05). Continuous variables are expressed as median (25th to 75th percentile). Categorical variables are n (%). * Laboratory data at time of $Mg^{2+}$ measurement. Subsections appear in boldface type. Abbreviations: BMI: body mass index, CMV: cytomegalovirus, CNI: calcineurin inhibitor, eGFR: estimated glomerular filtration rate, GN: glomerulonephritis, KT: kidney transplantation, mTOR: mammalian target of rapamycin.

Data of clinical and laboratory variables were available in >95% of all study participants, except for bicarbonate, which was just available in n = 339 (90.2%).

$Mg^{2+}$ sufficient patients had significantly worse kidney function based on creatinine and eGFR measurements (p = 0.001 and p = 0.009, respectively). A significant correlation between $Mg^{2+}$ levels and eGFR (Pearson correlation coefficient r = −0.138, p = 0.008) was observed.

At the time of the $Mg^{2+}$ measurement, most patients (88.0%) used tacrolimus. The percentage of patients with an $Mg^{2+}$ deficiency was significantly higher in patients with higher tacrolimus levels (percentages in the respective tacrolimus tertials: 56.1% vs. 61.7% vs. 74.6%; $p = 0.014$).

In total, $n = 52$ (15.2%) KTRs received oral $Mg^{2+}$ supplementation (Table 1). No statistically significant difference in $Mg^{2+}$ supplementation between patients with and without $Mg^{2+}$ deficiency was observed ($p = 0.424$).

### 3.2. Percentage and Incidence of Infections

During the first year after KT, UTI was observed in 204 (54.3%), CMV in 182 (48.4%), EBV in 96 (25.5%), and polyoma in 33 (8.8%) patients. In total, 236 (62.8%) patients had a viral infection. The incidences for all infection types were lower during the second year after KT: 69 (19.0%) patients had UTI, 31 (8.5%) CMV, 5 (1.4%) polyoma, and 10 (2.7%) EBV for a total of 41 (11.3%) patients with a viral infection.

There was significant difference in UTI incidence rates within the first 12 months between the patients with and without $Mg^{2+}$ deficiency (58.5% vs. 47.6%, $p = 0.039$). Moreover, a significantly higher incidence rate of viral infections was observed in patients with $Mg^{2+}$ deficiency in the first year after KT (69.9% vs. 51.7%, $p < 0.001$). There were significantly lower incidence rates of UTI and viral infections between the 12–24-month time period as compared to the first year after KT (McNemar test: both $p < 0.001$). However, no significant differences in the infection incidence rates 12–24 months post-transplant were observed between the study groups ($p = 0.807$ for UTI and $p = 0.474$ for viral infections). A total of 72 (19.1%) patients required hospital admission due to infection during the first year, and 27 (7.4%) in the second year after KT. The incidence rates of UTI and viral infections at the two time intervals after KT according to the study groups ($Mg^{2+} < 0.7$ mmol/L and $Mg^{2+}$ levels $\geq 0.7$ mmol/L) are shown in Table 2.

**Table 2.** Incidence rate of all urinary tract and viral infections during the first and second year after KT.

| Type of Infections | Whole Study Population | | Patients without $Mg^{2+}$ Deficiency | | Patients with $Mg^{2+}$ Deficiency | |
|---|---|---|---|---|---|---|
| | N = 376 | N = 364 | N = 147 | N = 143 | N = 229 | N = 221 |
| | 0–12 Months | 12–24 Months | 0–12 Months | 12–24 Months | 0–12 Months | 12–24 Months |
| Urinary tract infections | 204 (54.3) | 69 (19.0) | 70 (47.6) | 28 (19.6) | 134 (58.5) | 41 (18.6) |
| Viral infections | 236 (62.8) | 41 (11.3) | 76 (51.7) | 14 (9.8) | 160 (69.9) | 27 (12.2) |
| Detailed viral infections | | | | | | |
| CMV | 182 (48.4) | 31 (8.5) | 57 (38.8) | 10 (7.0) | 125 (54.6) | 21 (9.5) |
| Polyoma | 33 (8.8) | 5 (1.4) | 8 (5.4) | 1 (0.7) | 25 (10.9) | 4 (1.8) |
| EBV | 96 (25.5) | 10 (2.7) | 37 (25.2) | 4 (2.8) | 59 (25.8) | 6 (2.7) |

Categorical variables are n (%). Abbreviations: CMV: cytomegalovirus, EBV: Epstein–Barr virus.

### 3.3. Risk Factors for Infections

Regarding $Mg^{2+}$ deficiency, the univariable logistic regression analysis 12 months after KT revealed significant associations with both UTIs and viral infections compared to $Mg^{2+}$ sufficiency (OR: 1.55, 95% CI: 1.02–2.35, $p = 0.039$ and OR: 2.17, 95% CI: 1.41–3.33, $p < 0.001$). Age at the time of KT, female sex, BMI, and delayed graft function were significantly associated with higher UTI incidence, while higher eGFR, living kidney donation, and serum albumin levels were associated with a decreased incidence of UTIs within 12 months after KT (Table S1). On the other hand, age at the time of KT and donor CMV serostatus showed a significant higher association with the incidence of viral infections, while a higher eGFR and serum albumin levels also showed an association with decreased viral infection incidence rates within 12 months after KT (Table S2).

In multivariable logistic regression analyses, including all parameters, serum $Mg^{2+}$ deficiency remained a significant predictor of UTIs and viral infections during the first year after KT (OR: 1.73, 95% CI: 1.04–2.86, $p$ = 0.035 and OR: 2.05, 95% CI: 1.23–3.41, $p$ = 0.006, respectively) (Table 3).

**Table 3.** Multivariable logistic regression analysis of confounders for urinary tract and viral infections incidence during the first year after KT.

| Test Variable | Urinary Tract Infections | | | Viral Infections | | |
|---|---|---|---|---|---|---|
| | OR | 95% CI | $p$-Value | OR | 95% CI | $p$-Value |
| Age at the time of KT (years) | 1.01 | 0.99–1.04 | 0.179 | 1.00 | 0.98–1.02 | 0.982 |
| Gender (female) | **4.57** | **2.56–8.16** | **<0.001** | 1.14 | 0.66–1.98 | 0.635 |
| BMI (kg/m$^2$) | 1.06 | 0.99–1.13 | 0.090 | 1.05 | 0.98–1.12 | 0.179 |
| Nicotine abuse | 1.36 | 0.83–2.22 | 0.217 | 1.65 | 1.00–2.74 | 0.051 |
| Serum Mg$^{2+}$ (deficiency) | **1.73** | **1.04–2.86** | **0.035** | **2.05** | **1.23–3.41** | **0.006** |
| eGFR | 1.00 | 0.98–1.01 | 0.603 | 0.99 | 0.98–1.00 | 0.150 |
| Albumin | 0.74 | 0.49–1.12 | 0.155 | 0.73 | 0.48–1.12 | 0.146 |
| CNI serum level (highest tertile) | 0.98 | 0.59–1.63 | 0.940 | 1.68 | 0.98–2.86 | 0.059 |
| Dialysis vintage (<1 year) | 1.06 | 0.43–2.59 | 0.898 | 0.66 | 0.27–1.61 | 0.362 |
| Hemodialysis | 0.85 | 0.47–1.53 | 0.587 | 0.91 | 0.50–1.66 | 0.760 |
| Previous KT | 0.84 | 0.46–1.54 | 0.573 | 0.88 | 0.47–1.64 | 0.681 |
| Living kidney donation | 0.53 | 0.21–1.38 | 0.194 | 1.57 | 0.61–4.07 | 0.349 |
| Donor CMV seropositivity | 1.12 | 0.69–1.82 | 0.650 | **2.54** | **1.54–4.18** | **<0.001** |
| Recipient CMV seropositivity | 0.67 | 0.41–1.11 | 0.123 | 1.66 | 1.00–2.76 | 0.051 |
| Delayed graft function | 1.47 | 0.83–2.60 | 0.190 | 1.19 | 0.97–2.13 | 0.552 |
| Diabetes mellitus | 1.06 | 0.52–2.18 | 0.877 | 0.82 | 0.40–1.69 | 0.589 |
| GN as primary kidney disease | 0.72 | 0.44–1.20 | 0.205 | 0.80 | 0.48–1.34 | 0.387 |

Statistically significant $p$-values appear in boldface type ($p < 0.05$). Abbreviations: BMI: body mass index, CI: confidence interval; CMV: cytomegalovirus, CNI: calcineurin inhibitor, eGFR: estimated glomerular filtration rate, GN: glomerulonephritis, KT: kidney transplantation, mTOR: mammalian target of rapamycin, OR: odds ratio.

In a subgroup analysis for patients receiving IL2Ra ($n$ = 242), Mg$^{2+}$ deficiency remained significantly associated with higher incidences of viral infections during the first year after KT (OR: 2.484, 95% CI: 1.28–4.80, $p$ = 0.007) but not for UTIs (OR: 1.22, 95% CI: 0.66–2.27, $p$ = 0.529). A similar analysis in patients receiving ATG ($n$ = 50) did not show significant associations, likely due to the small number of observations.

*3.4. Risk Factors for Magnesium Deficiency*

In the univariable analysis, serum phosphorus level < 2.6 mg/dL (OR: 2.54, 95% CI: 1.64–3.93, $p$ < 0.001), CRP levels > 5 mg/dL (OR: 0.61, 95% CI: 0.38–0.96, $p$ = 0.035), high CNI serum levels (OR: 2.29, 95% CI: 1.35–3.89, $p$ = 0.002 for comparison of highest vs. lowest tertial), and donor CMV seropositivity (OR: 1.68, 95% CI: 1.09–2.58, $p$ = 0.018) were associated with the presence of Mg$^{2+}$ deficiency. All factors investigated for the risk of hypomagnesemia are shown in Table S3.

*3.5. Mortality*

During the first year after transplantation, 12 (3.2%) patients died, five of which had an infection as a likely cause of death. During a median follow-up period of 6.7 (IQR 4.9–9.3) years, 67 (17.8%) patients died due to any cause. In Cox regression analysis, Mg$^{2+}$ deficiency was not significantly associated with all-cause mortality (HR: 1.15, 95% CI: 0.70–1.89, $p$ = 0.577; Figure S1). However, age at the time of the transplantation (HR: 1.08, 95% CI: 1.05–1.10, $p$ < 0.001), BMI (HR: 1.07, 95% CI: 1.01–1.13, $p$ = 0.017), DM (HR: 2.04, 95% CI: 1.18–3.50, $p$ = 0.010), delayed graft function (HR: 1.88, 95% CI: 1.16–3.03, $p$ = 0.010) and donor CMV seropositivity (HR: 1.66, 95% CI: 1.02–2.73, $p$ = 0.044) were associated with an increased risk of all-cause mortality. In contrast, glomerulonephritis as primary renal disease (HR: 0.42, 95% CI: 0.23–0.76, $p$ = 0.004), serum albumin level (HR: 0.57, 95% CI: 0.39–0.84, $p$ = 0.005) and eGFR at the time of the Mg$^{2+}$ measurement (HR: 0.98, 95% CI: 0.97–0.99, $p$ = 0.002) were associated with decreased risk of all-cause mortality. Sex, maintenance immunosuppression, hemodialysis, living kidney donation, recipient CMV serostatus, and smoking status were not associated with an increased risk of all-cause mortality.

## 4. Discussion

Opportunistic infections are a common and significant cause of morbidity, graft loss, reduced quality of life, and mortality among KTRs [15,24,25]. Although the incidence of these infections is highest shortly after transplantation, these infections continue to have a significant impact on outcomes after this period. In this study, we observed $Mg^{2+}$ deficiency in more than sixty percent of KTRs within three months post-transplant and found that $Mg^{2+}$ deficiency is associated with an increased incidence of UTIs and viral infections within the first year after KT, independent of potential confounders.

Hypomagnesaemia is a known risk factor for new-onset DM after transplantation (NODAT) [26] and seems to play a role in cardiovascular (CV) morbidity and mortality after KT [27–29]. Moreover, there has been an increasing interest in the potential role of $Mg^{2+}$ in the immune system and a possible regulatory function in acquired immunity by regulating the proliferation and development of lymphocytes [30]. Several early studies provided evidence on a close relationship between $Mg^{2+}$ and the inflammatory response in animal models [31–33]. In a mouse model, reduced serum $Mg^{2+}$ concentration led to impaired $CD8^+$ T-cell response to influenza A virus infection, reduced T-cell activity, and exacerbated mortality [34]. In humans, intracellular free $Mg^{2+}$ controls the expression of the activating receptor natural killer group 2 member D (NKG2D) and is required for the cytotoxic activity of NK and $CD8^+$ T-cells [18]. Moreover, a case report suggested that in vitro addition of $Mg^{2+}$ may restore the cytotoxicity of $CD8^+$ T-cells in patients with mutations in interleukin-2-inducible T-cell kinase (ITK) and magnesium transporter 1 (MAGT1) [35]. Additionally, supplemental $Mg^{2+}$ might also indirectly influence T-cell receptor signaling by binding several protein kinases [36].

Data from clinical studies on metabolic syndrome revealed a direct link between hypomagnesemia and inflammation, indicating its role in more complex inflammatory processes [37,38]. T-cell activation is an energy-dependent process driven by a switch from oxidative phosphorylation to aerobic glycolysis [39]. T-cells upregulate insulin receptors, which is necessary for their effective function [40]. Consequently, impaired insulin responsiveness may lead to impaired adaptive immunity. Several studies have revealed an association between hypomagnesemia and type 2 DM and NODAT [41,42], while oral $Mg^{2+}$ supplementation increases insulin sensitivity and metabolic control in type 2 DM patients [37], but not in those with NODAT [43,44]. Importantly, DM as a cause of end-stage renal disease (ESRD) is associated with an increased risk of infectious death during the first post-transplant year in KTRs [45]. In our patient cohort, 15.4% of the patients had DM prior to KT, which was not a risk factor for UTIs or viral infections during the first year after KT. Nevertheless, insufficient glycemic control in KTRs might be an essential aspect of post-transplant infections since insulin plays a pivotal role in the activation of T-cells, and this link with $Mg^{2+}$ should be explored further in mechanistic studies.

Recent clinical data revealed worse mortality rates in patients with pneumonia and hypomagnesemia admitted to the intensive care unit (ICU) [46]. In pediatric liver transplant recipients with pre-transplant hypomagnesemia, increased mortality risk due to sepsis was observed [47]. Despite the importance of opportunistic infections for outcomes in KTRs and the high prevalence of $Mg^{2+}$ deficiency in this setting, to our knowledge, only one study has investigated the potential impact of $Mg^{2+}$ on infection complications after KT. In their well-designed, single-center prospective cohort study, Van Laecke and colleagues investigated 873 KTRs and found a dose-dependent association between a single baseline serum $Mg^{2+}$ concentration and incidence of severe infections in KTRs [19]. However, in our study, we mainly focused on mild transplant-related infections managed in the daily routine practice in the transplant out- and inpatient setting with different primary endpoints. Our findings reflect an observation in a cohort of KTRs managed in an ambulatory setting or admitted to the ward due to reasons primarily not necessarily associated with UTIs or viral infections. This is an important aspect since an efficient strategy to prevent severe infections after KT is required. Moreover, apart from severe infections, such as CMV viral syndrome and tissue invasive disease, a number of indirect immunomodulatory effects of viral infections on long-term kidney function have been postulated [48]. This indirect connection may lead to an increased incidence of acute and chronic rejection after KT, which may be caused by a bystander activation of alloreactive T-cells during an antiviral response of the organ recipient. Additionally, the incidence of other opportunistic infections may also be influenced by these effects [48]. Nevertheless, the results of these studies support each other regarding the observation between $Mg^{2+}$ deficiency and infections among KTRs and provide additional evidence on the role of $Mg^{2+}$ on infections in an independent cohort of KTRs.

Current immunosuppression strategies block T-lymphocytes primarily to prevent cellular rejection. The use of mTORi among KTRs seems to be associated with a reduced risk of CMV infections compared to those treated with a regular dose of CNI alone. Moreover, a combination of mTORi and a reduced dose of CNI also revealed the same effect. Interestingly, polyoma infections were not influenced by the different immunosuppression regimens [49,50]. In our cohort, $Mg^{2+}$ deficiency remained an independent risk factor of UTIs and viral infections in the univariable analysis and after adjustment for possible confounders (even including maintenance immunosuppression therapy). However, the percentage of $Mg^{2+}$ deficiency was higher in patients with the highest tertile of serum tacrolimus concentrations. Adverse events of CNIs include renal $Mg^{2+}$ wasting leading to $Mg^{2+}$ deficiency [51]. Thus, CNIs might indirectly further increase the susceptibility of KTRs for viral infections. On the contrary, CNI avoidance and withdrawal might lead to acute graft rejection, while a reduced dose of CNIs (particularly low-dose tacrolimus regimen in combination with an interleukin (IL)-2 receptor blocker) in induction regimes seems to be appropriate to reduce acute rejection [49,52]. Since T-cells are responsible for controlling viral infections and a direct link between T-cell function and $Mg^{2+}$ exists, a comprehensive approach investigating the associations between immunosuppression, $Mg^{2+}$, and particularly viral infections after KT is warranted.

Importantly, there was also a significant difference in the incidence of UTIs during the first year after KT between the patients with and without $Mg^{2+}$ deficiency. This observation might also be in line with the present knowledge on the effect of $Mg^{2+}$ on the immune function, significantly affecting the NK and CD8+ T-cell function [7,18]. Basically, in response to UTIs, a wide range of cells of the innate immune system, such as neutrophils, macrophages, and mast cells, are involved. The possible role of NK cells on UTIs (or bacterial infections) via tumor necrosis factor (TNF) production was reported [53]. However, their exact role in the pathogenesis of UTIs remains unclear. In addition, adaptive immune responses seem to be also limited in the immune response for UTIs [54].

In our cohort, patients with $Mg^{2+}$ deficiency had lower serum phosphate levels and better kidney function compared with patients without $Mg^{2+}$ deficiency. Hypophosphatemia is a common finding in KTRs, especially in those with immediate graft function and a high pre-transplant serum PTH level due to the significant urinary phosphorus loss driven by the effects of high levels of PTH and FGF-23 [55]. The constellation of lower serum $Mg^{2+}$ and phosphate levels may be a marker of a better tubular graft function. Nevertheless, serum phosphate levels start to normalize within the first few months after KT due to the reduced FGF-23 levels [56], while hypomagnesemia might persist for several years after KT. The relationship between decreased serum $Mg^{2+}$ levels and accelerated graft function decline or development of renal lesions involving innate immune pathways has been discussed [42]. However—until now—no clear association for these relationships was found. Dietary and supplementary interventions containing $Mg^{2+}$ and phosphate may lead to better nutritional status and indirectly improve immune function, especially within the first year after KT. However, prospective studies on this issue are needed.

Hypomagnesaemia is a known predictor of CV and all-cause mortality in dialysis patients [17,57,58]. Among KTRs, a possible relationship between the accelerated decline of graft function and hypomagnesemia was suggested [42]. Garnier and colleagues hypothesized an indirect positive effect of $Mg^{2+}$ on CV-related morbidity and mortality through decreased CV risk as a beneficial effect of $Mg^{2+}$ supplementation [42]. To our knowledge, data on long-term all-cause mortality and $Mg^{2+}$ status among KTRs are lacking. In our analysis, no statistically significant association between baseline serum $Mg^{2+}$ concentration and all-cause mortality was observed.

Some limitations should be considered when interpreting the results. First, given the design of the study as a single-center analysis based on retrospectively collected data, missing data were unavoidable. In the early years of the observation period, CMV PCR testing was not widely available, and a pp65 antigenemia assay was frequently used to identify CMV infections. However, in this period, this semi-quantitative fluorescent assay based on the detection of CMV infected cells in peripheral blood was the standard diagnostic approach to identify CMV infections in KTRs [59]. Notably, this assay is comparable in sensitivity to CMV PCR [60]. Second, induction therapies might represent an essential aspect of incidence rates of opportunistic infections after KT [61,62]. Good quality systematic review data provided clear evidence on the increased risk of CMV infections in patients treated by ATG [63]. In our subgroup analysis, $Mg^{2+}$ deficiency was significantly associated with higher incidence of viral infections, but not with UTIs during the first year after KT in patients receiving IL2Ra in a multivariate analysis. This might be explained due to the high incidence rate of UTIs in the first 6 months after KT, which is a time period with a higher effect rate of IL2Ra. This result may allude to the significant effect of IL2Ra on UTIs. In contrast, a similar analysis in patients receiving ATG did not show these results in our patient cohort. However, these observations need to be interpreted with caution due to small number of observations (n = 50 in the ATG group) as well as missing data, and a possible link should be investigated more extensively. On the other hand, recent evidence shows the decline of infection risk in KTRs that received lower ATG doses [64]. Additionally, previous data suggested no influence of induction therapy on severe infections among KTRs with hypomagnesemia [19]. Third, most $Mg^{2+}$ is found intracellular, and only around 1% is present in the blood, representing a small fraction of the total body reserves. Thus, serum $Mg^{2+}$ concentration may not represent intracellular $Mg^{2+}$ availability, which is an overall limitation on studies interpreting data using serum $Mg^{2+}$ measurements. Current methods estimating intracellular $Mg^{2+}$ concentration are invasive and expensive with low evidence level on their efficacy. Nevertheless, serum $Mg^{2+}$ measurement is the most available and commonly used test to access $Mg^{2+}$ status [65]. In addition, in blood, 20–30% of $Mg^{2+}$ is bound to albumin and other serum proteins. In our cohort, only a small proportion of KTRs (24.3%) had a serum albumin level < 3.5 g/dL. Additionally, serum $Mg^{2+}$ concentration can be influenced by many factors, including pH, azotemia, insulin resistance, post-transplantation volume expansion, low dietary $Mg^{2+}$ intake, or time of blood sample taken [42,66]. In this analysis, we used a single $Mg^{2+}$ measurement, and the question arises as to the variability within an individual patient. However, hypomagnesemia is an extensively described phenomenon in KTRs due to several pathophysiological and clinical factors [9,42], and it is rather unlikely that these factors potentially move patients by one to another study group. Finally, we do not have data on proton-pump inhibitor (PPI) therapies, which are a possible risk factor for hypomagnesemia and frequently prescribed for KTRs [67]. The possible association between the use of PPIs and $Mg^{2+}$ in link with infection complications should be addressed in future studies. Nevertheless, to the best of our knowledge, this is the first study addressing the impact of serum $Mg^{2+}$ on opportunistic infections and UTIs among KTRs managed in an ambulatory setting or admitted to the ward due to reasons not associated with infection-related complications.

## 5. Conclusions

In our study involving KTRs, $Mg^{2+}$ deficiency was independently associated with UTIs and viral infections in the early phase after KT. Our findings have implications for both research and clinical practice. The independent association between $Mg^{2+}$ deficiency and UTIs and viral infections highlights the need to explore the immunological effects of $Mg^{2+}$ in KTRs in more detail. Specific risk factors for $Mg^{2+}$ deficiency, particularly different immunosuppressive strategies, may further be characterized and used for more intensive serum $Mg^{2+}$ controlling to improve its potential effects not only on infection risk but on CV risk factors as well. In clinical praxis, the critical evaluation and potential use of reduced CNI regimes and correction of serum $Mg^{2+}$ level might be beneficial to preventing viral infections among KTRs. The current results support the hypothesis that $Mg^{2+}$ plays an important role in adaptive immunity among KTRs. Further, especially interventional studies on $Mg^{2+}$ and opportunistic infections in KTRs are warranted, at best designed as randomized placebo-controlled trials on $Mg^{2+}$ supplementation in KTRs with $Mg^{2+}$ deficiency.

**Supplementary Materials:** The following are available online at https://www.mdpi.com/article/10.3390/nu13041296/s1, Figure S1: All-cause mortality comparing patients with a serum $Mg^{2+}$ level $\geq$ 0.7 mmol/L and < 0.7 mmol/L, Table S1: Unadjusted logistic regression analysis of risk factors for hypomagnesemia, Table S2: Unadjusted logistic regression analysis of risk factors for viral infections, Table S3: Unadjusted logistic regression analysis of risk factors for hypomagnesemia.

**Author Contributions:** B.O. and A.R.R. conceived the project and wrote the manuscript. G.P., R.R. and A.B. performed data analysis and wrote the manuscript. B.O., A.R.R., G.P. and A.B. interpreted the data. B.O., J.B., A.P., L.S., C.H., M.K., M.S.B. and L.H. performed data abstraction. A.T.D., P.S., S.P. and C.T. critically revised the manuscript for important intellectual content. A.R.R. was responsible for final manuscript approval. All authors have read and agreed to the published version of the manuscript.

**Funding:** This research received no external funding.

**Institutional Review Board Statement:** The study was conducted according to the guidelines of the Declaration of Helsinki, and approved by the Institutional Ethics Committee of the Medical University of Graz (EK-Number: 31-226 ex 18/19; 1 March 2019).

**Informed Consent Statement:** Patient consent was waived due to the retrospective design of the study.

**Data Availability Statement:** The data presented in this study are available from the corresponding author on reasonable request.

**Conflicts of Interest:** The authors declare no conflict of interest.

# References

1. De Baaij, J.H.F.; Hoenderop, J.G.J.; Bindels, R.J.M. Magnesium in man: Implications for health and disease. *Physiol. Rev.* **2015**, *95*, 1–46. [CrossRef]
2. Maguire, M.E. Magnesium and cell proliferation. *Ann. N. Y. Acad. Sci.* **1988**, *551*, 201–215. [CrossRef] [PubMed]
3. Mooren, F.C.; Krüger, K.; Völker, K.; Golf, S.W.; Wadepuhl, M.; Kraus, A. Oral magnesium supplementation reduces insulin resistance in non-diabetic subjects—A double-blind, placebo-controlled, randomized trial. *Diabetes Obes. Metab.* **2011**, *13*, 281–284. [CrossRef] [PubMed]
4. Pilchova, I.; Klacanova, K.; Tatarkova, Z.; Kaplan, P.; Racay, P. The involvement of $Mg^{2+}$ in regulation of cellular and mitochondrial functions. *Oxid. Med. Cell. Longev.* **2017**, *2017*, 6797460. [CrossRef] [PubMed]
5. Galland, L. Magnesium and immune function: An overview. *Magnesium* **1988**, *7*, 290–299.
6. Liang, R.Y.; Wu, W.; Huang, J.; Jiang, S.P.; Lin, Y. Magnesium affects the cytokine secretion of $CD4^+$ T lymphocytes in asthma. *J. Asthma.* **2012**, *49*, 1012–1015. [CrossRef]
7. Brandao, K.; Deason-Towne, F.; Perraud, A.L.; Schmitz, C. The role of $Mg^{2+}$ in immune cells. *Immunol. Res.* **2013**, *55*, 261–269. [CrossRef]
8. Li, F.Y.; Chaigne-Delalande, B.; Kanellopoulou, C.; Davis, J.C.; Matthews, H.F.; Douek, D.C.; Cohen, J.I.; Uzel, G.; Su, H.C.; Lenardo, M.J. Second messenger role for $Mg^{2+}$ revealed by human T-cell immunodeficiency. *Nature* **2011**, *475*, 471–476. [CrossRef]
9. Van Laecke, S.; Van Biesen, W. Hypomagnesaemia in kidney transplantation. *Transplant. Rev.* **2015**, *29*, 154–160. [CrossRef]
10. Barton, C.H.; Vaziri, N.D.; Martin, D.C.; Choi, S.; Alikhani, S. Hypomagnesemia and renal magnesium wasting in renal transplant recipients receiving cyclosporine. *Am. J. Med.* **1987**, *83*, 693–699. [CrossRef]
11. Markell, M.S.; Altura, B.T.; Sarn, Y.; Barbour, R.; Friedman, E.A.; Altura, B.M. Relationship of ionized magnesium and cyclosporine level in renal transplant recipients. *Ann. N. Y. Acad. Sci.* **1993**, *696*, 408–411. [CrossRef]
12. Nijenhuis, T.; Hoenderop, J.G.; Bindels, R.J. Downregulation of $Ca^{2+}$ and $Mg^{2+}$ transport proteins in the kidney explains tacrolimus (FK506)-induced hypercalciuria and hypomagnesemia. *J. Am. Soc. Nephrol.* **2004**, *15*, 549–557. [CrossRef] [PubMed]
13. Navaneethan, S.D.; Sankarasubbaiyan, S.; Gross, M.D.; Jeevanantham, V.; Monk, R.D. Tacrolimus associated hypomagnesemia in renal transplant recipients. *Transplant. Proc.* **2006**, *38*, 1320–1322. [CrossRef] [PubMed]

14. Fishman, J.A. Infection in solid-organ transplant recipients. *N. Eng. J. Med.* **2007**, *357*, 2601–2614. [CrossRef] [PubMed]
15. Karuthu, S.; Blumberg, E.A. Common infections in kidney transplant recipients. *Clin. J. Am. Soc. Nephrol.* **2012**, *7*, 2058–2070. [CrossRef] [PubMed]
16. Cippà, P.E.; Schiesser, M.; Ekberg, H.; van Gelder, T.; Mueller, N.J.; Cao, C.A.; Fehr, T.; Bernasconi, C. Risk stratification for rejection and infection after kidney transplantation. *Clin. J. Am. Soc. Nephrol.* **2015**, *10*, 2213–2220. [CrossRef] [PubMed]
17. Sakaguchi, Y.; Fujii, N.; Shoji, T.; Hayashi, T.; Rakugi, H.; Isaka, Y. Hypomagnesemia is a significant predictor of cardiovascular and non-cardiovascular mortality in patients undergoing hemodialysis. *Kidney. Int.* **2014**, *85*, 174–181. [CrossRef] [PubMed]
18. Chaigne-Delalande, B.; Li, F.Y.; O'Connor, G.M.; Lukacs, M.J.; Jiang, P.; Zheng, L.; Shatzer, A.; Biancalana, M.; Pittaluga, S.; Matthews, H.F.; et al. $Mg^{2+}$ regulates cytotoxic functions of NK and CD8 T cells in chronic EBV infection through NKG2D. *Science* **2013**, *341*, 186–191. [CrossRef]
19. Van Laecke, S.; Vermeiren, P.; Nagler, E.V.; Caluwe, R.; De Wilde, M.; Van der Vennet, M.; Peeters, P.; Randon, C.; Vermassen, F.; Vanholder, R.; et al. Magnesium and infection risk after kidney transplantation: An observational cohort study. *J. Infect.* **2016**, *73*, 8–17. [CrossRef]
20. Deak, A.T.; Ionita, F.; Kirsch, A.H.; Odler, B.; Rainer, P.P.; Kramar, R.; Kubatzki, M.P.; Eberhard, K.; Berghold, A.; Rosenkranz, A.R. Impact of cardiovascular risk stratification strategies in kidney transplantation over time. *Nephrol. Dial. Transplant.* **2020**, *35*, 1810–1818. [CrossRef]
21. Levey, A.S.; Stevens, L.A.; Schmid, C.H.; Zhang, Y.L.; Castro, A.F.; Feldman, H.I.; Kusek, J.W.; Eggers, P.; Van Lente, F.; Greene, T.; et al. A new equation to estimate glomerular filtration rate. *Ann. Intern. Med.* **2009**, *150*, 604–612. [CrossRef] [PubMed]
22. American Diabetes Association. Diagnosis and classification of diabetes mellitus. *Diabetes Care* **2014**, *37*, S81–S90. [CrossRef]
23. Schröppel, B.; Legendre, C. Delayed kidney graft function: From mechanism to translation. *Kidney. Int.* **2014**, *86*, 251–258. [CrossRef]
24. Weinrauch, L.A.; D'Elia, J.A.; Weir, M.R.; Bunnapradist, S.; Finn, P.V.; Liu, J.; Claggett, B.; Monaco, A.P. Infection and Malignancy Outweigh Cardiovascular Mortality in Kidney Transplant Recipients: Post Hoc Analysis of the FAVORIT Trial. *Am. J. Med.* **2018**, *131*, 165–172. [CrossRef]
25. Bodro, M.; Linares, L.; Chiang, D.; Moreno, A.; Cervera, C. Managing recurrent urinary tract infections in kidney transplant recipients. *Expert. Rev. Anti-Infect. Ther.* **2018**, *16*, 723–732. [CrossRef] [PubMed]
26. Huang, J.W.; Famure, O.; Li, Y.; Kim, S.J. Hypomagnesemia and the risk of new-onset diabetes mellitus after kidney transplantation. *J. Am. Soc. Nephrol.* **2016**, *27*, 1793–1800. [CrossRef] [PubMed]
27. Van Laecke, S.; Maréchal, C.; Verbeke, F.; Peeters, P.; Van Biesen, W.; Devuyst, O.; Jadoul, M.; Vanholder, R. The relation between hypomagnesaemia and vascular stiffness in renal transplant recipients. *Nephrol. Dial. Transplant.* **2011**, *26*, 2362–2369. [CrossRef] [PubMed]
28. Kisters, K.; Gremmler, B.; Hausberg, M. Magnesium and arterial stiffness. *Hypertension* **2006**, *47*, e3. [CrossRef] [PubMed]
29. Shechter, M.; Sharir, M.; Labrador, M.J.; Forrester, J.; Silver, B.; Bairey Merz, C.N. Oral magnesium therapy improves endothelial function in patients with coronary artery disease. *Circulation* **2000**, *102*, 2353–2358. [CrossRef]
30. Feske, S.; Skolnik, E.Y.; Prakriya, M. Ion channels and transporters in lymphocyte function and immunity. *Nat. Rev. Immunol.* **2012**, *12*, 532–547. [CrossRef]
31. Weglicki, W.B.; Phillips, T.M.; Freedman, A.M.; Cassidy, M.M.; Dickens, B.F. Magnesium deficiency elevates circulating levels of inflammatory cytokines and endothelia. *Mol. Cell. Biochem.* **1992**, *110*, 169–173. [CrossRef] [PubMed]
32. Weglicki, W.B.; Dickens, B.F.; Wagner, T.L.; Chmielinska, J.J.; Phillips, T.M. Immunoregulation by neuropeptides in magnesium deficiency ex vivo effect of enhanced substance P production on circulating T lymphocytes from magnesium-deficient mice. *Magnes. Res.* **1996**, *9*, 3–11. [PubMed]
33. Malpuech-Brugère, C.; Nowacki, W.; Daveau, M.; Gueux, E.; Linard, C.; Rock, E.; Lebreton, J.P.; Mazur, A.; Rayssiguier, Y. Inflammatory response following acute magnesium deficiency in the rat. *Biochim. Biophys. Acta* **2000**, *1501*, 91–98. [CrossRef]
34. Kanellopoulou, C.; George, A.B.; Masutani, E.; Cannons, J.L.; Ravell, J.C.; Yamamoto, T.N.; Smelkinson, M.G.; Jiang, P.D.; Matsuda-Lennikov, M.; Reilley, J.; et al. $Mg^{2+}$ regulation of kinase signaling and immune function. *J. Exp. Med.* **2019**, *216*, 1828–1842. [CrossRef] [PubMed]
35. Howe, M.K.; Dowdell, K.; Roy, A.; Niemela, J.E.; Wilson, W.; McElwee, J.J.; Hughes, J.D.; Cohen, J.I. Magnesium restores activity to peripheral blood cells in a patient with functionally impaired interleukin-2-inducible T cell kinase. *Front. Immunol.* **2019**, *10*, 2000. [CrossRef]
36. Nolen, B.; Taylor, S.; Ghosh, G. Regulation of protein kinases; controlling activity through activation segment conformation. *Mol. Cell* **2004**, *15*, 661–675. [CrossRef]
37. Guerrero-Romero, F.; Rodriguez-Moran, M. Magnesium improves the beta-cell function to compensate variation of insulin sensitivity: Double-blind, randomized clinical trial. *Eur. J. Clin. Investig.* **2011**, *41*, 405–410. [CrossRef]
38. Mooren, F.C. Magnesium and disturbances in carbohydrate metabolism. *Diabetes Obes. Metab.* **2015**, *17*, 813–823. [CrossRef]
39. Chang, C.H.; Curtis, J.D.; Maggi, L.B., Jr.; Faubert, B.; Villarino, A.V.; O'Sullivan, D.; Huang, S.C.C.; Van Der Windt, G.J.; Blagih, J.; Qiu, J.; et al. Posttranscriptional control of T cell effector function by aerobic glycolysis. *Cell* **2013**, *153*, 1239–1251. [CrossRef]
40. Fischer, H.J.; Sie, C.; Schumann, E.; Witte, A.K.; Dressel, R.; van den Brandt, J.; Reichardt, H.M. The insulin receptor plays a critical role in T cell function and adaptive immunity. *J. Immunol.* **2017**, *198*, 1910–1920. [CrossRef]

41. Gommers, L.M.; Hoenderop, J.G.; Bindels, R.J.; de Baaij, J.H. Hypomagnesemia in type 2 diabetes: A vicious circle? *Diabetes* **2016**, *65*, 3–13. [CrossRef] [PubMed]
42. Garnier, A.S.; Duveau, A.; Planchais, M.; Subra, J.F.; Sayegh, J.; Augusto, J.F. Serum magnesium after kidney transplantation: A systematic review. *Nutrients* **2018**, *10*, 729. [CrossRef]
43. Van Laecke, S.; Nagler, E.V.; Taes, Y.; Biesen, W.V.; Peetres, P.; Vanholder, R. The effect of magnesium supplements on early post-transplantation glucose metabolism: A randomized controlled trial. *Transplant. Int.* **2014**, *27*, 895–902. [CrossRef] [PubMed]
44. Van Laecke, S.; Caluwe, R.; Huybrechts, I.; Nagler, E.V.; Vanholder, R.; Peeters, P.; Van Vlem, B.; Van Biesen, W. Effects of magnesium supplements on insulin secretion after kidney transplantation: A randomized controlled trial. *Ann. Transplant.* **2017**, *22*, 524–531. [CrossRef]
45. Kinnunen, S.; Karhapää, P.; Juutilainen, A.; Finne, P.; Helanterä, I. Secular trends in infection-related mortality after kidney transplantation. *Clin. J. Am. Soc. Nephrol.* **2018**, *13*, 755–762. [CrossRef] [PubMed]
46. Nasser, R.; Mohammad, E.N.; Mashiach, T.; Azzam, Z.S.; Braun, E. The association between serum magnesium levels and community-acquired pneumonia 30-day mortality. *BMC. Infect. Dis.* **2018**, *18*, 698. [CrossRef]
47. Elgendy, H.M.; El Moghazy, W.M.; Uemoto, S.; Fukuda, K. Pre transplant serum magnesium level predicts outcome after pediatric living donor liver transplantation. *Ann. Transplant.* **2012**, *17*, 29–37.
48. Helanterä, I.; Egli, A.; Koskinen, P.; Lautenschlager, I.; Hirsch, H.H. Viral impact on long-term kidney graft function. *Infect. Dis. Clin. N. Am.* **2010**, *24*, 339–371. [CrossRef]
49. Karpe, K.M.; Talaulikar, G.S.; Walters, G.D. Calcineurin inhibitor withdrawal or tapering for kidney transplant recipients. *Cochrane Database Syst. Rev.* **2017**, *7*, CD006750. [CrossRef] [PubMed]
50. Mallat, S.G.; Tanios, B.Y.; Itani, H.S.; Lotfi, T.; McMullan, C.; Gabardi, S.; Akl, E.A.; Azzi, J.R. CMV and BKPyV infections in renal transplant recipients receiving an mTOR inhibitor-based versus a CNI-based regimen: A systematic review and meta-analysis of randomized, controlled trials. *Clin. J. Am. Soc. Nephrol.* **2017**, *12*, 1321–1336. [CrossRef]
51. Gratreak, B.D.; Swanson, E.A.; Lazelle, R.A.; Jelen, S.K.; Hoenderop, J.; Bindels, R.J.; Yang, C.L.; Ellison, D.H. Tacrolimus-induced hypomagnesemia and hypercalciuria requires FKBP12 suggesting a role for calcineurin. *Physiol. Rep.* **2020**, *8*, e14316. [CrossRef]
52. Ekberg, H.; Tedesco-Silva, H.; Demirbas, A.; Vítko, Š.; Nashan, B.; Gürkan, A.; Margreiter, R.; Hugo, C.; Grinyó, J.M.; Frei, U.; et al. Reduced exposure to calcineurin inhibitors in renal transplantation. *N. Eng. J. Med.* **2007**, *357*, 2562–2575. [CrossRef] [PubMed]
53. Gur, C.; Coppenhagen-Glazer, S.; Rosenberg, S.; Yamin, R.; Enk, J.; Glasner, A.; Bar-On, Y.; Fleissig, O.; Naor, R.; Abed, J.; et al. Natural killer cell-mediated host defense against uropathogenic E. coli is counteracted by bacterial hymolysinA-dependent killing of NK cells. *Cell. Host Microbe* **2013**, *14*, 664–674. [CrossRef] [PubMed]
54. Abraham, S.N.; Miao, Y. The nature of immune responses to urinary tract infections. *Nat. Rev. Immunol.* **2015**, *10*, 655–663. [CrossRef]
55. Vangala, C.; Pan, J.; Cotton, R.T.; Ramanathan, V. Mineral and bone disorders after kidney transplantation. *Front. Med.* **2018**, *5*, 211. [CrossRef]
56. Wolf, M.; Weir, M.R.; Kopyt, N.; Mannon, R.B.; Von Visger, J.; Deng, H.; Yue, S.; Vincenti, F. A prospective cohort study of mineral metablism after kidney transplantation. *Transplantation* **2016**, *100*, 184–193. [CrossRef]
57. Ishimura, E.; Okuno, S.; Yamakawa, T.; Inaba, M.; Nishizawa, Y. Serum magnesium concentration is a significant predictor of mortality in maintenance hemodialysis patients. *Magnes. Res.* **2007**, *20*, 237–244.
58. Lacson, E., Jr.; Wang, W.; Ma, L.; Passlick-Deetjen, J. Serum magnesium and mortality in hemodialysis patients in the United States: A cohort study. *Am. J. Kidney Dis.* **2015**, *66*, 1056–1066. [CrossRef] [PubMed]
59. Kidney Disease: Improving Global Outcomes (KDIGO) Transplant Work Group. *Am. J. Transplant.* **2009**, *9*, S1–S155.
60. Humar, A.; Snydman, D. AST Infectious Diseases Community of Practice. Cytomegalovirus in solid organ transplant recipients. *Am. J. Transplant.* **2009**, *9*, S78–S86. [CrossRef]
61. Bayraktar, A.; Catma, Y.; Akyildiz, A.; Demir, E.; Bakkaloglu, H.; Ucar, A.R.; Dirim, A.B.; Usta Akgul, S.; Temurhan, S.; Gok, A.F.K.; et al. Infectious complications of induction therapies in kidney transplantation. *Ann. Transplant.* **2019**, *24*, 412–417. [CrossRef]
62. Bertrand, D.; Chavarot, N.; Gatault, P.; Garrouste, C.; Bouvier, N.; Grall-Jezequel, A.; Jaureguy, M.; Caillard, S.; Lemoine, M.; Colosio, C.; et al. Opportunistic infections after conversion to belatacept in kidney transplantation. *Nephrol. Dial. Transplant.* **2020**, *35*, 336–345. [CrossRef] [PubMed]
63. Hill, P.; Cross, N.B.; Barnett, A.N.R.; Palmer, S.C.; Webster, A.C. Polyclonal and monoclonal antibodies for induction therapy in kidney transplant recipients. *Cochrane. Database Syst. Rev.* **2017**, *1*, CD004759. [CrossRef] [PubMed]
64. Hellemans, R.; Bosmans, J.L.; Abramowicz, D. Induction therapy for kidney transplant recipients: Do we still need anti-IL2 receptor monoclonal antibodies? *Am. J. Transplant.* **2017**, *17*, 22–27. [CrossRef]
65. Reddy, S.T.; Soman, S.S.; Yee, J. Magnesium balance and measurement. *Adv. Chronic Kidney Dis.* **2018**, *25*, 224–229. [CrossRef] [PubMed]
66. Kanbay, M.; Goldsmith, D.; Uyar, M.E.; Turgut, F.; Covic, A. Magnesium in chronic kidney disease: Challenges and opportunities. *Blood Purif.* **2010**, *29*, 280–292. [CrossRef] [PubMed]
67. Al-Aly, Z.; Maddukuri, G.; Xie, Y. Proton pump inhibitors and kidney: Implications of current evidence for clinical practice and when and how to describe. *Am. J. Kidney Dis.* **2020**, *75*, 497–507. [CrossRef]

Article

# Magnesium Absorption in Intestinal Cells: Evidence of Cross-Talk between EGF and TRPM6 and Novel Implications for Cetuximab Therapy

Giuseppe Pietropaolo [1,†], Daniela Pugliese [2], Alessandro Armuzzi [2], Luisa Guidi [2], Antonio Gasbarrini [2], Gian Lodovico Rapaccini [2], Federica I. Wolf [1,\*] and Valentina Trapani [1,\*]

1 Sezione di Patologia Generale, Dipartimento di Medicina e Chirurgia Traslazionale, Fondazione Policlinico Universitario A. Gemelli IRCCS—Università Cattolica del Sacro Cuore, 00168 Rome, Italy; giuseppe.pietropaolo@uniroma1.it
2 UOC Medicina Interna e Gastroenterologia, Dipartimento di Medicina e Chirurgia Traslazionale, Fondazione Policlinico Universitario A. Gemelli IRCCS—Università Cattolica del Sacro Cuore, 00168 Rome, Italy; daniela.pugliese@policlinicogemelli.it (D.P.); alessandro.armuzzi@unicatt.it (A.A.); luisa.guidi@unicatt.it (L.G.); antonio.gasbarrini@unicatt.it (A.G.); gianludovico.rapaccini@policlinicogemelli.it (G.L.R.)
\* Correspondence: federica.wolf@unicatt.it (F.I.W.); valentina.trapani@unicatt.it (V.T.)
† Current address: Dipartimento di Medicina Molecolare, laboratory affiliated to Istituto Pasteur Italia-Fondazione Cenci Bolognetti, Sapienza Università di Roma, 00161 Rome, Italy.

Received: 25 September 2020; Accepted: 22 October 2020; Published: 26 October 2020

**Abstract:** Hypomagnesemia is very commonly observed in cancer patients, most frequently in association with therapy with cetuximab (CTX), a monoclonal antibody targeting the epithelial growth factor receptor (EGFR). CTX-induced hypomagnesemia has been ascribed to renal magnesium (Mg) wasting. Here, we sought to clarify whether CTX may also influence intestinal Mg absorption and if Mg supplementation may interfere with CTX activity. We used human colon carcinoma CaCo-2 cells as an in vitro model to study the mechanisms underlying Mg transport and CTX activity. Our findings demonstrate that TRPM6 is the key channel that mediates Mg influx in intestinal cells and that EGF stimulates such influx; consequently, CTX downregulates TRPM6-mediated Mg influx by interfering with EGF signaling. Moreover, we show that Mg supplementation does not modify either the CTX IC50 or CTX-dependent inhibition of ERK1/2 phosphorylation. Our results suggest that reduced Mg absorption in the intestine may contribute to the severe hypomagnesemia that occurs in CTX-treated patients, and Mg supplementation may represent a safe and effective nutritional intervention to restore Mg status without impairing the CTX efficacy.

**Keywords:** biomarker; colorectal cancer; EGFR; hypomagnesemia; magnesium supplementation; monoclonal antibodies; targeted therapy

## 1. Introduction

Nutritional deficits, defined as an imbalance between intake and metabolic requirements, are very common in cancer patients and may be caused by both the tumor itself and its treatment [1]. Appropriate nutritional interventions that correct such an imbalance reduce the risk of interruption or discontinuous treatment and improve quality of life [2]. Hypomagnesemia is frequent in oncologic patients, especially in those subjected to cisplatin-based therapies [3]. More recently, hypomagnesemia has emerged as the most notable adverse effect of the anti-EGFR monoclonal antibody cetuximab (CTX), which is widely used for advanced colorectal cancer (CRC) [4]. Several meta-analyses have shown that the incidence of all-grade hypomagnesemia in CTX-treated patients could be as high as about 35%; in about 5% of cases, hypomagnesemia can be severe (grade 3–4) and cause symptoms that require magnesium (Mg) supplementation [5–8]. On the other hand, early hypomagnesemia seems to act as a good predictor of the efficacy and outcome of CTX in *KRAS* wild-type CRC patients [9]. Although opposing results have also been reported [10], a recent meta-analysis confirmed that hypomagnesemia is associated with better progression-free survival, overall survival, and overall relative risk in CTX-treated *KRAS* wild-type CRC patients [11]. In addition to the clinical relevance of these findings, Vincenzi et al. [12] went as far as proposing that reduced serum Mg levels might potentiate the chemotherapeutic effects of CTX, which raised an intense debate among the scientific community [13,14].

Mg is a micronutrient involved in a plethora of cell functions, acting as a cofactor for a multitude of enzymes [15]. Systemic Mg homeostasis depends on the concomitant action of the intestine, responsible for Mg uptake from food, and the kidneys, which regulate Mg excretion. Magnesium is absorbed through different mechanisms, including passive paracellular transport, which is driven by the electrochemical gradient, and active transcellular transport, which is mediated by two highly homologous Mg channels—transient receptor potential melastatin (TRPM) channels type 6 and 7. TRPM7 is ubiquitously expressed, while TRPM6 is mainly expressed in the kidneys, the distal small intestine, and the colon [16]. Although the distal convoluted tubule of the kidney has long been considered the key gatekeeper of systemic Mg, the latest findings challenged such a view and suggested that intestinal Mg uptake might be of primary relevance [17]. Recent results corroborate the view that TRPM6, rather than TRPM7, modulates magnesium homeostasis in the colon. Ferioli et al. reported that TRPM6 function cannot be replaced by other channels [18]. Likewise, our group demonstrated that, in colon mucosa, TRPM6 is responsible for Mg influx and cell proliferation leading to mucosal healing [19,20].

The present view is that CTX-induced hypomagnesemia originates from molecular cross-talk between the EGF pathway and the regulatory mechanism for systemic Mg homeostasis in the kidneys. Such interaction was elucidated through the discovery of a mutation in the *EGF* gene in a rare genetic condition characterized by renal Mg wasting [21] and the following molecular characterization [22]. It was demonstrated that EGF acts as a magnesiotropic hormone by stimulating the surface expression and activity of the TRPM6 channel on the apical membrane of kidney epithelial cells, which in turn mediates Mg uptake. Therefore, by antagonizing EGF, ultimately CTX inhibits renal Mg reabsorption by TRPM6 and alters the whole-body Mg balance.

In addition to its well-established role in tumor growth and progression, EGF is also an important actor in intestinal development and mucosal repair [23]. Furthermore, EGF has been shown to be an important regulator of the expression, trafficking, and activity of epithelial transport proteins in the intestine [24]. However, despite the current emphasis on the importance of gut absorption for Mg homeostasis, the molecular cross-talk between EGF and TRPM6 has never been investigated in intestinal epithelial cells so far, nor is it known whether CTX can also affect the intestinal Mg absorption. Moreover, evidence that hypomagnesemia may serve as a positive prognostic factor in CTX-treated patients strongly contrasts with the fact that hypomagnesemia may cause severe discomfort and may even pose a serious threat to their lives. In this context, a crucial issue remains unanswered regarding the possibility that Mg supplementation may interfere with CTX efficacy. In the present paper, we sought to clarify two pressing issues that might have important clinical implications for CTX therapy: (1) the role of altered intestinal Mg absorption in the development of CTX-induced hypomagnesaemia; (2) the effect of Mg supplementation on the efficacy of CTX treatment.

## 2. Materials and Methods

### 2.1. Cell Culture

The constitutive activation of the MAPK pathway may limit the effectiveness of CTX treatment [25]. We screened different colon cancer cell lines and chose human colon carcinoma CaCo-2 cells, which harbor wild-type forms of critical genes, such as *KRAS*, *BRAF*, *PI3K3CA*, and *PTEN* [26]. CaCo-2 cells also express EGFR [27]. Cells were routinely grown in Dulbecco's modified Eagle's medium (DMEM) supplemented with 20% fetal bovine serum (FBS), 2mM of glutamine, 100 U/mL of penicillin, and 100 µg/mL of streptomycin in a 5% $CO_2$ humidified atmosphere at 37 °C. The reagents for cell culture were from Euroclone (Pero, Milan, Italy). Recombinant EGF was purchased from PeproTech (London, UK) and used at a concentration of 10 ng/mL, as previously reported [28]. Before EGF stimulation, the cells were starved in FBS-free medium for 24 h. Cetuximab was kindly provided by the Oncology Pharmacy Unit, "Agostino Gemelli" University Hospital, and used at an optimal concentration of 70 µg/mL, as inferred from cytotoxicity assays (see Sections 2.2 and 3.1). To obtain a transient downregulation of TRPM6, predesigned siRNA against human TRPM6 was purchased from Qiagen. Specific siRNAs were transfected into cells (1300 ng per 400,000 cells) using HiPerFect Transfection Reagent (Qiagen, Milan, Italy) following the manufacturer's protocol (https://www.qiagen.com/it/transfectionprotocols/transfectionprotocol/). Non-silencing scrambled sequences were used as controls.

## 2.2. MTT Cytotoxicity Assay

Cells were seeded in 24-well plates at a density of 40,000 cells/well and allowed to adhere for 24 h before drug treatment. CTX (concentration range: 7.5 µg/mL to 240 µg/mL) was added to the culture medium in triplicates. After 24 h, the culture medium was replaced with serum-free medium containing 3-[4,5-dimethylthiazol-2-yl]-2,5-diphenyltetrazolium bromide (MTT, 1 mg/mL), and cells were incubated for 90 min at 37 °C. Finally, formazan crystals were dissolved in acidified isopropanol (0.04 N HCl in isopropanol), and the absorbance was read at $\lambda = 565$ nm. Data were analyzed using Prism software (version 5.01, GraphPad Software Inc., La Jolla, CA, USA) and dose–response curves were obtained by nonlinear regression (sigmoidal curve, variable slope).

## 2.3. Western Blotting

Cells were lysed in RIPA buffer (50 mM of Tris, pH 8, 150 mM of NaCl, 1 mM of EDTA, 1% NP-40, 0.05% sodium deoxycholate, 0.1% SDS) supplemented with protease and phosphatase inhibitors (Halt™ inhibitor cocktail, ThermoFisher Scientific, Milan, Italy). Protein concentrations were determined using the Bradford protein assay (Bio-Rad). Cell extracts were resolved by SDS-PAGE; transferred to PVDF membranes; and probed with rabbit polyclonal anti-TRPM6 (1:500, Biorbyt), anti-ERK1/2 (1:1000, Cell Signaling Technology), anti phospho-ERK1/2 (1:1000, Cell Signaling Technology), or anti-β-actin (1:1000, Sigma-Aldrich) primary antibodies. Horseradish peroxidase-conjugated secondary antibodies (GE Healthcare) were detected by the ECL Prime Western Blotting Detection Reagent (GE Healthcare) and the ChemiDoc XRS system (Bio-Rad). Densitometric analysis was performed by the ImageJ software (NIH, http://imagej.nih.gov/ij/).

## 2.4. Mg Influx Measurements

Subconfluent cells grown on 35 mm microscopy dishes (µ-dish, ibidi GmbH) were loaded with 3 µM of Mag-Fluo-4-AM (ThermoFisher Scientific), and imaged in a $Na^+$, $Ca^{2+}$, and $Mg^{2+}$-free buffer at a confocal laser scanning microscope (Nikon A1 MP), as previously described [28]. Cytosolic fluorescence signals were recorded as time series at a sampling frequency of 30 frames/min. The baseline was monitored for 30 s, then $MgSO_4$ was added drop-wise to a final concentration of 5 mM. Changes in the intracellular Mg levels at the single-cell level were estimated by the mean fluorescent increment $\Delta F/F$ [29]. Image analysis was performed by the NIS-Elements Confocal Software on 10 representative cells in each microscopic field, and experiments were repeated independently at least three times.

## 2.5. Statistical Analyses

All the experiments were repeated independently three times. The Prism software (version 5.01, GraphPad Software Inc., La Jolla, CA, USA) was used for all the statistical analyses. Statistical significance was evaluated using Student's t-test, when comparing two groups; one-way ANOVA, when comparing more groups in relation to only one variable; and two-way ANOVA, when comparing more groups in relation to two variables. ANOVA was followed by Bonferroni's test. Differences were considered statistically significant for *p*-values < 0.05, and significance levels were assigned as follows: * for $p < 0.05$, ** for $p < 0.01$.

## 3. Results

### 3.1. TRPM6 Mediates Mg Influx in CTX-Sensitive CaCo-2 Cells

First, we assessed the sensitivity of CaCo-2 cells to CTX with an MTT assay. The IC50 at 24h was (66 ± 16) µg/mL; therefore, in the following experiments, a CTX dose of 70 µg/mL was used. Next, we characterized the basal Mg influx capacity by the live imaging of CaCo-2 cells loaded with the Mg-specific fluorescent probe Mag-Fluo-4. The addition of 5 mM of $MgSO_4$ to the extracellular medium induced a rapid increase in fluorescence (i.e., intracellular Mg concentration) up to about 10% of the basal level; the fluorescence then gradually decreased to basal levels within about 3 min (Figure 1a, solid circles). CaCo-2 cells express the TRPM6 channel [19]. To determine whether the detected Mg influx is mediated by TRPM6, we repeated the same experiment in *TRPM6*-silenced CaCo-2 cells. Silencing by transient siRNA transfection significantly decreased the TRPM6 protein levels, as assessed by Western blot analysis at 48h (Figure 1b,c). After 48 h from siRNA transfection, the Mg influx upon the addition of extracellular $MgSO_4$ was nearly abolished in *TRPM6*-silenced cells in comparison with the control cells (Figure 1a, open circles). These results prove that TRPM6 is the key channel that mediates the Mg influx in CaCo-2 cells.

### 3.2. Cetuximab and EGF Modulate Mg Influx

Molecular crosstalk between the EFGR pathway and TRPM6 has been described in kidney epithelial cells [21,22]. We sought to assess whether the same mechanisms may modulate the Mg influx in intestinal epithelial cells. As shown in Figure 2a, EGF stimulation (10 ng/mL, 24 h) induced an increase in the basal Mg influx capacity of CaCo-2 cells (open circles vs. open squares), while CTX reduced the EGF-dependent increase in Mg influx (solid circles vs. open circles) and completely abrogated the basal Mg influx (solid squares vs. open squares). Western blot analysis proved that 24h of treatment with EGF (10 ng/mL) upregulated the TRPM6 expression, while concomitant exposure to CTX (70 µg/mL) resulted in TRPM6 levels comparable to the basal expression (Figure 2b,c). We conclude that EGF signaling leads to increased levels of the TRPM6 channel in intestinal cells, and CTX, by interfering with this signaling, downregulates the TRPM6-mediated Mg influx.

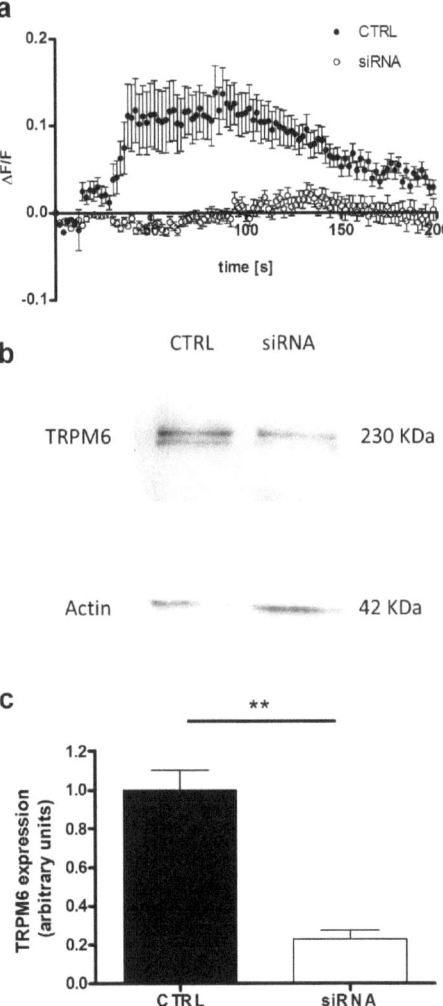

**Figure 1.** The TRPM6 channel mediates the Mg influx in human colon carcinoma cells. CaCo-2 cells were transiently silenced for *TRPM6* and assessed 48h after transfection. (**a**) Mg influx capacity in *TRPM6*-silenced (open circles) vs. control cells (solid circles), as assessed by the live imaging of Mag-Fluo-4-loaded cells; a representative experiment is shown. (**b**) TRPM6 protein expression in *TRPM6*-silenced and control cells, as evaluated by Western blot analysis; a representative blot is shown. (**c**) Quantification of TRPM6 protein expression by Western blot densitometry normalized to β-actin levels ($n = 3$, mean ± SD) in *TRPM6*-silenced (white bar) and control (black bar) cells. ** $p < 0.01$ by paired Student's t-test. Full scans of original blots are available in Figure S1.

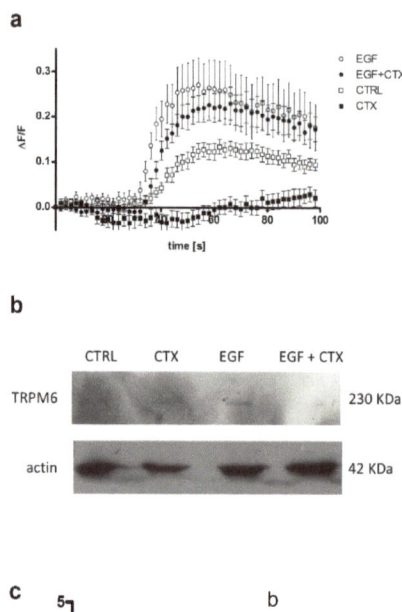

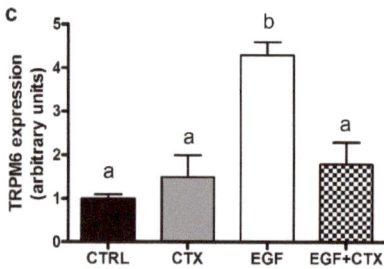

**Figure 2.** Epidermal growth factor (EGF) stimulates TRPM6 channel expression and Mg influx in human colon carcinoma cells. CaCo-2 cells were serum-starved for 24 h and exposed to EGF (10 ng/mL) and cetuximab (CTX, 70 µg/mL), either alone or in combination, for a further 24h. (**a**) Mg influx capacity, as assessed by the live imaging of Mag-Fluo-4-loaded cells; a representative experiment is shown. (**b**) TRPM6 protein expression, as evaluated by Western blot analysis; a representative blot is shown. (**c**) Quantification of TRPM6 protein expression by Western blot densitometry normalized to β-actin levels ($n = 3$, mean ± SD). Data sharing the same letter are not significantly different ($p > 0.05$) according to one-way ANOVA, followed by Bonferroni's test. Full scans of original blots are available in Figure S2.

### 3.3. Mg Supplementation Does Not Affect Cetuximab Efficacy

To evaluate whether the extracellular Mg availability may alter sensitivity to the growth inhibitory effects of CTX, we challenged CaCo-2 cells with CTX in Mg-supplemented (5 mM of MgSO$_4$) medium. As shown in Figure 3a, the IC50 for CTX at 24h did not change significantly in the Mg-supplemented cells in comparison to the control cells (50 ± 10 vs. 66 ± 16 µg/mL, respectively). CTX prevents the dimerization of the EGFR and the activation of downstream pathways; primary or acquired resistance to CTX is mainly due to the constitutive activation of MEK signaling with subsequent MAPK activation [30]. Therefore, we also assessed whether Mg supplementation could interfere with the CTX-dependent inhibition of ERK1/2 phosphorylation. Western blot analysis showed that the presence of 5 mM of MgSO$_4$ did not substantially change the amount of phospho-ERK1/2 in CTX-treated cells (Figure 3b); two-way ANOVA confirmed that the Mg supplementation had no effect ($p = 0.70$), while CTX treatment had a very significant effect ($p < 0.001$, Figure 3c) on the ERK1/2 phosphorylation.

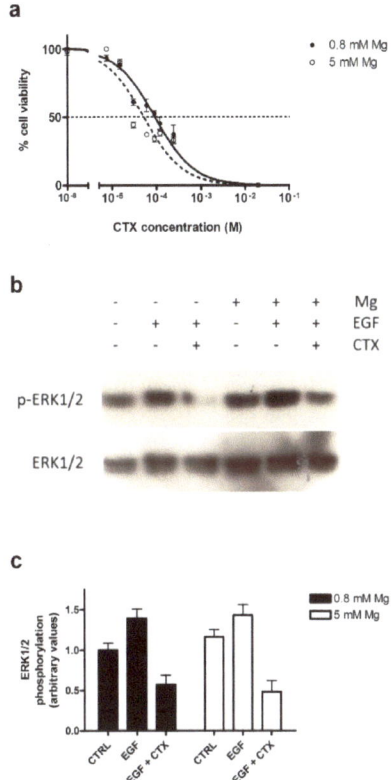

**Figure 3.** Mg supplementation does not affect cancer cell sensitivity to CTX. (**a**) A representative dose-response curve of CaCo-2 cells exposed to CTX (24 h) in control medium (0.8 mM of MgSO$_4$, solid line and circles) or Mg-supplemented medium (5 mM of MgSO$_4$, dotted line and open circles). (**b**) Phosphorylation of extracellular signal-regulated kinase (ERK) 1/2 in CaCo2 cells challenged for 24h with EGF (10 ng/mL) with or without CTX (70 µg/mL) in control (0.8 mM of MgSO$_4$) or Mg-supplemented (5 mM of MgSO$_4$) medium, as assessed by Western blot analysis; a representative blot is shown. (**c**) Quantification of ERK1/2 phosphorylation by Western blot densitometry, normalized to total ERK1/2 levels ($n = 3$, mean ± SD). Two-way ANOVA indicated a very significant effect ($p < 0.001$) of EGF/CTX treatment, while Mg concentration had no significant effect ($p = 0.70$). The effect of CTX on unstimulated cells in control or Mg-supplemented medium is reported in Figure S3.

## 4. Discussion

Cancer-associated hypomagnesemia has long been recognized, and was originally attributed to the metabolic demands of tumor growth; now, it has become clear that also cancer therapies play an important role [13,14]. In contrast to cisplatin, which damages renal tubules and hence causes a generalized electrolyte wasting [3], CTX induces hypomagnesemia by a specific antagonistic effect on renal Mg reabsorption. However, the interference between the mode of action of CTX and the homeostatic mechanisms of Mg has not received the deserved attention to its clinical implications. In this paper, we provide evidence supporting two relevant issues: (1) CTX-induced hypomagnesemia is not just due to renal wasting, but also to impaired intestinal absorption; (2) Mg supplementation does not modify CTX cytotoxicity.

Until recently, renal Mg reabsorption was thought to play a pivotal role in maintaining systemic Mg homeostasis [16]. However, conditional knockout murine models have proved that wild-type kidneys are not able to compensate for the ablation of intestinal TRPM6 and have pointed to an indispensable function of gut Mg absorption in the maintenance of proper Mg status [17]. The data we present here are in line with this view, confirming that cross-talk between EGF and TRPM6 occurs also in intestinal cells. In our work, we focus on the pathological setting of colon cancer and the mechanisms of CTX-induced hypomagnesemia. We are aware that our results do not conclusively prove that the activation of the EGFR pathway regulates the TRPM6-mediated intestinal Mg absorption in a more physiological context; future studies in more appropriate models will address this issue. On the other hand, it could be speculated that, in cancer cells, autocrine signaling by EGF may potentiate the TRPM6-mediated magnesium influx or compete with CTX. However, the EGF production by CaCo-2 cells does not undermine the value of our results, since in our model CTX does inhibit cellular growth as well as both basal and EGF-stimulated magnesium influx.

We propose that the impaired Mg absorption in the intestine heavily contributes to the severe hypomagnesemia that occurs in many CTX-treated patients. In the kidneys, the EGF-dependent modulation of TRPM6 has been ascribed to two different mechanisms: (1) the altered endomembrane trafficking of TRPM6, which results in increased TRPM6 channel activity on the cell surface [22], and (2) a transcriptional effect on the *TRPM6* mRNA expression via the MAPK/ERK pathway [31]. In our colon cellular model, we confirm that EGF increases the TRPM6 protein expression, while we have no evidence that acute EGF stimulation affects the number of recycling channels in favor of increased plasma membrane expression. We are aware of the limitations of our approach resulting from analyzing the TRPM6 protein expression by Western blot. However, the mRNA expression levels might not necessarily translate into a functional channel protein. Indeed, post-translational modifications may affect the protein levels, regardless of (or in addition to) transcriptional events; for example, this has been demonstrated for the sister channel TRPM7 [32]. On the other hand, ion channels are notoriously challenging to study for two main reasons: (a) the paucity of channel molecules per cell—a rough estimate obtained by electrophysiological studies is about 70 molecules/cell (Prof. Andrea Fleig, personal communication); (b) the difficulty of producing specific antibodies, due to structural constraints and sequence similarity. The commercial anti-TRPM6 antibody that we use is the only known antibody that does not display cross-reactivity with TRPM7 [19]. Ultimately, we are interested in the functional effects of cross-talk between EGF and TRPM6; in this respect, by using a functional assay we provide definitive proof that, in intestinal cells, (a) TRPM6 is indispensable for mediating the Mg influx (Figure 1a), and (b) CTX modulates the TRPM6-mediated Mg influx (Figure 2a).

The symptoms of hypomagnesemia range from depression and muscle spasms to arrhythmias and seizures, and significantly worsen the quality of life of patients [16]. Severe hypomagnesemia warrants treatment by intravenous and/or oral magnesium supplementation for the duration of CTX therapy, and cases of CTX dose reduction and discontinuation have been documented [33]. No evidence-based guidelines have currently been developed for the management of hypomagnesaemia in the context of CTX cancer therapy [33]. We demonstrate that CTX inhibits the TRPM6-mediated transcellular Mg transport. However, oral Mg supplementation, by increasing the intraluminal Mg concentration and thus favoring the paracellular route, may represent an effective strategy to restore the Mg status in CTX-treated patients, even in the presence of limited renal transcellular reabsorption. Most importantly, such approach would be feasible on an outpatient basis.

Although most medical oncologists agree on the necessity of restoring the Mg status in symptomatic patients [31], the relationship between Mg and cancer remains highly controversial [34]. Early reports found that, in murine models, a low Mg availability resulted in the inhibition of primary tumor growth, but at the same time enhanced metastasis formation [35,36]. On the other hand, more recently Mg status was reported to have no influence on tumor progression in two different animal models [37,38]. Notably, Mg supplementation protected against cisplatin-induced acute kidney injury without compromising the cisplatin-mediated killing of an ovarian tumor xenograft in mice [37]. Despite the limitations of an in vitro model, our data oppose the view that the effect of Mg deficiency on cell proliferation is synergistic with that of CTX: Mg supplementation did not significantly affect cell growth (Figure S4), nor did it alter the effect of CTX inhibitory activity on cell growth or on MAPK signaling (Figure 3).

In conclusion, we present evidence that Mg supplementation does not compromise CTX efficacy, and suggest that nutritional intervention may be a safe and cost-effective approach by which to maximize patient wellbeing and improve CRC management. Further preclinical and clinical research will be necessary to clarify the relationship between Mg and CTX in various experimental tumor models and the potential of Mg supplementation in CTX-treated patients.

**Supplementary Materials:** The following are available online at http://www.mdpi.com/2072-6643/12/11/3277/s1: Figure S1: full scan of the original blot for TRPM6 shown in Figure 1; Figure S2: full scan of the original blot for TRPM6 shown in Figure 2; Figure S3: phosphorylation of ERK1/2 in unstimulated CaCo2 treated with CTX in control or Mg-supplemented medium; Figure S4: growth curve of CaCo-2 cells in control or Mg-supplemented medium.

**Author Contributions:** Conceptualization, F.I.W. and V.T.; formal analysis, G.P. and V.T.; investigation, G.P., D.P. and V.T.; resources, A.A. and L.G.; writing—original draft preparation, V.T.; writing—review and editing, G.P., D.P., A.A., L.G., A.G., G.L.R., F.I.W. and V.T.; supervision, F.I.W. and V.T.; funding acquisition, F.I.W. All authors have read and agreed to the published version of the manuscript.

**Funding:** This research was funded by MIUR (Italian Ministry of University and Research) D.3.2-2015 and D.1.2017.

**Acknowledgments:** Confocal imaging was performed at the LABCEMI (Laboratorio Centralizzato di Microscopia Ottica ed Elettronica), Università Cattolica del Sacro Cuore, Fondazione Policlinico Universitario "Agostino Gemelli" IRCCS, Rome, Italy.

**Conflicts of Interest:** The authors declare no conflict of interest.

# References

1. Martin, L.; Senesse, P.; Gioulbasanis, I.; Antoun, S.; Bozzetti, F.; Deans, C.; Strasser, F.; Thoresen, L.; Jagoe, R.T.; Chasen, M.; et al. Diagnostic criteria for the classification of cancer-associated weight loss. *J. Clin. Oncol.* **2015**, *33*, 90–99. [CrossRef] [PubMed]
2. Arends, J.; Bachmann, P.; Baracos, V.; Barthelemy, N.; Bertz, H.; Bozzetti, F.; Fearon, K.; Hütterer, E.; Isenring, E.; Kaasa, S.; et al. ESPEN guidelines on nutrition in cancer patients. *Clin. Nutr.* **2017**, *36*, 11–48. [CrossRef] [PubMed]
3. Finkel, M.; Goldstein, A.; Steinberg, Y.; Granowetter, L.; Trachtman, H. Cisplatinum nephrotoxicity in oncology therapeutics: Retrospective review of patients treated between 2005 and 2012. *Pediatr. Nephrol.* **2014**, *29*, 2421–2424. [CrossRef] [PubMed]
4. Hofheinz, R.D.; Segaert, S.; Safont, M.J.; Demonty, G.; Prenen, H. Management of adverse events during treatment of gastrointestinal cancers with epidermal growth factor inhibitors. *Crit. Rev. Oncol. Hematol.* **2017**, *114*, 102–113. [CrossRef]
5. Cao, Y.; Liao, C.; Tan, A.; Liu, L.; Gao, F. Meta-analysis of incidence and risk of hypomagnesemia with cetuximab for advanced cancer. *Chemotherapy* **2010**, *56*, 459–465. [CrossRef]
6. Petrelli, F.; Borgonovo, K.; Cabiddu, M.; Ghilardi, M.; Barni, S. Risk of anti-EGFR monoclonal antibody-related hypomagnesemia: Systematic review and pooled analysis of randomized studies. *Expert Opin. Drug Saf.* **2012**, *11*, S9–S19. [CrossRef]
7. Chen, P.; Wang, L.; Li, H.; Liu, B.; Zou, Z. Incidence and risk of hypomagnesemia in advanced cancer patients treated with cetuximab: A meta-analysis. *Oncol. Lett.* **2013**, *5*, 1915–1920. [CrossRef]
8. Wang, Q.; Qi, Y.; Zhang, D.; Gong, C.; Yao, A.; Xiao, Y.; Yang, J.; Zhou, F.; Zhou, Y. Electrolyte disorders assessment in solid tumor patients treated with anti-EGFR monoclonal antibodies: A pooled analysis of 25 randomized clinical trials. *Tumour Biol.* **2015**, *36*, 3471–3482. [CrossRef]
9. Vincenzi, B.; Galluzzo, S.; Santini, D.; Rocci, L.; Loupakis, F.; Correale, P.; Addeo, R.; Zoccoli, A.; Napolitano, A.; Graziano, F.; et al. Early magnesium modifications as a surrogate marker of efficacy of cetuximab-based anticancer treatment in KRAS wild-type advanced colorectal cancer patients. *Ann. Oncol.* **2011**, *22*, 1141–1146. [CrossRef]

10. Vickers, M.M.; Karapetis, C.S.; Tu, D.; O'Callaghan, C.J.; Price, T.J.; Tebbutt, N.C.; Van Hazel, G.; Shapiro, J.D.; Pavlakis, N.; Gibbs, P.; et al. Association of hypomagnesemia with inferior survival in a phase III, randomized study of cetuximab plus best supportive care versus best supportive care alone: NCIC CTG/AGITG CO.17. *Ann. Oncol.* **2013**, *24*, 953–960. [CrossRef]
11. Hsieh, M.C.; Wu, C.F.; Chen, C.W.; Shi, C.S.; Huang, W.S.; Kuan, F.C. Hypomagnesemia and clinical benefits of anti-EGFR monoclonal antibodies in wild-type KRAS metastatic colorectal cancer: A systematic review and meta-analysis. *Sci. Rep.* **2018**, *8*, 2047. [CrossRef] [PubMed]
12. Vincenzi, B.; Santini, D.; Tonini, G. Biological interaction between anti-epidermal growth factor receptor agent cetuximab and magnesium. *Exp. Opin. Pharmacother.* **2008**, *9*, 1267–1269. [CrossRef] [PubMed]
13. Wolf, F.I.; Trapani, V.; Cittadini, A.; Maier, J.A. Hypomagnesaemia in oncologic patients: To treat or not to treat? *Magnes. Res.* **2009**, *22*, 5–9. [CrossRef] [PubMed]
14. Wolf, F.I.; Cittadini, A.R.; Maier, J.A. Magnesium and tumors: Ally or foe? *Cancer Treat. Rev.* **2009**, *35*, 378–382. [CrossRef]
15. Bairoch, A. The ENZYME database in 2000. *Nucleic Acids Res.* **2000**, *28*, 304–305. [CrossRef]
16. de Baaij, J.H.; Hoenderop, J.G.; Bindels, R.J. Magnesium in man: Implications for health and disease. *Physiol. Rev.* **2015**, *95*, 1–46. [CrossRef]
17. Chubanov, V.; Ferioli, S.; Wisnowsky, A.; Simmons, D.G.; Leitzinger, C.; Einer, C.; Jonas, W.; Shymkiv, Y.; Bartsch, H.; Braun, A.; et al. Epithelial magnesium transport by TRPM6 is essential for prenatal development and adult survival. *Elife* **2016**, *5*, e20914. [CrossRef]
18. Ferioli, S.; Zierler, S.; Zaißerer, J.; Schredelseker, J.; Gudermann, T.; Chubanov, V. TRPM6 and TRPM7 differentially contribute to the relief of heteromeric TRPM6/7 channels from inhibition by cytosolic $Mg^{2+}$ and Mg•ATP. *Sci. Rep.* **2017**, *7*, 8806. [CrossRef]
19. Luongo, F.; Pietropaolo, G.; Gautier, M.; Dhennin-Duthille, I.; Ouadid-Ahidouch, H.; Wolf, F.I.; Trapani, V. TRPM6 is Essential for Magnesium Uptake and Epithelial Cell Function in the Colon. *Nutrients* **2018**, *10*, 784. [CrossRef]
20. Trapani, V.; Petito, V.; Di Agostini, A.; Arduini, D.; Hamersma, W.; Pietropaolo, G.; Luongo, F.; Arena, V.; Stigliano, E.; Lopetuso, L.R.; et al. Dietary Magnesium Alleviates Experimental Murine Colitis Through Upregulation of the Transient Receptor Potential Melastatin 6 Channel. *Inflamm. Bowel Dis.* **2018**, *24*, 2198–2210. [CrossRef]
21. Groenestege, W.M.; Thébault, S.; van der Wijst, J.; van den Berg, D.; Janssen, R.; Tejpar, S.; van den Heuvel, L.P.; van Cutsem, E.; Hoenderop, J.G.; Knoers, N.V.; et al. Impaired basolateral sorting of pro-EGF causes isolated recessive renal hypomagnesemia. *J. Clin. Invest.* **2007**, *117*, 2260–2267. [CrossRef]
22. Thebault, S.; Alexander, R.T.; Tiel Groenestege, W.M.; Hoenderop, J.G.; Bindels, R.J. EGF increases TRPM6 activity and surface expression. *J. Am. Soc. Nephrol.* **2009**, *20*, 78–85. [CrossRef] [PubMed]
23. Tang, X.; Liu, H.; Yang, S.; Li, Z.; Zhong, J.; Fang, R. Epidermal Growth Factor and Intestinal Barrier Function. *Med. Inflamm.* **2016**, *2016*, 1927348. [CrossRef]
24. Mroz, M.S.; Keely, S.J. Epidermal growth factor chronically upregulates $Ca^{2+}$-dependent $Cl^-$ conductance and TMEM16A expression in intestinal epithelial cells. *J. Physiol.* **2012**, *590*, 1907–1920. [CrossRef]
25. Parseghian, C.M.; Napolitano, S.; Loree, J.M.; Kopetz, S. Mechanisms of Innate and Acquired Resistance to Anti-EGFR Therapy: A Review of Current Knowledge with a Focus on Rechallenge Therapies. *Clin. Cancer Res.* **2019**, *25*, 6899–6908. [CrossRef]
26. Ahmed, D.; Eide, P.W.; Eilertsen, I.A.; Danielsen, S.A.; Eknæs, M.; Hektoen, M.; Lind, G.E.; Lothe, R.A. Epigenetic and genetic features of 24 colon cancer cell lines. *Oncogenesis* **2013**, *2*, e71. [CrossRef] [PubMed]
27. Shigeta, K.; Hayashida, T.; Hoshino, Y.; Okabayashi, K.; Endo, T.; Ishii, Y.; Hasegawa, H.; Kitagawa, Y. Expression of Epidermal Growth Factor Receptor Detected by Cetuximab Indicates Its Efficacy to Inhibit In Vitro and In Vivo Proliferation of Colorectal Cancer Cells. *PLoS ONE* **2013**, *8*, e66302. [CrossRef]
28. Trapani, V.; Arduini, D.; Luongo, F.; Wolf, F.I. EGF stimulates $Mg^{2+}$ influx in mammary epithelial cells. *Biochem. Biophys. Res. Commun.* **2014**, *454*, 572–575. [CrossRef] [PubMed]
29. Trapani, V.; Schweigel-Röntgen, M.; Cittadini, A.; Wolf, F.I. Intracellular magnesium detection by fluorescent indicators. *Methods Enzymol.* **2012**, *505*, 421–444.

30. Troiani, T.; Napolitano, S.; Vitagliano, D.; Morgillo, F.; Capasso, A.; Sforza, V.; Nappi, A.; Ciardiello, D.; Ciardiello, F.; Martinelli, E. Primary and acquired resistance of colorectal cancer cells to anti-EGFR antibodies converge on MEK/ERK pathway activation and can be overcome by combined MEK/EGFR inhibition. *Clin. Cancer Res.* **2014**, *20*, 3775–3786. [CrossRef]
31. Ikari, A.; Sanada, A.; Okude, C.; Sawada, H.; Yamazaki, Y.; Sugatani, J.; Miwa, M. Up-regulation of TRPM6 transcriptional activity by AP-1 in renal epithelial cells. *J. Cell. Physiol.* **2010**, *222*, 481–487. [CrossRef] [PubMed]
32. Castiglioni, S.; Cazzaniga, A.; Trapani, V.; Cappadone, C.; Farruggia, G.; Merolle, L.; Wolf, F.I.; Iotti, S.; Maier, J. Magnesium homeostasis in colon carcinoma LoVo cells sensitive or resistant to doxorubicin. *Sci. Rep.* **2015**, *5*, 16538. [CrossRef] [PubMed]
33. Thangarasa, T.; Gotfrit, J.; Goodwin, R.A.; Tang, P.A.; Clemons, M.; Imbulgoda, A.; Vickers, M.M. Epidermal growth factor receptor inhibitor-induced hypomagnesemia: A survey of practice patterns among Canadian gastrointestinal medical oncologists. *Curr. Oncol.* **2019**, *26*, e162–e166. [CrossRef]
34. Trapani, V.; Wolf, F.I. Dysregulation of $Mg^{2+}$ homeostasis contributes to acquisition of cancer hallmarks. *Cell Calcium.* **2019**, *83*, 102078. [CrossRef]
35. Maier, J.A.; Nasulewicz-Goldeman, A.; Simonacci, M.; Boninsegna, A.; Mazur, A.; Wolf, F.I. Insights into the mechanisms involved in magnesium-dependent inhibition of primary tumor growth. *Nutr. Cancer.* **2007**, *59*, 192–198. [CrossRef] [PubMed]
36. Nasulewicz, A.; Wietrzyk, J.; Wolf, F.I.; Dzimira, S.; Madej, J.; Maier, J.A.; Rayssiguier, Y.; Mazur, A.; Opolski, A. Magnesium deficiency inhibits primary tumor growth but favors metastasis in mice. *Biochim. Biophys. Acta.* **2004**, *1739*, 26–32. [CrossRef]
37. Solanki, M.H.; Chatterjee, P.K.; Xue, X.; Gupta, M.; Rosales, I.; Yeboah, M.M.; Kohn, N.; Metz, C.N. Magnesium protects against cisplatin-induced acute kidney injury without compromising cisplatin-mediated killing of an ovarian tumor xenograft in mice. *Am. J. Physiol. Renal Physiol.* **2015**, *309*, F35–F47. [CrossRef]
38. Huang, J.; Furuya, H.; Faouzi, M.; Zhang, Z.; Monteilh-Zoller, M.; Kawabata, K.G.; Horgen, F.D.; Kawamori, T.; Penner, R.; Fleig, A. Inhibition of TRPM7 suppresses cell proliferation of colon adenocarcinoma in vitro and induces hypomagnesemia in vivo without affecting azoxymethane-induced early colon cancer in mice. *Cell Commun. Signal.* **2017**, *15*, 30. [CrossRef]

**Publisher's Note:** MDPI stays neutral with regard to jurisdictional claims in published maps and institutional affiliations.

© 2020 by the authors. Licensee MDPI, Basel, Switzerland. This article is an open access article distributed under the terms and conditions of the Creative Commons Attribution (CC BY) license (http://creativecommons.org/licenses/by/4.0/).

*Review*

# Mg²⁺ Transporters in Digestive Cancers

Julie Auwercx [1], Pierre Rybarczyk [1,2], Philippe Kischel [1], Isabelle Dhennin-Duthille [1], Denis Chatelain [2], Henri Sevestre [1,2], Isabelle Van Seuningen [3], Halima Ouadid-Ahidouch [1], Nicolas Jonckheere [3,†] and Mathieu Gautier [1,*,†]

1. Université de Picardie Jules Verne, UFR des Sciences, UR-UPJV 4667, F-80000 Amiens, France; julie.auwercx@etud.u-picardie.fr (J.A.); Rybarczyk.Pierre@chu-amiens.fr (P.R.); philippe.kischel@u-picardie.fr (P.K.); isabelle.dhennin@u-picardie.fr (I.D.-D.); sevestre.henri@chu-amiens.fr (H.S.); halima.ahidouch-ouadid@u-picardie.fr (H.O.-A.)
2. Service d'Anatomie et Cytologie Pathologique, CHU Amiens-Picardie, F-80000 Amiens, France; chatelain.denis@chu-amiens.fr
3. University Lille, CNRS, Inserm, CHU Lille, UMR9020-U1277—CANTHER—Cancer Heterogeneity Plasticity and Resistance to Therapies, F-59000 Lille, France; isabelle.vanseuningen@inserm.fr (I.V.S.); nicolas.jonckheere@inserm.fr (N.J.)
* Correspondence: mathieu.gautier@u-picardie.fr; Tel.: +33-3-22-82-76-42
† These authors contributed equally to this work.

**Abstract:** Despite magnesium ($Mg^{2+}$) representing the second most abundant cation in the cell, its role in cellular physiology and pathology is far from being elucidated. $Mg^{2+}$ homeostasis is regulated by $Mg^{2+}$ transporters including Mitochondrial RNA Splicing Protein 2 (MRS2), Transient Receptor Potential Cation Channel Subfamily M, Member 6/7 (TRPM6/7), Magnesium Transporter 1 (MAGT1), Solute Carrier Family 41 Member 1 (SCL41A1), and Cyclin and CBS Domain Divalent Metal Cation Transport Mediator (CNNM) proteins. Recent data show that $Mg^{2+}$ transporters may regulate several cancer cell hallmarks. In this review, we describe the expression of $Mg^{2+}$ transporters in digestive cancers, the most common and deadliest malignancies worldwide. Moreover, $Mg^{2+}$ transporters' expression, correlation and impact on patient overall and disease-free survival is analyzed using Genotype Tissue Expression (GTEx) and The Cancer Genome Atlas (TCGA) datasets. Finally, we discuss the role of these $Mg^{2+}$ transporters in the regulation of cancer cell fates and oncogenic signaling pathways.

**Keywords:** magnesium transporters; digestive cancers; TCGA; overall survival

**Citation:** Auwercx, J.; Rybarczyk, P.; Kischel, P.; Dhennin-Duthille, I.; Chatelain, D.; Sevestre, H.; Van Seuningen, I.; Ouadid-Ahidouch, H.; Jonckheere, N.; Gautier, M. Mg²⁺ Transporters in Digestive Cancers. *Nutrients* **2021**, *13*, 210. https://doi.org/10.3390/nu13010210

Received: 22 December 2020
Accepted: 8 January 2021
Published: 13 January 2021

**Publisher's Note:** MDPI stays neutral with regard to jurisdictional claims in published maps and institutional affiliations.

**Copyright:** © 2021 by the authors. Licensee MDPI, Basel, Switzerland. This article is an open access article distributed under the terms and conditions of the Creative Commons Attribution (CC BY) license (https://creativecommons.org/licenses/by/4.0/).

## 1. Introduction

According to the International Agency for Research, digestive cancers are the most common and deadliest malignancies worldwide [1]. In this review, we choose to focus on the main digestive cancers namely esophageal adenocarcinoma, gastric cancer, pancreatic ductal adenocarcinoma and colorectal cancer.

Esophageal cancer (ESAC) is ranked in the seventh position in terms of incidence and in the sixth in terms of mortality [1]. ESAC is the most common type of esophageal cancer in industrialized countries [2]. It is among the most lethal digestive malignancies with only 16% of patients surviving 5 years after diagnosis and a median survival that is less than 1 year [3]. The main risk factor for ESAC is the gastroesophageal reflux disease, that leads to inflammation of esophageal and remodeling of tissue into a metaplastic, specialized intestinal epithelium named Barrett's esophagus. Tobacco smoking and obesity have been also identified as others strong risk factors for ESAC [3].

Gastric cancer (GC) is the fifth most common cancer worldwide and the third deadliest [1]. The 5-year survival rate is dependent of the stage of disease at the diagnosis. GCs detected at early stage have a 5-year survival rate around 80% [4]. There is strong evidence that *Helicobacter pylori* infection is a risk factor for GC development, therefore *Helicobacter pylori* has been classified as a class I carcinogen by International Agency for Research on Cancer [5]. As for many cancers, dietary factors play also a role in stomach carcinogenesis. Fruits, vegetables, and vitamins intake seem to have a protective role, while alcohol, coffee, meat and high salt consumption seem to increase the risk of developing GC.

Pancreatic ductal adenocarcinoma (PDAC) is the 7th leading cause of global cancer deaths in industrialized countries and the 3rd in USA, while it is ranked in the 11th position in term of incidence [6]. Unlike most cancers, the PDAC incidence is in constant progression and it is estimated that it will become the second deadliest cancer in 2030 [7]. The poor prognosis associated with PDAC is because this malignancy is mainly diagnosed too late in an advanced and metastatic stage. To date, carbohydrate antigen 19-9 (CA 19-9) is the only diagnostic marker for PDAC approved by the U.S. Food and Drug Administration (FDA). However, other cancers and benign diseases can cause CA 19-9 overexpression which can explain the poor specificity. Thus, there is an urgent need for specific biomarkers for PDAC [8]. To date, cigarette smoking and family history are the main risk factors but dietary style and obesity have been also considered [6].

Colorectal cancer (CRC) is at the third rank in term of incidence and at the second rank in term of mortality [1]. The incidence of CRC is country-dependent and the main factor risks for CRC are family hereditary, red and processed meat consumption, alcohol drinking, obesity, and inflammatory bowel disease [9]. Surprisingly, while the overall incidence and mortality are decreasing, the incidence of early-onset CRC, generally diagnosed before 50 years old, is increasing worldwide. The increase in early-onset CRC incidence associated with a higher mortality rate for young adults may be associated with Western lifestyle, including diet [9]. Consequently, there is an important role of nutrition in cause and prevention of CRC [10].

Nutrients are transported through the gastrointestinal tract and nutrient deficiency could be associated with digestive cancer initiation and/or promotion. Among these nutrients, low magnesium intake is observed in a large part of the population, especially in industrialized countries. The aim of this review is to present the current knowledge on magnesium levels and digestive cancer development. Firstly, we will focus on magnesium transporter expression in digestive cancers by analyzing the Cancer Genome Atlas (TCGA). In the last part, the role of these magnesium transporters in cancer cell fate and their potential importance as new biomarkers in digestive cancers will be discussed.

## 2. Magnesium

Magnesium ($Mg^{2+}$) is one of the most important ions in health and is the second most abundant cation in the cell with a concentration estimated between 10 and 30 mM. Due to the binding to different partners like ATP, ribosomes, or nucleotides, the free intracellular $Mg^{2+}$ levels lower to 0.5 to 1.2 mM [11]. $Mg^{2+}$ is essential in almost all cellular processes, acting as a cofactor and activator for various enzymes [11]. For example, $Mg^{2+}$ is essential in DNA stabilization, DNA repair mechanisms, or even protein synthesis [12–15]. New interactions are still being discovered, expanding the importance of this cation [16].

Normal $Mg^{2+}$ in blood serum levels for healthy people is about 0.7–1 mM, corresponding to an average daily intake (ADI) of 320–420 mg/day [17,18]. This $Mg^{2+}$ intake is absorbed mostly in the small intestine by two mechanisms: paracellular transport and via the expression of membrane transporters (Figure 1). Paracellular transport is predominant, mainly because of low expression of claudins in the small intestine [19,20]. Numerous $Mg^{2+}$ transporters are also present in the plasma membrane of intestine cells for $Mg^{2+}$ absorption. An average of 100 mg is absorbed in the intestine, depending on the daily $Mg^{2+}$ intake [11]. Kidneys filters around 2400 mg of $Mg^{2+}$ per day in the glomeruli, where most of the $Mg^{2+}$ (2300 mg) is reabsorbed in the thick ascending limb of Henle's loop. $Mg^{2+}$ is mainly stored in bones but also in muscles and soft tissues. [11,21]. This organization allows the $Mg^{2+}$ homeostasis balance, maintaining a constant 0.7–1 mM $Mg^{2+}$ serum level in normal conditions.

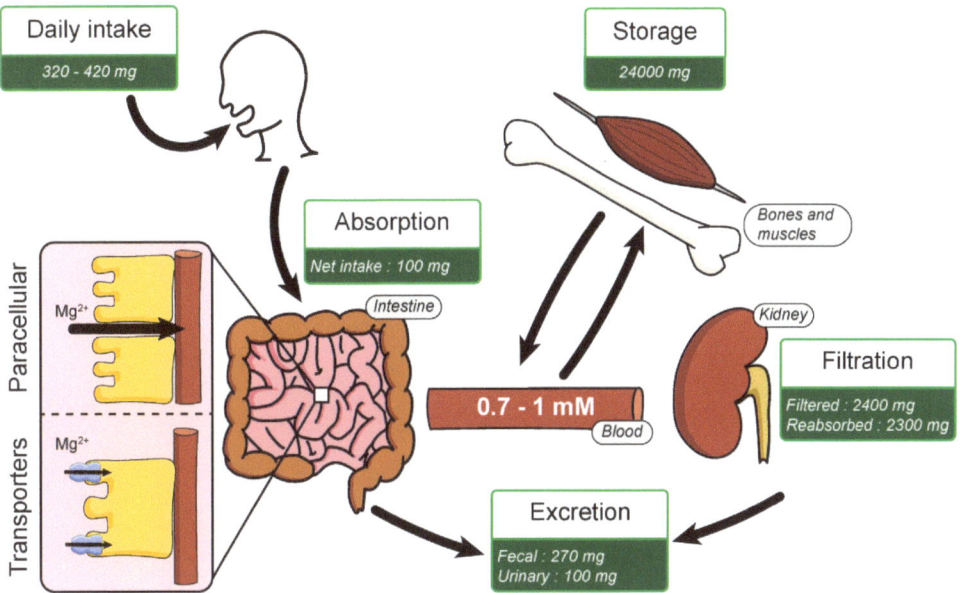

**Figure 1.** Summary of Mg$^{2+}$ homeostasis.

Unfortunately, our alimentation contains nowadays less Mg$^{2+}$ because of the development of the food industry and changes in soils due to intensive farming [22,23]. Along with the modifications of our eating habits and the prevalence of processed foods, it is shown that a large number of adults do not reach the recommended Mg$^{2+}$ average daily intake [24]. Hypomagnesemia is characterized by Mg$^{2+}$ serum levels <0.7 mM, but it is often underestimated because the serum levels are not representative of the whole Mg$^{2+}$ availability [25]. Hypomagnesemia is associated with several health issues such as epilepsy, cystic fibrosis, atherosclerosis, and type 2 diabetes [26–29].

Several studies suggest that calcium (Ca$^{2+}$) and Mg$^{2+}$ can compete during intestinal absorption, leading to the consideration also of the Ca$^{2+}$/Mg$^{2+}$ ratio for assessing Ca$^{2+}$ and Mg$^{2+}$ intakes [30].

Due to its importance, Mg$^{2+}$, requires a specific transport system. The first magnesium transporters were identified in prokaryotes, with the identification of the proteins magnesium/cobalt transporter (CorA), magnesium-transporting ATPase (MgtA/B/E) [31]. Subsequently, Mg$^{2+}$ transporters were identified and cloned in other models (Figure 2). In Mammals, several transporters have been identified and will be described in this manuscript.

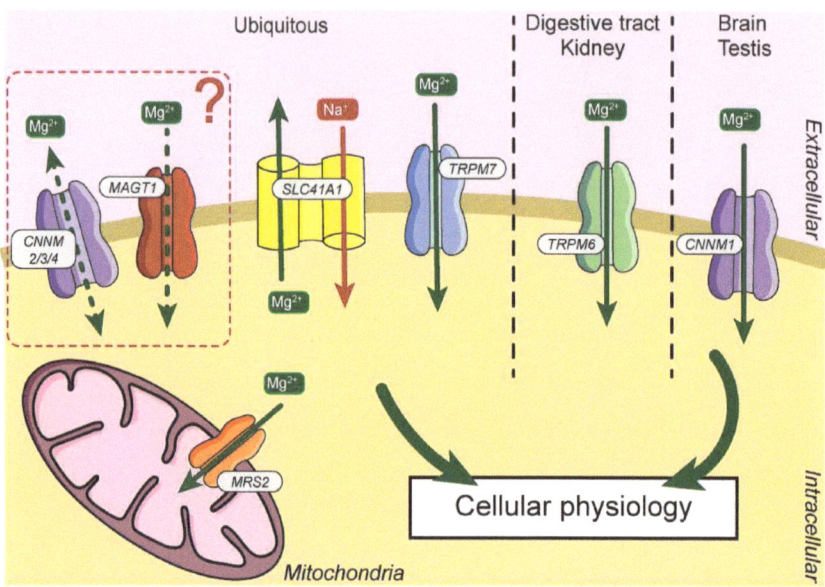

**Figure 2.** General distribution and localization of $Mg^{2+}$ transporters in cells. $Mg^{2+}$, magnesium; $Na^+$, sodium; CNNM2/3/4, Cyclin and CBS Domain Divalent Metal Cation Transport Mediator2/3/4; MAGT1, Magnesium Transporter 1; SLC41A1, Solute Carrier Family 41, Member 1; TRPM7, Transient Receptor Potential Cation Channel Subfamily M Member 7; TRPM6, Transient Receptor Potential Cation Channel Subfamily M Member 6; CNNM1, Cyclin and CBS Domain Divalent Metal Cation Transport Mediator1; MRS2, Mitochondrial RNA Splicing Protein 2.

### 2.1. MRS2

The first $Mg^{2+}$ transporter characterized in Metazoa is the Mitochondrial RNA splicing protein 2 (MRS2/MRS2p). It was discovered as a CorA homolog, localized in the mitochondrial inner membrane, and involved in $Mg^{2+}$ mitochondrial uptake. A ubiquitous mRNA expression of MRS2 was found in rat tissues [32]. With the use of a MagFura-2 fluorescent probe in yeast, MRS2 overexpression was shown to enhance mitochondrial $Mg^{2+}$ influx [33]. On the other hand, the mitochondrial $Mg^{2+}$ influx was abolished upon MRS2 gene deletion. The MRS2 protein is therefore described as an essential magnesium transporter in the mitochondria.

### 2.2. TRPM7 and TRPM6

The transient receptor potential cation channel subfamily M member 7 (TRPM7) was discovered and cloned by two teams under different names, Long TRP Channel 7 (LTRPC7) and TRP-Phospholipase C Interacting Kinase (TRP-PLIK). TRPM7 is first known as the long transient receptor potential channel 7 (LTRPC7), a member of LTRPCs by its similarity with the first 1200 amino-acids [34]. Its carboxy-terminal tail is pretty unique as it contains a kinase domain, with significant homology to the protein-kinase family of Myosin Heavy Chain Kinase/Eukaryotic Elongation Factor 2 Kinase (MHCK/eEF-2) [35]. In the DT-40 lymphoma cell line, TRPM7 has a role in viability and proliferation [34]. Using the patch-clamp technique in the HEK-293 cell line, it was shown that LTRPC7 was permeable to $Ca^{2+}$ and $Mg^{2+}$ and was inhibited by cytosolic free $Mg^{2+}$ and magnesium bound to ATP (MgATP) [34]. In the same year, the TRP-PLIK was described, with similarity with the LTRPC family [36]. TRP-PLIK, with its kinase domain, is suspected to have autophosphorylation properties. Using patch-clamp on CHO-K1 cells, it was shown that TRP-PLIK is permeable to $Ca^{2+}$ and monovalent cations like sodium ($Na^+$) or potassium ($K^+$). Other electrophysiological studies on HEK-293 cells have shown that the

TRPM7 channel is also permeable to other cations like zinc, nickel, baryum, cobalt, strontium, and cadmium [37]. TRPM7 expression was found to be ubiquitous in mouse and human tissues [36,38]. TRPM7 is now proposed as an essential actor in magnesium homeostasis, embryonic development, and mineral absorption [39–41].

The Transient Receptor Potential Cation Channel Subfamily M, Member 6 (TRPM6) is the second TRPM channel involved in $Mg^{2+}$ transport. The mutated gene is associated with "hypomagnesemia with secondary hypocalcemia" (HSH) [42,43]. TRPM6 shares strong homology with TRPM7, and also has an alpha-kinase domain at its C-terminus [35]. Strong TRPM6 mRNA expression was found in the intestine and the distal convoluted tube (DCT) in mouse kidney tissues, and this expression was confirmed in human tissues [44]. The protein was detected at the apical membrane of DCT in mouse kidney and the brush-border membrane of the small intestine. Using a patch-clamp, it was shown that TRPM6 is responsible for $Mg^{2+}$ currents [44] and TRPM6 is now considered as an essential actor in $Mg^{2+}$ (re)absorption in the kidney.

### 2.3. SLC41A1

The human transporter solute carrier family 41, member 1 (SLC41A1) was also identified by homology with a prokaryote $Mg^{2+}$ transporter, the Magnesium Transporter E (MgtE). Both transporters share similarities on two transmembrane domains [45]. SLC41A1 expression is ubiquitous, with highest expression in heart and testis tissues. By using a patch-clamp in Xenopus oocytes, SLC41A1 was identified as a voltage-dependent $Mg^{2+}$ transporter and is also permeable to other divalent cations such as cobalt, copper, and zinc [46]. Interestingly, upregulation of SLC41A1 transcripts has been observed following hypomagnesemia in mouse kidney tissues [46]. Based on $Mg^{2+}$ imaging on HEK293 cells, SLC41A1 is now identified as a $Na^+/Mg^{2+}$ exchanger that allows $Mg^{2+}$ efflux [47].

### 2.4. MAGT1

The Magnesium Transporter 1 (MAGT1) was identified as a gene upregulated in conditions of $Mg^{2+}$ deficiency [48]. This transporter is voltage-dependent and involved in $Mg^{2+}$ uptake when expressed in Xenopus oocytes. Unlike the other $Mg^{2+}$ transporters, MAGT1 is able to achieve a specific transport. Its expression is ubiquitous in all human tissues. It is also essential for the development of zebrafish, underlying a role in vertebrate embryonic development [49]. $Mg^{2+}$ was suspected as a second messenger in the X-linked human immunodeficiency with $Mg^{2+}$ defect and Epstein–Barr virus infection and neoplasia (XMEN): it appeared that mutations in the MAGT1 gene were actually involved [50]. In disorders like XMEN and congenital disorders of glycosylation (CDG), these mutations caused N-glycosylation defects [51]. MAGT1 has been recently identified as a member of glycoside complexes, regulated by $Mg^{2+}$ [52].

### 2.5. CNNM Family

The Cyclin and CBS Domain Divalent Metal Cation Transport Mediator (CNNM) family was first known as the Ancient Conserved Domain Protein (ACDP) family by the conserved domain structures among different species like yeasts, bacteria, and others like *Drosophilia Melanogaster* [53]. The ACDP family has four members, ACDP1/2/3/4 (corresponding to CNNM1/2/3/4), that share a minimum of 62.8% nucleotide similarity. It has been shown by Northern-blotting that ACDP2/3/4 are found in all human tissues, while ACDP1 is found mostly in brain and testis tissues. Other works found out that mouse and human ACDP were similar in structure and tissue distribution [54].

The two most studied members of the ACDP/CNNM family are ACDP2/CNNM2 and ACDP4/CNNM4. By studying CNNM2 in Xenopus oocytes, it has been defined that this protein was a cation transporter for magnesium, cobalt, manganese, strontium, baryum, and copper [55]. CNNM2 mRNA was also regulated by $Mg^{2+}$ deficiency in distal convoluted tubule (MDCT) epithelial cells. However, the role of CNNM2 as a $Mg^{2+}$ transporter is still debated because of its $Mg^{2+}$ sensitivity and transport capacity in HEK293 cells [56].

Th role of CNNM4 was firstly studied in rat spinal cord dorsal horn neurons, where it interacts with the Cytochrome C Oxidase Copper Chaperone 11 (COX11) [57]. Since its overexpression in HEK293 cells causes $Cu^{2+}$, $Mn^{2+}$, and $Co^{2+}$ toxicity, CNNM4 was suggested as a divalent cation transporter. Other studies localized CNNM4 on the basolateral side of intestinal epithelial cells, where it extrudes $Mg^{2+}$. Mice lacking CNNM4 also show hypomagnesemia and Jalili syndrome, characterized by cone–rod dystrophy and amelogenesis defect [58]. However, the role of CNNM4 as a $Mg^{2+}$ transporter or a $Na^+/Mg^{2+}$ is still discussed [58,59].

## 3. $Mg^{2+}$ Intake and Digestive Cancers

There is much evidence suggesting an association between $Mg^{2+}$ intake and digestive cancer risk and/or development. For example, high $Mg^{2+}$ intake and particularly low $Ca^{2+}/Mg^{2+}$ ratio protects against reflux esophagitis and Barret's esophagus, two precursors of ESAC. However, no significant associations were observed between $Mg^{2+}$ intake and ESAC incidence [60]. However, the association is less evident for GC because there is only a suggestive trend for a preventive effect of high $Mg^{2+}$ intake in non-cardia GC depending of gender and dietary source of $Mg^{2+}$ [61].

In PDAC, a first study from 2012 in a large cohort (142,203 men and 334,999 women) recruited between 1992 and 2000 shows no association between $Mg^{2+}$ intake and cancer risk [62]. Another study has investigated the association between nutrients intake from fruit and vegetable and PDAC risk [63]. The results show an inverse association between PDAC risk and nutrient intake, including $Mg^{2+}$, in a dose-dependent manner. Importantly, Dibaba et al. have shown in a large cohort, followed from 2000 to 2008, that every 100 mg per day decrement in $Mg^{2+}$ intake was associated with a 24% increase in PDAC incidence [64]. Moreover, analysis of metallomics in PDAC reveals a lower concentration of $Mg^{2+}$ in urine of patients with PDAC [65].

$Mg^{2+}$ intake was associated with a lower risk for CRC, particularly in people with low $Ca^{2+}/Mg^{2+}$ intake ratio [66]. Importantly, Dai et al. also show that the *Thr1482Ile* polymorphism in the *TRPM7* gene increases the risk for adenomatous and hyperplastic polyps [66]. It was also shown that $Mg^{2+}$ intake around 400 mg per day has a protective effect for CRC incidence in postmenopausal women [67]. A meta-analysis from 29 studies published on PubMed, Web of Science and the Chinese National Knowledge Infrastructure confirms that the high intake of $Mg^{2+}$ is inversely associated with the risk of CRC [68]. Assessment of $Mg^{2+}$ concentration in serum showed an inverse association with CRC risk in female but not in male. Moreover, no significant association was detected between dietary $Mg^{2+}$ and CRC risk in this study [69]. Finally, Wesselink et al. suggested that an interaction between normal 25-hydroxyvitamin $D_3$ concentration and high $Mg^{2+}$ intake is essential for reducing the risk of mortality by CRC [70].

To summarize, these epidemiologic studies suggested that high $Mg^{2+}$ intake by diet and/or supplemental compounds is inversely associated with CRC, PDAC and possibly ESAC risk, but not with GC risk.

## 4. Expression of $Mg^{2+}$ Transporters in Digestive Cancers

Ion channels are essential for physiological function of the digestive system. Although some of these (e.g., chloride, potassium, calcium, sodium and zinc) are dysregulated in cancer [71,72], the expression of $Mg^{2+}$ transporters in digestive cancers is less extensively studied.

### 4.1. Analysis of the Literature

In ESAC, immunohistochemistry (IHC) analyses have shown that TRPM7 protein was expressed in the cytoplasm of carcinoma cells while not detected in the non-cancerous esophageal epithelia. High TRPM7 staining was associated with better 5-year survival in patients with ESAC [73] (Table 1).

Table 1. Expression of $Mg^{2+}$ Transporters in Digestive Cancer Tissues.

| Cancer | Transporter | Expression in Cancerous Tissues (Compared to Normal Tissues) | Technique | Reference |
|---|---|---|---|---|
| ESAC | TRPM7 | Upregulated | IHC | [73] |
| PDAC | TRPM7 | Upregulated | IHC In silico + Protein Atlas | [74–77] |
| | SLC41A1 | Downregulated | qRT-PCR Western-Blot | [78] |
| CRC | TRPM6 | Downregulated (RNA) Upregulated (Protein) | In silico qRT-PCR IHC | [79–81] |
| | TRPM7 | Upregulated | In silico qRT-PCR IF IHC | [81,82] |
| | MAGT1 | Upregulated | In silico qRT-PCR | [83] |
| | CNNM4 | Downregulated | IHC | [84] |

ESAC, Esophageal Cancer; PDAC, Pancreatic Ductal Adenocarcinoma; CRC, Colorectal Cancer; MAGT1, Magnesium Transporter 1; SLC41A1, Solute Carrier Family 41 Member 1; TRPM7, Transient Receptor Potential Cation Channel Subfamily M Member 7; TRPM6, Transient Receptor Potential Cation Channel Subfamily M Member 6; CNNM4, Cyclin and CBS Domain Divalent Metal Cation Transport Mediator4; IHC, Immunohistochemistry; qRT-PCR, quantitative RT-PCR; IF, Immunofluorescence.

In PDAC patients, TRPM7 protein was overexpressed in cancerous tissues when compared to normal adjacent tissues [74–77]. TRPM7 expression correlates with tumor size, grade, and a high expression of this protein associates with a poor prognosis in patients [75,76]. On the other hand, SLC41A1 protein and mRNA were downregulated in PDAC patients compared to normal adjacent tissues, using quantitative real time-PCR (qRT-PCR), IHC, and in silico studies [78]. SLC41A1 expression is inversely correlated with tumor grade and was positively associated with a better outcome for patients [78].

In CRC, the expression of $Mg^{2+}$ transporters has been investigated in numerous studies. Evaluation of TRPM6 expression at the mRNA level (by qRT-PCR and in TCGA datasets), it has been found that TRPM6 was downregulated in cancerous colorectal tissues and that a high TRPM6 expression correlates with better survival in patients [79,80]. However, overexpression of TRPM6 protein was observed by IHC on several colorectal cancerous tissues when compared to matched normal tissues [81]. TRPM7 was also found upregulated in CRC using in silico datasets but also qRT-PCR, immunofluorescence, and IHC on tissues. TRPM7 expression was also associated with tumor infiltration, tumor grade, and the presence of distant metastasis [81,82]. In qRT-PCR-based and in silico studies, the $Mg^{2+}$ transporter MAGT1 was found to be overexpressed in colorectal cancerous tissues [83]. High MAGT1 expression also correlates with chemotherapeutic resistance, metastatic status, and tumor stage [83]. CNNM4 protein was also found downregulated in an IHC analysis of cancerous colorectal tissues, and inversely correlates with tumor malignancy [84].

*4.2. Analysis of the Human Protein Atlas*

The analysis of the Protein Atlas program provides some interesting data on IHC staining of $Mg^{2+}$ transporters on paraffin-embedded tumoral tissue sections (Table 2). Most transporters are expressed in PDAC, CRC and GC. However, TRPM7 is only found in GC, while TRPM6 and CNNM2 are not detected. A homogenous moderate to strong staining is found for SCL41A1 and MRS2, whereas MagT1 staining appears more heterogeneous. These results need to be confirmed in larger cohorts, because the number of studied cases varies currently from 8 to 12. Although staining intensity of cancer cells is analyzed, the difference between normal and peritumoral tissues is not systematically considered. These data provide preliminary results on $Mg^{2+}$ transporters in digestive cancer tissues, but they still need to be confirmed in larger cohorts and by a comparative study with non-cancerous or healthy tissues.

**Table 2.** Expression of Magnesium transporters using immunohistochemistry (IHC) in some digestive cancers based on The Human Protein Atlas data.

| Transporter | GC | PDAC | CRC |
|---|---|---|---|
| TRPM7 | 33% | 0% | 0% |
| MAGT1 | 44.4% | 22.2% | 83.3% |
| SLC41A1 | 91.7% | 100% | 100% |
| MRS2 | 75% | 91.7% | 75% |
| CNNM: | | | |
| CNNM1 | 12.5% | 36.4% | 16.7% |
| CNNM3 | 72.7% | 100% | 100% |
| CNNM4 | 50% | 58.3% | 83.3% |

Percentages of tumoral tissue samples with strong and moderate staining using IHC were calculated from the protein atlas data [85]. Only data for Pancreatic Ductal Adenocarcinoma (PDAC), Colorectal Cancer (CRC) and Gastric Cancer (GC) were available and for each transporters a number of patients ranging between 8 and 12 was tested. CNNM1/2/3/4, Cyclin and CBS Domain Divalent Metal Cation Transport Mediator1/2/3/4; MAGT1, Magnesium Transporter 1; SLC41A1, Solute Carrier Family 41 Member 1; TRPM7, Transient Receptor Potential Cation Channel Subfamily M Member 7; TRPM6, Transient Receptor Potential Cation Channel Subfamily M Member 6; CNNM1, Cyclin And CBS Domain Divalent Metal Cation Transport Mediator1; MRS2, Mitochondrial RNA Splicing Protein 2. No data were found for TRPM6 and CNNM2 expression.

*4.3. Transcriptome Analysis in Datasets*

We analyzed $Mg^{2+}$ transporters' expression in digestive cancers, correlation and their impact on patient overall and disease-free survival using Genotype Tissue Expression (GTEx) and The Cancer Genome Atlas (TCGA) datasets using GEPIA2 and RStudio tools, as previously performed [86].

4.3.1. $Mg^{2+}$ Transporters Expression in Digestive Cancers

We investigated $Mg^{2+}$ transporters expression using available datasets. Whisker boxplots of $Mg^{2+}$ transporters mRNA (TRPM6, TRPM7, MAGT1, SLC41A1, MRS2, CNNM1, CNNM2, CNNM3, CNNM4) were generated using GEPIA2, that compiles GTEx and TCGA datasets of normal and tumoral samples from the different digestive organs of interest (Supplementary Figure S1).

We observed a statistically significant overexpression of the transporters TRPM7, MAGT1, SLC41A1, CNNM2, CNNM3, and CNNM4 in the esophageal carcinoma (ESCA) samples when compared to corresponding normal tissues ($p < 0.01$) (Figure 3A). MAGT1, CNNM2 and CNNM4 mRNA were increased in tumoral tissues of stomach adenocarcinoma (STAD) when compared to normal stomach tissues ($p < 0.01$) (Figure 3B). In pancreatic adenocarcinoma (PAAD), colon adenocarcinoma (COAD) and rectum adenocarcinoma (READ), MAGT1 and CNNM4 mRNA relative levels were increased when compared to their normal corresponding samples ($p < 0.01$) (Figure 3C–E). A limitation of this type of transcriptome analysis is the homogeneity variances and related robustness of the analysis. This is why it will be important to perform additional studies on independent datasets for each digestive cancer as well as analyzing formalin-fixed paraffin-embedded (FFPE) samples by IHC. Moreover, characterization of gene of interest expression in pathological stages or other clinical features might reinforce the involvement of each magnesium transporter during carcinogenesis progression.

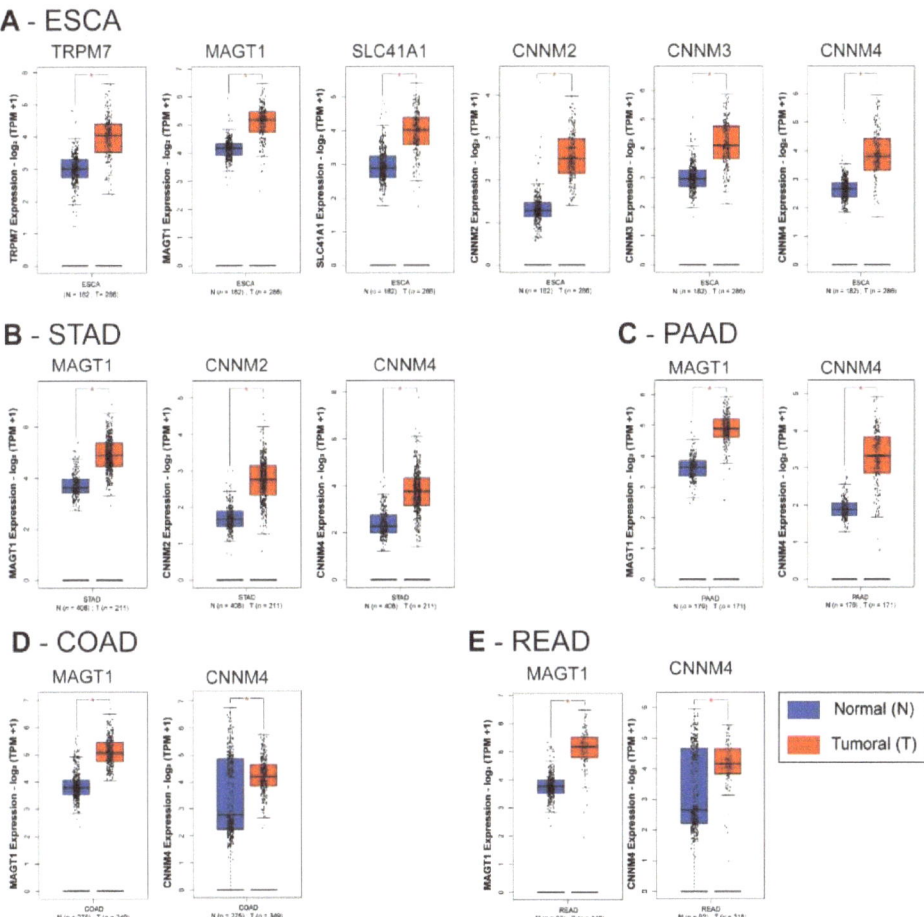

**Figure 3.** Relative mRNA expression of magnesium transporters in digestive cancers and normal tissues. Whiskers boxplots for $Mg^{2+}$ transporters mRNA (TRPM6, TRPM7, MAGT1, SLC41A1, MRS2, CNNM1, CNNM2, CNNM3, CNNM4) were generated using GEPIA2 from The Cancer Genome Atlas (TCGA) and Genotype-Tissue Expression (GTEx) samples. TCGA datasets were (**A**) Esophageal carcinoma (ESCA), (**B**) Stomach Adenocarcinoma (STAD), (**C**) Pancreatic Adenocarcinoma (PAAD), (**D**) Colon Adenocarcinoma (COAD) and (**E**). Rectum Adenocarcinoma (READ). TRPM7, Transient Receptor Potential Cation Channel Subfamily M, Member 7; MAGT1, Magnesium Transporter 1; SLC41A1, Solute Carrier Family 41, Member 1; MRS2, Mitochondrial RNA Splicing Protein 2; CNNM4, Cyclin and CBS Domain Divalent Metal Cation Transport Mediator 4; $n$ = number of samples; N, normal; T, tumoral. Relative mRNA levels are expressed as log2 transcripts per million bases (TPM). Only significant results (\* $p < 0.01$) are presented. The whole dataset analysis is provided as supplementary data (Supplementary Figure S1).

### 4.3.2. $Mg^{2+}$ Transporters and Patient Survival

We then searched for a possible association between the Mg transporters' expression and patient survival. We generated survival heatmaps and Kaplan–Meier curves for overall survival (OS) and disease-free survival (DFS) in TCGA datasets using the GEPIA2 tool (Figure 4).

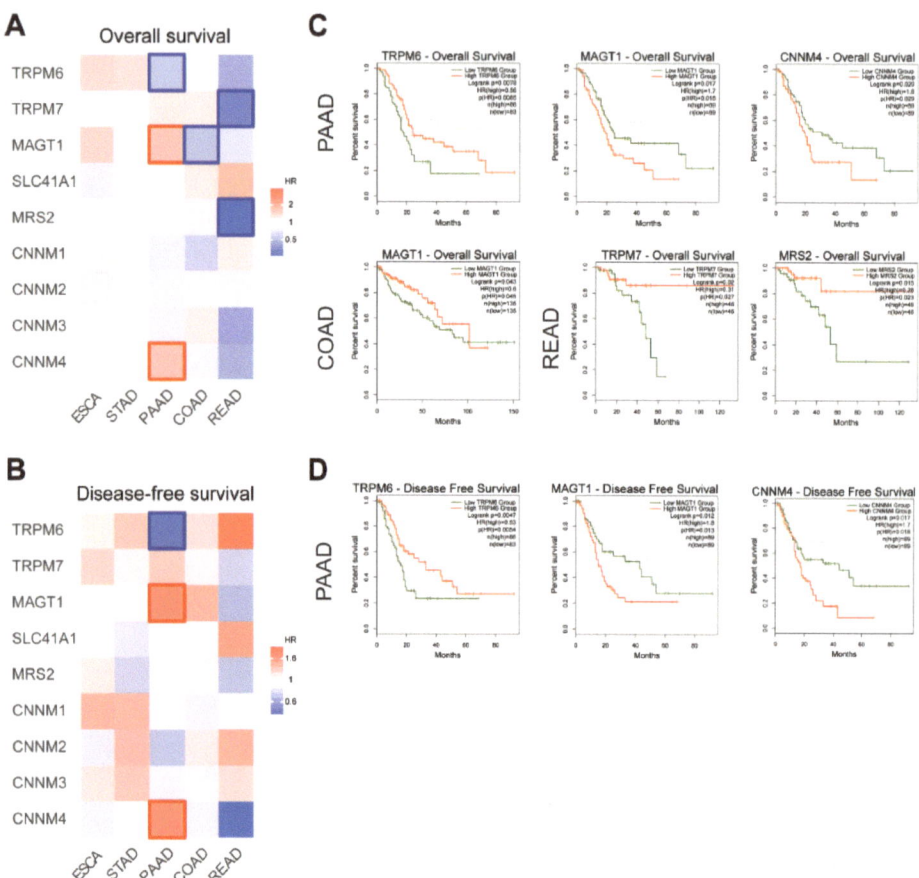

**Figure 4.** Analysis of patient survival in digestive cancers. Survival heatmaps were generated using GEPIA2 with TCGA data for overall survival (**A**) and disease-free survival (**B**) for Esophageal Cancer (ESCA), Stomach Adenocarcinoma (STAD), Pancreatic Adenocarcinoma (PAAD), Colon Adenocarcinoma (COAD) and Rectum Adenocarcinoma (READ) datasets. Survival is expressed as hazard ratio (HR). Framed squares represent significative statistical values ($p < 0.05$). Kaplan–Meier curves for overall survival (**C**) and disease-free survival (**D**) were analyzed using GEPIA2. TRPM6, Transient Receptor Potential Cation Channel Subfamily M, Member 6. TRPM7, Transient Receptor Potential Cation Channel Subfamily M, Member 7; MAGT1, Magnesium Transporter 1; SLC41A1, Solute Carrier Family 41, Member 1; MRS2, Mitochondrial RNA Splicing Protein 2; CNNM4, Cyclin and CBS Domain Divalent Metal Cation Transport Mediator 4. Only statistically significant curves (* $p < 0.05$) are presented.

For PAAD patients, we observed that high expression MAGT1 and CNNM4 mRNA were associated with shorter patient overall survival, while high TRPM6 mRNA expression was correlated with better outcome in those patients. Similar correlations were observed in patients for disease-free survival.

For COAD patients, a high expression of MAGT1 mRNA is associated with a better outcome in patient overall survival.

For READ patients, a longer overall survival is observed when patients have a high expression of TRPM7 and MRS2 mRNA.

Survival analysis of TCGA datasets provides many interesting clues for future research. However, additional analyses of independent cohorts remain mandatory as well as more advanced statistical analysis using R package such as "regnet" in order to increase the robustness of the in silico analysis [87].

#### 4.3.3. $Mg^{2+}$ Transporters and Patient Survival

For each dataset, we studied the co-expression of $Mg^{2+}$ transporters by performing Pearson's correlation analysis and principal component analysis (PCA) using RStudio (R scripts were previously described in [86]) (Supplementary Tables S1–S4 for the whole analysis). In esophageal cancer (TCGA-ESCA dataset), we observed a strong positive correlation among CNNM transporters. TRPM7 was also positively correlated with CNNM transporters and MAGT1 (0.19 < R < 0.29). SLC41A1 is positively correlated with CNNM1 (R = 0.19) and CNNM2 (R = 0.16). Relationships between these variables were further confirmed in our PCA plot in which we observed a proximity among CNNM transporters, TRPM7 and MAGT1 (Figure 5A).

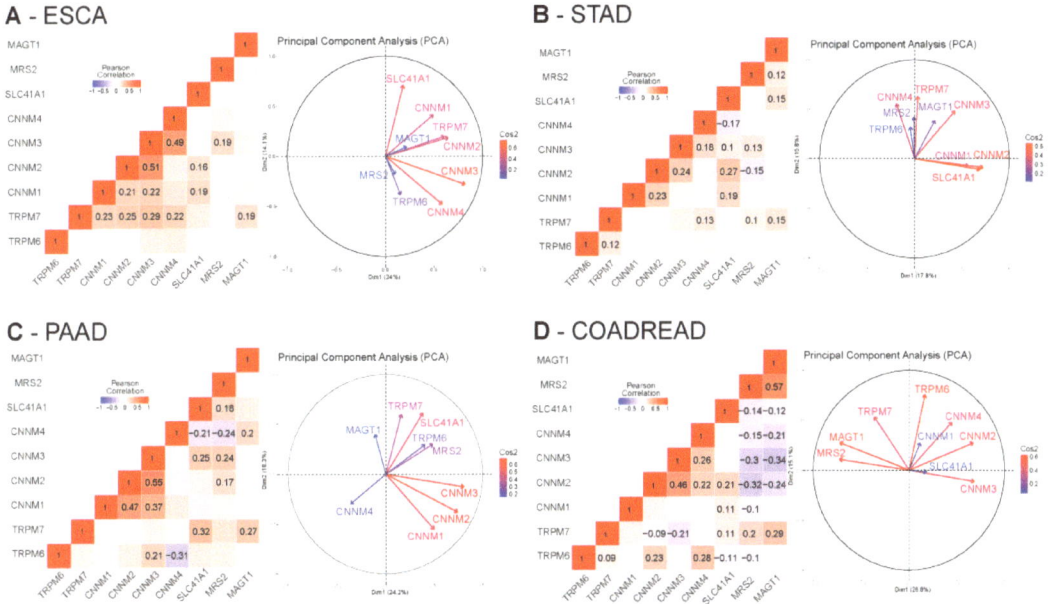

**Figure 5.** Correlation analysis of relative magnesium transporters mRNA levels in digestive cancers TCGA cohorts. Pearson cor-relation R values were calculated for each magnesium transporter mRNA using RStudio. All queries for TRPM6, TRPM7, MAGT1, SLC41A1, MRS2, CNNM1, CNNM2, CNNM3 and CNNM4 genes were realized in ESCA (**A**), STAD (**B**), PAAD (**C**) and COAD-READ (**D**) datasets from TCGA using the cBioPortal website. mRNA expression values were retrieved as RNA-Seq by Expectation Maximization RSEM (Batch normalized from Illumina HiSeq_RNASeqV2). Only significative correlations values ($p < 0.05$) are presented. Principal component analysis (PCA) of $Mg^{2+}$ transporters mRNA relative expression in TCGA datasets are also represented. Grouped variables are positively correlated whereas opposed variables are negatively correlated. Independency of the variables is formed by a 90° angle formed by two arrows. The quality of the variables on the Principal Component Analysis (PCA) are designated by Cos2 (square cosine, squared coordinates) values.

In gastric cancer (TCGA-STAD dataset), we observed a mild positive correlation of TRPM7 with TRPM6, MAGT1, CNNM3 and CNNM4 (=0.1< R < 0.22). SLC41A1 is also positively correlated to MAGT1, CNNM1, CNNM2, CNNM3 (R = 0.1–0.27) but negatively correlated to CNNM4 (R = −0.17). MRS2 was positively correlated with TRPM7, CNNM3 and SLC41A1 (R = 0.1–0.13) but negatively correlated to CNNM2 (R = −0.15). CNNM2 is positively correlated with CNNM3 (R = 0.24). Those positive and negative correlations are confirmed in the PCA plot, for example a close proximity of SLC41A1 with CNNM1 and CNNM2 (Figure 5B).

In pancreatic cancer (PAAD dataset), MAGT1 is positively correlated to TRPM7 (R = 0.27) and CNNM4 (R = 0.2). CNNM1, CNNM2 and CNNM3 are also positively correlated with each other (0.17 < R < 0.25). CNNM2, CNNM3, and SLC41A1 are positively correlated to MRS2. SLC41A1 is also positively correlated to CNNM3, and TRPM7. CNNM4 is negatively correlated with SLC41A1 (R = −0.21) and MRS2 (R = −0.24) On the PCA, MAGT1, CNNM4 and TRPM6 are independent (Figure 5C).

In colorectal cancer (COAD-READ dataset), MAGT1 and MRS2 expression are negatively correlated to CNNM2, CNNM3, CNNM4 and SLC41A1 (−0.1 < R < −0.32) but are positively correlated between each other (R = 0.57). TRPM7 is also positively correlated to MAGT1 (R = 0.29), MRS2 (R = 0.2) and SLC41A1 (R = 0.11) but negatively correlated with CNNM2 (R = −0.09) and CNNM3 (R = −0.21). TRPM6 is positively correlated to TRPM7 (R = 0.09), CNNM2 (R = 0.23) and CNNM4 (R = 0.28) but is negatively correlated to SLC41A1 (R = −0.11) and MRS2 (R = −0.1). SLC41A1 is positively correlated to CNNM1 (R = 0.11), CNNM2 (R = 0.21) and TRPM7 (R = 0.11) but negatively correlated to TRPM6 (R = −0.11). On PCA analysis, the strong positive correlation expression of MAGT1 and MRS2 is confirmed and we observed an independency between TRPM6 and CNNM3 and MRS2 (Figure 5D).

## 5. Regulation of Digestive Cancer Cell Fates by Magnesium Transporters

Dysregulation of $Mg^{2+}$ transporters could contribute to cancer hallmarks by regulating malignant cell proliferation, migration or invasion [88].

As mentioned above, TRPM7 overexpression has been proposed as an independent good prognosis biomarker in ESAC. In the TE5 human ESAC cell line, TRPM7 silencing by siRNA enhances cell proliferation as well as migration and invasion [73]. Although the molecular mechanisms involved in TRPM7-mediated ESAC cell proliferation, migration and invasion are far from being fully elucidated, TRPM7 channel expression may prevent the oncogenic properties of ESAC cells.

TRPM7 expression is also detected in GC cell lines but TRPM7 silencing decreases cell viability probably by inducing apoptosis. Interestingly, $Mg^{2+}$ supplementation (10 mM) maintains cell growth when TRPM7 expression is inhibited [89]. Moreover, Mrs2 is upregulated in the multidrug-resistant GC cell line, SGC7901/ADR, increasing adriamycin release and promoting cell growth by p27 downregulation and cyclin D1 upregulation [90].

Similar results are found in PDAC cell lines BxPC-3 and PANC-1 where TRPM7 silencing reduces the cell proliferation by accumulating the cells in G0-G1 phases without affecting the number of apoptotic cells. In high $Mg^{2+}$ culture media, the proliferation of TRPM7-deficient cells is restored [74]. Yee et al. further show that TRPM7 expression is required for preventing BxPC-3 and PANC-1 cells from senescence, and that TRPM7 silencing enhances cytotoxicity induced by gemcitabine treatment [91]. However, it has been also shown that TRPM7 inhibition by using small interfering RNA (siRNA) decreases BxPC-3 cell migration without affecting cell viability [76]. In this study, we show that TRPM7 regulates cation constitutive entry and cytosolic $Mg^{2+}$ levels. Interestingly, external $Mg^{2+}$ supplementation maintains normal cytosolic $Mg^{2+}$ levels in TRPM7-deficient cells, suggesting that TRPM7 silencing may be compensated by others' $Mg^{2+}$ entry pathways in BxPC-3 cells. Importantly, cell migration is also restored by $Mg^{2+}$ supplementation in TRPM7-deficient BxPC-3, indicating that $Mg^{2+}$ is essential for PDAC cell migration. TRPM7 channels are also required for PDAC cell invasion [75]. We have recently shown that TRPM7 was involved in the secretion of both heat-shock protein 90α (Hsp90α), urokinase plasminogen activator (uPA) and matrix metalloproteinase-2 (MMP-2), leading to enhanced PDAC cell invasion [77]. Moreover, TRPM7 regulates $Mg^{2+}$ constitutive entry and $Mg^{2+}$-dependent cell invasion in the MIA PaCa-2 PDAC cell line. TRPM7 channels are linked to pancreatic carcinogenesis, since they are overexpressed in epithelial pancreatic cells chronically exposed to cadmium pollutant, leading to cytosolic $Mg^{2+}$ accumulation and enhanced cell invasion [92]. Finally, TRPM7 channels are also involved in the interaction between PDAC cells and the tumoral microenvironment because TRPM7 membrane currents and TRPM7-dependent cell migration are both stimulated following treatment with elastin-derived peptides (EDP) that are released by the degradation of the extracellular matrix. Our study suggests that TRPM7 is involved in the response to EDP through its interaction with the ribosomal protein SA (RPSA) [93].

In a colon carcinoma LoVo cell model, cytosolic $Mg^{2+}$ levels are higher in doxorubicin-resistant cells when compared to the doxorubicin-sensitive ones [94]. Nevertheless, resistant cells express less TRPM6 and TRPM7 channels, leading to lower $Mg^{2+}$ influx. TRPM7 silencing induces the acquisition of a more resistant phenotype in sensitive cells, indicating that TRPM7 channel expression is associated with chemoresistance in CRC. Cazzaniga et al. have shown that TRPM7 downregulation was accompanied by the upregulation of MagT1 in doxorubicin-resistant LoVo cells [95]. Moreover, MagT1 silencing strongly inhibits resistant cell proliferation without affecting total intracellular $Mg^{2+}$. Luongo et al. further suggest that TRPM6 and TRPM7 assemblage as heterotetrameric channels regulates $Mg^{2+}$ influx and cell proliferation in the HT-29 CRC cell line [96]. In an azoxymethane-induced colorectal cancer mouse model, the use of the TRPM7 inhibitor waixenicin A induces hypomagnesemia via insufficient $Mg^{2+}$ absorption by the colon. However, neither waixenicin A, nor low $Mg^{2+}$ diet affect the formation of pre-neoplastic lesions in the colon [97]. Furthermore, Huang et al. show that the TRPM7 channel is the main transporter for $Mg^{2+}$ influx in both the HT-29 CRC cell line and in primary mouse colon epithelial cells. Su et al. also show that TRPM7 silencing decreases both HT-29 and SW-480 cell proliferation, migration and invasion [82]. Inhibition of TRPM7 induces the upregulation of E-cadherin and the downregulation of N-cadherin in CRC cells suggesting that this channel may regulate epithelial-to-mesenchymal transition (EMT). Taken together, these results strongly suggest that TRPM7 and $Mg^{2+}$ are not involved in early stages of colon carcinogenesis. On the other hand, TRPM7 is involved in processes occurring at late stages of colon carcinogenesis like EMT, cell migration, invasion and chemoresistance. Recently, Yamazaki et al. have demonstrated that CNNM4 deficiency accelerates epithelial colon cell proliferation in mice [98]. Moreover, primary organoids growth is also enhanced in CNNM4-deficient mice. Intriguingly, capsaicin-stimulated $Ca^{2+}$ influx is also reduced in colonic organoids derived from CNNM4-deficient mice while intracellular $[Mg^{2+}]$ is increased. These results suggest a functional interaction between TRPV1 $Ca^{2+}$-channels and CNNM4 in colon. Finally, CNNM4-deficient mice treated with azoxymethane followed by dextran sodium sulfate form more polyps. Importantly, histological analyses of CNNM4-deficient polyps reveal the presence of invading cancer cells. These data clearly show a protective role of CNNM4 expression against colon carcinogenesis.

To resume, $Mg^{2+}$ transporter expression is altered in numerous digestive cancers, leading to dysregulation of cell fates. To date, most of these studies are focused on the TRPM7 channel, but the role of other $Mg^{2+}$ transporters cannot be excluded. However, it still remains unclear if these cancer cell fates may be regulated by intracellular $Mg^{2+}$ homeostasis disturbance, since the role of $Mg^{2+}$ as a second messenger is still being debated.

## 6. $Mg^{2+}$-Regulated Signaling Pathways in Digestive Cancer Cells

In cell-based studies, high $Mg^{2+}$ concentration can cause increased tyrosine-kinases activities. $Mg^{2+}$ is indeed a crucial divalent cation required for the activity of kinases, including non-receptor tyrosine kinases (e.g., Src and Abl Proto-Oncogenes, Janus Kinase (JAK), Focal Adhesion Kinase (FAK), Suppressor Of Cytokine Signaling (SOCS)) and receptor tyrosine kinases (RTKs, such as Growth Factor Receptors (GFR) including VEGFR, EGFR, FGFR, and PDGFR) [99]. Following binding of growth factors to their respective RTKs, cytoplasmic proteins containing Src homology region 2 or phospho-tyrosine-binding domains are recruited to the cell membrane. These recruited proteins either have intrinsic enzymatic activity (such as Src and Phospholipase C (PLC)), or serve as docking proteins that are capable to bind additional enzymes [100,101]. Activated RTKs are able to trigger a wide range of downstream signaling pathways, including RAS/RAF/MEK/MAPK, PLC/PKC, PI3K/AKT/mTOR, and JAK/STAT [99].

Although there is scarce information regarding activation of signaling pathways in digestive cancer, recent papers show evidence for the involvement of at least two pathways in these cancers cells: the AKT/mTOR pathway on one hand, and the JAK/STAT pathway on the other hand.

Xie and collaborators were able to show a direct relationship between expression of the SLC41A1 transporter and activation of the AKT/mTOR pathway [78]. SLC41A1 mediates both $Mg^{2+}$ uptake and extrusion [102]. SLC41A1 expression is correlated with clinical outcomes in patients with pancreatic ductal adenocarcinoma, with SLC41A1 being down-regulated in tumors. Overexpression of SLC41A1 suppressed orthotopic tumor growth in a mouse model and reduced the cell proliferation, colony formation, and invasiveness of KP3 and Panc-1 cell lines. Overexpression of SLC41A1 promoted $Mg^{2+}$ efflux and suppressed AKT/mTOR activity, which is the upstream regulator of Bax and Bcl-2. An increase in AKT activity and supplementation with $Mg^{2+}$ abolished SLC41A1-induced tumor suppression [78].

At least another $Mg^{2+}$ transporter was found to be associated with the mTOR pathway: CNNM4. This transporter is highly expressed in the colon epithelia, and also strongly expressed in the intestine [58]. In this latter tissue, CNNM4 is localized at the basolateral membrane of epithelial cells and mediates intestinal $Mg^{2+}$ absorption from the tubular lumen across the epithelial sheet, by extruding intracellular $Mg^{2+}$ to the body inside. CNNM4-deficient mice can grow without severe defects but have moderately lowered levels of magnesium in the blood when compared to control wild-type mice due to malabsorption of magnesium [58].

In ApcΔ14/+ mice, which spontaneously form benign polyps in the intestine, deletion of Cnnm4 promoted malignant progression of intestinal polyps to adenocarcinomas. IHC analyses of tissues from patients with colon cancer demonstrated an inverse relationship between CNNM4 expression and colon cancer malignancy, thereby supporting the notion that CNNM4 suppresses the progression of cancer malignancy in humans [84].

CNNM4-dependent $Mg^{2+}$ efflux is apparently able to suppress tumor progression by regulating energy metabolism [84]: $Mg^{2+}$ is able to bind several biomolecules, with ATP being most probably the utmost important. CNNM4 knockdown is able to increase intracellular $Mg^{2+}$ levels, and to significantly increase ATP levels in HEK293 cells. Moreover, CNNM4 knockdown (or $Mg^{2+}$ supplementation) is able to selectively abrogate AMPK hyperphosphorylation (AMPK being phosphorylated and activated under energy-deficient conditions [103]). mTOR is known to be a major downstream target of AMPK signaling [104] and has significant roles in cancer development. Through monitoring of S6K (a well-known substrate of mTOR), Funato and collaborators have clearly identified CNNM4 as a modulator of mTOR signaling [84].

Moreover, it has been shown that Cnnm4 deficiency suppresses $Ca^{2+}$ signaling and promotes cell proliferation in the colon epithelia. These results establish the functional interplay between $Mg^{2+}$ and $Ca^{2+}$ in the colon epithelia, which is crucial for maintaining the dynamic homeostasis of the epithelial tissue [98].

The second signaling pathway influenced by magnesium levels is the STAT pathway. This influence occurs through Magnesium-dependent Phosphatase (MDP)-1. This enzyme, belonging to the haloacid dehalogenase family, has a protein–tyrosine phosphatase function and is potentially involved in glycation repair (Fortpied et al., 2006).

Forced expression of MDP1 in the gastric cancer cell line BGC-823 inhibited cell proliferation, whereas the knockdown of MDP1 protein promoted cell growth. Overexpression of MDP1 in BGC-823 cells also enhanced cell senescence and apoptosis. Signal transducer and activator of transcription 3 (*Stat3*), as well as the c-Jun N-terminal kinase (JNK) were found to mediate the biological function of MDP1 [105].

TRPM7 has also been linked to Stat3: disrupted expression of Trpm7 in mice causes downregulated expression of *Stat3* mRNA [106].

Finally, it has to be mentioned that anti-RTK EGFR antibodies are able to dramatically reduce serum magnesium concentration [107,108]. Although EGFR tyrosine kinase inhibitors can also potentially induce hypomagnesaemia, typical concentrations used in the clinic seem to be insufficient to induce this side-effect [108]. Inhibition of the EGFR induces a mutated-like TRPM6 syndrome [108], while stimulation of the EGFR increase current through TRPM6 (but not TRPM7) [109]. The α-kinase domain of TRPM6 is not involved in the EGF receptor-mediated increase in channel activity: the activation relies on both Src and the downstream effector Rac1, the latter being able to increase TRPM6 mobility and increase cell surface abundance by redistributing endomembrane TRPM6 to the plasma membrane [109] (Figure 6).

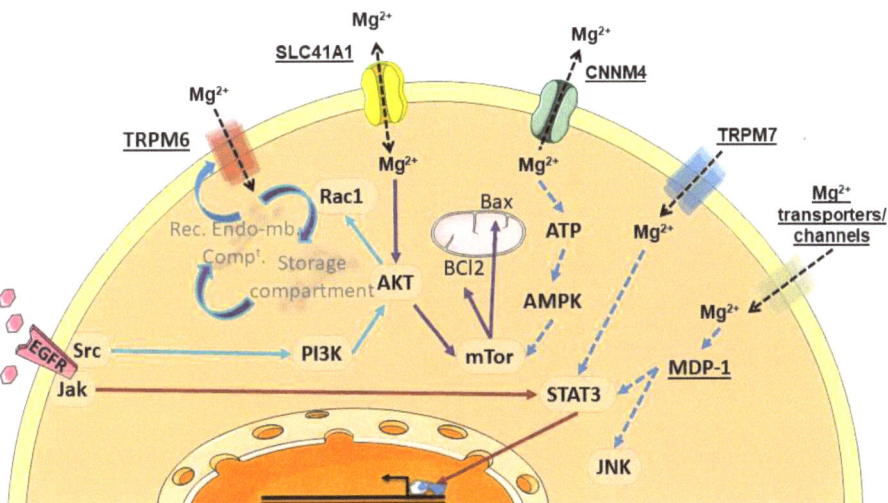

**Figure 6.** Summary of signaling pathways associated with $Mg^{2+}$ transporters. $Mg^{2+}$ transporter, channels and $Mg^{2+}$-dependent proteins are underlined. These proteins activate kinases, signaling proteins and/or transcription factors (framed in ovals). As a result, some processes, such as protein redistribution to the plasma membrane, control of effector proteins and transcriptional regulation can occur. Rec. Endo-mb. Comp$^t$.: recycling endomembrane compartment. $Mg^{2+}$, Magnesium; EGFR, Epidermal Growth Factor Receptor; TRPM6, Transient Receptor Potential Cation Channel Subfamily M, Member 6; TRPM7, Transient Receptor Potential Cation Channel Subfamily M, Member 7; SLC41A1, Solute Carrier Family 41, Member 1; CNNM4, Cyclin and CBS Domain Divalent Metal Cation Transport Mediator 4; Jak, Janus Kinase; Src, Src Proto-Oncogene; PI3K, Phosphatidylinositol-4,5-Biphosphate 3-Kinase; AKT, AKT Serine/Threonine Kinase; Rac1, Rac-Family Small GTPase 1; mTor, Mechanistic Target of Rapamycin Kinase; BCl2, BCL2 Apoptosis Regulator; Bax, BCL2 Associated X Apoptosis Regulator; ATP, Adenosine Tri-Phosphate; AMPK, AMP-Activated Protein Kinase, STAT3, Signal Transducer And Activator Of Transcription 3; MDP-1, Magnesium-Dependent Phosphatase-1; JNK, c-Jun *N*-terminal Kinase.

Other transporters have been demonstrated to activate magnesium-dependent signaling pathways, such as MAGT1. Indeed, Li and collaborators have identified mutations in this magnesium transporter gene, in a novel X-linked human immunodeficiency characterized by CD4 lymphopenia, severe chronic viral infections, and defective T lymphocyte activation [50]. MAGT1 deficiency was shown to abrogate $Mg^{2+}$ influx, leading to defective activation of phospholipase Cγ and consequently impaired responses to antigen receptor engagement in patients harboring this XMEN (X-link immunodeficiency with Magnesium defect and EBV infection and Neoplasia) disease. However, it must be noted (i) that $Mg^{2+}$ supplementation has not proven successful [110], and (ii) that XMEN disease has been recently shown to be a congenital disorder of glycosylation that affects a restricted subset of glycoproteins. MAGT1 is actually a non-catalytic subunit required for N-glycosylation of key immune cells receptors [111]. The mechanism by which MAGT1 is involved in the magnesium homeostasis and how magnesium affects glycosylation requires further investigation [110].

## 7. Conclusions and Perspectives

The aim of this work was to make an overview of $Mg^{2+}$ transporter expression and their role in esophageal adenocarcinoma (ESAC), gastric cancer (GC), pancreatic ductal adenocarcinoma (PDAC), and colorectal cancer (CRC). These digestive cancers are the most common and the deadliest malignancies worldwide. Numerous epidemiologic studies strongly suggest that digestive cancer incidence and mortality may be dependent on lifestyle, and particularly diet. $Mg^{2+}$ content is continuously decreased in alimentation of industrialized countries, leading to nutritional deficiency in $Mg^{2+}$ for a large part of the population, estimated to ~60% in the USA [112]. While $Mg^{2+}$ is the second most abundant cation in the cell, its role in physiology and pathology is less extensively studied than others such as $Ca^{2+}$, $Na^+$ or $K^+$. Cellular $Mg^{2+}$ homeostasis is regulated by membrane transporters. Among these transporters, TRPM7 has been clearly identified as the main $Mg^{2+}$ gatekeeper for cell intake in both non-cancer and cancer cells. On the other hand, the functional characterization of other candidates such as MAGT1 or CNNM4 as $Mg^{2+}$ transporter is still under debate. The reviewed literature, as well as the Human Protein Atlas analyses indicate that $Mg^{2+}$ transporter expression is altered in most digestive cancers. The discrepancies between the data could be explained by the low number of cohorts and/or by the methodology used. In particular, antibodies targeting ion channels and transporters often display a poor specificity, inducing potential cross-reactivity with other molecules [113]. Analyses of $Mg^{2+}$ transporter expression in cells and tissues by immunochemistry should be systematically completed by other methods of detection such as functional (e.g., electrophysiology and $Mg^{2+}$-imaging) and transcriptomic analysis. In this work, we analyzed $Mg^{2+}$ transporters expression in digestive cancers, correlation and their impact on patient overall and disease-free survival using Genotype Tissue Expression (GTEx) and The Cancer Genome Atlas (TCGA), as previously described [86]. Interestingly, our data reveal that the MAGT1 is overexpressed in all digestive cancers. Moreover, MAGT1 expression is associated with a poor survival in PDAC patients and with a better survival in CRC patients. Cazzaniga et al. show that MAGT1 is overexpressed in a CRC cell line resistant to doxorubicin compared to the sensitive ones, and it regulates cell proliferation [95]. To our knowledge, the role of MAGT1 has not been yet studied in other digestive cancer cell models and further investigations are needed to better understand the mechanisms involving this protein in digestive cancers. It has been shown that MAGT1 expression can restore TRPM7 deficiency, intracellular $Mg^{2+}$ homeostasis and cell viability in some cellular models suggesting a possible transcriptomic regulation [95,114]. Interestingly, our data show a positive correlation between TRPM7/MAGT1/CNNM4 expression in both ESAC and PDAC, and also between TRPM7 and MAGT1 expression in CRC. This suggests a possible interaction between these three proteins in ESAC and PDAC cancer cells. In our opinion, such complex may be of great interest for the research of new biomarkers, especially for PDAC as high expression of TRPM7/MAGT1/CNNM4 clearly discriminates a unique profile of patients with the poorest survival (Figure 7). We found that TRPM7/MAGT1/CNNM4 high expression is also associated with a poor survival in the other digestive cancers (Supplementary Figure S2). However, these results are only preliminary and should be confirmed by additional analyses on independent datasets, as well as IHC on FFPE. TRPM7 possess a functional kinase domain able to phosphorylate serine or threonine residues, while MAGT1 is implicated in protein glycosylation. Additionally, it has been suggested that CNNM4 expression could regulate TRP channel expression such as TRPV1 [98]. Therefore, it is possible that transcriptomic regulations and/or post-translational modifications can occur between TRPM7, MAGT1 and CNNM4 in digestive cancer cells.

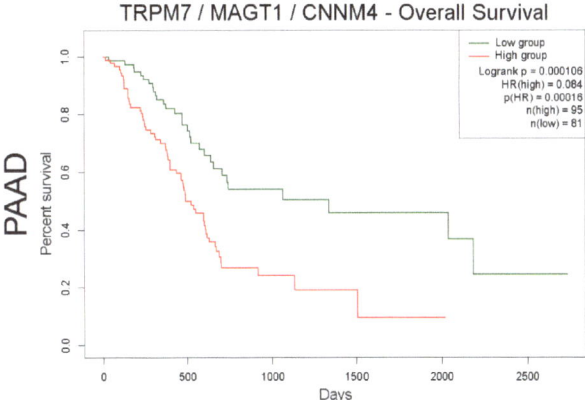

**Figure 7.** Analysis of overall survival of the TRPM7/MAGT1/CNNM4 signature in pancreatic cancer datasets using SurvExpress. PAAD patients were stratified using a gene signature combining TRPM7, MAGT1, and CNNM4. Kaplan–Meier curves were analyzed using the optimized SurvExpress Maximize algorithm. The number of analyzed patients across time (days) is indicated below the horizontal axis for both conditions, as previously described [86].

To conclude, this review highlights $Mg^{2+}$ transporters and their associated signaling pathways as promising biomarkers in digestive cancers. The transcriptomic analysis of datasets reveals a $Mg^{2+}$ transporter signature involving TRPM7/MAGT1/CNNM4 in ESAC and PDAC. In PDAC, the correlation between TRPM7, MAGT1 and CNNM4 expression is associated with a low survival. Further studies are needed to better understand the physiological significance of this complex in cancer cell models.

**Supplementary Materials:** The following are available online at https://www.mdpi.com/2072-6643/13/1/210/s1, Figure S1: Relative mRNA expression of magnesium transporters in digestive cancers and normal tissues. Figure S2: Analysis of overall survival of the TRPM7/MAGT1/CNNM4 signature in digestive cancer datasets using SurExpress (ESCA for esophageal carcinoma (A); STAD for stomach adenocarcinoma (B); PAAD for pancreatic adenocarcinoma (C); COADREAD for colorectal adenocarcinoma (D). Table S1. Pearson correlation of magnesium transporters combination in esophageal cancer (ESCA) dataset. Table S2. Pearson correlation of magnesium transporters combination in stomach cancer (STAD) dataset. Table S3. Pearson correlation of magnesium transporters combination in pancreatic cancer (PAAD) dataset. Table S4. Pearson correlation of magnesium transporters combination in colorectal cancer (COADREAD) dataset.

**Author Contributions:** Conceptualization M.G. and N.J.; formal analysis N.J. and J.A.; project administration M.G. and N.J.; writing—original draft J.A., P.R., P.K., I.D.-D., N.J., and M.G.; writing—review and editing D.C., H.S., I.V.S., and H.O.-A. All authors have read and agreed to the published version of the manuscript.

**Funding:** This research was funded by the "Ministère de l'Enseignement Supérieur de la Recherche et de l'Innovation" (J.A.), the "Université de Picardie Jules Verne (UPJV)" (J.A., P.K., I.D.D., H.O.A., and M.G.), the "Université de Lille" (I.V.S. and N.J.), "le CHU Amiens-Picardie" (P.R., D.C., and H.S.), the "Institut National de la Santé et de la Recherche Médicale (Inserm) " (I.V.S. and NJ), the "Centre National de la Recherche Scientifique (CNRS)" (I.V.S. and N.J.), "La Ligue Nationale contre le Cancer" (comité Septentrion N.J. and M.G.), and "l'Agence Nationale de Sécurité Sanitaire, de l'Alimentation, de l'Environnement et du Travail" (N.J. and M.G.).

**Institutional Review Board Statement:** Not applicable.

**Informed Consent Statement:** Not applicable.

**Data Availability Statement:** All data are available and are based upon public data extracted from TCGA Research Network (http://cancergenome.nih.gov/), Genome Tissue Expression (GTEX) project (http://www.GTEXportal.org/), and Gene Expression Omnibus (GEO) database (http://www.ncbi.nml.nih.gov/geo/) using GEPIA2 (http://gepia2.cancer-pku.cn/) and SurvExpress (SurvExpress—Web resource for Biomarker comparison and validation of Survival gene eXpression data in cancer (itesm.mx)).

**Acknowledgments:** In this section, you can acknowledge any support given which is not covered by the author contribution or funding sections. This may include administrative and technical support, or donations in kind (e.g., materials used for experiments).

**Conflicts of Interest:** The authors declare no conflict of interest.

## References

1. Bray, F.; Ferlay, J.; Soerjomataram, I.; Siegel, R.L.; Torre, L.A.; Jemal, A. Global cancer statistics 2018: GLOBOCAN estimates of incidence and mortality worldwide for 36 cancers in 185 countries. *CA Cancer J. Clin.* **2018**, *68*, 394–424. [CrossRef] [PubMed]
2. Arnold, M.; Soerjomataram, I.; Ferlay, J.; Forman, D. Global incidence of oesophageal cancer by histological subtype in 2012. *Gut* **2015**, *64*, 381–387. [CrossRef] [PubMed]
3. Rubenstein, J.H.; Shaheen, N.J. Epidemiology, diagnosis, and management of esophageal adenocarcinoma. *Gastroenterology* **2015**, *149*, 302–317. [CrossRef] [PubMed]
4. Eusebi, L.H.; Telese, A.; Marasco, G.; Bazzoli, F.; Zagari, R.M. Gastric cancer prevention strategies: A global perspective. *J. Gastroenterol. Hepatol.* **2020**. [CrossRef]
5. De Martel, C.; Georges, D.; Bray, F.; Ferlay, J.; Clifford, G.M. Global burden of cancer attributable to infections in 2018: A worldwide incidence analysis. *Lancet Glob. Health* **2020**, *8*, e180–e190. [CrossRef]
6. Rawla, P.; Sunkara, T.; Gaduputi, V. Epidemiology of pancreatic cancer: Global trends, etiology and risk factors. *World J. Oncol.* **2019**, *10*, 10–27. [CrossRef]
7. Rahib, L.; Smith, B.D.; Aizenberg, R.; Rosenzweig, A.B.; Fleshman, J.M.; Matrisian, L.M. Projecting cancer incidence and deaths to 2030: The unexpected burden of thyroid, liver, and pancreas cancers in the United States. *Cancer Res.* **2014**, *74*, 2913–2921. [CrossRef]
8. Khomiak, A.; Brunner, M.; Kordes, M.; Lindblad, S.; Miksch, R.C.; Ohlund, D.; Regel, I. Recent discoveries of diagnostic, prognostic and predictive biomarkers for pancreatic cancer. *Cancers* **2020**, *12*, 3234. [CrossRef]
9. Akimoto, N.; Ugai, T.; Zhong, R.; Hamada, T.; Fujiyoshi, K.; Giannakis, M.; Wu, K.; Cao, Y.; Ng, K.; Ogino, S. Rising incidence of early-onset colorectal cancer—A call to action. *Nat. Rev. Clin. Oncol.* **2020**. [CrossRef]
10. Thanikachalam, K.; Khan, G. Colorectal cancer and nutrition. *Nutrients* **2019**, *11*, 164. [CrossRef]
11. De Baaij, J.H.; Hoenderop, J.G.; Bindels, R.J. Magnesium in man: Implications for health and disease. *Physiol. Rev.* **2015**, *95*, 1–46. [CrossRef] [PubMed]
12. Rubin, H. Central role for magnesium in coordinate control of metabolism and growth in animal cells. *Proc. Natl. Acad. Sci. USA* **1975**, *72*, 3551–3555. [CrossRef] [PubMed]
13. Chiu, T.K.; Dickerson, R.E. 1 A crystal structures of B-DNA reveal sequence-specific binding and groove-specific bending of DNA by magnesium and calcium. *J. Mol. Biol.* **2000**, *301*, 915–945. [CrossRef] [PubMed]
14. Ban, C.; Junop, M.; Yang, W. Transformation of MutL by ATP binding and hydrolysis: A switch in DNA mismatch repair. *Cell* **1999**, *97*, 85–97. [CrossRef]
15. Calsou, P.; Salles, B. Properties of damage-dependent DNA incision by nucleotide excision repair in human cell-free extracts. *Nucleic Acids Res.* **1994**, *22*, 4937–4942. [CrossRef]
16. Caspi, R.; Billington, R.; Keseler, I.M.; Kothari, A.; Krummenacker, M.; Midford, P.E.; Ong, W.K.; Paley, S.; Subhraveti, P.; Karp, P.D. The MetaCyc database of metabolic pathways and enzymes—A 2019 update. *Nucleic Acids Res.* **2020**, *48*, D445–D453. [CrossRef]
17. Lowenstein, F.W.; Stanton, M.F. Serum magnesium levels in the United States, 1971–1974. *J. Am. Coll. Nutr.* **1986**, *5*, 399–414. [CrossRef]
18. IMSC. Dietary reference intakes for calcium, phosphorus, magnesium, vitamin d, and fluoride. In *Dietary Reference Intakes for Calcium, Phosphorus, Magnesium, Vitamin D, and Fluoride*; National Academies Press: Washington, DC, USA, 1997. [CrossRef]
19. Lameris, A.L.; Huybers, S.; Kaukinen, K.; Makela, T.H.; Bindels, R.J.; Hoenderop, J.G.; Nevalainen, P.I. Expression profiling of claudins in the human gastrointestinal tract in health and during inflammatory bowel disease. *Scand. J. Gastroenterol.* **2013**, *48*, 58–69. [CrossRef]
20. Amasheh, S.; Fromm, M.; Gunzel, D. Claudins of intestine and nephron—A correlation of molecular tight junction structure and barrier function. *Acta Physiol.* **2011**, *201*, 133–140. [CrossRef]
21. Elin, R.J. Magnesium: The fifth but forgotten electrolyte. *Am. J. Clin. Pathol.* **1994**, *102*, 616–622. [CrossRef]
22. Worthington, V. Nutritional quality of organic versus conventional fruits, vegetables, and grains. *J. Altern. Complement. Med.* **2001**, *7*, 161–173. [CrossRef] [PubMed]
23. Rosanoff, A. Changing crop magnesium concentrations: Impact on human health. *Plant Soil* **2012**, *368*, 139–153. [CrossRef]

24. Fulgoni, V.L., 3rd; Keast, D.R.; Bailey, R.L.; Dwyer, J. Foods, fortificants, and supplements: Where do Americans get their nutrients? *J. Nutr.* **2011**, *141*, 1847–1854. [CrossRef] [PubMed]
25. Workinger, J.L.; Doyle, R.P.; Bortz, J. Challenges in the diagnosis of magnesium status. *Nutrients* **2018**, *10*, 1202. [CrossRef]
26. Sinert, R.; Zehtabchi, S.; Desai, S.; Peacock, P.; Altura, B.T.; Altura, B.M. Serum ionized magnesium and calcium levels in adult patients with seizures. *Scand. J. Clin. Lab Invest.* **2007**, *67*, 317–326. [CrossRef] [PubMed]
27. Gupta, A.; Eastham, K.M.; Wrightson, N.; Spencer, D.A. Hypomagnesaemia in cystic fibrosis patients referred for lung transplant assessment. *J. Cyst. Fibros.* **2007**, *6*, 360–362. [CrossRef]
28. Liao, F.; Folsom, A.R.; Brancati, F.L. Is low magnesium concentration a risk factor for coronary heart disease? The Atherosclerosis Risk in Communities (ARIC) Study. *Am. Heart J.* **1998**, *136*, 480–490. [CrossRef]
29. Barbagallo, M.; Dominguez, L.J. Magnesium metabolism in type 2 diabetes mellitus, metabolic syndrome and insulin resistance. *Arch. Biochem. Biophys.* **2007**, *458*, 40–47. [CrossRef]
30. Rosanoff, A.; Dai, Q.; Shapses, S.A. Essential nutrient interactions: Does low or suboptimal magnesium status interact with vitamin d and/or calcium status? *Adv. Nutr.* **2016**, *7*, 25–43. [CrossRef]
31. Smith, R.L.; Maguire, M.E. Microbial magnesium transport: Unusual transporters searching for identity. *Mol. Microbiol.* **1998**, *28*, 217–226. [CrossRef]
32. Zsurka, G.; Gregan, J.; Schweyen, R.J. The human mitochondrial Mrs2 protein functionally substitutes for its yeast homologue, a candidate magnesium transporter. *Genomics* **2001**, *72*, 158–168. [CrossRef] [PubMed]
33. Kolisek, M.; Zsurka, G.; Samaj, J.; Weghuber, J.; Schweyen, R.J.; Schweigel, M. Mrs2p is an essential component of the major electrophoretic $Mg^{2+}$ influx system in mitochondria. *EMBO J.* **2003**, *22*, 1235–1244. [CrossRef] [PubMed]
34. Nadler, M.J.; Hermosura, M.C.; Inabe, K.; Perraud, A.L.; Zhu, Q.; Stokes, A.J.; Kurosaki, T.; Kinet, J.P.; Penner, R.; Scharenberg, A.M.; et al. LTRPC7 is a Mg. ATP-regulated divalent cation channel required for cell viability. *Nature* **2001**, *411*, 590–595. [CrossRef] [PubMed]
35. Ryazanov, A.G.; Pavur, K.S.; Dorovkov, M.V. Alpha-kinases: A new class of protein kinases with a novel catalytic domain. *Curr. Biol.* **1999**, *9*, R43–R45. [CrossRef]
36. Runnels, L.W.; Yue, L.; Clapham, D.E. TRP-PLIK, a bifunctional protein with kinase and ion channel activities. *Science* **2001**, *291*, 1043–1047. [CrossRef]
37. Monteilh-Zoller, M.K.; Hermosura, M.C.; Nadler, M.J.; Scharenberg, A.M.; Penner, R.; Fleig, A. TRPM7 provides an ion channel mechanism for cellular entry of trace metal ions. *J. Gen. Physiol.* **2003**, *121*, 49–60. [CrossRef]
38. Fonfria, E.; Murdock, P.R.; Cusdin, F.S.; Benham, C.D.; Kelsell, R.E.; McNulty, S. Tissue distribution profiles of the human TRPM cation channel family. *J. Recept Signal Transduct. Res.* **2006**, *26*, 159–178. [CrossRef]
39. Schmitz, C.; Perraud, A.L.; Johnson, C.O.; Inabe, K.; Smith, M.K.; Penner, R.; Kurosaki, T.; Fleig, A.; Scharenberg, A.M. Regulation of vertebrate cellular $Mg^{2+}$ homeostasis by TRPM7. *Cell* **2003**, *114*, 191–200. [CrossRef]
40. Ryazanova, L.V.; Rondon, L.J.; Zierler, S.; Hu, Z.; Galli, J.; Yamaguchi, T.P.; Mazur, A.; Fleig, A.; Ryazanov, A.G. TRPM7 is essential for $Mg^{(2+)}$ homeostasis in mammals. *Nat. Commun.* **2010**, *1*, 109. [CrossRef]
41. Mittermeier, L.; Demirkhanyan, L.; Stadlbauer, B.; Breit, A.; Recordati, C.; Hilgendorff, A.; Matsushita, M.; Braun, A.; Simmons, D.G.; Zakharian, E.; et al. TRPM7 is the central gatekeeper of intestinal mineral absorption essential for postnatal survival. *Proc. Natl. Acad. Sci. USA* **2019**, *116*, 4706–4715. [CrossRef]
42. Schlingmann, K.P.; Weber, S.; Peters, M.; Niemann Nejsum, L.; Vitzthum, H.; Klingel, K.; Kratz, M.; Haddad, E.; Ristoff, E.; Dinour, D.; et al. Hypomagnesemia with secondary hypocalcemia is caused by mutations in TRPM6, a new member of the TRPM gene family. *Nat. Genet.* **2002**, *31*, 166–170. [CrossRef] [PubMed]
43. Walder, R.Y.; Landau, D.; Meyer, P.; Shalev, H.; Tsolia, M.; Borochowitz, Z.; Boettger, M.B.; Beck, G.E.; Englehardt, R.K.; Carmi, R.; et al. Mutation of TRPM6 causes familial hypomagnesemia with secondary hypocalcemia. *Nat. Genet.* **2002**, *31*, 171–174. [CrossRef] [PubMed]
44. Voets, T.; Nilius, B.; Hoefs, S.; van der Kemp, A.W.; Droogmans, G.; Bindels, R.J.; Hoenderop, J.G. TRPM6 forms the $Mg^{2+}$ influx channel involved in intestinal and renal $Mg^{2+}$ absorption. *J. Biol. Chem.* **2004**, *279*, 19–25. [CrossRef] [PubMed]
45. Wabakken, T.; Rian, E.; Kveine, M.; Aasheim, H.-C. The human solute carrier SLC41A1 belongs to a novel eukaryotic subfamily with homology to prokaryotic MgtE $Mg^{2+}$ transporters. *Biochem. Biophys. Res. Commun.* **2003**, *306*, 718–724. [CrossRef]
46. Goytain, A.; Quamme, G.A. Functional characterization of human SLC41A1, a $Mg^{2+}$ transporter with similarity to prokaryotic MgtE $Mg^{2+}$ transporters. *Physiol. Genom.* **2005**, *21*, 337–342. [CrossRef]
47. Kolisek, M.; Nestler, A.; Vormann, J.; Schweigel-Rontgen, M. Human gene SLC41A1 encodes for the $Na^{(+)}/Mg^{(2+)}$ exchanger. *Am. J. Physiol. Cell. Physiol.* **2012**, *302*, C318–C326. [CrossRef]
48. Goytain, A.; Quamme, G.A. Identification and characterization of a novel mammalian $Mg^{2+}$ transporter with channel-like properties. *BMC Genom.* **2005**, *6*, 48. [CrossRef]
49. Zhou, H.; Clapham, D.E. Mammalian MagT1 and TUSC3 are required for cellular magnesium uptake and vertebrate embryonic development. *Proc. Natl. Acad. Sci. USA* **2009**, *106*, 15750–15755. [CrossRef]
50. Li, F.Y.; Chaigne-Delalande, B.; Kanellopoulou, C.; Davis, J.C.; Matthews, H.F.; Douek, D.C.; Cohen, J.I.; Uzel, G.; Su, H.C.; Lenardo, M.J. Second messenger role for $Mg^{2+}$ revealed by human T-cell immunodeficiency. *Nature* **2011**, *475*, 471–476. [CrossRef]

51. Blommaert, E.; Peanne, R.; Cherepanova, N.A.; Rymen, D.; Staels, F.; Jaeken, J.; Race, V.; Keldermans, L.; Souche, E.; Corveleyn, A.; et al. Mutations in MAGT1 lead to a glycosylation disorder with a variable phenotype. *Proc. Natl. Acad. Sci. USA* **2019**, *116*, 9865–9870. [CrossRef]
52. Matsuda-Lennikov, M.; Biancalana, M.; Zou, J.; Ravell, J.C.; Zheng, L.; Kanellopoulou, C.; Jiang, P.; Notarangelo, G.; Jing, H.; Masutani, E.; et al. Magnesium transporter 1 (MAGT1) deficiency causes selective defects in N-linked glycosylation and expression of immune-response genes. *J. Biol. Chem.* **2019**, *294*, 13638–13656. [CrossRef] [PubMed]
53. Wang, C.-Y.; Shi, J.-D.; Yang, P.; Kumar, P.G.; Li, Q.-Z.; Run, Q.-G.; Su, Y.-C.; Scott, H.S.; Kao, K.-J.; She, J.-X. Molecular cloning and characterization of a novel gene family of four ancient conserved domain proteins (ACDP). *Gene* **2003**, *306*, 37–44. [CrossRef]
54. Wang, C.Y.; Yang, P.; Shi, J.D.; Purohit, S.; Guo, D.; An, H.; Gu, J.G.; Ling, J.; Dong, Z.; She, J.X. Molecular cloning and characterization of the mouse Acdp gene family. *BMC Genom.* **2004**, *5*, 7. [CrossRef] [PubMed]
55. Goytain, A.; Quamme, G.A. Functional characterization of ACDP2 (ancient conserved domain protein), a divalent metal transporter. *Physiol. Genom.* **2005**, *22*, 382–389. [CrossRef]
56. Sponder, G.; Mastrototaro, L.; Kurth, K.; Merolle, L.; Zhang, Z.; Abdulhanan, N.; Smorodchenko, A.; Wolf, K.; Fleig, A.; Penner, R.; et al. Human CNNM2 is not a $Mg^{(2+)}$ transporter per se. *Pflugers Arch.* **2016**, *468*, 1223–1240. [CrossRef]
57. Guo, D.; Ling, J.; Wang, M.H.; She, J.X.; Gu, J.; Wang, C.Y. Physical interaction and functional coupling between ACDP4 and the intracellular ion chaperone COX11, an implication of the role of ACDP4 in essential metal ion transport and homeostasis. *Mol. Pain* **2005**, *1*, 15. [CrossRef]
58. Yamazaki, D.; Funato, Y.; Miura, J.; Sato, S.; Toyosawa, S.; Furutani, K.; Kurachi, Y.; Omori, Y.; Furukawa, T.; Tsuda, T.; et al. Basolateral $Mg^{2+}$ extrusion via CNNM4 mediates transcellular $Mg^{2+}$ transport across epithelia: A mouse model. *PLoS Genet.* **2013**, *9*, e1003983. [CrossRef]
59. Arjona, F.J.; de Baaij, J.H.F. CrossTalk opposing view: CNNM proteins are not $Na^{(+)}/Mg^{(2+)}$ exchangers but $Mg^{(2+)}$ transport regulators playing a central role in transepithelial $Mg^{(2+)}$ (re)absorption. *J. Physiol.* **2018**, *596*, 747–750. [CrossRef]
60. Dai, Q.; Cantwell, M.M.; Murray, L.J.; Zheng, W.; Anderson, L.A.; Coleman, H.G. Dietary magnesium, calcium:magnesium ratio and risk of reflux oesophagitis, Barrett's oesophagus and oesophageal adenocarcinoma: A population-based case-control study. *Br. J. Nutr.* **2016**, *115*, 342–350. [CrossRef]
61. Shah, S.C.; Dai, Q.; Zhu, X.; Peek, R.M., Jr.; Smalley, W.; Roumie, C.; Shrubsole, M.J. Associations between calcium and magnesium intake and the risk of incident gastric cancer: A prospective cohort analysis of the National Institutes of Health-American Association of Retired Persons (NIH-AARP) Diet and Health Study. *Int. J. Cancer* **2019**. [CrossRef]
62. Molina-Montes, E.; Wark, P.A.; Sánchez, M.-J.; Norat, T.; Jakszyn, P.; Luján-Barroso, L.; Michaud, D.S.; Crowe, F.; Allen, N.; Khaw, K.-T.; et al. Dietary intake of iron, heme-iron and magnesium and pancreatic cancer risk in the European prospective investigation into cancer and nutrition cohort. *Int. J. Cancer* **2012**, *131*, E1134–E1147. [CrossRef]
63. Jansen, R.J.; Robinson, D.P.; Stolzenberg-Solomon, R.Z.; Bamlet, W.R.; de Andrade, M.; Oberg, A.L.; Rabe, K.G.; Anderson, K.E.; Olson, J.E.; Sinha, R.; et al. Nutrients from fruit and vegetable consumption reduce the risk of pancreatic cancer. *J. Gastrointest. Cancer* **2013**, *44*, 152–161. [CrossRef] [PubMed]
64. Dibaba, D.; Xun, P.; Yokota, K.; White, E.; He, K. Magnesium intake and incidence of pancreatic cancer: The VITamins and Lifestyle study. *Br. J. Cancer* **2015**, *113*, 1615–1621. [CrossRef]
65. Schilling, K.; Larner, F.; Saad, A.; Roberts, R.; Kocher, H.M.; Blyuss, O.; Halliday, A.N.; Crnogorac-Jurcevic, T. Urine metallomics signature as an indicator of pancreatic cancer. *Metallomics* **2020**. [CrossRef] [PubMed]
66. Dai, Q.; Shrubsole, M.J.; Ness, R.M.; Schlundt, D.; Cai, Q.; Smalley, W.E.; Li, M.; Shyr, Y.; Zheng, W. The relation of magnesium and calcium intakes and a genetic polymorphism in the magnesium transporter to colorectal neoplasia risk. *Am. J. Clin. Nutr.* **2007**, *86*, 743–751. [CrossRef] [PubMed]
67. Gorczyca, A.M.; He, K.; Xun, P.; Margolis, K.L.; Wallace, J.P.; Lane, D.; Thomson, C.; Ho, G.Y.; Shikany, J.M.; Luo, J. Association between magnesium intake and risk of colorectal cancer among postmenopausal women. *Cancer Causes Control* **2015**, *26*, 1761–1769. [CrossRef] [PubMed]
68. Meng, Y.; Sun, J.; Yu, J.; Wang, C.; Su, J. Dietary intakes of calcium, iron, magnesium, and potassium elements and the risk of colorectal cancer: A meta-analysis. *Biol. Trace Elem. Res.* **2019**, *189*, 325–335. [CrossRef]
69. Polter, E.; Onyeaghala, G.C.; Lutsey, P.L.; Folsom, A.R.; Joshu, C.E.; Platz, E.A.; Prizment, A.E. Prospective association of serum and dietary magnesium with colorectal cancer incidence. *Cancer Epidemiol. Biomark. Prev.* **2019**. [CrossRef]
70. Wesselink, E.; Kok, D.E.; Bours, M.J.L.; de Wilt, J.H.; van Baar, H.; van Zutphen, M.; Geijsen, A.M.J.R.; Keulen, E.T.P.; Hansson, B.M.E.; van den Ouweland, J.; et al. Vitamin D, magnesium, calcium, and their interaction in relation to colorectal cancer recurrence and all-cause mortality. *Am. J. Clin. Nutr.* **2020**. [CrossRef]
71. Anderson, K.J.; Cormier, R.T.; Scott, P.M. Role of ion channels in gastrointestinal cancer. *World J. Gastroenterol.* **2019**, *25*, 5732–5772. [CrossRef]
72. Stokłosa, P.; Borgström, A.; Kappel, S.; Peinelt, C. TRP channels in digestive tract cancers. *Int. J. Mol. Sci.* **2020**, *21*, 1877. [CrossRef] [PubMed]
73. Nakashima, S.; Shiozaki, A.; Ichikawa, D.; Hikami, S.; Kosuga, T.; Konishi, H.; Komatsu, S.; Fujiwara, H.; Okamoto, K.; Kishimoto, M.; et al. Transient receptor potential melastatin 7 as an independent prognostic factor in human esophageal squamous cell carcinoma. *Anticancer Res.* **2017**, *37*, 1161–1167. [CrossRef] [PubMed]

74. Yee, N.S.; Zhou, W.; Liang, I.C. Transient receptor potential ion channel Trpm7 regulates exocrine pancreatic epithelial proliferation by $Mg^{2+}$-sensitive Socs3a signaling in development and cancer. *Dis. Model Mech.* **2011**, *4*, 240–254. [CrossRef] [PubMed]
75. Yee, N.S.; Kazi, A.A.; Li, Q.; Yang, Z.; Berg, A.; Yee, R.K. Aberrant over-expression of TRPM7 ion channels in pancreatic cancer: Required for cancer cell invasion and implicated in tumor growth and metastasis. *Biol. Open* **2015**, *4*, 507–514. [CrossRef] [PubMed]
76. Rybarczyk, P.; Gautier, M.; Hague, F.; Dhennin-Duthille, I.; Chatelain, D.; Kerr-Conte, J.; Pattou, F.; Regimbeau, J.M.; Sevestre, H.; Ouadid-Ahidouch, H. Transient receptor potential melastatin-related 7 channel is overexpressed in human pancreatic ductal adenocarcinomas and regulates human pancreatic cancer cell migration. *Int. J. Cancer* **2012**, *131*, E851–E861. [CrossRef] [PubMed]
77. Rybarczyk, P.; Vanlaeys, A.; Brassart, B.; Dhennin-Duthille, I.; Chatelain, D.; Sevestre, H.; Ouadid-Ahidouch, H.; Gautier, M. The transient receptor potential melastatin 7 channel regulates pancreatic cancer cell invasion through the Hsp90alpha/uPA/MMP2 pathway. *Neoplasia* **2017**, *19*, 288–300. [CrossRef]
78. Xie, J.; Cheng, C.S.; Zhu, X.Y.; Shen, Y.H.; Song, L.B.; Chen, H.; Chen, Z.; Liu, L.M.; Meng, Z.Q. Magnesium transporter protein solute carrier family 41 member 1 suppresses human pancreatic ductal adenocarcinoma through magnesium-dependent Akt/mTOR inhibition and bax-associated mitochondrial apoptosis. *Aging* **2019**, *11*, 2681–2698. [CrossRef]
79. Xie, B.; Zhao, R.; Bai, B.; Wu, Y.; Xu, Y.; Lu, S.; Fang, Y.; Wang, Z.; Maswikiti, E.P.; Zhou, X.; et al. Identification of key tumorigenesis-related genes and their microRNAs in colon cancer. *Oncol. Rep.* **2018**, *40*, 3551–3560. [CrossRef]
80. Ibrahim, S.; Dakik, H.; Vandier, C.; Chautard, R.; Paintaud, G.; Mazurier, F.; Lecomte, T.; Gueguinou, M.; Raoul, W. Expression profiling of calcium channels and calcium-activated potassium channels in colorectal cancer. *Cancers* **2019**, *11*, 561. [CrossRef]
81. Pugliese, D.; Armuzzi, A.; Castri, F.; Benvenuto, R.; Mangoni, A.; Guidi, L.; Gasbarrini, A.; Rapaccini, G.L.; Wolf, F.I.; Trapani, V. TRPM7 is overexpressed in human IBD-related and sporadic colorectal cancer and correlates with tumor grade. *Dig. Liver Dis.* **2020**. [CrossRef]
82. Su, F.; Wang, B.F.; Zhang, T.; Hou, X.M.; Feng, M.H. TRPM7 deficiency suppresses cell proliferation, migration, and invasion in human colorectal cancer via regulation of epithelial-mesenchymal transition. *Cancer Biomark.* **2019**, *26*, 451–460. [CrossRef]
83. Zheng, K.; Yang, Q.; Xie, L.; Qiu, Z.; Huang, Y.; Lin, Y.; Tu, L.; Cui, C. Overexpression of MAGT1 is associated with aggressiveness and poor prognosis of colorectal cancer. *Oncol. Lett.* **2019**, *18*, 3857–3862. [CrossRef] [PubMed]
84. Funato, Y.; Yamazaki, D.; Mizukami, S.; Du, L.; Kikuchi, K.; Miki, H. Membrane protein CNNM4-dependent $Mg^{2+}$ efflux suppresses tumor progression. *J. Clin. Investig.* **2014**, *124*, 5398–5410. [CrossRef] [PubMed]
85. Digre, A.; Lindskog, C. The Human Protein Atlas-Spatial localization of the human proteome in health and disease. *Protein Sci.* **2021**, *30*, 218–233. [CrossRef] [PubMed]
86. Jonckheere, N.; Auwercx, J.; Hadj Bachir, E.; Coppin, L.; Boukrout, N.; Vincent, A.; Neve, B.; Gautier, M.; Trevino, V.; van Seuningen, I. Unsupervised hierarchical clustering of pancreatic adenocarcinoma dataset from TCGA defines a mucin expression profile that impacts overall survival. *Cancers* **2020**, *12*, 3309. [CrossRef]
87. Ren, J.; Du, Y.; Li, S.; Ma, S.; Jiang, Y.; Wu, C. Robust network-based regularization and variable selection for high-dimensional genomic data in cancer prognosis. *Genet. Epidemiol.* **2019**, *43*, 276–291. [CrossRef]
88. Trapani, V.; Wolf, F.I. Dysregulation of $Mg^{(2+)}$ homeostasis contributes to acquisition of cancer hallmarks. *Cell Calcium* **2019**, *83*, 102078. [CrossRef]
89. Kim, B.J.; Park, E.J.; Lee, J.H.; Jeon, J.-H.; Kim, S.J.; So, I. Suppression of transient receptor potential melastatin 7 channel induces cell death in gastric cancer. *Cancer Sci.* **2008**, *99*, 2502–2509. [CrossRef]
90. Chen, Y.; Wei, X.; Yan, P.; Han, Y.; Sun, S.; Wu, K.; Fan, D. Human mitochondrial Mrs2 protein promotes multidrug resistance in gastric cancer cells by regulating p27, cyclin D1 expression and cytochrome C release. *Cancer Biol. Ther.* **2009**, *8*, 607–614. [CrossRef]
91. Yee, N.S.; Zhou, W.; Lee, M.; Yee, R.K. Targeted silencing of TRPM7 ion channel induces replicative senescence and produces enhanced cytotoxicity with gemcitabine in pancreatic adenocarcinoma. *Cancer Lett.* **2012**, *318*, 99–105. [CrossRef]
92. Vanlaeys, A.; Fouquet, G.; Kischel, P.; Hague, F.; Pasco-Brassart, S.; Lefebvre, T.; Rybarczyk, P.; Dhennin-Duthille, I.; Brassart, B.; Ouadid-Ahidouch, H.; et al. Cadmium exposure enhances cell migration and invasion through modulated TRPM7 channel expression. *Arch. Toxicol.* **2020**, *94*, 735–747. [CrossRef] [PubMed]
93. Lefebvre, T.; Rybarczyk, P.; Bretaudeau, C.; Vanlaeys, A.; Cousin, R.; Brassart-Pasco, S.; Chatelain, D.; Dhennin-Duthille, I.; Ouadid-Ahidouch, H.; Brassart, B.; et al. TRPM7/RPSA complex regulates pancreatic cancer cell migration. *Front. Cell Dev. Biol.* **2020**, *8*, 549. [CrossRef] [PubMed]
94. Castiglioni, S.; Cazzaniga, A.; Trapani, V.; Cappadone, C.; Farruggia, G.; Merolle, L.; Wolf, F.I.; Iotti, S.; Maier, J.A.M. Magnesium homeostasis in colon carcinoma LoVo cells sensitive or resistant to doxorubicin. *Sci. Rep.* **2015**, *5*, 16538. [CrossRef] [PubMed]
95. Cazzaniga, A.; Moscheni, C.; Trapani, V.; Wolf, F.I.; Farruggia, G.; Sargenti, A.; Iotti, S.; Maier, J.A.M.; Castiglioni, S. The different expression of TRPM7 and MagT1 impacts on the proliferation of colon carcinoma cells sensitive or resistant to doxorubicin. *Sci. Rep.* **2017**, *7*, 40538. [CrossRef]
96. Luongo, F.; Pietropaolo, G.; Gautier, M.; Dhennin-Duthille, I.; Ouadid-Ahidouch, H.; Wolf, F.I.; Trapani, V. TRPM6 is essential for magnesium uptake and epithelial cell function in the colon. *Nutrients* **2018**, *10*, 784. [CrossRef]
97. Huang, J.; Furuya, H.; Faouzi, M.; Zhang, Z.; Monteilh-Zoller, M.; Kawabata, K.G.; Horgen, F.D.; Kawamori, T.; Penner, R.; Fleig, A. Inhibition of TRPM7 suppresses cell proliferation of colon adenocarcinoma in vitro and induces hypomagnesemia in vivo without affecting azoxymethane-induced early colon cancer in mice. *Cell Commun. Signal.* **2017**, *15*. [CrossRef]

98. Yamazaki, D.; Hasegawa, A.; Funato, Y.; Tran, H.N.; Mori, M.X.; Mori, Y.; Sato, T.; Miki, H. Cnnm4 deficiency suppresses Ca$^{(2+)}$ signaling and promotes cell proliferation in the colon epithelia. *Oncogene* **2019**, *38*, 3962–3969. [CrossRef]
99. Zou, Z.G.; Rios, F.J.; Montezano, A.C.; Touyz, R.M. TRPM7, magnesium, and signaling. *Int. J. Mol. Sci.* **2019**, *20*, 1877. [CrossRef]
100. Lemmon, M.A.; Schlessinger, J. Cell signaling by receptor tyrosine kinases. *Cell* **2010**, *141*, 1117–1134. [CrossRef]
101. Du, Z.; Lovly, C.M. Mechanisms of receptor tyrosine kinase activation in cancer. *Mol. Cancer* **2018**, *17*, 58. [CrossRef]
102. Arjona, F.J.; Latta, F.; Mohammed, S.G.; Thomassen, M.; van Wijk, E.; Bindels, R.J.M.; Hoenderop, J.G.J.; de Baaij, J.H.F. SLC41A1 is essential for magnesium homeostasis in vivo. *Pflügers Arch. Eur. J. Physiol.* **2019**, *471*, 845–860. [CrossRef] [PubMed]
103. Hardie, D.G.; Ross, F.A.; Hawley, S.A. AMPK: A nutrient and energy sensor that maintains energy homeostasis. *Nat. Rev. Mol. Cell. Biol.* **2012**, *13*, 251–262. [CrossRef] [PubMed]
104. Yuan, H.X.; Xiong, Y.; Guan, K.L. Nutrient sensing, metabolism, and cell growth control. *Mol. Cell* **2013**, *49*, 379–387. [CrossRef] [PubMed]
105. Zhu, J.; Deng, L.; Chen, B.; Huang, W.; Lin, X.; Chen, G.; Tzeng, C.-M.; Ying, M.; Lu, Z. Magnesium-dependent phosphatase (MDP) 1 is a potential suppressor of gastric cancer. *Curr. Cancer Drug Targets* **2019**, *19*, 817–827. [CrossRef] [PubMed]
106. Jin, J.; Desai, B.N.; Navarro, B.; Donovan, A.; Andrews, N.C.; Clapham, D.E. Deletion of Trpm7 disrupts embryonic development and thymopoiesis without altering Mg$^{2+}$ homeostasis. *Science* **2008**, *322*, 756–760. [CrossRef]
107. Tejpar, S.; Piessevaux, H.; Claes, K.; Piront, P.; Hoenderop, J.G.J.; Verslype, C.; Van Cutsem, E. Magnesium wasting associated with epidermal-growth-factor receptor-targeting antibodies in colorectal cancer: A prospective study. *Lancet Oncol.* **2007**, *8*, 387–394. [CrossRef]
108. Costa, A.; Tejpar, S.; Prenen, H.; Van Cutsem, E. Hypomagnesaemia and targeted anti-epidermal growth factor receptor (EGFR) agents. *Target Oncol.* **2011**, *6*, 227–233. [CrossRef]
109. Thebault, S.; Alexander, R.T.; Tiel Groenestege, W.M.; Hoenderop, J.G.; Bindels, R.J. EGF increases TRPM6 activity and surface expression. *J. Am. Soc. Nephrol.* **2009**, *20*, 78–85. [CrossRef]
110. Ravell, J.C.; Chauvin, S.D.; He, T.; Lenardo, M. An update on XMEN disease. *J. Clin. Immunol.* **2020**, *40*, 671–681. [CrossRef]
111. Ravell, J.C.; Matsuda-Lennikov, M.; Chauvin, S.D.; Zou, J.; Biancalana, M.; Deeb, S.J.; Price, S.; Su, H.C.; Notarangelo, G.; Jiang, P.; et al. Defective glycosylation and multisystem abnormalities characterize the primary immunodeficiency XMEN disease. *J. Clin. Investig.* **2020**, *130*, 507–522. [CrossRef]
112. King, D.E.; Mainous, A.G., 3rd; Geesey, M.E.; Woolson, R.F. Dietary magnesium and C-reactive protein levels. *J. Am. Coll. Nutr.* **2005**, *24*, 166–171. [CrossRef] [PubMed]
113. Danbolt, N.C.; Zhou, Y.; Furness, D.N.; Holmseth, S. Strategies for immunohistochemical protein localization using antibodies: What did we learn from neurotransmitter transporters in glial cells and neurons. *Glia* **2016**, *64*, 2045–2064. [CrossRef] [PubMed]
114. Deason-Towne, F.; Perraud, A.L.; Schmitz, C. The Mg$^{2+}$ transporter MagT1 partially rescues cell growth and Mg$^{2+}$ uptake in cells lacking the channel-kinase TRPM7. *FEBS Lett.* **2011**, *585*, 2275–2278. [CrossRef] [PubMed]

# Article

# Magnesium Deficiency Alters Expression of Genes Critical for Muscle Magnesium Homeostasis and Physiology in Mice

Dominique Bayle [1], Cécile Coudy-Gandilhon [1], Marine Gueugneau [1], Sara Castiglioni [2], Monica Zocchi [2], Magdalena Maj-Zurawska [3,4], Adriana Palinska-Saadi [3,4], André Mazur [1], Daniel Béchet [1,*] and Jeanette A. Maier [2,5]

1. UNH, Unité de Nutrition Humaine, Université Clermont Auvergne, INRAE, F-63000 Clermont-Ferrand, France; dominique.bayle@inrae.fr (D.B.); cecile.coudy-gandilhon@inrae.fr (C.C.-G.); marine.gueugneau@inrae.fr (M.G.); andre.mazur@inrae.fr (A.M.)
2. Department of Biomedical and Clinical Sciences Luigi Sacco, Università di Milano, 20157 Milano, Italy; sara.castiglioni@unimi.it (S.C.); monica.zocchi@unimi.it (M.Z.); jeanette.maier@unimi.it (J.A.M.)
3. Biological and Chemical Research Centre, University of Warsaw, PL-02-089 Warsaw, Poland; mmajzur@chem.uw.edu.pl (M.M.-Z.); adusp@cnbc.uw.edu.pl (A.P.-S.)
4. Faculty of Chemistry, University of Warsaw, PL-02-093 Warsaw, Poland
5. Interdisciplinary Centre for Nanostructured Materials and Interfaces (CIMaINa), Università di Milano, 20133 Milano, Italy
* Correspondence: daniel.bechet@inrae.fr

**Abstract:** Chronic $Mg^{2+}$ deficiency is the underlying cause of a broad range of health dysfunctions. As 25% of body $Mg^{2+}$ is located in the skeletal muscle, $Mg^{2+}$ transport and homeostasis systems (MgTHs) in the muscle are critical for whole-body $Mg^{2+}$ homeostasis. In the present study, we assessed whether $Mg^{2+}$ deficiency alters muscle fiber characteristics and major pathways regulating muscle physiology. C57BL/6J mice received either a control, mildly, or severely $Mg^{2+}$-deficient diet (0.1%; 0.01%; and 0.003% $Mg^{2+}$ wt/wt, respectively) for 14 days. $Mg^{2+}$ deficiency slightly decreased body weight gain and muscle $Mg^{2+}$ concentrations but was not associated with detectable variations in gastrocnemius muscle weight, fiber morphometry, and capillarization. Nonetheless, muscles exhibited decreased expression of several MgTHs (*MagT1*, *CNNM2*, *CNNM4*, and *TRPM6*). Moreover, TaqMan low-density array (TLDA) analyses further revealed that, before the emergence of major muscle dysfunctions, even a mild $Mg^{2+}$ deficiency was sufficient to alter the expression of genes critical for muscle physiology, including energy metabolism, muscle regeneration, proteostasis, mitochondrial dynamics, and excitation–contraction coupling.

**Keywords:** skeletal muscle; magnesium; magnesium transporters; transcriptome

## 1. Introduction

Magnesium ($Mg^{2+}$) intake is suboptimal in the population of Western countries, which results in an increased risk of latent $Mg^{2+}$ deficiency with Western diet behavior [1]. In addition, there is an increased risk of low $Mg^{2+}$ status in the elderly and after several current pharmacological treatments, such as proton pump inhibitors, thiazides, cetuximab, cisplatin, and some antibiotics [2,3]. In hospitalized patients, hypomagnesemia is a frequent finding often associated with other electrolyte disorders [4]. Clinical manifestations, depending on the severity and chronicity of deficiency, include a large variety of symptoms, e.g., neuromuscular symptoms (hyperexcitability, tetany, cramps, fasciculation, tremor, spasms, weakness), fatigue, tachycardia, anorexia, apathy, and behavioral alterations. In comparison to severe acute $Mg^{2+}$ deficiency, the diagnosis of chronic latent $Mg^{2+}$ deficiency is difficult, because magnesemia is often within reference intervals and results in nonspecific clinical symptoms [5,6]. However, it is well recognized that chronic $Mg^{2+}$ deficiency contributes to a broad range of metabolic, cardiovascular, immune, and neurological disorders [7].

$Mg^{2+}$ is the second (after $K^+$) most abundant intracellular cation and $Mg^{2+}$ is critical for a number of biological processes. $Mg^{2+}$ is a natural $Ca^{2+}$ antagonist, the activator of more than 200 enzymes, and the direct cofactor of over 600 enzymes [7]. In addition, intracellular $Mg^{2+}$ is buffered by many biological molecules, including proteins, RNAs, DNA, and ATP. ATP is mainly bound to $Mg^{2+}$, and $MgATP^{2-}$ is the active species in enzyme binding and energy production. Therefore, the intracellular concentration of $Mg^{2+}$ must be tightly regulated. This is achieved through the activity of $Mg^{2+}$ permeable channels and transporters. The last few years have seen rapid progress in the identification and characterization of $Mg^{2+}$ transport and homeostasis systems (MgTHs), including transient receptor potential cation channel subfamily M member 6 (TRPM6) and 7 (TRPM7), magnesium transporter 1 (MagT1), magnesium transporter MRS2, solute carrier family 41 member 1 (Slc41a1) and 3 (Slc41a3), cyclin, CBS domain divalent metal cation transport mediator 1 (CNNM1) and 4 (CNNM4) [7–9]. Although the precise function of MgTHs is still under investigation, current knowledge suggests that cellular $Mg^{2+}$ homeostasis is regulated by the combined action of several ubiquitous $Mg^{2+}$ transporters.

Besides its role in energy production, in the skeletal muscle $Mg^{2+}$ controls contraction by acting as a $Ca^{2+}$ antagonist on $Ca^{2+}$-permeable channels and $Ca^{2+}$-binding proteins. Accordingly, $Mg^{2+}$ deficiency decreases muscle strength [10]. Moreover, aging, frequently associated with low $Mg^{2+}$ status, is characterized by the gradual decline of muscle mass and performance. About 25% of body $Mg^{2+}$ is located in the skeletal muscle, which indicates that the expression of MgTHs is relevant to whole-body $Mg^{2+}$ homeostasis. Gene and/or protein expression of ubiquitous MgTHs have been demonstrated in the skeletal muscle (http://www.proteinatlas.org/ accessed on 10 April 2021), but their specific functions in this tissue have not been elucidated. Moreover, to our knowledge, few studies on the regulation of MgTHs in skeletal muscle under pathophysiological conditions, including $Mg^{2+}$ status, have been published [11].

We were interested in unveiling the early events occurring in the skeletal muscle in response to a low $Mg^{2+}$-containing diet. The principal aim of this study was to individuate whether and how a short-term $Mg^{2+}$-deficient diet modulates muscle fiber characteristics and cellular pathways critical for muscle physiology.

## 2. Materials and Methods

### 2.1. Animals

The present study (APAFIS#14025-201803121538803) was approved by the Ethics Committee C2EA-02 and was conducted in accordance with the National Research Council Guide for the Care and Use of Laboratory Animals. All animals were maintained in a temperature-controlled room (22 ± 1 °C) with a 12:12 h light:dark cycle and handled according to the recommendations of the Institutional Ethics Committee. Two-month-old male C57BL/6J mice were housed for one week in a standard environment with a control diet (0.1% $Mg^{2+}$ wt/wt). The mice were then randomly divided into three groups, and over the following two weeks, each group ($n$ = 12) was fed one of the three following diets: control diet (0.1% $Mg^{2+}$ wt/wt), mildly $Mg^{2+}$-deficient diet (0.01% $Mg^{2+}$ wt/wt), or severely $Mg^{2+}$-deficient diet (0.003% $Mg^{2+}$ wt/wt). The $Ca^{2+}$ content of the diets was 0.4% (wt/wt). All diets were prepared in our laboratory. Distilled water and food were available ad libitum. Quantitative magnetic resonance of live mice was carried out to estimate body fat and lean mass using an EcoMRI-100 analyzer (Echo Medical Systems LLC, Houston, TX, USA). At the end of the experiment, the animals were sacrificed, blood was collected from the heart in heparin-containing tubes, and gastrocnemius muscles were excised. Plasma was obtained by centrifugation (10 min, 3500 rpm, 4 °C) and frozen for later analysis. Muscles were weighed and (i) maintained in RNAlater (Qiagen, Courtaboeuf, France) overnight at 4 °C before extracting RNA, (ii) frozen in isopentane cooled on liquid $N_2$ and stored at −80 °C for histology, or (iii) snap-frozen in liquid $N_2$ for muscle $Mg^{2+}$ measurements.

## 2.2. Mg$^{2+}$ Analysis

Plasma magnesium was quantified using a Magnesium Calgamite kit according to the manufacturer's instructions (Biolabo, Maizy, France). Erythrocytes were washed 3 times with a saline solution, hemolyzed in water, and centrifuged. Muscle samples were mineralized in 65% HNO$_3$ for 48 h, and then 1.5 mL of deionized H$_2$O and 0.8 mL of 18 mol/L NaOH were added. Magnesium analyses were performed on a chemistry analyzer (Indiko Plus, Thermo Fisher Scientific, Vantaa, Finland).

## 2.3. Fiber Morphometry and Capillary Network

Serial cross sections (10 µm thick) were obtained using a cryostat (Microm, Francheville, France) at −25 °C. Myofiber morphometry and capillarization were assessed on cross sections after labeling with anti-laminin-α1 (Sigma, Saint-Quentin-Fallavier, France) and anti-CD31 (M0823 from Dako, Glostrup, Denmark), respectively, and capturing images by a BX-51 microscope (Olympus, Rungis, France) according to [12,13]. On average, 710 ± 48 fibers were analyzed per subject. Fiber cross-sectional area (CSA) and perimeter were determined for each fiber, using the image processing software Visilog-6.9 (Noesis, Gif-sur-Yvette, France) as previously described [12]. A shape factor (perimeter$^2$/4π CSA) was calculated, with a value of 1.0 indicating a circle and >1.0 an increasingly elongated ellipse. A mean of 229 ± 13 capillaries was analyzed per subject. Capillary density (CD) was expressed as the number of capillaries counted per square mm. Capillary-to-fiber ratio (C/F) was calculated as the ratio between the number of capillaries and the number of fibers present in the same area [13].

## 2.4. Quantitative Real-Time Polymerase Chain Reaction (qRT-PCR) Analysis

Independent RNA isolations were carried out for each gastrocnemius muscle sample. Total RNAs were extracted using the RNeasy Fibrous Tissue Mini Kit (Qiagen) following the manufacturer's conditions. RNA concentrations were measured using a NanoDrop ND-1000 (LabTech, Ringmer, UK), and RNA quality was verified by 1% agarose gel electrophoresis. One µg total RNA was used as a template for single-strand cDNA synthesis using High-Capacity cDNA RT Kit (Applied Biosystems, Foster City, CA, USA) in a total volume of 20 µL containing 1 X RT buffer, 4 mM dNTP mix, 1 X random primers, 50 U reverse transcriptase and 20 U RNase inhibitor. The primer sequence is reported in Table 1. The reverse transcription reactions were run as follows: 25 °C for 10 min, 37 °C for 120 min, and 85 °C for 5 s. PCR was carried out in a final volume of 20 µL containing 10 µL Power SYBR Green PCR Master Mix (Applied Biosystems), 0.4 µL of each primer at 10 pmol/µL, and 2 µL of the cDNA solution. qRT-PCR amplification was performed using a CFX96 Real-Time PCR Detection System (Bio-Rad, Marnes-la-Coquette, France) with the following thermal cycler conditions: 15 min at 95 °C, followed by 45 cycles of 15 s at 95 °C, and 1 min at 60 °C. Raw data were analyzed using CFX Maestro (Bio-Rad) and compared by the ΔΔCt method. Results are expressed relative to the housekeeping gene (*Actb*) transcript quantity.

## 2.5. TaqMan Low-Density Array (TLDA)

A total of 500 ng (10 µL) cDNA of each sample was combined with 95 µL of nuclease-free water and 105 µL 2X TaqMan™ Fast Advanced Master Mix (Applied Biosystems) for the quantitative real-time PCR (qPCR) measurements. This mixture was divided equally over two sample-loading ports of the TLDA. The arrays were centrifuged once (1 min, 1300 rpm at room temperature) to equally distribute the sample over the wells. Subsequently, the card was sealed to prevent exchange between wells. qPCR amplification was performed using an Applied Biosystems 7900HT system with the following thermal cycler conditions: 2 min at 50 °C and 10 min at 94.5 °C, followed by 40 cycles of 30 s at 97 °C and 30 s at 59.7 °C. Raw data were analyzed using Sequence Detection System (SDS) Software v2.4 (Applied Biosystems). The expression of β-actin (*Actb*), β-glucuronidase (*Gusb*), and hypoxanthine phosphoribosyltransferase (*Hprt*) were used as controls. The genes analyzed are reported in Table 2.

Table 1. Sequence of primers used for qRT-PCR.

| Primers | Forward | Reverse |
|---|---|---|
| TRPM6 | GACAGTCTAAGCACCTTTTC | AAGCTTGTACCCTTCAGTAG |
| MagT1 | GATGGGCTTTTGCAGCTTTGT | GCAATACACATCATCCTTCGCT |
| MRS2 | GGTGATGTGCTCCGGTTTAGA | TGGCCTGGAGTGCTAACTCAT |
| Slc41a1 | TACTGGCCCTACTCCTTCTCC | GGGACTCAATCACTACCACCTC |
| Slc41a2 | CTTCAGCAAGAGATCAGAGCC | CCAGCATAGTCATCGTACTTGG |
| Slc41a3 | CTCAGCCTTGAGTTCCGCTTT | GCAGGATAGGTATGGCGACC |
| CNNM1 | GTAGGGTCACAACCTACATC | CATGACATACACAGAAGAGG |
| CNNM2 | AAGTGGCCCACCGTGAAAG | CGCTTCTACTTCTGTTGCTAGG |
| CNNM3 | GACTCCGGCACTGTCCTAGA | AGTGGATGGTTGTAGAAGCGG |
| CNNM4 | CTGCACATCCTTCTCGTTATGG | TGCGAGCATACTTTCTCTCCTT |

Table 2. Gene analyzed.

| Acvr2a | Activin A Receptor Type 2a | Map1lc3b | Microtubule-Associated Protein 1 Light Chain 3 Beta |
|---|---|---|---|
| Acvr2b | Activin A Receptor Type 2b | Mdm2 | Mouse double minute 2 homolog |
| Asb2 | Ankyrin Repeat And SOCS Box Containing 2 | Mef2c | Myocyte Enhancer Factor 2C |
| Atf4 | Activating Transcription Factor 4 | Mfn1 | Mitofusin 1 |
| Atg10 | Autophagy Related 10 | Mfn2 | Mitofusin 2 |
| Atg12 | Autophagy Related 12 | Mief1 | Mitochondrial Elongation Factor 1 |
| Atg16l1 | Autophagy Related 16 Like 1 | Mstn | Myostatin |
| Atg3 | Autophagy Related 3 | Myf5 | Myogenic Factor 5 |
| Atg5 | Autophagy Related 5 | Myod | Myogenic Differentiation 1 |
| Atg7 | Autophagy Related 7 | Myog | Myogenin |
| Atp2a1 | ATPase Sarcoplasmic/Endoplasmic Reticulum $Ca^{2+}$ Transporting 1 | Nbr1 | NBR1 Autophagy Cargo Receptor |
| Becn1 | Beclin 1 | Nrf1 | Nuclear Respiratory Factor 1 |
| Bnip3 | BCL2/adenovirus E1B Interacting Protein 3 | Opa1 | Mitochondrial dynamin like GTPase |
| Capn1 | Calpain 1 | Park2 | Parkin RBR E3 Ubiquitin Protein Ligase |
| Capn2 | Calpain 2 | Pik3c3 | Phosphatidylinositol 3-Kinase Catalytic Subunit Type 3 |
| Capn3 | Calpain 3 | Pink1 | PTEN Induced Kinase 1 |
| Casq1 | Calsequestrin 1 | Plin2 | Perilipin 2 |
| Cdkn1a | Cyclin Dependent Kinase Inhibitor 1A | Plin5 | Perilipin 5 |
| Chac1 | Glutathione-specific gamma-glutamylcyclotransferase 1 | Ppargc1a | Peroxisome Proliferator Activated Receptor Gamma Coactivator 1 Alpha |
| CNNM2 | Cyclin and CBS Domain Divalent Metal Cation Transport Mediator 2 | Rhot1 | Ras Homolog Family Member T1 |
| CNNM4 | Cyclin and CBS Domain Divalent Metal Cation Transport Mediator 4 | Ryr1 | Ryanodine Receptor 1 |
| Cox1 | Mitochondrially Encoded Cytochrome C Oxidase I | Slc27a1 | Solute Carrier Family 27 Member 1 |
| Creb1 | CAMP Responsive Element Binding Protein 1 | Slc2a1 | Solute Carrier Family 2 Member 1 |

Table 2. Cont.

| | | | |
|---|---|---|---|
| Cs | citrate synthase | Slc2a4 | Solute Carrier Family 2 Member 4 |
| Ctsl | Cathepsin L | Slc6a8 | Solute Carrier Family 6 Member 8 |
| Ddit3 | DNA Damage Inducible Transcript 3 | Slc41a1 | Solute Carrier family 41 member 1 |
| Dnm1l | Dynamin 1 Like | Sqstm1 | Sequestosome 1 |
| Eif4ebp1 | Eukaryotic Translation Initiation Factor 4E Binding Protein 1 | Srebf1 | Sterol Regulatory Element Binding Transcription Factor 1 |
| Fabp3 | Fatty Acid Binding Protein 3 | Srebf2 | Sterol Regulatory Element Binding Transcription Factor 2 |
| Fbxo21 | F-Box Protein 21 | Srl | Sarcalumenin |
| Fbxo30 | F-Box Protein 30 | Tfam | Transcription Factor A, Mitochondrial |
| Fbxo31 | F-Box Protein 31 | Trim63 | Tripartite Motif-containing 63 |
| Fbxo32 | F-Box Protein 32 | Trim72 | Tripartite Motif-containing 72 |
| Fis1 | Fission, Mitochondrial 1 | TRPM6 | Transient Receptor Potential cation channel subfamily M member 6 |
| Foxo3 | Forkhead Box O3 | Ube2b | Ubiquitin Conjugating Enzyme E2 B |
| Fst | Follistatin | Ube2e1 | Ubiquitin-Conjugating Enzyme E2 E1 |
| Fundc1 | FUN14 Domain Containing 1 | Ube2g1 | Ubiquitin-Conjugating Enzyme E2G 1 |
| Gabarapl1v | Gamma-Aminobutyric Acid A Receptor-Associated Protein-Like 1 | Ube2j1 | Ubiquitin-Conjugating Enzyme E2J 1 |
| Gadd45a | Growth Arrest and DNA Damage Inducible Alpha | Ube2j2 | Ubiquitin-Conjugating Enzyme E2J 2 |
| Gapdh | Glyceraldehyde-3-Phosphate Dehydrogenase | Ube2l3 | Ubiquitin-Conjugating Enzyme E2L 3 |
| Lamp2 | Lysosomal-Associated Membrane Protein 2 | Ulk1 | Unc-51 Like Autophagy Activating Kinase 1 |
| MagT1 | Magnesium Transporter 1 | Zeb1 | Zinc Finger E-Box Binding Homeobox 1 |

### 2.6. Western Blot

Gastrocnemius muscles were mechanically shredded in a Potter homogenizer with lysis buffer (50 mM Tris-HCl pH 7.4, 150 mM NaCl, 1% NP-40, 0.25% Na-deoxycholate) containing protease inhibitors. Total proteins were quantified using the Bradford reagent (Sigma-Aldrich, St. Louis, MO, USA). Equal amounts of proteins were separated by SDS–PAGE on 4–20% Mini-PROTEAN TGX Stain-free Gels (Bio-Rad, Hercules, CA, USA) and transferred to nitrocellulose membranes by using Trans-Blot® TurboTM Transfer Pack (Bio-Rad). After blocking with bovine serum albumin (BSA), Western blot analysis was performed using primary antibodies against Myog, Opa1 (BD Biosciences, St. Diego, CA, USA), Gapdh, and Mfn2 (Santa-Cruz Biotechnology, Dallas, TX, USA). The filters were washed and incubated with secondary antibodies conjugated to horseradish peroxidase (Amersham Pharmacia Biotech Italia, Cologno Monzese, Italy) were used. The immunoreactive proteins were detected with ClarityTM Western ECL substrate (Bio-Rad) and images were captured with a ChemiDoc MP Imaging System (Bio-Rad). The nitrocellulose sheets were used as control loading. Densitometry of the bands was performed with the software ImageLab (Bio-Rad). The Western blots shown are representative and the densitometric analysis was performed on three independent experiments.

### 2.7. Statistical Analysis

Data are presented as means ±SE. To determine whether or not data sets were normally distributed, the Shapiro–Wilk and Kolmogorov–Smirnov normality tests were performed. When data were normally distributed, statistical comparisons between groups were performed applying either Student's $t$ test or one-way ANOVA, followed by a Tukey's post hoc test, as appropriate. Mann–Whitney U tests were performed when data in at least one group were not normally distributed. Correction for multiple testing was performed with R according to [14], and TLDA significance was set at q-value < 0.05. Univariate linear Pearson's regression was carried out to investigate relationships between MgTHs mRNA levels and body weight gain. Statistical analyses were performed using XLSTAT (Addinsoft, Paris, France), and significance was set at $P < 0.05$.

## 3. Results

### 3.1. $Mg^{2+}$-Deficient Diet Reduces Muscle $Mg^{2+}$ Concentrations but Does Not Affect Fiber Characteristics

To investigate the modulation of the expression of $Mg^{2+}$ transport and homeostasis systems (MgTHs) in response to $Mg^{2+}$ status, we used a model of C57BL/6J mice receiving a mildly or severely $Mg^{2+}$-deficient diet and compared it to mice under an $Mg^{2+}$-sufficient diet [15,16]. After 14 days of diet, mice fed a mildly or severely $Mg^{2+}$-deficient diet exhibited a 26% and 75% reduction in plasma $Mg^{2+}$ concentration, respectively (Figure 1a). However, no significant change in erythrocyte $Mg^{2+}$ concentration was detected (Figure 1b). In parallel, moderate and severe $Mg^{2+}$ deficiencies resulted in a 4.8% and 5.4% decrease in intramuscular $Mg^{2+}$ concentrations, respectively, compared to the control $Mg^{2+}$-sufficient diet (Figure 1c).

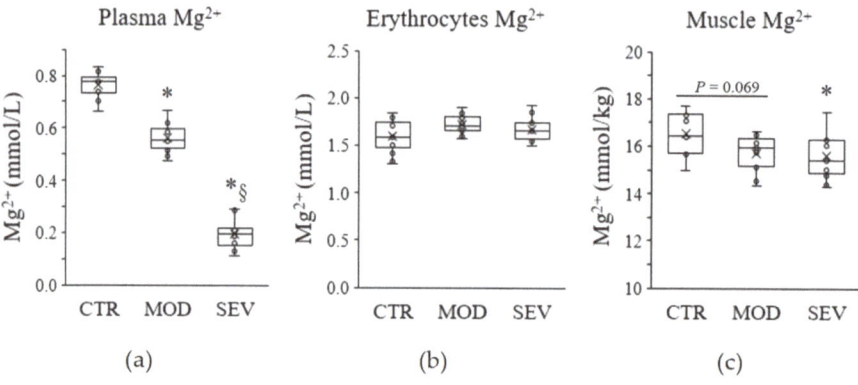

**Figure 1.** $Mg^{2+}$ status in mice. $Mg^{2+}$ concentrations were measured in (**a**) plasma, (**b**) erythrocytes, and (**c**) gastrocnemius muscle of mice fed either a control (CTR), a mildly (MOD), or a severely (SEV) $Mg^{2+}$-deficient diet. Data (N = 12 per group) are presented as box-and-whisker plots (centerline, median; box limits, first and third quartiles; whiskers, 1.5 x interquartile range; points, outliers; x in the box, mean). * Significant difference ($P < 0.05$) from the CTR group. § Significant difference ($P < 0.05$) from the MOD group.

Moderate and severe $Mg^{2+}$ deficiencies were associated with a significant decline in body weight gain (Figure 2a), and magnetic resonance imaging (EchoMRI) indicated modest trends for whole-body fat and lean mass reductions (Figure 2b,c). Nonetheless, gastrocnemius muscle weight did not differ between $Mg^{2+}$-deficient and $Mg^{2+}$-sufficient diets (Figure 3a). Semiquantitative histology further indicated that $Mg^{2+}$ deficiency was not associated with a detectable variation in muscle fiber cross-sectional area, shape, capillary density (CD), or capillary-to-fiber ratio (C/F) (Figure 3b–d).

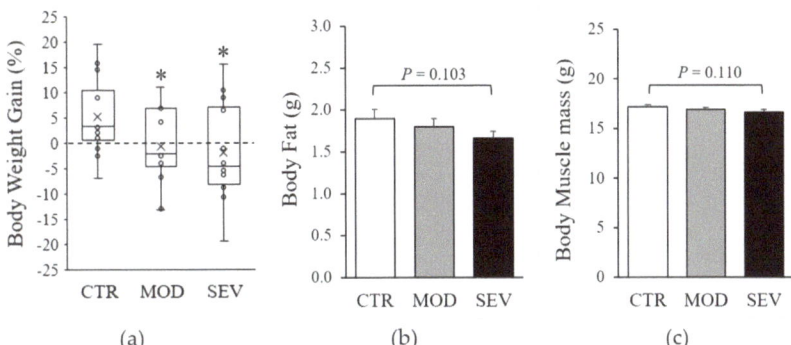

**Figure 2.** Body composition of mice fed Mg$^{2+}$-deficient diet. Mice were fed either a control (CTR), a mildly (MOD), or a severely (SEV) Mg$^{2+}$-deficient diet: (**a**) body weight, data (N = 12) presented as box-and-whisker plots, (**b**) body fat, and (**c**) body muscle mass were determined by EcoMRI; results are means + SE (N = 12). * Significant difference ($P < 0.05$) from the CTR group.

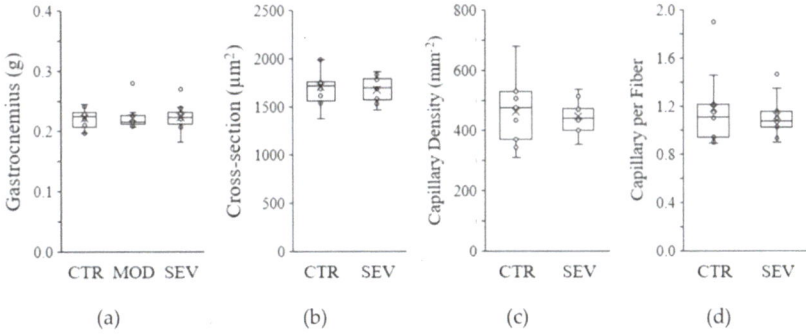

**Figure 3.** Mg$^{2+}$-deficient diet and muscle characteristics. Mice were fed either a control (CTR), a mildly (MOD), or a severely (SEV) Mg$^{2+}$-deficient diet: (**a**) gastrocnemius muscle weight; (**b**) muscle fiber cross-sectional area; (**c**) muscle capillary density; (**d**) number of capillaries per fiber. Data (N = 12 per group) are presented as box-and-whisker plots.

### 3.2. Mg$^{2+}$-Deficient Diet Alters Expression of Muscle MgTHs

Although no evidence emerged in terms of atrophy or altered morphology or capillarization of muscle fibers, modest declines in muscular Mg$^{2+}$ concentrations might nonetheless modulate specific mechanisms to maintain cellular Mg$^{2+}$ homeostasis. We assessed whether Mg$^{2+}$ deficiency could be associated with differential expression of MgTHs. qRT-PCR analyses performed on gastrocnemius muscle revealed significant declines in *MagT1*, *CNNM2*, *CNNM4*, and *TRPM6* mRNAs in mice under Mg$^{2+}$-deficient diets, compared to Mg$^{2+}$-sufficient diets (Figure 4a). MagT1 is a controversial protein initially isolated as a critical mediator of Mg$^{2+}$ homeostasis in eukaryotes [17] and then as an integral part of the *N*-linked glycosylation complex [18]. As MagT1 mutations associate with hypomagnesemia, we included MagT1 in the list of MgTHs. The mRNA levels of the other MgTHs, i.e., *CNNM1*, *CNNM3*, *MRS2*, *Slc41a1*, *Slc41a2*, and *Slc41a3*, were not altered by Mg$^{2+}$ deficiency.

Interestingly, regression analyses performed with all mice indicated positive correlation between body weight gain and several MgTHs mRNAs, i.e., *MagT1* (r = 0.35, $P = 0.038$), *CNNM2* (r = 0.38, $P = 0.022$), *CNNM3* (r = 0.40, $P = 0.015$), *CNNM4* (r = 0.48, $P = 0.003$), *MRS2* (r = 0.45, $P = 0.006$), *Slc41a1* (r = 0.50, $P = 0.002$) (Figure 4b).

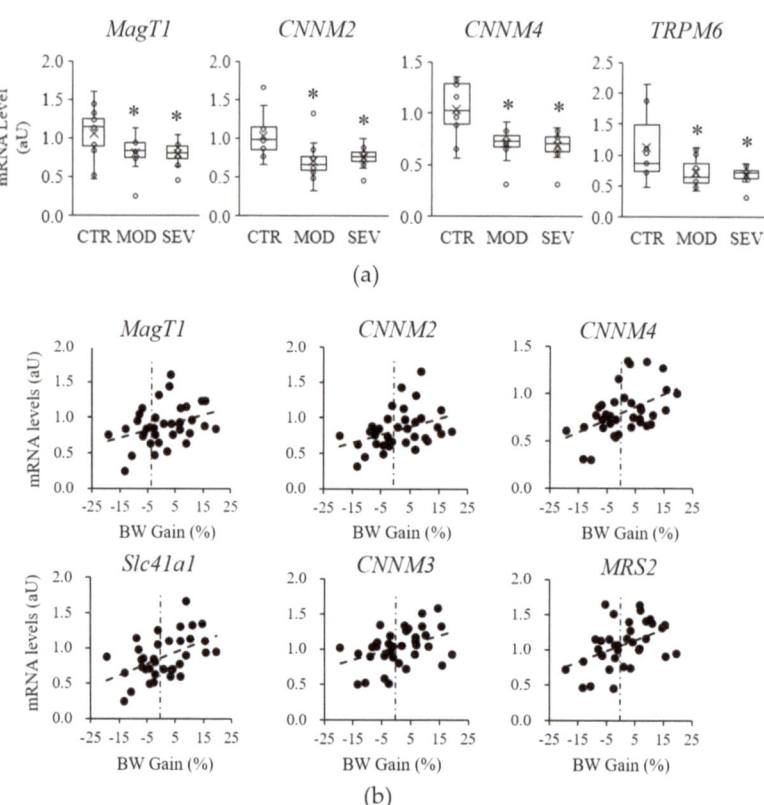

**Figure 4.** $Mg^{2+}$-deficient diet and muscle $Mg^{2+}$ transport and homeostatic systems (MgTHs) mRNA levels. Mice were fed either a control (CTR), a mildly (MOD), or a severely (SEV) $Mg^{2+}$-deficient diet: (**a**) gastrocnemius MgTHs mRNA levels; data (N = 12) are presented as box-and-whisker plots; * indicates significant difference ($P < 0.05$) from the CTR group; (**b**) examples of linear Pearson's regressions between muscle MgTHs mRNA levels and body weight (BW) gain.

*3.3. Mild $Mg^{2+}$ Deficiency Alters the Expression of Genes Important for Muscle Energy Metabolism and Regeneration*

As $Mg^{2+}$ is the activator or cofactor of a number of enzymes, modest declines in intramuscular $Mg^{2+}$ concentrations might modulate biological processes that are critical for muscle physiology. Initially, to assess muscle stress, we evaluated the expression of *Trim72*, *Chac1*, and *Ddit3*. *Trim72* codes for a protein specifically located in the sarcolemma and involved in membrane repair [19]. *Chac1* encodes a protein acting downstream of ATF4 implicated in muscle atrophy [20], and *Ddit3* encodes a member of the CCAAT/enhancer-binding protein (C/EBP) family of transcription factors and inhibits myogenesis [21]. We found that severely and mildly $Mg^{2+}$-deficient diets did not modulate *Trim72* (Figure 5). Moreover, severe $Mg^{2+}$ deficiency, but not mild $Mg^{2+}$ deficiency, upregulated the stress genes *Chac1* and *Ddit3* (Figure 5).

To reduce the potential interferences due to the activation of the stress response and to mimic chronic latent deficiency conditions, further investigations were focused on mild $Mg^{2+}$ deficiency. Moreover, TaqMan low-density array (TLDA) analyses revealed that mild $Mg^{2+}$ deficiency was sufficient to rapidly alter the expression of genes important for lipid and carbohydrate metabolism (Figure 6a). These included *Slc2a4* and *Slc6a8*, coding for Glut4 and creatine transporter (CT)-1, respectively, *Gapdh*, *Cs* (citrate synthase), *Plin2*, a marker of intramyocellular lipid droplets, and the transcription factors *Creb1*, *Srebf1*, and *Srebf2*. Mild $Mg^{2+}$ deficiency also rapidly altered the expression of genes involved in muscle regeneration (Figure 6b), i.e., *Myog* (myogenin), *Mef2c* (myocyte enhancer factor 2C), *Mstn* (myostatin), and its receptors (*Acvr2a* and *Acvr2b*).

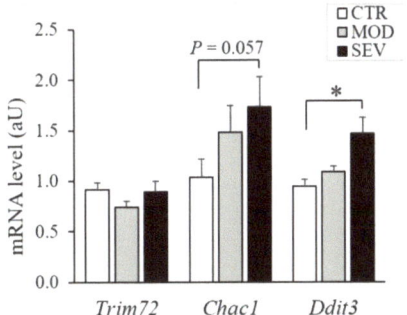

**Figure 5.** Muscle mRNA levels of stress genes in mice fed $Mg^{2+}$-deficient diet. TaqMan low-density array (TLDA) was used to measure mRNA levels in gastrocnemius muscle of mice fed either a control (CTR; white bars), a mildly (MOD; grey bars), or a severely $Mg^{2+}$-deficient diet (SEV; black bars). * indicates significant difference ($P < 0.05$) from the CTR group.

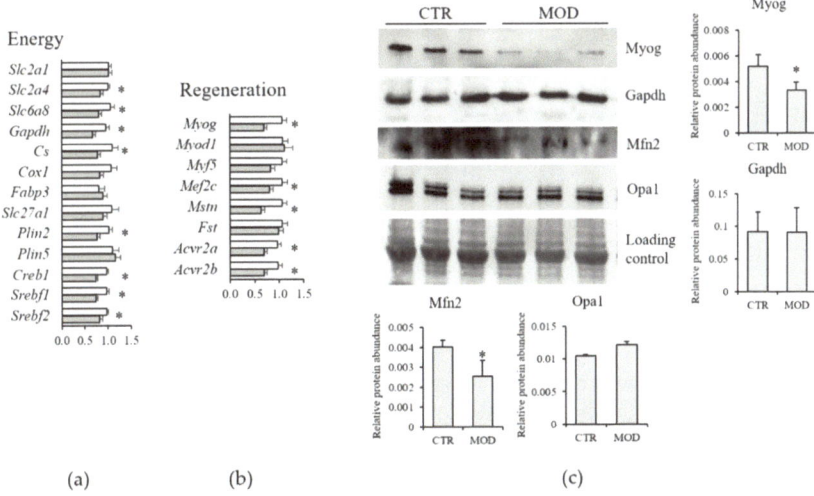

**Figure 6.** Muscle mRNA levels of genes involved in energy metabolism (**a**) and muscle regeneration (**b**) in mice fed a mildly $Mg^{2+}$-deficient diet. Results are means ± SE ($N = 12$). White columns CTR, grey columns mild $Mg^{2+}$ deficiency. (**c**) Western blot was performed on 40 μg of lysates using specific antibodies against Myog, Gapdh, Mfn1, and Opa1. A representative western is shown. Densitometry was performed on three different blots. * indicates significant difference ($P < 0.05$) from the CTR group.

We tested the total amounts of some proteins by Western blot on lysates from gastrocnemius muscles of 12 animals under control or mildly $Mg^{2+}$-deficient diet. As shown in Figure 6c, we found a significant reduction of Myog and no modulation of Gapdh.

## 3.4. Mild $Mg^{2+}$ Deficiency Alters Expression of Genes Important for Muscle Proteostasis

Amongst the transduction pathways implicated in the regulation of muscle atrophy [22,23], mild $Mg^{2+}$ deficiency was associated with a decreased expression of *Fbxo32* (MAFbx), *Zeb1*, *Fbxo31*, *Atf4*, and *Eif4ebp1*, while the mRNA levels of other major regulators, e.g., *Trim63* (MuRF1), *Foxo3*, *Mdm2*, *Fbxo30* (Musa1), *Fbxo21* (Smart), *Gadd45a*, and *Cdkn1a* (p21) did not significantly change (Figure 7a).

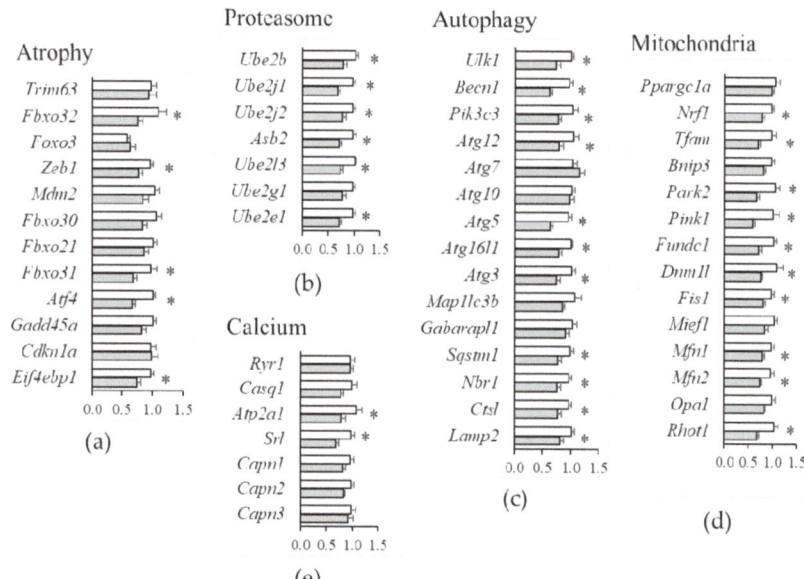

**Figure 7.** Muscle gene expression in mice fed a mildly $Mg^{2+}$-deficient diet. TaqMan low-density array (TLDA) was used to measure mRNA levels of genes critical for muscle physiology of mice fed either a control (white bars) or a mildly (grey bars) $Mg^{2+}$-deficient diet. Genes involved in muscle atrophy (**a**), proteostasis (**b**), autophagy (**c**), mitochondrial dynamics (**d**) and calcium homeostasis (**e**) were analyzed. Results are means ± SE (N = 12). * indicates significant difference ($P < 0.05$) from the CTR group.

In agreement with the downregulation of major atrogenes by mild $Mg^{2+}$ deficiency, e.g., *Fbxo32* (MAFbx), there was also evidence for altered expression of many genes involved in protein catabolism. This encompassed the downregulation of (i) E2-activating genes (*Ube2b*, *Ube2j1*, *Ube2j2*, *Asb2*, *Ube2l3*, *Ube2e1*) of the ubiquitin–proteasome system (Figure 7b); (ii) genes regulating initiation (*Ulk1*, *Becn1*, *Pik3c3*), elongation (*Atg12*, *Atg5*, *Atg16l1*, *Atg3*), substrate/cargo recruitment (*Sqstm1*, *Nbr1*) and lysosomal proteolysis (*Ctsl*, *Lamp2*) in autophagy (Figure 7c). However, $Mg^{2+}$ deficiency did not affect the expression of $Ca^{2+}$-dependent proteases (*Capn1*, *Capn2*, *Capn3* calpains) (Figure 7e).

## 3.5. Mild $Mg^{2+}$ Deficiency, Mitochondria, and $Ca^{2+}$ Homeostasis

Mitochondria play multifaceted roles in essential aspects of skeletal muscle cell physiology [24]. Our TLDA analyses indicated that a mild $Mg^{2+}$ deficiency is sufficient to downregulate genes regulating mitophagy (*Park2*, *Pink1*, *Fundc1*) (Figure 7d), in addition to reducing the expression of genes implicated in mitogenesis (*Nrf1*, *Tfam*), mitochondrial fission (*Dnm1l*, *Fis1*), and fusion (*Mfn1*, *Mfn2*, *Rhot1*) (Figure 7d). By Western blot, we confirmed the downregulation of mitofusin (Mfn)2 (Figure 6c). Opa1, instead, was not modulated, both at the RNA and the protein levels (Figures 6c and 7d).

Mg$^{2+}$ may also modulate Ca$^{2+}$-permeable channels and Ca$^{2+}$-binding proteins. Ca$^{2+}$ handling by the sarcoplasmic reticulum is a key feature in muscle contraction. Action potentials elicit contraction by the release of Ca$^{2+}$ from the sarcoplasmic reticulum through the ryanodine receptors (*Ryr1*) that are regulated by calsequestrin-1 (*Casq1*). For muscle relaxation, Ca$^{2+}$ is transported back to the sarcoplasmic reticulum by sarco/endoplasmic reticulum Ca$^{2+}$-ATPases (Serca), regulated by sarcalumenins (*Srl*) [25,26]. It is noteworthy that our TLDA study emphasized that a mild Mg$^{2+}$ deficiency did not affect the expression of genes involved in contraction (*Ryr1*, *Casq1*), while it downregulated the expression of those implicated in relaxation, *Atp2a1* (Serca), and *Srl* (Figure 7e).

## 4. Discussion

Appropriate nutrition is indispensable for normal muscle metabolism and function [7]. In humans, cross-sectional associations between low Mg$^{2+}$ intake and loss of skeletal mass and function were reported in large population cohorts [27–29].

To get insight into the role of Mg$^{2+}$ in muscle health, we developed an experimental model in which mice were fed a moderately or severely Mg$^{2+}$-deficient diet. Both these dietetic regimens significantly reduced serum Mg$^{2+}$ levels, a useful biomarker of Mg$^{2+}$ status [6]. Unexpectedly, we did not detect a reduction of erythrocytes Mg$^{2+}$ concentration, differently from a previous report demonstrating the concomitant reduction of serum and erythrocyte Mg$^2$ in animals fed the severely deficient diet [15]. We have no explanation at the moment for this discrepancy. Interestingly, we found that some Mg$^{2+}$ transporters were significantly downregulated in the muscle. This finding explains the reduced amounts of intracellular Mg$^{2+}$ in the muscle and might also represent an initial, adaptive response aimed at maintaining circulating Mg$^{2+}$ as close to physiological levels as possible. In this perspective, it is noteworthy that 12–13 week-old *Trpm6*-deficient adult mice are hypomagnesemic and sarcopenic, events reversible upon supplementation with Mg$^{2+}$ [30]. Additionally, *CNNM2+/−* mice were hypomagnesemic, but no data are available on the skeletal muscle yet [31]. In our experimental model, Mg$^{2+}$ deficiency was associated with a significant decline in body weight gain. Interestingly, regression analysis revealed a positive correlation between body weight gain and the downregulation of *MagT1*, *CNNM2*, *CNNM3*, *CNNM4*, *MRS2*, *Slc41a1* in the muscle. Indeed, redundancy among members of Mg$^{2+}$ transporters likely enables functional compensation to maintain sufficient Mg$^{2+}$ homeostasis resulting in normal body weight.

We found no structural signs of muscle atrophy after 14 days in Mg$^{2+}$ deficient regimen. This finding might be due to the downregulation of the genes coding for myostatin and its receptors, which activate the ubiquitin–proteasome and autophagy pathways, thus resulting in muscle wasting [32]. Accordingly, TLDA analysis revealed the reduced expression of genes involved in the regulation of the proteasome and autophagy in mice fed a moderately low Mg$^{2+}$ diet.

By TLDA analysis, we found significant differences in the expression of genes coding for proteins involved in energy metabolisms. Glucose transport into cells is the first step in glucose metabolism. We here describe the downregulation of the gene encoding the glucose transporter Glut4. Interestingly, in the gastrocnemius of type 2 diabetic rats, Glut4 was reduced, and Mg$^{2+}$ supplementation was sufficient to revert it [33]. The low expression of Glut4 in Mg$^{2+}$-deficient mice indicates that less glucose might be available for energy production, further impaired by the downregulation of *Citrate synthase*, involved in the first reaction of the Krebs cycle. Reduced amounts of *Plin2*, *Srebf1*, and *Srebf2* transcripts might be predictive of altered lipid metabolism. It is noteworthy that *Srebf1* and *Srebf2* are downregulated in the skeletal muscle of diabetic individuals [34] and that the overexpression of *Plin2* ameliorates insulin sensitivity in skeletal muscle [35]. Additionally, *Slc6a8*, coding for creatine transporter (CT)-1, is reduced in moderately Mg$^{2+}$-deficient mice. In the skeletal muscle, the creatine system is fundamental for optimal energy utilization, especially at the beginning of exercise and during intense physical activity, because it serves as a first-line energy buffer that maintains ATP levels constant. CT1 is the major route for creatine entry in skeletal muscle cells and has a central role in ensuring high intracellular creatine content [36]. Consistently, CT1-deficient mice feature muscle atrophy, reduced strength, and endurance [37].

TLDA also disclosed the reduced expression of genes involved in mitophagy, fusion, and fission, thus indicating alterations of mitochondrial dynamics that might be accompanied by impaired energy production in $Mg^{2+}$-deficient mice. Moreover, we found a marked reduction of the total amounts of Mfn2, a GTPase located on the outer mitochondrial membrane which is critical for mitochondrial fusion [38]. It is noteworthy that altered mitochondria have been implicated in sarcopenia in the elderly, in muscle atrophy associated with disuse, in muscular dystrophies, and in insulin resistance [24]. TLDA also shows the downregulation of genes that coordinate the different steps of autophagy in $Mg^{2+}$-deficient mice. This result is in agreement with data from cultured cells. Indeed, both *TRPM7* or *MagT1* silencing and $Mg^{2+}$ deficiency activate autophagy in human mesenchymal stem cells induced osteogenic differentiation [39]. Consistently, high concentrations of extracellular $Mg^{2+}$ inhibit autophagy in chondrocyte ATDC5 cells [40].

Our study presents some limitations. First of all, the experiments were performed using young growing animals. It will be relevant to extend our studies using animals of different ages and to evaluate metabolic parameters related to glucose metabolism and lipid metabolism to obtain a complete overview of the fundamental function exerted by skeletal muscle. We also highlight that our study was performed on male mice. To identify potential gender-related differences, the same experiments should be performed on females. Another limitation is that only some proteins were evaluated. Further experiments are required to confirm the altered expressions at the protein level and/or their functional activity.

## 5. Conclusions

Our results emphasize that even a mild $Mg^{2+}$ deficiency, as found in the Western population, is sufficient to modulate the gene expression of major pathways, mostly related to energy metabolism, proteostasis, autophagy, and mitochondrial dynamics in the skeletal muscle. Consequently, supplementing $Mg^{2+}$ in $Mg^{2+}$-deficient individuals might be a simple and costless countermeasure to maintain healthy muscles and metabolic balance.

**Author Contributions:** Conceptualization, A.M., D.B. (Daniel Béchet), and J.A.M.; data curation, D.B. (Dominique Bayle), A.M., and D.B. (Daniel Béchet); formal analysis, D.B. (Dominique Bayle), C.C.-G., M.Z., and M.M.-Z.; funding acquisition, S.C. and D.B. (Daniel Béchet); investigation, D.B. (Dominique Bayle) and C.C.-G.; Methodology, D.B. (Dominique Bayle), C.C.-G., M.G., S.C., M.Z., A.P.-S., and D.B. (Daniel Béchet); project administration, A.M., D.B. (Daniel Béchet), and J.A.M.; supervision, A.M., D.B. (Daniel Béchet), and J.A.M.; validation, S.C., A.M., D.B. (Daniel Béchet), and J.A.M.; writing—original draft preparation, D.B. (Daniel Béchet); writing—review and editing, A.M., D.B. (Daniel Béchet), and J.A.M. All authors have read and agreed to the published version of the manuscript.

**Funding:** This research was funded by the French government's IDEX-ISITE initiative 16-IDEX-0001 (CAP 20-25) and, in part, by Università di Milano (Fondi del Piano di Sviluppo di Ricerca 2020).

**Institutional Review Board Statement:** The study was conducted according to the guidelines of the Declaration of Helsinki and approved by the Ethics Committee C2EA-02 of INRAE (Protocol Code 14025-201803121538803 of 25 April 2018).

**Data Availability Statement:** The data presented in this study are openly available in INRA Dataverse at https://data.inrae.fr/dataset.xhtml?persistentId=doi:10.15454/IVBODD (accessed on 24 June 2021).

**Acknowledgments:** Acknowledgements to the animal facility staff in charge of the experimental protocol and to Séverine Valero for the technical assistance and plasma magnesium analysis. The authors acknowledge support from the University of Milan through the APC initiative.

**Conflicts of Interest:** The authors declare no conflict of interest.

## References

1. Cazzola, R.; Della Porta, M.; Manoni, M.; Iotti, S.; Pinotti, L.; Maier, J.A. Going to the roots of reduced magnesium dietary intake: A tradeoff between climate changes and sources. *Heliyon* **2020**, *6*, e05390. [CrossRef] [PubMed]
2. Lo Piano, F.; Corsonello, A.; Corica, F. Magnesium and elderly patient: The explored paths and the ones to be explored: A review. *Magnes. Res.* **2019**, *32*, 1–15. [CrossRef] [PubMed]
3. Kisters, K.; Gröber, U. Magnesium and thiazide diuretics. *Magnes. Res.* **2018**, *31*, 143–145. [CrossRef]
4. Hansen, B.A.; Bruserud, Ø. Hypomagnesemia in critically ill patients. *J. Intensive Care* **2018**, *6*, 1–11. [CrossRef]
5. Arnaud, M.J. Update on the assessment of magnesium status. *Br. J. Nutr.* **2008**, *99*, S24–S36. [CrossRef]
6. Witkowski, M.; Hubert, J.; Mazur, A. Methods of assessment of magnesium status in humans: A systematic review. *Magnes. Res.* **2011**, *24*, 163–180. [CrossRef]
7. de Baaij, J.H.F.; Hoenderop, J.G.J.; Bindels, R.J.M. Magnesium in Man: Implications for Health and Disease. *Physiol. Rev.* **2015**, *95*, 1–46. [CrossRef] [PubMed]
8. Giménez-Mascarell, P.; González-Recio, I.; Fernández-Rodríguez, C.; Oyenarte, I.; Müller, D.; Martínez-Chantar, M.L.; Martínez-Cruz, L.A. Current Structural Knowledge on the CNNM Family of Magnesium Transport Mediators. *Int. J. Mol. Sci.* **2019**, *20*, 1135. [CrossRef] [PubMed]
9. Kolisek, M.; Sponder, G.; Pilchova, I.; Cibulka, M.; Tatarkova, Z.; Werner, T.; Racay, P. Magnesium Extravaganza: A Critical Compendium of Current Research into Cellular Mg(2+) Transporters Other than TRPM6/7. *Rev. Physiol. Biochem. Pharmacol.* **2019**, *176*, 65–105. [CrossRef] [PubMed]
10. Carvil, P.; Cronin, J. Magnesium and implications on muscle function. *Strength Cond. J.* **2010**, *32*, 48–54. [CrossRef]
11. Coudy-Gandilhon, C.; Gueugneau, M.; Taillandier, D.; Combaret, L.; Polge, C.; Roche, F.; Barthélémy, J.-C.; Féasson, L.; Maier, J.A.; Mazur, A.; et al. Magnesium transport and homeostasis-related gene expression in skeletal muscle of young and old adults: Analysis of the transcriptomic data from the PROOF cohort Study. *Magnes. Res.* **2019**, *32*, 72–82. [CrossRef] [PubMed]
12. Gueugneau, M.; Coudy-Gandilhon, C.; Théron, L.; Meunier, B.; Barboiron, C.; Combaret, L.; Taillandier, D.; Polge, C.; Attaix, D.; Picard, B.; et al. Skeletal muscle lipid content and oxidative activity in relation to muscle fiber type in aging and metabolic syndrome. *J. Gerontol. A Biol. Sci. Med. Sci.* **2015**, *70*, 566–576. [CrossRef]
13. Gueugneau, M.; Coudy-Gandilhon, C.; Meunier, B.; Combaret, L.; Taillandier, D.; Polge, C.; Attaix, D.; Roche, F.; Féasson, L.; Barthélémy, J.-C.; et al. Lower skeletal muscle capillarization in hypertensive elderly men. *Exp. Gerontol.* **2016**, *76*, 80–88. [CrossRef] [PubMed]
14. Storey, J.D.; Tibshirani, R. Statistical significance for genomewide studies. *Proc. Natl. Acad. Sci. USA* **2003**, *100*, 9440–9445. [CrossRef] [PubMed]
15. Rondón, L.J.; Groenestege, W.M.T.; Rayssiguier, Y.; Mazur, A. Relationship between low magnesium status and TRPM6 expression in the kidney and large intestine. *Am. J. Physiol. Regul. Integr. Comp. Physiol.* **2008**, *294*, R2001–R2007. [CrossRef] [PubMed]
16. Ryazanova, L.V.; Rondon, L.J.; Zierler, S.; Hu, Z.; Galli, J.; Yamaguchi, T.P.; Mazur, A.; Fleig, A.; Ryazanov, A.G. TRPM7 is essential for Mg(2+) homeostasis in mammals. *Nat. Commun.* **2010**, *1*, 109. [CrossRef] [PubMed]
17. Li, F.Y.; Chaigne-Delalande, B.; Kanellopoulou, C.; Davis, J.C.; Matthews, H.F.; Douek, D.C.; Cohen, J.I.; Uzel, G.; Su, H.C.; Lenardo, M.J. Second messenger role for Mg2+ revealed by human T-cell immunodeficiency. *Nature* **2011**, *475*, 471–476. [CrossRef]
18. Blommaert, E.; Péanne, R.; Cherepanova, N.A.; Rymen, D.; Staels, F.; Jaeken, J.; Race, V.; Keldermans, L.; Souche, E.; Corveleyn, A.; et al. Mutations in MAGT1 lead to a glycosylation disorder with a variable phenotype. *Proc. Natl. Acad. Sci. USA* **2019**, *116*, 9865–9870. [CrossRef] [PubMed]
19. Cai, Y.; Masumiya, H.; Weisleder, N.; Matsuda, N.; Nishi, M.; Hwang, M.; Ko, J.-K.; Lin, P.; Thornton, A.; Zhao, X.; et al. MG53 nucleates assembly of cell membrane repair machinery. *Nat. Cell Biol.* **2009**, *11*, 56–64. [CrossRef] [PubMed]
20. Ebert, S.M.; Bullard, S.A.; Basisty, N.; Marcotte, G.R.; Skopec, Z.P.; Dierdorff, J.M.; Al-Zougbi, A.; Tomcheck, K.C.; DeLau, A.D.; Rathmacher, J.A.; et al. Activating transcription factor 4 (ATF4) promotes skeletal muscle atrophy by forming a heterodimer with the transcriptional regulator C/EBPβ. *J. Biol. Chem.* **2020**, *295*, 2787–2803. [CrossRef]
21. AlSudais, H.; Lala-Tabbert, N.; Wiper-Bergeron, N. CCAAT/Enhancer Binding Protein β inhibits myogenic differentiation via ID3. *Sci. Rep.* **2018**, *8*, 16613. [CrossRef]
22. Ebert, S.M.; Al-Zougbi, A.; Bodine, S.C.; Adams, C.M. Skeletal Muscle Atrophy: Discovery of Mechanisms and Potential Therapies. *Physiology* **2019**, *34*, 232–239. [CrossRef] [PubMed]
23. Larsson, L.; Degens, H.; Li, M.; Salviati, L.; Lee, Y., II; Thompson, W.; Kirkland, J.L.; Sandri, M. Sarcopenia: Aging-Related Loss of Muscle Mass and Function. *Physiol. Rev.* **2019**, *99*, 427–511. [CrossRef] [PubMed]
24. Gouspillou, G.; Hepple, R.T. Editorial: Mitochondria in Skeletal Muscle Health, Aging and Diseases. *Front. Physiol.* **2016**, *7*, 446. [CrossRef]
25. Leberer, E.; Timms, B.G.; Campbell, K.P.; MacLennan, D.H. Purification, calcium binding properties, and ultrastructural localization of the 53,000- and 160,000 (sarcalumenin)-dalton glycoproteins of the sarcoplasmic reticulum. *J. Biol. Chem.* **1990**, *265*, 10118–10124. [CrossRef]
26. Gueugneau, M.; Coudy-Gandilhon, C.; Gourbeyre, O.; Chambon, C.; Combaret, L.; Polge, C.; Taillandier, D.; Attaix, D.; Friguet, B.; Maier, A.B.; et al. Proteomics of muscle chronological ageing in post-menopausal women. *BMC Genom.* **2014**, *15*, 1165. [CrossRef] [PubMed]

27. Petermann-Rocha, F.; Chen, M.; Gray, S.R.; Ho, F.K.; Pell, J.P.; Celis-Morales, C. Factors associated with sarcopenia: A cross-sectional analysis using UK Biobank. *Maturitas* **2020**, *133*, 60–67. [CrossRef] [PubMed]
28. Ter Borg, S.; de Groot, L.C.P.G.M.; Mijnarends, D.M.; de Vries, J.H.M.; Verlaan, S.; Meijboom, S.; Luiking, Y.C.; Schols, J.M.G.A. Differences in Nutrient Intake and Biochemical Nutrient Status Between Sarcopenic and Nonsarcopenic Older Adults-Results From the Maastricht Sarcopenia Study. *J. Am. Med. Dir. Assoc.* **2016**, *17*, 393–401. [CrossRef]
29. Orsso, C.E.; Tibaes, J.R.B.; Oliveira, C.L.P.; Rubin, D.A.; Field, C.J.; Heymsfield, S.B.; Prado, C.M.; Haqq, A.M. Low muscle mass and strength in pediatrics patients: Why should we care? *Clin. Nutr.* **2019**, *38*, 2002–2015. [CrossRef]
30. Chubanov, V.; Ferioli, S.; Wisnowsky, A.; Simmons, D.G.; Leitzinger, C.; Einer, C.; Jonas, W.; Shymkiv, Y.; Bartsch, H.; Braun, A.; et al. Epithelial magnesium transport by TRPM6 is essential for prenatal development and adult survival. *Elife* **2016**, *5*. [CrossRef]
31. Franken, G.A.C.; Seker, M.; Bos, C.; Siemons, L.A.H.; van der Eerden, B.C.J.; Christ, A.; Hoenderop, J.G.J.; Bindels, R.J.M.; Müller, D.; Breiderhoff, T.; et al. Cyclin M2 (CNNM2) knockout mice show mild hypomagnesaemia and developmental defects. *Sci. Rep.* **2021**, *11*, 1–12. [CrossRef] [PubMed]
32. Milan, G.; Romanello, V.; Pescatore, F.; Armani, A.; Paik, J.H.; Frasson, L.; Seydel, A.; Zhao, J.; Abraham, R.; Goldberg, A.L.; et al. Regulation of autophagy and the ubiquitin-proteasome system by the FoxO transcriptional network during muscle atrophy. *Nat. Commun.* **2015**, *6*, 1–14. [CrossRef]
33. Morakinyo, A.O.; Samuel, T.A.; Adekunbi, D.A. Magnesium upregulates insulin receptor and glucose transporter-4 in streptozotocin-nicotinamide-induced type-2 diabetic rats. *Endocr. Regul.* **2018**, *52*, 6–16. [CrossRef] [PubMed]
34. Sewter, C.; Berger, D.; Considine, R.V.; Medina, G.; Rochford, J.; Ciaraldi, T.; Henry, R.; Dohm, L.; Flier, J.S.; O'Rahilly, S.; et al. Human obesity and type 2 diabetes are associated with alterations in SREBP1 isoform expression that are reproduced ex vivo by tumor necrosis factor-alpha. *Diabetes* **2002**, *51*, 1035–1041. [CrossRef]
35. Bosma, M.; Hesselink, M.K.C.; Sparks, L.M.; Timmers, S.; Ferraz, M.J.; Mattijssen, F.; van Beurden, D.; Schaart, G.; de Baets, M.H.; Verheyen, F.K.; et al. Perilipin 2 improves insulin sensitivity in skeletal muscle despite elevated intramuscular lipid levels. *Diabetes* **2012**, *61*, 2679–2690. [CrossRef] [PubMed]
36. Brault, J.J.; Abraham, K.A.; Terjung, R.L. Muscle creatine uptake and creatine transporter expression in response to creatine supplementation and depletion. *J. Appl. Physiol.* **2003**, *94*, 2173–2180. [CrossRef]
37. Stockebrand, M.; Sasani, A.; Das, D.; Hornig, S.; Hermans-Borgmeyer, I.; Lake, H.A.; Isbrandt, D.; Lygate, C.A.; Heerschap, A.; Neu, A.; et al. A Mouse Model of Creatine Transporter Deficiency Reveals Impaired Motor Function and Muscle Energy Metabolism. *Front. Physiol.* **2018**, *9*, 773. [CrossRef]
38. Filadi, R.; Pendin, D.; Pizzo, P. Mitofusin 2: From functions to disease. *Cell Death Dis.* **2018**, *9*, 1–13. [CrossRef] [PubMed]
39. Castiglioni, S.; Romeo, V.; Locatelli, L.; Cazzaniga, A.; Maier, J.A.M. TRPM7 and MagT1 in the osteogenic differentiation of human mesenchymal stem cells in vitro. *Sci. Rep.* **2018**, *8*, 1–10. [CrossRef]
40. Yue, J.; Jin, S.; Gu, S.; Sun, R.; Liang, Q. High concentration magnesium inhibits extracellular matrix calcification and protects articular cartilage via Erk/autophagy pathway. *J. Cell. Physiol.* **2019**, *234*, 23190–23201. [CrossRef]

Article

# Magnesium Influences Membrane Fusion during Myogenesis by Modulating Oxidative Stress in C2C12 Myoblasts

Monica Zocchi [1], Daniel Béchet [2], André Mazur [2], Jeanette A. Maier [1,3] and Sara Castiglioni [1,*]

1. Department of Biomedical and Clinical Sciences L. Sacco, Università di Milano, Via G.B. Grassi 74, 20157 Milano, Italy; monica.zocchi@unimi.it (M.Z.); jeanette.maier@unimi.it (J.A.M.)
2. INRAE, UNH, Unitéde Nutrition Humaine, Université Clermont Auvergne, 63001 Clermont-Ferrand, France; daniel.bechet@inrae.fr (D.B.); andre.mazur@inrae.fr (A.M.)
3. Interdisciplinary Centre for Nanostructured Materials and Interfaces (CIMaINa), Università di Milano, 20133 Milan, Italy
* Correspondence: sara.castiglioni@unimi.it

**Abstract:** Magnesium (Mg) is essential to skeletal muscle where it plays a key role in myofiber relaxation. Although the importance of Mg in the mature skeletal muscle is well established, little is known about the role of Mg in myogenesis. We studied the effects of low and high extracellular Mg in C2C12 myogenic differentiation. Non-physiological Mg concentrations induce oxidative stress in myoblasts. The increase of reactive oxygen species, which occurs during the early phase of the differentiation process, inhibits myoblast membrane fusion, thus impairing myogenesis. Therefore, correct Mg homeostasis, also maintained through a correct dietary intake, is essential to assure the regenerative capacity of skeletal muscle fibers.

**Keywords:** magnesium; myogenesis; oxidative stress; membrane fusion

## 1. Introduction

Magnesium (Mg) is the second most abundant cation within the intracellular compartment of the human body and is essential to all living cells, including skeletal myocytes [1–3]. Almost 25% of total Mg is contained in skeletal muscles where it plays a key role in myofiber relaxation by acting as a calcium (Ca) antagonist on Ca-permeable channels and Ca-binding proteins. It is known that a correct dietary Mg intake is positively associated with muscle strength and health [2] and that a Mg-deficient condition is associated with muscle cramps, spasms, weakness, and a higher risk of developing age-related sarcopenia [1,4]. Although the importance of Mg in the mature skeletal muscle is well established, little is known about the role of Mg in myoblasts and during myogenesis, i.e., process leading to muscle generation [5].

Skeletal muscle is a highly complex and heterogeneous tissue, the largest in the body. In the embryo, the first muscle fibers arise from the somites, transient structures which originate from the paraxial mesoderm. Additional fibers are generated from myogenic progenitors, which initially proliferate extensively and, once the muscle has matured, enter quiescence and reside as satellite cells between basal lamina and sarcolemma. Satellite cells use asymmetric divisions for self-maintenance and, at the same time, for generating a myogenic progeny with the potential to differentiate into new fibers. The differentiation process is coordinated by the sequential and coordinated expression of myogenic regulatory factors (MRFs), including MyoD and Myf5, which commit the cells to the myogenic program, and myogenin (Myog) and MRF4, which are responsible for the terminal stages of differentiation. Finally, differentiated myoblasts fuse with each other to form multiple nuclear myotubes [6,7]. In the adult, the satellite cells guarantee tissue renewal and repair.

Reactive oxygen species (ROS) are important players in the regulation of myogenesis [8,9]. ROS act in a hormetic fashion because they may exert beneficial or detrimental

action depending on their amount and the duration of their production. A moderate increase of ROS is necessary for myogenesis. On the other hand, higher ROS levels can affect the efficiency of myogenic differentiation.

In this study we analyzed the effects of physiological (1 mM), low (0.1 mM) and high (3, 6, and 10 mM) extracellular Mg concentrations on myogenesis in C2C12 cells, an in vitro model of murine myoblasts which are able to efficiently differentiate in myotubes under specific culture conditions [10,11].

## 2. Materials and Methods

### 2.1. Cell Culture

C2C12 cells are murine proliferating myoblasts that, after serum depletion, differentiate to generate multinucleated myotubes [12]. The C2C12 cell line was purchased from American Type Culture Collection (ATCC). The cells were serially passaged in culture medium (CM) composed of Dulbecco's Modified Eagle Medium (DMEM) containing high glucose and 20% of heat-inactivated fetal bovine serum (FBS), glutamine (2 mM), and 1% penicillin/streptomycin according to the manufacture's instruction. To induce myogenic differentiation, 25,000 cells/cm$^2$ were seeded. 24 h later, they were exposed to a differentiation medium (DM) consisting of DMEM with high glucose supplemented with 2% horse serum. In our experiments, the cells were cultured in DM containing low (0.1 mM), physiological (1 mM) or high (3, 6, and 10 mM) MgSO$_4$ concentrations.

N-Acetylcysteine (NAC) (Sigma-Aldrich, St. Louis, MO, USA) was used to inhibit ROS production by pre-treating the cells for 24 h at the final concentration of 10 mM. The concentration was defined by performing a dose-response cell viability assay (data not shown). Images of cultured cells were acquired by optical microscopy with a Zeiss Primovert phase-contrast microscope (10× objective).

### 2.2. SDS-PAGE and Western Blot

C2C12 cells were lyzed in lysis buffer (50 mM Tris-HCl pH 7.4, 150 mM NaCl, 1% NP-40, 0.25% Na-deoxycholate) containing protease inhibitors. A syringe was used to better homogenize the lysates. Total proteins were quantified using the Bradford reagent (Sigma-Aldrich, St. Louis, MO, USA). Equal amounts of proteins were separated by SDS-PAGE on 4–20% Mini-PROTEAN TGX Stain-free Gels (Bio-Rad, Hercules, CA, USA) and transferred to nitrocellulose membranes by using Trans-Blot® TurboTM Transfer Pack (Bio-Rad, Hercules, CA, USA). After blocking with bovine serum albumin (BSA), Western blot analysis was performed using primary antibodies against Myosin Heavy Chain (MHC) (R&D Systems, Minneapolis, MN, USA), Caveolin-3 (BD Biosciences, San Diego, California USA), Myomixer (R&D Systems, Minneapolis, MN, USA), and β-actin (Santa-Cruz Biotechnology, Dallas, TX, USA). After extensive washing, secondary antibodies conjugated with horseradish peroxidase (Amersham Pharmacia Biotech Italia, Cologno Monzese, Italy) were used. The immunoreactive proteins were detected with ClarityTM Western ECL substrate (Bio-Rad, Hercules, CA, USA) and images were captured with a ChemiDoc MP Imaging System (Bio-Rad, Hercules, CA, USA). Densitometry of the bands was performed with the software ImageLab (Bio-Rad, Hercules, CA, USA). The Western blots shown are representative and the densitometric analysis was performed on three independent experiments.

### 2.3. Immunofluorescence

The immunofluorescence staining and imaging were performed directly in culture wells. Cells were seeded in 24-well plates and, after 144 h of differentiation, fixed for 15 min in phosphate-buffered saline (PBS) containing 4% paraformaldehyde and 2% sucrose (pH 7.6). Cells were permeabilized and blocked for 30 min in a PBS solution containing 2% BSA and 0.3% Triton. To stain myotubes, the cells were incubated with anti-MHC primary antibody (R&D Systems, Minneapolis, MN, USA) and with an Alexa Fluor 488 secondary antibody (Thermo Fisher Scientific, Waltham, MA, USA). The nuclei were stained with 4′,6-diamidino-2-phenylindole (DAPI). Images were acquired using FLoid™ Cell Imaging Station (Thermo Fisher Scientific, Waltham, MA, USA).

### 2.4. ROS Production Analysis

For the detection of ROS [13], C2C12 cells were cultured in 96-well black plates (Greiner bio-one, Frickenhausen, Germany). At the end of the experiments, cells were incubated with 10 mM 2′-7′-dichlorofluorescein diacetate (DCFDA) (Thermo Fisher Scientific, Waltham, MA, USA) solution for 30 min at 37 °C. DCFDA was deacetylated by cellular esterases to a non-fluorescent compound which was then oxidized by ROS into the fluorescent molecule 2′,7′–dichlorofluorescein (DCF) ($\lambda$exc = 495 nm, $\lambda$emm = 529 nm). The dye fluorescent emission was measured using Varioskan LUX Multimode Microplate Reader (Thermo Fisher Scientific, Waltham, MA, USA). DCF fluorescence was normalized on cell counts. Images of DCF fluorescence emission were acquired on living cells cultured in 24-well plates. At the end of the experiment, cells were incubated for 30 min with the dye and images were captured with FLoid™ Cell Imaging Station (Thermo Fisher Scientific, Waltham, MA, USA). The results shown are the mean of three independent experiments performed in triplicate.

### 2.5. Statistical Analysis

Data are expressed as the mean ± standard deviation. The data were non-parametric and normally distributed and were analyzed using one-way ANOVA. The $p$-values deriving from multiple pairwise comparisons were corrected using the Bonferroni method. Statistical significance was defined as $p$-value $\leq 0.05$. In the figures, * $p \leq 0.05$; ** $p \leq 0.01$; *** $p \leq 0.001$; **** $p \leq 0.0001$.

## 3. Results
### 3.1. Low and High Extracellular Mg Impair Myogenesis

After 144 h of differentiation in low (0.1 mM), physiological (1 mM), and high (3, 6, and 10 mM) Mg concentrations, we analyzed C2C12 cells by optical microscope and by immunofluorescence using antibodies against Myosin Heavy Chain (MHC), a contractile protein only expressed in differentiated myotubes. As shown in Figure 1, we detected a significant reduction in the number of myotubes (Figure 1a) and in the fusion index values (nuclei in myotubes vs. total nuclei) (Figure 1b) both in low and high extracellular Mg vs. the control cells differentiated in physiological Mg concentrations. By Western blot (Figure 1c), we confirmed the reduction of the MHC expression in low and high Mg, suggesting that non-physiological extracellular concentrations of Mg impair myogenesis.

### 3.2. Low and High Mg Induce ROS Accumulation during Myogenesis

Because ROS are critical signaling molecules involved in muscle differentiation [8], we measured their amounts by DCF assay during myogenesis of C2C12 cells in low- and high-Mg conditions. We found an increase of ROS after 24 h of serum deprivation both in low- and high-Mg conditions compared to the physiological condition, while we did not detect ROS modulation after 72 and 144 h of cell differentiation (Figure 2).

### 3.3. ROS Accumulation Is Involved in the Impairment of Myogenesis

To understand if the early ROS increase in low and high Mg is involved in the impairment of myogenesis, we analyzed the differentiation process after treating the cells with the antioxidant N-Acetylcysteine (NAC), a glutathione precursor, before inducing the cells to differentiate. By DCF assay, we demonstrated that NAC efficiently prevents ROS increase at 24 h of culture in low and high Mg (Figure 3a). After 144 h, the myotubes were analyzed by optical microscopy and by immunofluorescence of MHC. No significant differences were observed in C2C12 cells exposed to different Mg concentrations in the presence of NAC (Figure 3b). Moreover, after NAC pre-treatment, the fusion index values in low and high Mg were comparable to the controls (Figure 3c). Western blot in Figure 3d shows no appreciable differences in MHC total amounts in the presence of NAC under all the Mg conditions tested.

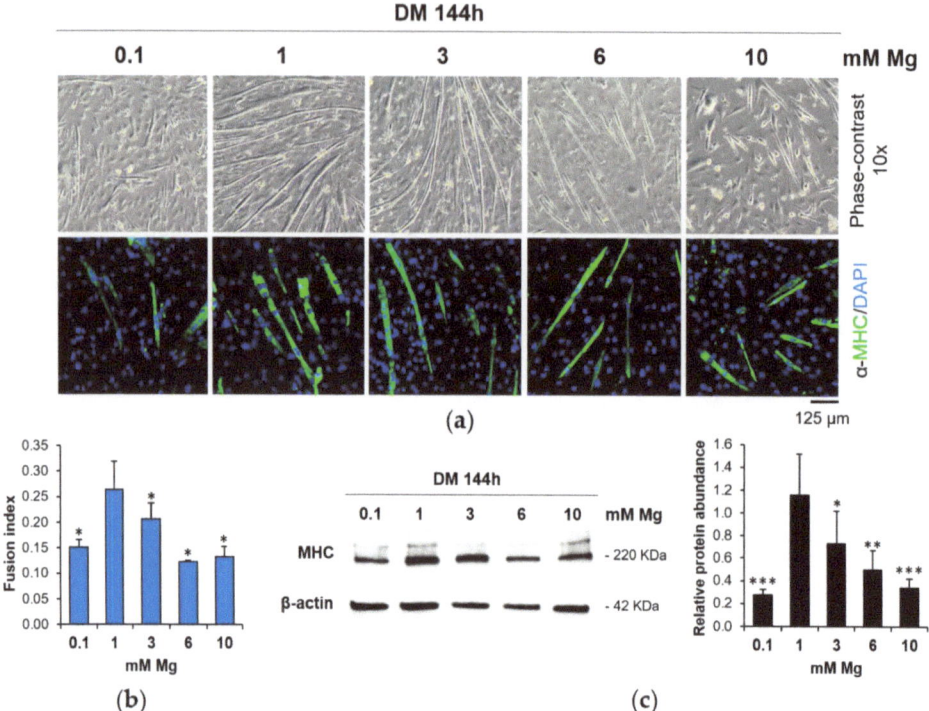

**Figure 1.** Low and high Mg concentrations inhibit myogenesis. C2C12 cells were cultured in differentiation medium (DM) for 144 h in the presence of different extracellular concentrations of Mg. (**a**) Pictures were taken with optical microscope (10× magnification, upper panels). After immunofluorescence with antibodies against Myosin Heavy Chain (MHC; green fluorescence), images were acquired using a fluorescence microscope (10× magnification, lower panels). The nuclei were stained with 4′,6-diamidino-2-phenylindole (DAPI). (**b**) Fusion index was calculated as the ratio of the number of nuclei within myotubes (>2 nuclei) to the total number of nuclei in the field and quantified based on (**a**). (**c**) MHC levels were analyzed by Western blot. β-actin was used as control of loading. A representative blot (left) and densitometry performed on three independent experiments and obtained by ImageLab (right) are shown. * Indicates significance with respect to 1 mM Mg (* $p \leq 0.05$; ** $p \leq 0.01$; *** $p \leq 0.001$).

*3.4. Low and High Mg Impair Myoblasts Fusion*

Because our data demonstrate that low and high Mg concentrations decrease the fusion index value, we investigated the levels of two key proteins involved in myoblast membrane fusion, Caveolin-3 and Myomixer, after 144 h of differentiation in the presence of different Mg concentrations. As shown in Figure 4, low- and high-Mg conditions downregulate both Caveolin-3 and Myomixer. Importantly, this effect is prevented by treating the cells with NAC.

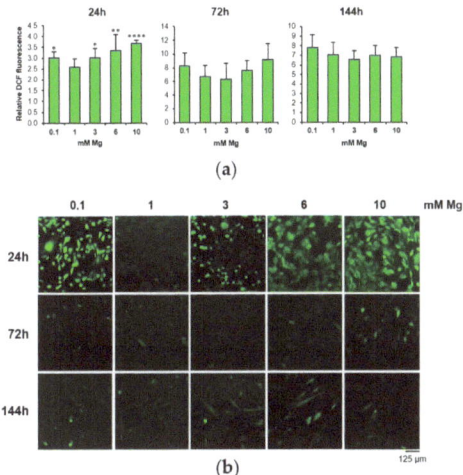

**Figure 2.** Low and high Mg induce ROS during myogenesis. After 24, 72 and 144 h of culture in DM in the presence of different extracellular concentrations of Mg, ROS accumulation was measured by 2′,7′–dichlorofluorescein (DCF) assay. Fluorescence was normalized to the cell number (**a**). Images of DCF fluorescence emission were acquired on living cells (10× magnification) (**b**). * Indicates significance with respect to 1 mM Mg (* $p \leq 0.05$; ** $p \leq 0.01$; **** $p \leq 0.0001$).

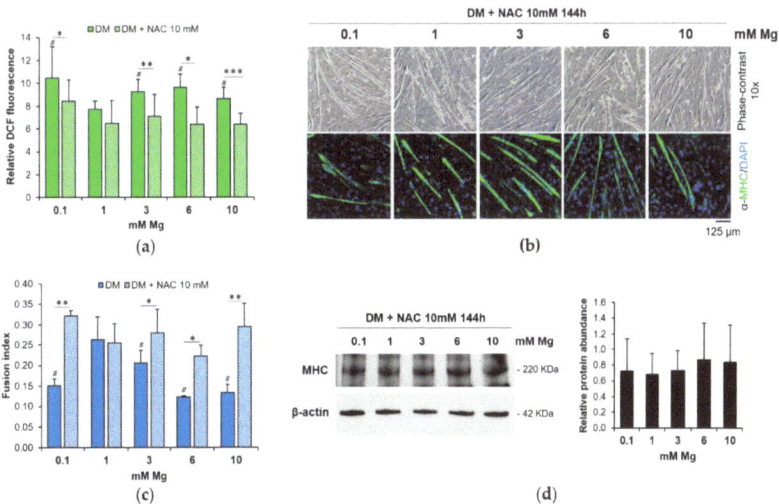

**Figure 3.** The antioxidant N-Acetylcysteine (NAC) prevents the impairment of myogenesis in low and high Mg. C2C12 cells were treated or not for 24 h with NAC in culture medium (CM) and then cultured in DM for 144 h with different extracellular concentrations of Mg. (**a**) ROS production was evaluated using DCF assay and the fluorescence was normalized to the cell number. (**b**) Upper panels: optical microscopy (10× magnification); lower panels: fluorescence microscopy after staining with antibodies against MHC (10× magnification). The nuclei were stained with DAPI. (**c**) Fusion index was quantified based on (**b**). (**d**) MHC total amounts were analyzed by Western blot. β-actin was used as control of loading. A representative blot (left) and the densitometry obtained by ImageLab (right) are shown. # Indicates significance with respect to DM 1 mM Mg (# $p \leq 0.05$). * indicates significance between DM and respective DM + NAC 10 mM (* $p \leq 0.05$; ** $p \leq 0.01$; *** $p \leq 0.001$).

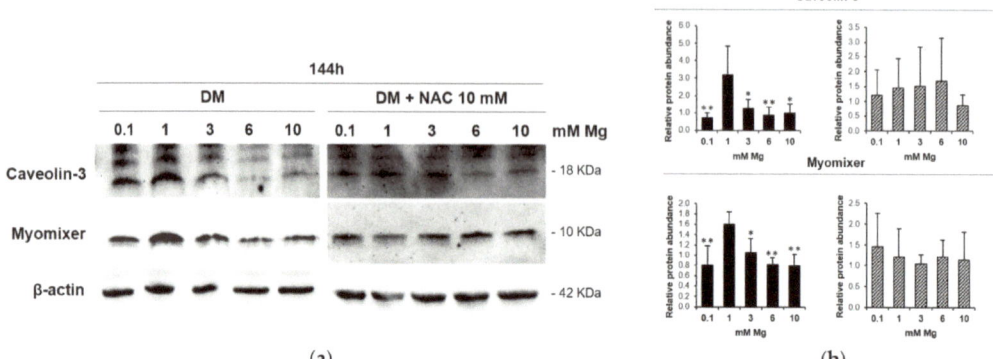

**Figure 4.** Low and high Mg inhibit myoblast fusion. C2C12 cells were treated or not for 24 h with NAC in CM and then cultured in DM for 144 h with different extracellular concentrations of Mg. Caveolin-3 and Myomixer expression was analysed by Western blot. β-actin was used as control of loading. A representative blot (**a**) and densitometry obtained by ImageLab (**b**) are shown. * Indicates significance with respect to 1 mM Mg (* $p \leq 0.05$; ** $p \leq 0.01$).

## 4. Discussion

Magnesium is an essential mineral for human health and its homeostasis is regulated by the balance between intestinal absorption and renal excretion [14]. An insufficient dietary Mg intake, a long-term alcohol abuse, the chronic use of some drugs, and pre-existing pathologies, such as diabetes, can lead to hypomagnesemia [1]. On the other hand, patients with acute or chronic kidney disease, hypothyroidism, and especially cortico-adrenal insufficiency can develop hypermagnesemia [15]. Both hypomagnesemia and hypermagnesemia impact on cell, tissue, and organ functions, leading to many disorders, among which neuromuscular disorders. However, very little is known about the molecular effects of variations in Mg homeostasis on skeletal muscle cells.

Here we studied how low and high extracellular Mg concentrations impact on the process of muscle fiber formation. C2C12 murine myoblasts were induced to differentiate into myotubes using media with different Mg conditions. We demonstrate that low or high extracellular concentrations of Mg impair myogenesis because we observed a reduction in the number of myotubes and in the fusion index values compared to the physiological condition. We also detected an increase of ROS in low and high Mg during the early phase of the differentiation process, while after 72 and 144 h of differentiation, no differences in the amounts of ROS were found. We speculate that some antioxidants' mechanisms are activated as an adaptive response to guarantee cell viability. It is known that ROS are implicated in different metabolic processes, including skeletal muscle differentiation. A moderate generation of ROS, in combination with growth factors and chemokines, is necessary for muscle regeneration and repair [16,17]. On the contrary, higher amounts of ROS might target mitochondria and mitochondrial DNA, inducing the block of myogenesis [8,18]. Moreover, higher ROS levels induce a depletion of intracellular glutathione which in turn further increases ROS accumulation. ROS increase induces NF-kB activation which contributes to lowering the expression of MyoD, thereby inhibiting myogenesis [19]. Coherently, we show that the ROS increase in low and high Mg is directly responsible for the impairment of myogenesis because the antioxidant NAC prevents the detrimental effect of low and high Mg on myogenesis, both in terms of myotube formation and fusion index. Moreover, after treatment with NAC, no significant differences in MHC expression were detected in all the Mg conditions tested.

Mg deficiency has been already demonstrated to induce oxidative stress in different cell types [20,21]. Interestingly, in myoblasts we also observed oxidative stress in high-Mg conditions. We can speculate that both low and high Mg could reduce the activity of some antioxidant enzymes, fundamental to maintaining ROS below a physiological level, and/or increase pro-oxidant molecules. Indeed, at low levels, ROS act as signaling molecules without exerting toxic effects, while at high levels they have detrimental effects.

To better understand the effects of extracellular Mg variations on the differentiation process, we focused on myoblast membrane fusion and detected a reduction in the amounts of two proteins involved in myoblast fusion, namely, Caveolin-3 and Myomixer. Indeed, lack of Caveolin-3 expression is sufficient to severely affect the fusion process during myogenesis [22,23]. Myomixer is a newly discovered muscle-specific membrane peptide with a fundamental role in the fusion pore formation [24,25]. Moreover, the antioxidant NAC rescues the levels of both Caveolin-3 and Myomixer in the cells induced to differentiate in low and high Mg. For this reason, we can assert that, in our model, ROS increase occurring in low- and high-Mg conditions impairs myogenesis by inhibiting the fusion process.

In conclusion, non-physiological extracellular Mg concentrations induce oxidative stress which affects myogenesis of C2C12 cells by inhibiting myoblasts' fusion. Our data suggest that adequate Mg dietary intake and the maintenance of Mg homeostasis are necessary to guarantee correct skeletal muscle plasticity and the regenerative capacity of fibers, thus contributing to skeletal muscle health.

**Author Contributions:** Conceptualization, S.C., A.M., D.B. and J.A.M.; methodology, M.Z.; formal analysis, S.C., J.A.M.; data curation, M.Z.; writing—original draft preparation, S.C. and J.A.M.; writing—review and editing, S.C., J.A.M., A.M. and D.B.; funding acquisition, S.C., J.A.M., A.M. and D.B. All authors have read and agreed to the published version of the manuscript.

**Funding:** This research was funded, in part, by Università di Milano (Fondi del Piano di sviluppo di ricerca 2020) and the French government's IDEX-ISITE initiative 16-IDEX-0001 (CAP 20–25).

**Institutional Review Board Statement:** Not applicable.

**Informed Consent Statement:** Not applicable.

**Data Availability Statement:** The data presented in this study are openly available in Dataverse at https://dataverse.unimi.it/dataverse/nutrients/ (accessed on 1 March 2021).

**Conflicts of Interest:** The authors declare no conflict of interest.

# References

1. de Baaij, J.H.F.; Hoenderop, J.G.J.; Bindels, R.J.M. Magnesium in Man: Implications for Health and Disease. *Physiol. Rev.* **2015**, *95*, 1–46. [CrossRef]
2. Welch, A.A.; Kelaiditi, E.; Jennings, A.; Steves, C.J.; Spector, T.D.; MacGregor, A. Dietary Magnesium Is Positively Associated with Skeletal Muscle Power and Indices of Muscle Mass and May Attenuate the Association between Circulating C-Reactive Protein and Muscle Mass in Women. *J. Bone Miner. Res.* **2016**, *31*, 317–325. [CrossRef]
3. Welch, A.A.; Skinner, J.; Hickson, M. Dietary magnesium may be protective for aging of bone and skeletal muscle in middle and younger older age men and women: Cross-sectional findings from the UK biobank cohort. *Nutrients* **2017**, *9*, 1189. [CrossRef]
4. Knochel, J.P.; Cronin, R.E. The myopathy of experimental magnesium deficiency. *Adv. Exp. Med. Biol.* **1984**, *178*, 351–361. [CrossRef]
5. Molkentin, J.D.; Olson, E.N. Defining the regulatory networks for muscle development. *Curr. Opin. Genet. Dev.* **1996**, *6*, 445–453. [CrossRef]
6. Berkes, C.A.; Tapscott, S.J. MyoD and the transcriptional control of myogenesis. *Semin. Cell Dev. Biol.* **2005**, *16*, 585–595. [CrossRef]
7. Taylor, M.V.; Hughes, S.M. Mef2 and the skeletal muscle differentiation program. *Semin. Cell Dev. Biol.* **2017**, *72*, 33–44. [CrossRef] [PubMed]
8. Barbieri, E.; Sestili, P. Reactive oxygen species in skeletal muscle signaling. *J. Signal Transduct.* **2012**, *2012*, 982794. [CrossRef] [PubMed]
9. Malinska, D.; Kudin, A.P.; Bejtka, M.; Kunz, W.S. Changes in mitochondrial reactive oxygen species synthesis during differentiation of skeletal muscle cells. *Mitochondrion* **2012**, *12*, 144–148. [CrossRef]
10. Andrés, V.; Walsh, K. Myogenin expression, cell cycle withdrawal, and phenotypic differentiation are temporally separable events that precede cell fusion upon myogenesis. *J. Cell Biol.* **1996**, *132*, 657–666. [CrossRef] [PubMed]
11. Burattini, S.; Ferri, R.; Battistelli, M.; Curci, R.; Luchetti, F.; Falcieri, E. C2C12 murine myoblasts as a model of skeletal muscle development: Morpho-functional characterization. *Eur. J. Histochem.* **2004**, *48*, 223–233. [CrossRef]
12. Cazzaniga, A.; Ille, F.; Wuest, S.; Haack, C.; Koller, A.; Giger-Lange, C.; Zocchi, M.; Egli, M.; Castiglioni, S.; Maier, J.A. Scalable microgravity simulator used for long-term musculoskeletal cells and tissue engineering. *Int. J. Mol. Sci.* **2020**, *21*, 8908. [CrossRef]

13. Mammoli, F.; Castiglioni, S.; Parenti, S.; Cappadone, C.; Farruggia, G.; Iotti, S.; Davalli, P.; Maier, J.A.M.; Grande, A.; Frassineti, C. Magnesium is a key regulator of the balance between osteoclast and osteoblast differentiation in the presence of vitamin D 3. *Int. J. Mol. Sci.* **2019**, *20*, 385. [CrossRef] [PubMed]
14. Al Alawi, A.M.; Majoni, S.W.; Falhammar, H. Magnesium and Human Health: Perspectives and Research Directions. *Int. J. Endocrinol.* **2018**, *2018*, 1–17. [CrossRef] [PubMed]
15. Galán, I.; Vega, A.; Goicoechea, M.; Shabaka, A.; Gatius, S.; Abad, S.; López-gómez, J.M. Hypermagnesemia is Associated with All-Cause Mortality in Patients with Chronic Kidney Disease. *J. Clin. Exp. Nephrol.* **2020**, 1–8. [CrossRef]
16. Kozakowska, M.; Pietraszek-Gremplewicz, K.; Jozkowicz, A.; Dulak, J. The role of oxidative stress in skeletal muscle injury and regeneration: Focus on antioxidant enzymes. *J. Muscle Res. Cell Motil.* **2015**, *36*, 377–393. [CrossRef]
17. Rajasekaran, N.S.; Shelar, S.B.; Jones, D.P.; Hoidal, J.R. Reductive stress impairs myogenic differentiation. *Redox Biol.* **2020**, *34*, 101492. [CrossRef]
18. Rochard, P.; Rodier, A.; Casas, F.; Cassar-Malek, I.; Marchal-Victorion, S.; Daury, L.; Wrutniak, C.; Cabello, G. Mitochondrial activity is involved in the regulation of myoblast differentiation through myogenin expression and activity of myogenic factors. *J. Biol. Chem.* **2000**, *275*, 2733–2744. [CrossRef]
19. Ardite, E.; Barbera, J.A.; Roca, J.; Fernández-Checa, J.C. Glutathione depletion impairs myogenic differentiation of murine skeletal muscle C2C12 cells through sustained NF-kappaB activation. *Am. J. Pathol.* **2004**, *165*, 719–728. [CrossRef]
20. Sargenti, A.; Castiglioni, S.; Olivi, E.; Bianchi, F.; Cazzaniga, A.; Farruggia, G.; Cappadone, C.; Merolle, L.; Malucelli, E.; Ventura, C.; et al. Magnesium deprivation potentiates human mesenchymal stem cell transcriptional remodeling. *Int. J. Mol. Sci.* **2018**, *19*, 1410. [CrossRef] [PubMed]
21. Locatelli, L.; Fedele, G.; Castiglioni, S.; Maier, J.A. Magnesium deficiency induces lipid accumulation in vascular endothelial cells via oxidative stress—the potential contribution of edf-1 and ppary. *Int. J. Mol. Sci.* **2021**, *22*, 1050. [CrossRef] [PubMed]
22. Galbiati, F.; Volonté, D.; Engelman, J.A.; Scherer, P.E.; Lisanti, M.P. Targeted down-regulation of caveolin-3 is sufficient to inhibit myotube formation in differentiating C2C12 myoblasts. Transient activation of p38 mitogen-activated protein kinase is required for induction of caveolin-3 expression and subsequent myotube fo. *J. Biol. Chem.* **1999**, *274*, 30315–30321. [CrossRef]
23. Volonte, D.; Peoples, A.J.; Galbiati, F. Modulation of Myoblast Fusion by Caveolin-3 in Dystrophic Skeletal Muscle Cells: Implications for Duchenne Muscular Dystrophy and Limb-Girdle Muscular Dystrophy-1C. *Mol. Biol. Cell* **2003**, *14*, 4075–4088. [CrossRef] [PubMed]
24. Bi, P.; Ramirez-Martinez, A.; Li, H.; Cannavino, J.; McAnally, J.R.; Shelton, J.M.; Sánchez-Ortiz, E.; Bassel-Duby, R.; Olson, E.N. Control of muscle formation by the fusogenic micropeptide myomixer. *Science* **2017**, *356*, 323–327. [CrossRef] [PubMed]
25. Chen, B.; You, W.; Wang, Y.; Shan, T. The regulatory role of Myomaker and Myomixer–Myomerger–Minion in muscle development and regeneration. *Cell. Mol. Life Sci.* **2020**, *77*, 1551–1569. [CrossRef]

Article

# Assessment and Imaging of Intracellular Magnesium in SaOS-2 Osteosarcoma Cells and Its Role in Proliferation

Concettina Cappadone [1,*,†], Emil Malucelli [1,†], Maddalena Zini [2], Giovanna Farruggia [1,3], Giovanna Picone [1], Alessandra Gianoncelli [4], Andrea Notargiacomo [5], Michela Fratini [6,7], Carla Pignatti [2], Stefano Iotti [1,3,‡] and Claudio Stefanelli [8,‡]

1. Department of Pharmacy and Biotechnology, University of Bologna, 33, 40127 Bologna, Italy; emil.malucelli@unibo.it (E.M.); giovanna.farruggia@unibo.it (G.F.); giovanna.picone2@unibo.it (G.P.); stefano.iotti@unibo.it (S.I.)
2. Department of Biomedical and Neuromotor Sciences, Alma Mater Studiorum, University of Bologna, 33, 40126 Bologna, Italy; maddalena.zini@unibo.it (M.Z.); carla.pignatti@unibo.it (C.P.)
3. INBB—Biostructures and Biosystems National Institute, 00136 Rome, Italy
4. Elettra—Sincrotrone Trieste, 34149 Trieste, Italy; alessandra.gianoncelli@elettra.eu
5. Institute for Photonics and Nanotechnologies, Consiglio Nazionale delle Ricerche, 00156 Rome, Italy; andrea.notargiacomo@ifn.cnr.it
6. Institute of Nanotechnology-CNR c/o Physics Department at 'Sapienza' University, 00185 Rome, Italy; michela.fratini@gmail.com
7. IRCCS Fondazione Santa Lucia, 00179 Rome, Italy
8. Department for Life Quality Studies, Alma Mater Studiorum, University of Bologna, 47921 Rimini, Italy; claudio.stefanelli@unibo.it
* Correspondence: concettina.cappadone@unibo.it
† These authors contributed equally to this work.
‡ These authors contributed equally to this work.

Citation: Cappadone, C.; Malucelli, E.; Zini, M.; Farruggia, G.; Picone, G.; Gianoncelli, A.; Notargiacomo, A.; Fratini, M.; Pignatti, C.; Iotti, S.; et al. Assessment and Imaging of Intracellular Magnesium in SaOS-2 Osteosarcoma Cells and Its Role in Proliferation. *Nutrients* 2021, 13, 1376. https://doi.org/10.3390/nu13041376

Academic Editor: Mario Barbagallo

Received: 14 March 2021
Accepted: 17 April 2021
Published: 20 April 2021

**Publisher's Note:** MDPI stays neutral with regard to jurisdictional claims in published maps and institutional affiliations.

Copyright: © 2021 by the authors. Licensee MDPI, Basel, Switzerland. This article is an open access article distributed under the terms and conditions of the Creative Commons Attribution (CC BY) license (https:// creativecommons.org/licenses/by/ 4.0/).

**Abstract:** Magnesium is an essential nutrient involved in many important processes in living organisms, including protein synthesis, cellular energy production and storage, cell growth and nucleic acid synthesis. In this study, we analysed the effect of magnesium deficiency on the proliferation of SaOS-2 osteosarcoma cells. When quiescent magnesium-starved cells were induced to proliferate by serum addition, the magnesium content was 2–3 times lower in cells maintained in a medium without magnesium compared with cells growing in the presence of the ion. Magnesium depletion inhibited cell cycle progression and caused the inhibition of cell proliferation, which was associated with mTOR hypophosphorylation at Serine 2448. In order to map the intracellular magnesium distribution, an analytical approach using synchrotron-based X-ray techniques was applied. When cell growth was stimulated, magnesium was mainly localized near the plasma membrane in cells maintained in a medium without magnesium. In non-proliferating cells growing in the presence of the ion, high concentration areas inside the cell were observed. These results support the role of magnesium in the control of cell proliferation, suggesting that mTOR may represent an important target for the antiproliferative effect of magnesium. Selective control of magnesium availability could be a useful strategy for inhibiting osteosarcoma cell growth.

**Keywords:** magnesium; osteosarcoma; cell cycle; mTOR

## 1. Introduction

Magnesium is an essential nutrient with a wide range of metabolic, structural and regulatory functions [1]. Despite its presence in several types of food, the absorption and elimination of magnesium might be easily hindered by several factors. Even after its absorption, many substances may increase the excretion of magnesium in the kidneys and cause reduced plasma levels, such as excessive alcohol intake, diuretics, coffee, salt, sugar and excess fat [2].

It is estimated that from 2.5% to 15% of the world's population experiences some form of hypomagnesemia. Magnesium deficiency is frequently observed in industrialized countries. The US National Health and Nutrition Examination Survey (NHANES) stated that approximately one-half of all American adults have an inadequate intake of magnesium [3].

Magnesium is the fourth most abundant mineral and the second most abundant intracellular divalent cation in the body. Approximately 50% of magnesium can be found in bone, and approximately 50% is inside body tissue cells and organs, while less than 1% is found in the blood. A great deal of evidence shows that magnesium acts primarily as a signalling element in cell metabolism, and the concept that $Mg^{2+}$ is simply an electrolyte is obsolete [4,5]. In fact, magnesium plays a crucial role in many cellular processes, such as energy metabolism [6,7], protein and DNA synthesis [8,9] and several studies show that magnesium content directly correlates to proliferation in normal and transformed cells [10,11].

Intracellular magnesium homeostasis is primarily maintained by the ubiquitously expressed ion channel transient receptor potential melastatin (TRPM)7, which is a member of the transient receptor potential (TRP) family, possessing both ion channel and kinase activities [12]. TRPM7 and its homolog TRPM6 are strictly associated with intracellular signalling. Upon mitogen stimuli, cells are able to increase their intracellular magnesium content, most likely activating its influx. In contrast, magnesium deprivation inhibits DNA and protein synthesis and promotes cell growth arrest. A wide range of literature provides evidence about the essential role of magnesium in the transduction of proliferative signals. This could be explained by considering that protein kinases are strictly dependent on the complex of magnesium and ATP ($MgATP^{2-}$), which is the biologically active form in all living organisms [5]. As regards the role of magnesium in the control of cell proliferation, Rubin [4] postulated that the binding of growth factors to their membrane receptors causes a perturbation of the plasma membrane and consequently a release of the magnesium bound to membrane phospholipids. This release leads to a significant rise in the cytosolic free magnesium concentration, allowing $Mg^{2+}$ to displace other cations from the ATP complexes, increasing the active form of $MgATP^{2-}$.

Despite the evident link between magnesium and cell proliferation, the role of magnesium in cancer cells is scarcely documented and often contradictory [13–15]; in particular, the role of magnesium in primary bone tumours has not yet been examined. It is worth noting that in physiological conditions, magnesium plays an important role in bone metabolism [16–18] and can influence osteoblast and osteoclast differentiation, affecting bone growth and remodelling [19,20]. This work aims to study the role of magnesium in the growth of cancer SaOS2 osteosarcoma cells, monitoring the cellular concentration and compartmentalization of the cation by means of an advanced cellular imaging technique, as well as the effect of magnesium on specific signal transduction pathways.

## 2. Materials and Methods

### 2.1. Reagents

All reagents were Ultrapure grade and, unless otherwise specified, were from Merck-Millipore.

Dulbecco's Phosphate-Buffer Saline (DPBS) without $Ca^{2+}$ and $Mg^{2+}$ (8 g $L^{-1}$ NaCl, 0.2 g $L^{-1}$ KCl, 0.2 g $L^{-1}$ $Na_2HPO_4$, 0.2 g $L^{-1}$ $KH_2PO_4$, pH 7.2) was prepared in doubly distilled water. The fluorescent probe DCHQ5 was synthesized as previously reported [21] and was dissolved in dimethyl sulfoxide (DMSO) to a final concentration of 1 mg $mL^{-1}$. Aliquots were kept in the dark at 4 °C.

Foetal Bovine Serum (FBS, Euroclone, Milan, Italy) was dialyzed by using the Spectra/Por 4 Molecular Porous Dialysis Membrane (Spectrum, Austin, TX, USA) against Puck Buffer (NaCl 136.9 mM, KCl 5.4 mM, $NaHCO_3$ 4.2 mM and D-glucose monohydrate 4.2 M) plus EDTA for 2 days, and only Puck Buffer for the last 3 days. Calcium content in dialyzed FBS (dFBS) was restored by adding $CaCl_2$ at a final concentration of 1.8 mM, and dFBS was sterilized by filtering through a 0.45 μm pore size membrane filter.

## 2.2. Cell Culture

The human osteosarcoma cell line SaOS-2 (American Type Culture Collection, Manassas, VA, USA) was cultured at 37 °C and 5% $CO_2$ in MEM medium (Invitrogen, Carlsbad, CA, USA), supplemented with 2 mM L-Glutamine, 10% FBS, 1000 units $mL^{-1}$ penicillin and 1 mg $mL^{-1}$ streptomycin. The cells were seeded at $10^4$ cell/$cm^2$ in complete MEM and after 24 h the medium was substituted by the custom-made medium MEM w/o $Mg^{2+}$ (Invitrogen, CA, USA), and where needed, $MgCl_2$ was added at 1 mM concentrations. The medium was supplemented with 2 mM L-Glutamine, 10% FBS, 1000 units $mL^{-1}$ penicillin and 1 mg $mL^{-1}$ streptomycin.

To synchronize cells in G0/G1 phase and reduce intracellular magnesium content, cells were cultured in the medium containing 0.5% dFBS in the absence of magnesium for 24 h. Then, in order to stimulate cell proliferation, the cells were grown in a medium containing 5% dFBS in the presence or absence of 1 mM $MgCl_2$. To determine the rate of cell proliferation, viable cells were counted after 24 h and 48 h by using a Bürker hemocytometer in the presence of erythrosine 0.1% in PBS.

## 2.3. Flow Cytometric Assays

Flow cytometric assays were performed on an Epics Elite flow cytometer (Beckman Coulter, Brea, CA, USA) equipped with an Argon Ion laser tuned at 488 nm.

Cell cycle. To perform the cell cycle analysis, cells were fixed by 70% ice-cold ethanol and left at −20 °C overnight. After centrifugation pellets were resuspended in an appropriate volume (0.5–2 mL) of staining solution (DPBS, 5 µg $mL^{-1}$ Propidium Iodide (PI) and 10 µg $mL^{-1}$ DNAse free RNase A). Samples were incubated in the dark for 30 min at 37 °C and analysed by acquiring the PI red fluorescence on a linear scale at 600 nm. Data analysis is performed using the software program "ModFit" (Verity, Carrollton, TX, USA).

$p27^{Kip1}$ induction. Detached cells were washed 2 times from the growth medium in DPBS by centrifuging at 240 g for 10 min. The samples were then fixed with 3% paraformaldehyde at room temperature for 15 min. To remove any residual formaldehyde, samples were washed 2 times in PBS-glycine 0.1 M, followed by two washes in DPBS-BSA 1% performed to block nonspecific sites. A solution 1:9 of DPBS-ethanol (70%) was added in order to permeabilize the cell membranes and samples are maintained at −20 °C for 3 min. Samples were then washed 3 times in DPBS-BSA 1% by centrifuging at 240 g for 5 min. and marked with a rabbit primary antibody anti-$p27^{Kip1}$ under stirring at 4 °C overnight. The samples were then washed in DPBS, and marked with a secondary antibody FITC conjugated, diluted 1:1000 at room temperature for 1 h. Finally, to verify the $p27^{Kip1}$ expression levels in the function of the cell cycle, the samples were counterstained for DNA content by PI 5 µg $mL^{-1}$ and analysed by flow cytometry. FITC green fluorescence is collected at 525 nm on a logarithmic scale and PI red fluorescence at 600 nm on a linear scale.

## 2.4. Lactate Dehydrogenase Assay

To verify the effect of magnesium deprivation on cell viability, released lactate dehydrogenase (LDH) activity was assayed in the culture medium. Briefly, 2 mL of medium were centrifuged at 4000 g for 10 min. The supernatant was preserved and constitutes the sample. Sequentially, 1.325 mL of phosphate buffer 0, 1 M at pH 7, 50 µL of sodium pyruvate 23 mM and 50 µL of NADH 14 mM dissolved in TRIS 0.1 M at pH 7 were added to the cuvette. The reaction starts after the addition of 100 µL of the sample. The absorbance was then measured at 340 nm and at intervals of 1 min against a blank prepared with 100 µL of fresh medium.

## 2.5. Western Blotting

The level of protein expression at the indicated time points was evaluated by Western blotting, as previously described [22]. Protein samples were run in 6% (for mTOR) or 15% (for LC3 and $p27^{kip1}$) SDS-polyacrylamide gels. Polyclonal primary antibodies of rabbit anti-mTOR, anti-phospho-mTOR, anti-$p27^{kip1}$, anti-LCRA and anti-LCRB (Cell

Signaling Beverly, MA, USA) were diluted 1:1000. The secondary antibody anti-rabbit IgG was diluted 1:2500 in PBS with 3% of skimmed milk powder. The intensity of the bands was evaluated with the densitometric software GelPro Analyzer 3.0 (Media Cybernetic, Rockville, MD, USA). In graphs, band intensity was normalized to the loading control β-actin.

*2.6. Magnesium Determination by DCHQ5 Spectrofluorimetric Assay*

Detached cells were washed 3 times in DPBS at 240 $g$ for 10 min, counted and resuspended at $10^6$ cell mL$^{-1}$, and stored at $-20$ °C until the spectrofluorimetric analysis. Total intracellular magnesium was assessed on sonicated cell samples by using the fluorescent chemosensor DCHQ5, according to Sargenti et al. [21]. Magnesium concentration was normalized for the amount of cells mL$^{-1}$ used during analysis, and reported as nmoles/$10^6$ cells. To obtain the mM concentration of magnesium, the detected nmoles are divided by the cell volume (µL) calculated according to Malucelli et al. [23].

*2.7. Synchrotron Based X-ray Microscopy*

To perform X-ray microscopy [24], the cells were plated on a 200 nm-thick silicon nitride membrane window (Silson UK), grown for 24 h with 5% dFBS in the presence or absence of magnesium, and then dehydrated and fixed by chemical fixing: after two washes in ammonium acetate 100 mM, they were immersed in methanol/acetone 1:1 for 2 min at $-20$ °C and then air-dried.

The scanning transmission X-ray microscopy (STXM) and the X-ray fluorescence microscopy (XRFM) measurements were carried out at the beamline TWINMIC at Elettra Synchrotron (Trieste, Italy) [25]. The dehydrated cells were carefully examined with an optical microscope and selected following the criteria of integrity, dimensions, and distance from other cells.

A Fresnel zone plate focused the incoming beam (1475 eV), monochromatized by a plane-grating monochromator, to a circular spot of about 600 nm in diameter. The sample was transversally scanned in the zone plate focus pixel per pixel and in steps of 500 nm. At each step, the fluorescence radiation intensity was measured by eight silicon drift detectors (active area 30 mm$^2$) concentrically mounted at a 20° grazing angle with respect to the specimen plane, at a detector-to-specimen distance of 28 mm [26]. Simultaneously, the transmitted intensity T was measured by a fast-readout electron-multiplying low-noise charge-coupled device (CCD) detector through an X-ray–visible light converting system [27]. Zone plate, sample, and detectors were in a vacuum, thus avoiding any absorption and scattering by air.

Five STXM images were acquired on whole cells with a step size of 500 nm. In sequence, XRFM and simultaneously, STXM were carried out with a range of 6–8 s dwell time per pixel, depending on the cell size. The total acquisition time was in the range of 6–10 h (field of view of at least 20 µm × 20 µm, and spatial resolution 500 nm). The measurement of I0 was made by acquiring 25 points and repeating the measure five times. Atomic Force Microscopy (AFM) measurements were performed on selected cells before and after XRFM and STXM measurements.

Thereby, maps of magnesium molar concentration can be calculated using the algorithm developed by Malucelli et al. [24].

*2.8. Statistical Analysis*

The experiments were repeated at least three times, and the values were reported as mean ± standard deviation. One-way ANOVA analysis was also performed and values of $p < 0.05$ were taken to be statistically significant.

## 3. Results

### 3.1. Effects of Magnesium Deprivation on Intracellular Magnesium Content

In order to study the effects of magnesium deficiency in human osteosarcoma SaOS-2 cells, the cells were firstly synchronized in G0/G1 phase with a reduced intracellular magnesium content by culturing them for 24 h in a medium without magnesium, containing 0.5% dFBS. Then, the medium was replaced with a medium containing 5% dFBS in the presence or absence of 1 mM MgCl$_2$, and grown for 24 and 48 h.

To evaluate the intracellular total magnesium, the fluorescent chemosensor DCHQ5 was used. It is a diaza-crown-hydroxyquinoline that allows the assessment of intracellular total magnesium in a much lower number of cells than compared to other techniques and to other commercial dyes [21].

Following the Rubin model which postulates a release of membrane-bound magnesium and a consequent increase in the MgATP required by the protein kinases after mitogenic stimulation [10], the total intracellular amount of the cation was measured 24 h after the addition of serum in the culture medium. Cells grown for 24 h in the presence or absence of magnesium were lysed by sonication and analysed by spectrofluorometry. Figure 1 shows that in cells grown in a medium without magnesium, the total intracellular content of the ion was about 50% with respect to cells cultured in the presence of magnesium, ranging from 31 to 16.7 nmol/10$^6$ cells, respectively. Thus, considering the SaOS-2 volume [23], the intracellular magnesium content was 13.7 mM in cells grown with magnesium and 8.4 mM in cells grown without magnesium.

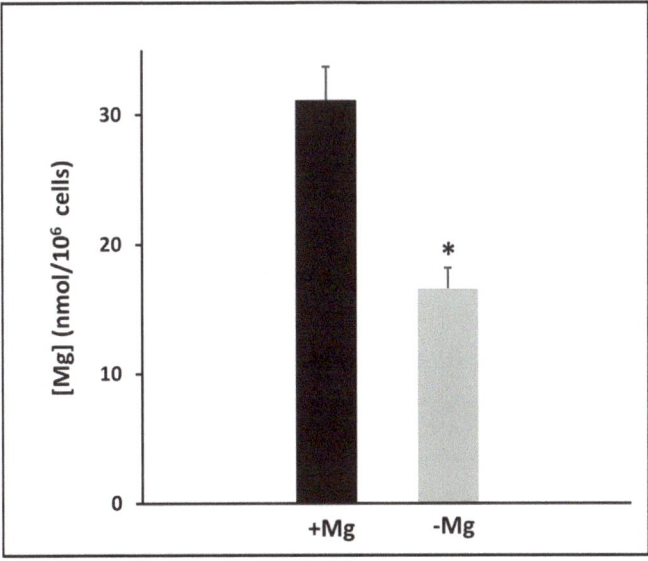

**Figure 1.** Intracellular content of total magnesium in SaOS-2 cells grown 24 h in the presence (+Mg) or absence (−Mg) of magnesium. Starved cells maintained in magnesium-free medium were stimulated to proliferate by adding 5% dFBS in the presence (+Mg) or absence (−Mg) of 1 mM MgCl$_2$. After 24 h, magnesium content was measured by a fluorescent probe. The data are reported as a mean ± SD of three independent experiments. * $p < 0.01$ vs. +Mg.

The dynamic of intracellular magnesium involves changes in its total amount and in cellular compartmentalization. Therefore, to understand the role of magnesium in signal transduction pathways linked to cell proliferation, it is important to evaluate not only its intracellular content but also its spatial distribution. To address this goal, we used an analytical approach of cellular imaging that combines techniques that are not widely utilized in biological studies [24], i.e., atomic force microscopy (AFM), scanning transmission X-ray microscopy (STXM) and X-ray fluorescence microscopy (XRFM). AFM allowed us to measure the thickness of the analysed cells. STXM records the light not absorbed (and then transmitted) by the X-ray irradiated sample, providing information about the local density and producing bidimensional images of the sample. XRFM generates X-ray fluorescence spectra of the cells, whose elaboration allowed us to draw the elemental map of the cell, which is complementary to the transmission map. Figure 2A shows the images of a single SaOS-2 cell obtained by these techniques.

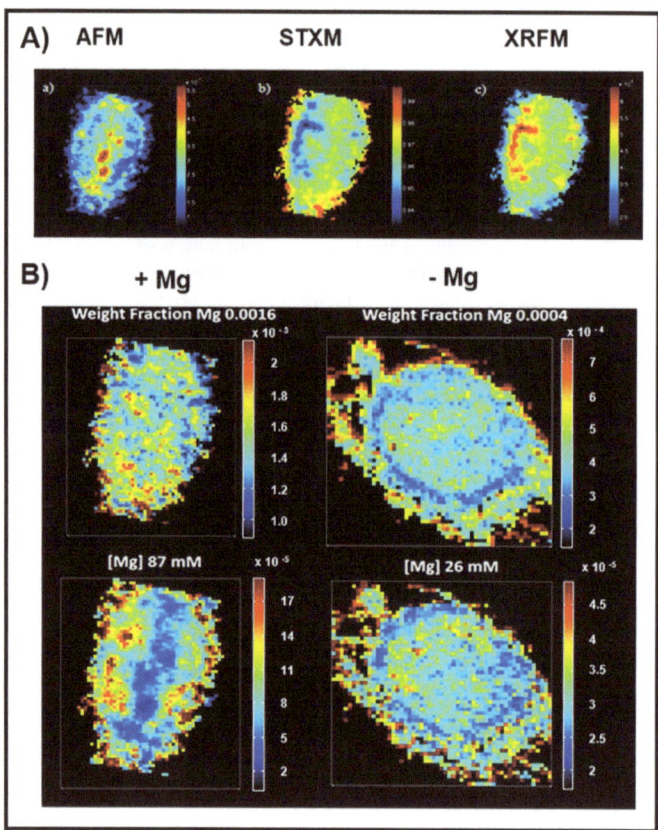

**Figure 2.** (**A**) Images of a single growing SaOS-2 cell obtained by three different microscopy techniques. (**a**) AFM analysis: the blue (red) colour indicates areas of lesser (greater) thickness; (**b**) STXM analysis: the blue (red) colour indicates areas in which the transmitted radiation is minimal (maximal); (**c**) XRFM analysis: the image is complementary to the STXM image because the incident radiation is more absorbed in the minimum transmission areas causing fluorescence. The blue (red) colour indicates areas of lesser (greater) fluorescence. (**B**) Distribution maps of the magnesium content in SaOS-2 cells cultured 24 h with 5% dFBS in the presence (left) or absence (right) of 1 mM $MgCl_2$. At the top, the weight fraction maps are reported. Below, the molar concentration maps are reported.

By combining STXM e XRFM images of a single cell, the elemental distributions expressed as weight fractions could be obtained. Malucelli et al. [24] proposed an algorithm which allowed us to combine the weight fraction map, obtained by STXM e XRFM analysis, with AFM data, providing a new elemental distribution map that merged local elemental composition and morphological information (volume). With this approach, it was possible to obtain a gross estimate of the molar concentration map in different zones inside a single cell [24,28]. Hence, the SaOS-2 cells synchronized in G0/G1 phase with reduced intracellular magnesium were stimulated to proliferate by adding 5% dFBS in the presence or absence of 1 mM $MgCl_2$. The cells were grown for 24 h on a silica frame and then fixed and analysed by AFM and X-ray microscopy. Figure 2B shows the magnesium maps evaluated as weight fraction and molar concentration of the cells grown in the presence or absence of magnesium. The analysis showed that magnesium concentration was estimated to be 87 mM in cells grown in the presence of $MgCl_2$ and decreased to 26 mM in magnesium-deprived cells, confirming the differences reported in Figure 1. In the reported maps, the magnesium concentration can be visually appreciated by a colour scale which goes from a very low signal in blue to a very high signal in red. In cells stimulated to proliferate in the presence of magnesium, the weight fraction map indicates a quite homogeneous distribution of magnesium within the cells, whereas the concentration map shows some "islets" with very high concentration. Furthermore, the central part of the cells shows a low concentration of magnesium, because this is the thickest part of the cells [24]. Interestingly, the cells maintained without magnesium show a somewhat more homogeneous intracellular distribution of the ion without high-concentration zones, and the red and yellow pixels (higher concentration) mainly remain confined in the area corresponding to the plasma membrane.

### 3.2. Effects of Magnesium Deficiency on Cell Growth and Cell Death

The effects of magnesium deficiency on the proliferation of human SaOS-2 osteosarcoma cells were investigated. The cells were firstly synchronized in G0/G1 phase with a reduced intracellular magnesium content, and afterward were grown for 24 and 48 h in a medium containing 5% dFBS in the presence or absence of 1 mM $MgCl_2$ as described. At the indicated time points, the cells were counted.

Figure 3A shows that cell proliferation was significantly decreased in cells grown in the absence of $MgCl_2$. In detail, the numbers of cells grown in the absence of $MgCl_2$ were 62% at 24 h and 53% at 48 h of those grown in the presence of 1 mM $MgCl_2$. Considering the SaOS-2 doubling time (36–38 h), the variations in proliferation are better highlighted at 48 h. However, following Rubin's hypothesis [10,29] we focused on early events and thus we also took into account the cell viability at 24 h.

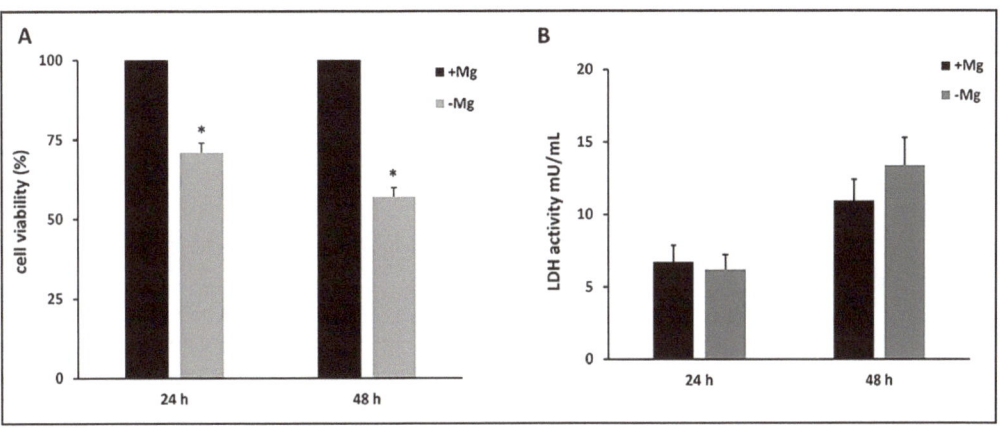

**Figure 3.** Effect of magnesium deficiency on SaOS-2 proliferation and viability. (**A**) Starved cells were stimulated to proliferate by adding 5% dFBS in presence (+Mg) or absence (−Mg) of 1 mM $MgCl_2$. After 24 h and 48 h the cell number was measured. The number of cells grown in the medium including $MgCl_2$ was arbitrarily taken as 100%. Data are means ± SD of three independent experiments. * $p < 0.05$ vs. +Mg. (**B**) LDH activity was measured in the culture medium of SaOS-2 cells grown 24 h and 48 h grown in the absence (−Mg) or presence (+Mg) of 1 mM $MgCl_2$. The data are reported as a mean ± SD of three determinations.

In order to evaluate the effect of magnesium deficiency on cellular viability, the released LDH activity was measured in a culture medium. Figure 3B shows that the deprivation of $MgCl_2$ did not induce any significant increase in LDH release at 24 and 48 h.

Taken together, these results indicate that magnesium deficiency reduced SaOS-2 cell proliferation without affecting cellular viability.

### 3.3. Effects of Magnesium Deficiency on Cell Cycle Progression

It is well known that different cell types have a different dependence on extracellular magnesium availability for their proliferation [10,11]. Hence, we examined the effect of magnesium availability on SaOS-2 cell cycle progression. As established in our experimental protocol, cells synchronized in G0/G1 phase with a low intracellular magnesium content were stimulated to grow in a medium containing 5% dFBS in the presence and absence of 1 mM $MgCl_2$ for 24 h and 48 h.

In cells cultured in the absence of $MgCl_2$, the percentage of cells in G0/G1 phase was markedly increased with respect to cells grown in the presence of the ion (Figure 4A), being 78% versus 53% at 24 h, and 84% versus 55% at 48 h (Figure 4B). On the other hand, the percentage of cells in S-phase was significantly lower in magnesium-deprived cells, while the cell distribution in the G2/M phase was not substantially influenced by magnesium deficiency. These results indicate that magnesium deficiency suppresses SaOS-2 cell cycle progression from G1 to S-phase, according to previous studies in kidney cells [30].

One feature of the role of magnesium in the control of cell growth pertains to the modulation of cell-cycle inhibitory proteins such as p27$^{Kip1}$ and p21$^{Cip1/WAF1}$ [31]. We investigated the effect of magnesium deficiency only on the expression of p27$^{Kip1}$ since SaOS-2 cells are p53-null and p21$^{Cip1/WAF1}$ protein is prevalently induced by p53 activation [32].

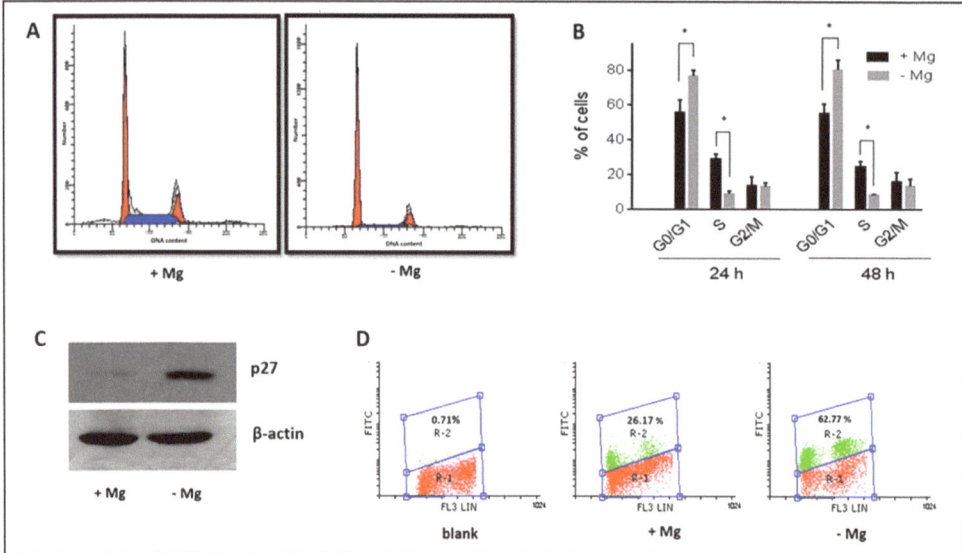

**Figure 4.** Effect of magnesium on the cell cycle progression of SaOS-2 cells. Starved cells were stimulated to proliferate by addition of 5% dFBS in the presence (+Mg) or absence (−Mg) of 1 mM $MgCl_2$ and analysed after the indicated times: (**A**) Typical cell cycle distribution after 24 h from serum addition, determined by flow cytometry. (**B**) percentage of cells in cell cycle phases after 24 h and 48 h; data are means ± SD obtained in three determinations; * $p < 0.05$. (**C**) Western blot analysis of p27$^{Kip1}$ protein in cells grown 24 h in the absence (left) and presence (right) of magnesium. The blot is representative of three experiments. (**D**) Expression of p27$^{Kip1}$ protein at 24 h in the function of cell cycle distribution determined by bi-parametric analysis: PI fluorescence (cell cycle) is shown on the X axis, while FITC fluorescence (p27$^{Kip1}$ protein) is reported on the Y axis.

A significant increase in the amount of p27$^{Kip1}$ protein was observed in magnesium-deprived cells stimulated to proliferate by dFBS addition (Figure 4C), suggesting that magnesium is involved in the regulation of p27$^{Kip1}$ in osteosarcoma SaOS-2 cells. Figure 4D shows the expression of p27$^{Kip1}$ in the function of the cell cycle phase. It is noteworthy that an increase in p27$^{Kip1}$ protein was associated with cells not resident in S-phase, as shown by the absence of green fluorescence corresponding to this phase. Overall, the percentage of p27$^{Kip1}$ positive cells ranges from 26% in Mg absence to 62% in Mg presence.

### 3.4. Effects of Magnesium Deficiency on mTOR Signaling

The activation of mTOR kinase represents a fundamental step in the initiation of protein synthesis and in cell growth. We assessed the expression of mTOR protein and mTOR phosphorylation at serine 2448 (S2448), since phosphorylated mTOR binds Raptor and becomes an active kinase [33].

Western blot analysis of the protein extracts from cells grown for 24h in the presence or absence of MgCl$_2$ revealed that magnesium deficiency did not alter mTOR protein level, but significantly reduced its phosphorylation at S2448 (Figure 5A).

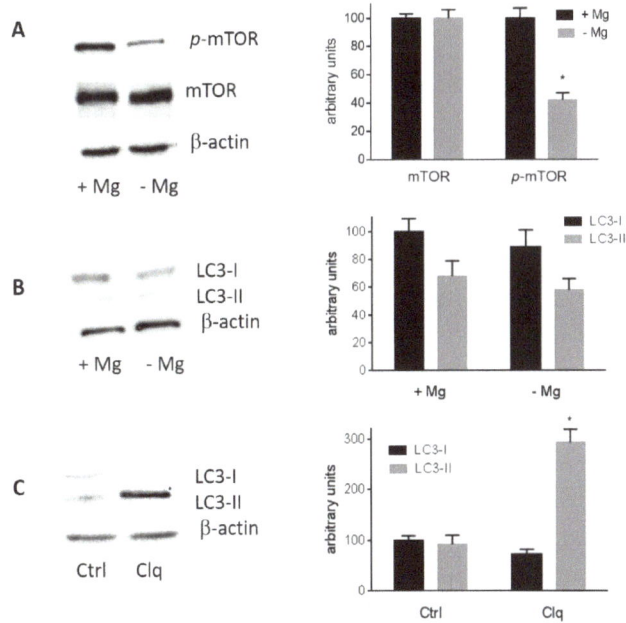

**Figure 5.** Effect of magnesium deficiency on mTOR level and phosphorylation and LC3 cleavage. Starved SaOS-2 cells were stimulated to proliferate by the addition of 5% dFBS in the presence (+Mg) or absence (−Mg) of 1 mM MgCl$_2$. After 24 h, cells were collected for protein analysis by Western blotting: (**A**) Left, Western blot analysis of total mTOR and phosphorylated mTOR (S2448). Right, densitometric analysis; the levels of mTOR and phosphorylated mTOR in the presence of magnesium are arbitrarily taken as 100; data are means ± SD of three determinations, * $p < 0.05$ vs. +Mg. (**B**) Left, Western blot analysis of LC3-I and LC3-II expression in cells grown in the presence or absence of 1 mM MgCl$_2$. Right, densitometric analysis; the level of LC3-I in the presence of magnesium is arbitrarily taken as 100; data are means ± SD of three determinations. (**C**) Left, the effect of 10 mM chloroquine (Clq) for 24 h in cells deprived of magnesium is shown as a positive control of LC3-II accumulation. Right, densitometric analysis; the level of LC3-I in control cells is arbitrarily taken as 100; data are means ± SD of three determinations, * $p < 0.05$ vs. control. Similar results were obtained in cells grown in the presence of 1 mM MgCl$_2$.

When nutrients are limited, mTOR in the mTORC1 complex is dephosphorylated and dissociates from the ULK complex, initiating the autophagy process [34], leading to the cleavage of LC3 protein which is considered a reliable marker of autophagy in mammalian cells [35]. However, even in the absence of magnesium, mTOR was found to be mainly dephosphorylated at S2448. Magnesium deficiency apparently did not induce LC3 cleavage and, consequently, did not cause a significant increase in the level of LC3-II protein, the cleaved and lipidated form of LC3 protein (Figure 5B). Treatment with 10 mM chloroquine as a positive control of LC3-II accumulation [36] caused a large increase in LC3-II level as expected in both magnesium-deprived cells and control cells (Figure 5C).

## 4. Discussion

Magnesium is a cofactor involved in more than 300 metabolic reactions in the body, including protein synthesis, cellular energy production and storage, cell growth and reproduction, and deoxyribonucleic acid and ribonucleic acid synthesis. Magnesium helps to maintain normal nerve and muscle function, cardiac excitability, vasomotor tone, blood pressure, immune system, bone integrity, and blood glucose levels. It also promotes intestinal calcium absorption. Based on its multiple functions within the human body, magnesium has been reported to play an important role in the prevention and treatment of many diseases [37], including cancer [38].

Within the cells, magnesium is present at a very high concentration, usually between 5 mM and 30 mM, and only a very small fraction, about 1 mM or less, is unbound. Some authors have suggested that the ionic form moves among cellular sub-compartments [39]. Nevertheless, magnesium intracellular compartmentalization has not yet been thoroughly elucidated, mainly because of the inadequacy of available techniques to map the intracellular distribution of this cation.

Knowledge of the intracellular concentration and distribution of the chemical elements in cells may reveal their function in a variety of cellular processes. The biological function of a chemical element in cells not only requires the determination of its intracellular quantity but also of the spatial distribution of its concentration [23,40]. In order to address this problem, we applied the multimodal fusion approach developed by Malucelli et al. [24] which combines synchrotron radiation microscopy techniques with off-line atomic force microscopy, and offers the possibility of achieving a detailed map of the intracellular concentration of magnesium [24,28]. This method requires the implementation of images obtained by AFM, XRFM e STXM and the utilization of a specifically elaborated algorithm, allowing an estimate of the molar concentration map of intracellular magnesium.

In this way, we observed that magnesium is mainly confined at the plasma membrane in quiescent cells. When cells are stimulated to grow with a medium containing 5% serum and normal magnesium concentration, proliferation starts and magnesium moves toward the inner areas of the cell, consistently with Rubin's model [10]. In contrast, when SaOS-2 cells were stimulated to grow in the absence of extracellular magnesium, the ion mainly remained confined in the area corresponding to the plasma membrane and the cells were unable to proliferate, as shown by the reduced number of cells and by the accumulation in G0/G1 phase of the cell cycle.

Many tumour cells proved to be resistant to magnesium deficiency [31,32], even if the reduction of magnesium to a very low level can modify cell cycle distribution in some tumour cell lines [18,31]. SaOS-2 cells appear to be sensitive to magnesium depletion, and the block of the cell cycle is associated with a significant increase in p27$^{Kip1}$ that in these p53-null cells represents the main inhibitor of cyclin-dependent kinases. A similar upregulation of p27$^{Kip1}$ in cancer cell lines grown in magnesium-deficient medium has been reported [41,42].

Magnesium deficiency was also shown to decrease mTOR phosphorylation at serine 2448. Phosphorylation of serine 2448 is associated with mTORC1 complex activation [33,43], whereas hypophosphorylation causes mTORC1 inhibition and can lead to the activation of autophagy [34]. However, we cannot detect any significant change in the amount of LC3-II, the cleaved and lipidated form of LC3 protein, whose increase represents a hallmark of autophagy [35], suggesting that autophagy is not a major mechanism associated with the reduced proliferation of magnesium-deficient cells.

Magnesium has an essential role in the transduction of proliferative signals. Rubin's theory about the role of magnesium in the control of proliferation [10,29] postulates that the release of membrane-bound magnesium leads to an increase in cytosolic free $Mg^{2+}$ with a consequent increase in MgATP, required by protein kinases. Among the kinases involved in proliferative pathways, mTOR has an unusually high $K_m$ for MgATP, about 1 mM. Thus, the MgATP complex is a limiting factor for the activation of mTOR kinase, initiation of protein synthesis and, consequently, the progression of the cell cycle from the G1 phase [10,29]. This consideration, together with our finding that magnesium deprivation causes mTOR hypophosphorylation at S2448, suggests that the antiproliferative effect of magnesium deficiency in osteosarcoma cells is mediated by mTOR. Further work is required to explore the effect of magnesium deficiency on other kinases of the signalling cascades involving mTOR.

Several epidemiological studies have provided evidence that a correlation exists between dietary magnesium and various types of cancer. In addition, impaired magnesium homeostasis is reported in cancer patients, and frequently complicates therapy with some anti-cancer drugs [38].

High levels of magnesium in drinking water protect against oesophageal and liver cancer, and it is inversely correlated with death from breast, prostate, and ovarian cancers, whereas no correlation existed for other tumours [14]. Dietary magnesium intake has been reported to have a statistically significant nonlinear inverse association with the risk of colorectal cancer. The greatest reduction for magnesium intake was a result of 200–270 mg/day [44]. Another study suggested that increasing the intake of magnesium-containing foods may help reduce the incidence and mortality of primary liver cancer [15].

Interestingly, a recent study found that cancer survivors used dietary supplements at a higher frequency and dose than individuals without cancer, but had an overall lower intake of nutrients from foods [45]. Our results indicate that the control of magnesium availability could be a useful strategy for inhibiting osteosarcoma cell growth, and support the hypothesis that mTOR may represent a target for the antiproliferative effect of magnesium deficiency.

Magnesium is also important for bone health. Interestingly, SaOS-2 cells display osteoblastic features similar to primary human osteoblastic cells and are often used as a model of osteogenic differentiation [20,46]. A recent review showed how optimal magnesium and vitamin D balance may improve bone metabolism and health outcomes [18]. Optimal magnesium levels contribute to the maintenance of skeletal health [47,48], and our results also suggest a mechanism that may be involved in the effects of magnesium deficiency in normal bone cells. This aspect is worthy of attention since there is a profound lack of awareness of the insufficient intake of magnesium in the population worldwide, and the decrease in magnesium content in processed foods and in newer varieties of grains, fruits, and vegetables poses a further challenge for adequate magnesium consumption.

## 5. Conclusions

Magnesium is an essential nutrient, but the links between magnesium, cell growth, and carcinogenesis still remain unclear and complex, with conflicting results being reported from many experimental, epidemiological and clinical studies.

It has been proposed that transformation causes a selective loss of the growth regulatory role of $Mg^{2+}$.

In view of the evidence that transformed cells have a diminished capacity to regulate their free $Mg^{2+}$, the effects of $Mg^{2+}$ deprivation on their behaviour were examined. In this study, we examined the effect of magnesium deficiency on the proliferation of SaOS-2 osteosarcoma cells. Magnesium depletion limited the ability of cells to progress in the cell cycle and caused the inhibition of cell proliferation, which was associated with mTOR hypophosphorylation at Serine 2448. In order to map the intracellular concentration and compartmentalization of the cation, an advanced cellular imaging technique using synchrotron-based X-ray techniques was applied. When cell growth was stimulated, magnesium was mainly localized near the plasma membrane in cells maintained in a medium without magnesium, whereas in non-proliferating cells growing in the presence of the ion, high concentration areas inside the cell were observed.

These results are compatible with Rubin's theory about the role of magnesium in the control of cell proliferation [29], and indicate that selective control of magnesium availability in osteosarcoma cells could be a useful strategy for inhibiting tumour growth.

**Author Contributions:** C.C. performed experimental design, cell culture and Mg quantification; E.M. supervised the experiment and data analysis; M.Z. performed Western blot experiments; G.F. performed the cytofluorimetric assay; G.P. performed Mg quantification; A.G. produced the X-Ray fluorescence microscopy measurements and analysis, A.N. produced the atomic force microscopy measurements and analysis; M.F. produced the atomic force microscopy measurements and analysis; C.P. and S.I. wrote the manuscript; C.S. supervised the project and wrote the manuscript. All authors have read and agreed to the published version of the manuscript.

**Funding:** This work was supported by the University of Bologna (RFO).

**Institutional Review Board Statement:** Not applicable.

**Informed Consent Statement:** Not applicable.

**Data Availability Statement:** Not applicable.

**Conflicts of Interest:** The authors declare no conflict of interest.

## References

1. de Baaij, J.H.F.; Hoenderop, J.G.J.; Bindels, R.J.M. Magnesium in Man: Implications for Health and Disease. *Physiol. Rev.* **2015**, *95*, 1–46. [CrossRef]
2. Johnson, S. The Multifaceted and Widespread Pathology of Magnesium Deficiency. *Med. Hypotheses* **2001**, *56*, 163–170. [CrossRef]
3. NHANES Questionnaires, Datasets, and Related Documentation. Available online: https://wwwn.cdc.gov/nchs/nhanes/continuousnhanes/default.aspx?BeginYear=2005 (accessed on 6 March 2021).
4. Rubin, H. Magnesium: The Missing Element in Molecular Views of Cell Proliferation Control. *Bioessays* **2005**, *27*, 311–320. [CrossRef] [PubMed]
5. Iotti, S.; Frassineti, C.; Sabatini, A.; Vacca, A.; Barbiroli, B. Quantitative Mathematical Expressions for Accurate in Vivo Assessment of Cytosolic [ADP] and ΔG of ATP Hydrolysis in the Human Brain and Skeletal Muscle. *Biochim. Biophys. Acta (BBA) Bioenerg.* **2005**, *1708*, 164–177. [CrossRef]
6. Feeney, K.A.; Hansen, L.L.; Putker, M.; Olivares-Yañez, C.; Day, J.; Eades, L.J.; Larrondo, L.F.; Hoyle, N.P.; O'Neill, J.S.; van Ooijen, G. Daily Magnesium Fluxes Regulate Cellular Timekeeping and Energy Balance. *Nature* **2016**, *532*, 375–379. [CrossRef]
7. Barbiroli, B.; Iotti, S.; Cortelli, P.; Martinelli, P.; Lodi, R.; Carelli, V.; Montagna, P. Low Brain Intracellular Free Magnesium in Mitochondrial Cytopathies. *J. Cereb. Blood Flow Metab.* **1999**, *19*, 528–532. [CrossRef] [PubMed]
8. Zieve, F.J.; Freude, K.A.; Zieve, L. Effects of Magnesium Deficiency on Protein and Nucleic Acid Synthesis in Vivo. *J. Nutr.* **1977**, *107*, 2178–2188. [CrossRef]
9. Mushegian, A.A. A Ribosomal Strategy for Magnesium Deficiency. *Sci. Signal.* **2016**, *9*, ec269. [CrossRef]
10. Rubin, H. The Logic of the Membrane, Magnesium, Mitosis (MMM) Model for the Regulation of Animal Cell Proliferation. *Arch. Biochem. Biophys.* **2007**, *458*, 16–23. [CrossRef] [PubMed]
11. Wolf, F.I.; Cittadini, A.R.M.; Maier, J.A.M. Magnesium and Tumors: Ally or Foe? *Cancer Treat. Rev.* **2009**, *35*, 378–382. [CrossRef] [PubMed]
12. Runnels, L.W.; Yue, L.; Clapham, D.E. TRP-PLIK, a Bifunctional Protein with Kinase and Ion Channel Activities. *Science* **2001**, *291*, 1043–1047. [CrossRef]
13. Zou, Z.-G.; Rios, F.J.; Montezano, A.C.; Touyz, R.M. TRPM7, Magnesium, and Signaling. *Int. J. Mol. Sci.* **2019**, *20*, 1877. [CrossRef] [PubMed]
14. Yang, C.-Y.; Chiu, H.-F.; Tsai, S.-S.; Wu, T.-N.; Chang, C.-C. Magnesium and Calcium in Drinking Water and the Risk of Death from Esophageal Cancer. *Magnes. Res.* **2002**, *15*, 215–222. [PubMed]
15. Zhong, G.-C.; Peng, Y.; Wang, K.; Wan, L.; Wu, Y.-Q.-L.; Hao, F.-B.; Hu, J.-J.; Gu, H.-T. Magnesium Intake and Primary Liver Cancer Incidence and Mortality in the Prostate, Lung, Colorectal and Ovarian Cancer Screening Trial. *Int. J. Cancer* **2020**, *147*, 1577–1586. [CrossRef]
16. Wolf, F.I.; Trapani, V. Magnesium and Its Transporters in Cancer: A Novel Paradigm in Tumour Development. *Clin. Sci.* **2012**, *123*, 417–427. [CrossRef]

17. Gaffney-Stomberg, E. The Impact of Trace Minerals on Bone Metabolism. *Biol. Trace Elem. Res.* **2019**, *188*, 26–34. [CrossRef]
18. Erem, S.; Atfi, A.; Razzaque, M.S. Anabolic Effects of Vitamin D and Magnesium in Aging Bone. *J. Steroid Biochem. Mol. Biol.* **2019**, *193*, 105400. [CrossRef]
19. Mammoli, F.; Castiglioni, S.; Parenti, S.; Cappadone, C.; Farruggia, G.; Iotti, S.; Davalli, P.; Maier, J.A.M.; Grande, A.; Frassineti, C. Magnesium Is a Key Regulator of the Balance between Osteoclast and Osteoblast Differentiation in the Presence of Vitamin $D_3$. *Int. J. Mol. Sci.* **2019**, *20*, 385. [CrossRef] [PubMed]
20. Picone, G.; Cappadone, C.; Pasini, A.; Lovecchio, J.; Cortesi, M.; Farruggia, G.; Lombardo, M.; Gianoncelli, A.; Mancini, L.; Ralf, H.M.; et al. Analysis of Intracellular Magnesium and Mineral Depositions during Osteogenic Commitment of 3D Cultured Saos2 Cells. *Int. J. Mol. Sci.* **2020**, *21*, 2368. [CrossRef] [PubMed]
21. Sargenti, A.; Farruggia, G.; Zaccheroni, N.; Marraccini, C.; Sgarzi, M.; Cappadone, C.; Malucelli, E.; Procopio, A.; Prodi, L.; Lombardo, M.; et al. Synthesis of a Highly $Mg^{2+}$-Selective Fluorescent Probe and Its Application to Quantifying and Imaging Total Intracellular Magnesium. *Nat. Protoc.* **2017**, *12*, 461–471. [CrossRef]
22. Andreani, A.; Burnelli, S.; Granaiola, M.; Leoni, A.; Locatelli, A.; Morigi, R.; Rambaldi, M.; Varoli, L.; Farruggia, G.; Stefanelli, C.; et al. Synthesis and Antitumor Activity of Guanylhydrazones from 6-(2,4-Dichloro-5-Nitrophenyl)Imidazo[2,1-*b*]Thiazoles and 6-Pyridylimidazo[2,1-*b*]Thiazoles. *J. Med. Chem.* **2006**, *49*, 7897–7901. [CrossRef] [PubMed]
23. Malucelli, E.; Procopio, A.; Fratini, M.; Gianoncelli, A.; Notargiacomo, A.; Merolle, L.; Sargenti, A.; Castiglioni, S.; Cappadone, C.; Farruggia, G.; et al. Single Cell versus Large Population Analysis: Cell Variability in Elemental Intracellular Concentration and Distribution. *Anal. Bioanal. Chem.* **2018**, *410*, 337–348. [CrossRef]
24. Malucelli, E.; Iotti, S.; Gianoncelli, A.; Fratini, M.; Merolle, L.; Notargiacomo, A.; Marraccini, C.; Sargenti, A.; Cappadone, C.; Farruggia, G.; et al. Quantitative Chemical Imaging of the Intracellular Spatial Distribution of Fundamental Elements and Light Metals in Single Cells. *Anal. Chem.* **2014**, *86*, 5108–5115. [CrossRef] [PubMed]
25. Gianoncelli, A.; Kourousias, G.; Merolle, L.; Altissimo, M.; Bianco, A. Current Status of the TwinMic Beamline at Elettra: A Soft X-Ray Transmission and Emission Microscopy Station. *J. Synchrotron Radiat.* **2016**, *23*, 1526–1537. [CrossRef] [PubMed]
26. Gianoncelli, A.; Kourousias, G.; Stolfa, A.; Kaulich, B. Recent Developments at the TwinMic Beamline at ELETTRA: An 8 SDD Detector Setup for Low Energy X-Ray Fluorescence. *J. Phys. Conf. Ser.* **2013**, *425*, 182001. [CrossRef]
27. Gianoncelli, A.; Morrison, G.R.; Kaulich, B.; Bacescu, D.; Kovac, J. Scanning Transmission X-Ray Microscopy with a Configurable Detector. *Appl. Phys. Lett.* **2006**, *89*, 251117. [CrossRef]
28. Picone, G.; Cappadone, C.; Farruggia, G.; Malucelli, E.; Iotti, S. The Assessment of Intracellular Magnesium: Different Strategies to Answer Different Questions. *Magnes. Res.* **2020**, *33*, 1–11. [CrossRef]
29. Rubin, H. Central Roles of $Mg^{2+}$ and $MgATP^{2-}$ in the Regulation of Protein Synthesis and Cell Proliferation: Significance for Neoplastic Transformation. *Adv. Cancer Res.* **2005**, *93*, 1–58. [CrossRef]
30. Ikari, A.; Sawada, H.; Sanada, A.; Tonegawa, C.; Yamazaki, Y.; Sugatani, J. Magnesium Deficiency Suppresses Cell Cycle Progression Mediated by Increase in Transcriptional Activity of P21(Cip1) and P27(Kip1) in Renal Epithelial NRK-52E Cells. *J. Cell Biochem.* **2011**, *112*, 3563–3572. [CrossRef]
31. Wolf, F.I.; Trapani, V. Cell (Patho)Physiology of Magnesium. *Clin. Sci.* **2008**, *114*, 27–35. [CrossRef]
32. Shamloo, B.; Usluer, S. P21 in Cancer Research. *Cancers* **2019**, *11*, 1178. [CrossRef]
33. Rosner, M.; Siegel, N.; Valli, A.; Fuchs, C.; Hengstschläger, M. MTOR Phosphorylated at S2448 Binds to Raptor and Rictor. *Amino Acids* **2010**, *38*, 223–228. [CrossRef]
34. Kuma, A.; Mizushima, N. Physiological Role of Autophagy as an Intracellular Recycling System: With an Emphasis on Nutrient Metabolism. *Semin. Cell Dev. Biol.* **2010**, *21*, 683–690. [CrossRef] [PubMed]
35. Yoshii, S.R.; Mizushima, N. Monitoring and Measuring Autophagy. *Int. J. Mol. Sci.* **2017**, *18*, 1865. [CrossRef] [PubMed]
36. Fourrier, C.; Bryksin, V.; Hattersley, K.; Hein, L.K.; Bensalem, J.; Sargeant, T.J. Comparison of Chloroquine-like Molecules for Lysosomal Inhibition and Measurement of Autophagic Flux in the Brain. *Biochem. Biophys. Res. Commun.* **2021**, *534*, 107–113. [CrossRef]
37. Gröber, U.; Schmidt, J.; Kisters, K. Magnesium in Prevention and Therapy. *Nutrients* **2015**, *7*, 8199–8226. [CrossRef] [PubMed]
38. Castiglioni, S.; Maier, J.A.M. Magnesium and Cancer: A Dangerous Liason. *Magnes. Res.* **2011**, *24*, 92–100. [CrossRef]
39. Romani, A.M.P. Magnesium in Health and Disease. *Met. Ions Life Sci.* **2013**, *13*, 49–79. [CrossRef]
40. Malucelli, E.; Fratini, M.; Notargiacomo, A.; Gianoncelli, A.; Merolle, L.; Sargenti, A.; Cappadone, C.; Farruggia, G.; Lagomarsino, S.; Iotti, S. Where Is It and How Much? Mapping and Quantifying Elements in Single Cells. *Analyst* **2016**, *141*, 5221–5235. [CrossRef]
41. Covacci, V.; Bruzzese, N.; Sgambato, A.; Di Francesco, A.; Russo, M.A.; Wolf, F.I.; Cittadini, A. Magnesium Restriction Induces Granulocytic Differentiation and Expression of P27Kip1 in Human Leukemic HL-60 Cells. *J. Cell Biochem.* **1998**, *70*, 313–322. [CrossRef]
42. Sgambato, A.; Wolf, F.I.; Faraglia, B.; Cittadini, A. Magnesium Depletion Causes Growth Inhibition, Reduced Expression of Cyclin D1, and Increased Expression of P27KIP1 in Normal but Not in Transformed Mammary Epithelial Cells. *J. Cell. Physiol.* **1999**, *180*, 245–254. [CrossRef]
43. Parrales, A.; López, E.; Lee-Rivera, I.; López-Colomé, A.M. ERK1/2-Dependent Activation of MTOR/MTORC1/P70S6K Regulates Thrombin-Induced RPE Cell Proliferation. *Cell Signal.* **2013**, *25*, 829–838. [CrossRef]

44. Qu, X.; Jin, F.; Hao, Y.; Zhu, Z.; Li, H.; Tang, T.; Dai, K. Nonlinear Association between Magnesium Intake and the Risk of Colorectal Cancer. *Eur. J. Gastroenterol. Hepatol.* **2013**, *25*, 309–318. [CrossRef]
45. Du, M.; Luo, H.; Blumberg, J.B.; Rogers, G.; Chen, F.; Ruan, M.; Shan, Z.; Biever, E.; Zhang, F.F. Dietary Supplement Use among Adult Cancer Survivors in the United States. *J. Nutr.* **2020**, *150*, 1499–1508. [CrossRef]
46. Zakłos-Szyda, M.; Nowak, A.; Pietrzyk, N.; Podsędek, A. Viburnum Opulus L. Juice Phenolic Compounds Influence Osteogenic Differentiation in Human Osteosarcoma Saos-2 Cells. *Int. J. Mol. Sci.* **2020**, *21*, 4909. [CrossRef]
47. Belluci, M.M.; Giro, G.; del Barrio, R.A.L.; Pereira, R.M.R.; Marcantonio, E.; Orrico, S.R.P. Effects of Magnesium Intake Deficiency on Bone Metabolism and Bone Tissue around Osseointegrated Implants: Effects of Magnesium Intake Deficiency on Bone Tissue. *Clin. Oral Implants Res.* **2011**, *22*, 716–721. [CrossRef]
48. Capozzi, A.; Scambia, G.; Lello, S. Calcium, Vitamin D, Vitamin K2, and Magnesium Supplementation and Skeletal Health. *Maturitas* **2020**, *140*, 55–63. [CrossRef]

Article

# Inhibition of $Mg^{2+}$ Extrusion Attenuates Glutamate Excitotoxicity in Cultured Rat Hippocampal Neurons

Yutaka Shindo [1], Ryu Yamanaka [1,2], Kohji Hotta [1] and Kotaro Oka [1,3,4,*]

1. Department of Bioscience and Informatics, Faculty of Science and Technology, Keio University, Yokohama, Kanagawa 223-8522, Japan; shindo@z5.keio.jp (Y.S.); 1024mbmail@gmail.com (R.Y.); khotta@bio.keio.ac.jp (K.H.)
2. Faculty of Pharmaceutical Sciences, Sanyo-Onoda City University, Sanyo-Onoda, Yamaguchi 756-0884, Japan
3. Waseda Research Institute for Science and Engineering, Waseda University, Shinjuku, Tokyo 162-8480, Japan
4. Graduate Institute of Medicine, College of Medicine, Kaohsiung Medical University, Kaohsiung 80708, Taiwan
* Correspondence: oka@bio.keio.ac.jp; Tel.: +81-45-563-1141

Received: 6 August 2020; Accepted: 8 September 2020; Published: 10 September 2020

**Abstract:** Magnesium plays important roles in the nervous system. An increase in the $Mg^{2+}$ concentration in cerebrospinal fluid enhances neural functions, while $Mg^{2+}$ deficiency is implicated in neuronal diseases in the central nervous system. We have previously demonstrated that high concentrations of glutamate induce excitotoxicity and elicit a transient increase in the intracellular concentration of $Mg^{2+}$ due to the release of $Mg^{2+}$ from mitochondria, followed by a decrease to below steady-state levels. Since $Mg^{2+}$ deficiency is involved in neuronal diseases, this decrease presumably affects neuronal survival under excitotoxic conditions. However, the mechanism of the $Mg^{2+}$ decrease and its effect on the excitotoxicity process have not been elucidated. In this study, we demonstrated that inhibitors of $Mg^{2+}$ extrusion, quinidine and amiloride, attenuated glutamate excitotoxicity in cultured rat hippocampal neurons. A toxic concentration of glutamate induced both $Mg^{2+}$ release from mitochondria and $Mg^{2+}$ extrusion from cytosol, and both quinidine and amiloride suppressed only the extrusion. This resulted in the maintenance of a higher $Mg^{2+}$ concentration in the cytosol than under steady-state conditions during the ten-minute exposure to glutamate. These inhibitors also attenuated the glutamate-induced depression of cellular energy metabolism. Our data indicate the importance of $Mg^{2+}$ regulation in neuronal survival under excitotoxicity.

**Keywords:** magnesium; excitotoxicity; fluorescence imaging; neuroprotection

## 1. Introduction

Magnesium plays important roles in the nervous system [1,2]. Elevation of the $Mg^{2+}$ concentration in cerebrospinal fluid (CSF) increases synapse formation, enhances recognition and learning abilities in rats [3], and causes neural stem cell proliferation in mice [4]. Further, a deficiency of $Mg^{2+}$ in the brain is implicated in neuronal diseases, and some researchers have reported lower $Mg^{2+}$ concentrations than normal in the brain of patients with neurodegenerative diseases [5,6]. In rats, significant loss of dopaminergic neurons in the substantia nigra, similar to the loss seen in Parkinson's disease, was elicited merely by feeding them a low-magnesium diet for two generations [7]. In contrast, $Mg^{2+}$ supplementation or overexpression of the $Mg^{2+}$ channel have neuro-protective effects on cellular and animal models of Parkinson's [8–10] and Alzheimer's diseases [11]. These evidences show that $Mg^{2+}$ is intricately involved in nervous system functioning and neuroprotection.

At the cellular level, $Mg^{2+}$ is essential for maintaining enzymatic activities and energy metabolism, and its intracellular concentration is regulated by a number of ion channels and transporters on cell and organelle membranes [12–16]. Recent researches have revealed that $Mg^{2+}$ also has more active roles in regulating intracellular signal transduction [17], cellular metabolism [18,19], and cell division [20]. These studies indicate that the concentration changes of $Mg^{2+}$ is in a small range from 0.5 mM to 1.2 mM but essential for cellular functions. In neurons, the most widely known role of $Mg^{2+}$ is the extracellular blockade and regulation of N-methyl-D-aspartate (NMDA) receptors, which occurs under physiologically normal extracellular concentrations [21,22]. In addition, intracellular $Mg^{2+}$ is thought to play an indispensable role in neuronal functions [1]. Previously, we demonstrated that some neurotransmitters, as well as neuronal excitation itself, elicit changes in the intracellular $Mg^{2+}$ concentration ($[Mg^{2+}]_i$) in cultured neurons [23–26]. In particular, glutamate induced interesting changes in $[Mg^{2+}]_i$. Although it is the most abundant neurotransmitter in the mammalian brain, excessive accumulation of glutamate around neurons induces neuronal cell death in a phenomenon known as excitotoxicity [27]. Toxic concentration of extracellular glutamate induced the release of $Mg^{2+}$ from mitochondria, which is triggered by excessive $Ca^{2+}$ accumulation in the mitochondria, leading to the increase in $[Mg^{2+}]_i$ [26]. Subsequently, the $[Mg^{2+}]_i$ turns to decrease. Given that $Mg^{2+}$ deficiency is involved in neuronal diseases, the decreasing $[Mg^{2+}]_i$ phase likely has detrimental effects on neuronal survival, and thus plays a key role in excitotoxicity. However, the mechanism causing the decreasing $[Mg^{2+}]_i$ phase and its effect on excitotoxicity has not yet been elucidated.

In this study, we examined whether changes in $[Mg^{2+}]_i$ are involved in neuronal cell death via excitotoxicity in rat hippocampal neurons. We investigated the mechanisms involved in the phase in which the transient $[Mg^{2+}]_i$ increase in response to glutamate is reversed, and the effects of $[Mg^{2+}]_i$ on cellular energy metabolism. Quinidine and amiloride were used to inhibit cellular $Mg^{2+}$ extrusion. Because those inhibitors are not specific for $Mg^{2+}$ transport, it is important to confirm that both inhibitors show similar effects and that the effects were mediated via the changes in $[Mg^{2+}]_i$. Glutamate excitotoxicity results from continuous activation of the NMDA receptors on neurons, which leads to excessive $Ca^{2+}$ influx into the cytosol and overload in the mitochondria, resulting in depolarization of the mitochondrial membrane potential and the release of cell death signals [27,28]. Mitochondria begin to release apoptosis signals within 10–20 min in response to toxic concentrations of glutamate [27]. Therefore, the $Mg^{2+}$ transient elicited during the first 10 min of exposure to the glutamate stimulus may affect intracellular signals prior to the release of cell death signals under excitotoxicity. We investigated the hypothesis that the intracellular $Mg^{2+}$ homeostasis and the intracellular $Mg^{2+}$ regulatory system are key to cell protection in neuronal pathology.

## 2. Materials and Methods

### 2.1. Ethical Approval

All animal procedures were approved by the ethics committee of Keio University (permit number 09106-(7)). All methods were carried out in accordance with the relevant guidelines and regulations.

### 2.2. Dissociation Culture of Rat Hippocampal Neurons

The primary cultures of hippocampal neurons were prepared from day 18 embryonic Wistar rats (Charles River Laboratories Japan, Tokyo, Japan). Extracted hippocampi were dissociated using a dissociation kit (Sumitomo Bakelite, Tokyo, Japan). Isolated hippocampal neurons were plated on glass bottom dishes (Iwaki, Tokyo, Japan) coated with poly-D-lysine (PDL; Sigma-Aldrich, St. Louis, MO, USA) for fluorescence imaging or in 96-well plates for MTT assay, and cultured in neurobasal medium (Thermo Fisher Scientific, Waltham, MA, USA) supplemented with B-27 (Thermo Fisher Scientific, Waltham, MA, USA), 2 mM L-glutamine, 50 U/mL penicillin, and 50 µg/mL streptomycin (Nacalai Tesuque, Kyoto, Japan). The neurons were cultured at 37 °C in a humidified atmosphere of 5% $CO_2$ for 7–9 days. Neuronal culture and synaptic formation among neurons were confirmed by immunofluorescence imaging of a neuron marker, βIII-tubulin (Sigma-Aldrich, St. Louis, MO, USA), and a synapse marker, synapsin I (Abcam, Cambridge, UK) (Figure S1).

*2.3. MTT Assay*

The MTT assay was conducted to measure cell viability. Neurons at the concentration of $8 \times 10^3$ cells/well were cultured on 96-well plates for 7 days. Neurons cultured in 96-well plates were incubated in culture medium with or without an inhibitor of $Mg^{2+}$ extrusion (quinidine at 200 µM or amiloride at 500 µM; Sigma-Aldrich) and/or an inhibitor of mechanistic target of rapamycin (mTOR) (Torin1 at 2.5 µM; Chem Scene, Monmouth Junction, NJ, USA) for 10 min. Next, 100 µM glutamate with 10 µM glycine or control culture medium was applied and cells were incubated for 10 min. The culture medium containing inhibitor and/or glutamate was then replaced with normal culture medium, and the neurons were incubated for 24 h. The neurons were then incubated in culture medium containing 0.5 mg/mL MTT (Nacalai Tesuque). The culture medium was removed and 100 µL of dimethyl sulfoxide (DMSO; Nacalai Tesuque) was added to each well to dissolve the precipitate, and absorbance at 570 nm was measured using a microplate reader, Fluoroskan Ascent FL (Thermo Fisher Scientific). Cell viability in the treatment cultures is expressed as a proportion of cell viability in the control cultures.

*2.4. Simultaneous Fluorescence Imaging of Intracellular $Mg^{2+}$ and $Ca^{2+}$*

Changes in $[Mg^{2+}]_i$ and intracellular $Ca^{2+}$ concentration ($[Ca^{2+}]_i$) were measured by simultaneous imaging using the probes KMG-104-AM [29] and Fura-Red-AM (Thermo Fisher Scientific), respectively. KMG-104 is sufficiently selective to measure the $Mg^{2+}$ signal without interference from the $Ca^{2+}$ signal [29,30]. Neurons cultured on glass-bottom dishes were washed with Hanks' balanced salt solution (HBSS; NaCl 137 mM, KCl 5.4 mM, $CaCl_2$ 1.3 mM, $MgCl_2$ 0.5 mM, $MgSO_4$ 0.4 mM, $Na_2HPO_4$ 0.3 mM, $KH_2PO_4$ 0.4 mM, $NaHCO_3$ 4.2 mM, D-glucose 5.6 mM, HEPES 10 mM, pH adjusted to 7.4 with NaOH) and incubated in HBSS containing 5 µM KMG-104-AM, 10 µM Fura-Red-AM and 0.02% pluronic F-127 (Thermo Fisher Scientific) at 37 °C for 30 min. The neurons were then washed twice and incubated in HBSS, in HBSS without $CaCl_2$ (nominally $Ca^{2+}$-free HBSS) for measurements under $Ca^{2+}$-free conditions, for 15 min at 37 °C to allow for complete hydrolysis of the acetoxymethyl (AM) ester.

Fluorescence measurements were performed using fluorescence microscope, ECLIPSE TE300 (Nikon, Tokyo, Japan) equipped with 10× and 20× objective lenses, S Fluor (Nikon). Excitation light with a wavelength of approximately 488 nm was selected from a 150W Xe lamp using a monochrometer unit (Hamamatsu photonics, Shizuoka, Japan). The fluorescence signals that passed through a 510 nm dichroic mirror were separated using a 590 nm dichroic mirror and detected with a CCD camera, HiSCA (Hamamatsu photonics) using a 535/55 nm band pass filter for KMG-104 and a 600 nm long-pass filter for Fura-Red, respectively. Time-lapse images were acquired every 5 sec. Fluorescence intensity (F) was calculated as the mean intensity in a region of interest (ROI) containing the entire cell body by using Aquacosmos software (Hamamatsu photonics). The values of F recorded during time-lapse imaging were normalized by the initial fluorescence intensity ($F_0$) of each cell. The KMG-104 data were presented and analyzed as $F/F_0$. Since a decrease in the fluorescence intensity of Fura-Red indicates an increase in $[Ca^{2+}]$, the Fura-Red data were presented and analyzed as $F_0/F$.

*2.5. Fluorescence Imaging of Mitochondrial Membrane Potential*

Mitochondrial membrane potential was measured using tetramethylrhodamine ethyl ester (TMRE; Thermo Fisher Scientific). The neurons were washed with HBSS and then incubated for 20 min in HBSS containing 25 nM TMRE. Next, the dye was diluted to 2.5 nM in HBSS, and the neurons were incubated for an additional 10 min to equilibrate the dye. For measurements under high $Mg^{2+}$ conditions, cells were stained and further incubated in HBSS with 8 mM $Mg^{2+}$.

Fluorescence of the TMRE was measured using the fluorescence microscope, ECLIPSE TE300. Excitation light with a wavelength of approximately 560 nm was selected from a 150W Xe lamp using the monochrometer unit. The fluorescence signals that passed through a 580 nm dichroic mirror and a

600 nm long-pass filter were detected with the CCD camera. F was calculated as the mean fluorescence intensity in a ROI containing the entire cell body. The values of F recorded during time-lapse imaging were normalized by the $F_0$ of each cell, and the data were presented and analyzed as $F/F_0$.

*2.6. Fluorescence Imaging of Intracellular ATP*

To compare intracellular ATP concentrations, genetically encoded ATP sensor, ATeam [31], was induced in neurons using an adeno-associated virus (AAV) vector. 30 μL of the AAV vector were added to every dish and neurons were incubated for 3–7 days to ensure sufficient expression of the sensor protein. After washing the cells twice with HBSS, or with HBSS with 8 mM $Mg^{2+}$ for measurements under high $Mg^{2+}$ conditions, the fluorescence was measured.

Fluorescence imaging of the neurons was performed using a confocal laser scanning microscope system, FV-1000 (Olympus, Tokyo, Japan) with a 40× oil-immersion objective lens. ATeam was excited at 440 nm from a diode laser. The fluorescence signals were separated using a 510 nm dichroic mirror and observed at 460–500 nm for cyan fluorescent protein (CFP) and 515–615 nm for yellow fluorescent protein (YFP). Fluorescence intensity was calculated as the mean intensity in a ROI containing the entire cell body. The intracellular ATP levels are represented by the YFP to CFP fluorescence ratio (ATeam ratio).

*2.7. Statistical Analysis*

For multiple testing, analysis of variance (ANOVA) was performed, and then Dunnett's test was used to compare the control group with the other groups or Tukey's test was used to compare all possible combinations of groups, assuming parametric distribution. $p < 0.05$ was considered significantly different.

## 3. Results

*3.1. Inhibitors of $Mg^{2+}$ Extrusion Attenuated the Excitotoxicity*

Previous studies have indicated that cells express several $Mg^{2+}$ extrusion mechanisms, and quinidine, amiloride, and imipramine are used to inhibit these $Mg^{2+}$ extrusion mechanism [32,33]. Because these inhibitors are not specific for $Mg^{2+}$ transport, we checked whether both quinidine and amiloride have similar effect on neurons. First, we examined whether inhibitors of $Mg^{2+}$ extrusion mechanisms had an effect on cell viability under excitotoxic conditions. A combination of 100 μM glutamate and 10 μM glycine was used as an excitotoxic glutamate stimulus. It has been demonstrated that even 10 min of exposure to toxic concentrations of glutamate induces, to some extent, neuronal cell death 24 h after the stimulation [27]. Here, we aimed to assess the impact of the glutamate stimulus for 10 min on cell viability 24 h after the stimulus and to evaluate the effects of inhibition of $Mg^{2+}$ extrusion on it. We therefore replaced the culture medium for a fresh one not containing glutamate or inhibitors after ten-minute exposure to glutamate stimulus in the presence or absence of an inhibitor to assess the effects of the glutamate stimulus and the inhibition of $Mg^{2+}$ extrusion only during the ten-minute glutamate stimulus. Then, the cell viability after 24 h was measured with MTT assay (Figure 1a). The ten-minute exposure to glutamate stimulus induced cell death in approximately 40% of the neurons after 24 h (controls with glutamate stimulus in Figure 1b,c). This experimental condition, which induces 40% cell death, was adopted in this study because it is easy to assess the effect of inhibitors on cell viability, whether to accelerate or suppress the toxicity. Both inhibitors partially attenuated this decrease in cell viability, while exposure to the inhibitors for 20 min without glutamate stimulus had no effect on viability (Figure 1b,c).

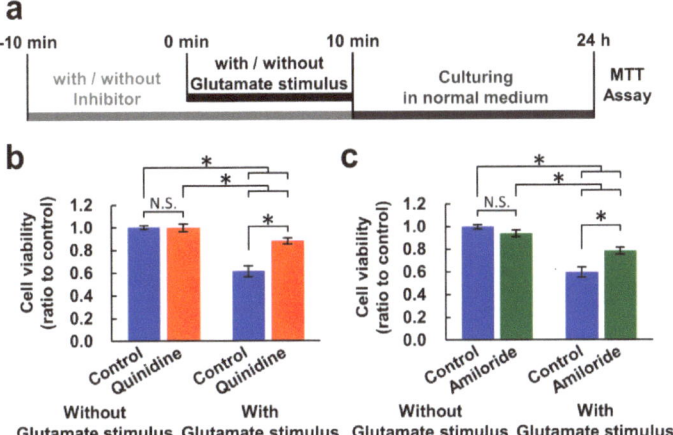

**Figure 1.** Inhibitors of $Mg^{2+}$ extrusion attenuated excitotoxicity. (**a**) Experimental procedure of the MTT assay for measuring the effect of a ten-minute glutamate stimulus. Comparison of cell viability in the presence or absence of quinidine (200 µM) (**b**) or amiloride (500 µM) (**c**) ($n$ = 30 for each). Error bars indicate SEM. * indicates $p < 0.05$ among all possible combinations by Tukey's test. N.S. indicates no statistically significant difference.

### 3.2. The Inhibitors Suppressed the Decreasing Phase of the Glutamate-Induced $Mg^{2+}$ Transient

To confirm the effects of the inhibitors on cellular ion transports immediately after the application of glutamate stimulus, $[Mg^{2+}]_i$ and $[Ca^{2+}]_i$ were simultaneously visualized using KMG-104 and Fura-Red, respectively. The glutamate stimulus elicited a gradual increase in $[Mg^{2+}]_i$ and a steep increase in $[Ca^{2+}]_i$, after which $[Mg^{2+}]_i$ began to decrease. Within 10 min, $[Mg^{2+}]_i$ in the control was below the initial concentration (Figure 2a). Quinidine (200 µM) and amiloride (500 µM) partially suppressed both the $[Mg^{2+}]_i$ decrease and the $[Ca^{2+}]_i$ increase, but did not suppress the $[Mg^{2+}]_i$ increase (Figure 2a,b). Distributions of the increases in $[Mg^{2+}]_i$ and $[Ca^{2+}]_i$ were different (Figure 2b). In addition, no morphological changes such as cell body rounding or neurite fragmentation were observed, indicating that the neurons are still alive at 10 min. Quinidine enhanced the $[Mg^{2+}]_i$ increase (Figure 2c), and both quinidine and amiloride suppressed the $Mg^{2+}$ decrease (Figure 2d). This indicates that glutamate stimulus simultaneously induced both $Mg^{2+}$ mobilization and $Mg^{2+}$ extrusion, which restricts amplitude of the increase in $[Mg^{2+}]_i$. As a result, neurons maintained a higher $[Mg^{2+}]_i$ level in the presence of the inhibitors than that in the control during the ten-minute exposure to glutamate stimulus. The inhibitors also partially attenuated the $[Ca^{2+}]_i$ peak observed within 1 min of the addition of the glutamate stimulus (Figure 2e). Following the initial peak, while the neurons in the control exhibited a further gradual increase in $[Ca^{2+}]_i$ to a level higher than the initial peak, in the presence of the inhibitors, $[Ca^{2+}]_i$ showed plateau after 4 min and did not exceed the initial peak level (Figure 2a middle panel,b,f). This was probably due to suppression of the $Ca^{2+}$ influx via the voltage-gated $Ca^{2+}$ channels (VGCC) by the higher $[Mg^{2+}]_i$ [34].

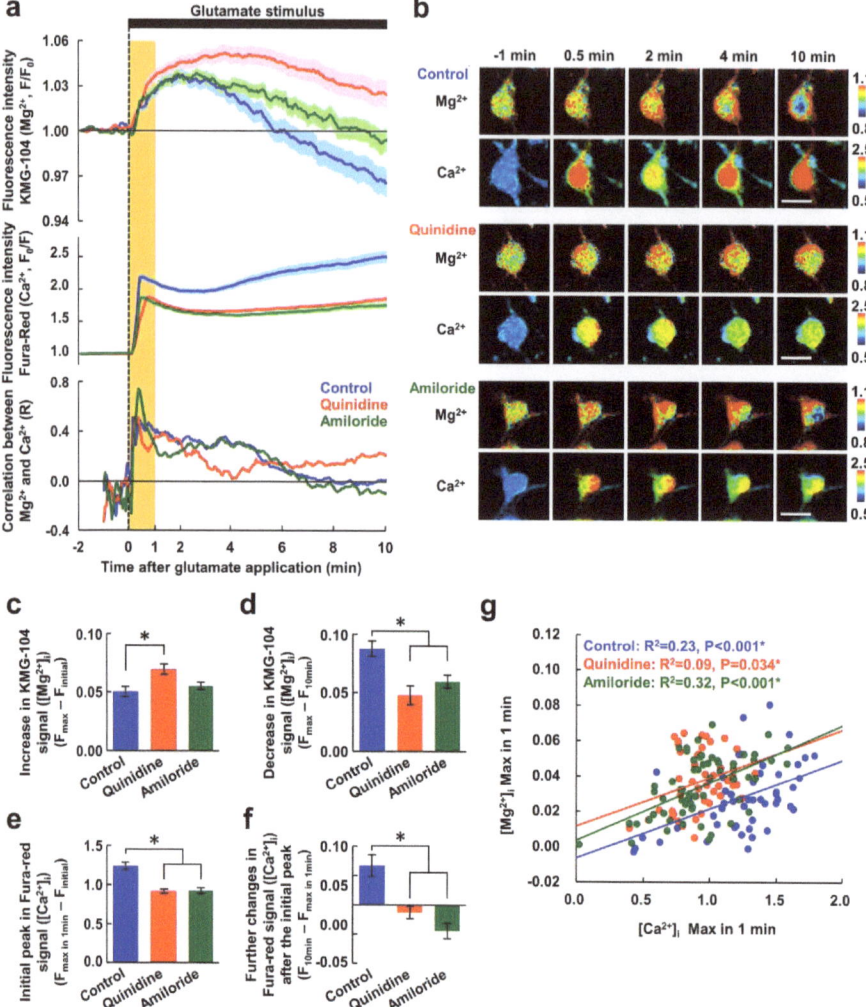

**Figure 2.** Inhibitors of $Mg^{2+}$ extrusion suppressed the decreasing phase of the glutamate-induced $Mg^{2+}$ transient. (**a**) Time-courses of $[Mg^{2+}]_i$ (upper) and $[Ca^{2+}]_i$ (middle) responses to the glutamate stimulus, and the correlation between them (bottom) in the presence or absence of inhibitor (blue: control, $n = 52$ cells; red: with quinidine at 200 µM, $n = 48$ cells; green: with amiloride at 500 µM, $n = 64$ cells; from 5 replicates each). $Mg^{2+}$ and $Ca^{2+}$ were simultaneously measured in the same neurons. Data are presented as the average (solid line) ± SEM (shaded area). The glutamate stimulus (100 µM glutamate with 10 µM glycine) was applied at 0 min (dotted line). The period 0–1 min is shaded orange. (**b**) Pseudo-colored images showing representative responses of $Mg^{2+}$ and $Ca^{2+}$ in each condition. Scale bars, 20 µm. Comparisons of amplitudes of increase in KMG-104 signal ($[Mg^{2+}]_i$) (**c**), decrease in KMG-104 signal ($[Mg^{2+}]_i$) (**d**), initial peak in Fura-red signal ($[Ca^{2+}]_i$) (**e**), and further changes in Fure-red signal ($[Ca^{2+}]_i$) after the initial peak (**f**) among the conditions extracted from the data shown in (**a**). Error bars indicate SEM. * indicates $p < 0.05$ compared with control by Dunnett's test. (**g**) Scatter plot of maximum values of $[Mg^{2+}]_i$ and $[Ca^{2+}]_i$ within the first minute (orange area in Figure 2a) and their regression lines (blue: control; red: with quinidine; green: with amiloride). $p$ values were determined from the correlation coefficient and the number of plots.

Under all three conditions, the changes in $[Mg^{2+}]_i$ and $[Ca^{2+}]_i$ were well correlated with each other immediately after the application of glutamate stimulus, and the correlation coefficient decreased with time (Figure 2a bottom panel). To further analyze the relationship between the $[Ca^{2+}]_i$ increase and the increasing phase of $[Mg^{2+}]_i$, we examined the correlation of the $[Mg^{2+}]_i$ and $[Ca^{2+}]_i$ maxima observed within 1 min of glutamate stimulus (Figure 2g). These correlated well in control neurons, because the increase in $[Mg^{2+}]_i$ is triggered by the increase in $[Ca^{2+}]_i$ as previously shown [26]. In the presence of the inhibitors, these maxima were still correlated, and although their regression lines shifted upward, they were almost parallel with each other (Figure 2g). This indicates that the inhibitors attenuated the $[Ca^{2+}]_i$ response, without affecting the $Mg^{2+}$ response, immediately after the stimulus.

### 3.3. Neurons Extruded $Mg^{2+}$ in Response to the Glutamate Stimulus

Next, we investigated whether the $Ca^{2+}$ signal is a prerequisite for the $Mg^{2+}$ extrusion. To do this, glutamate-induced responses were observed in nominally $Ca^{2+}$-free conditions. Under this conditions, there was only a small $[Ca^{2+}]_i$ response, and the $[Mg^{2+}]_i$ increase was completely abolished. However the decrease in $[Mg^{2+}]_i$ remained (Figure 3a). Combined with Figure 2g, this suggests that the $[Mg^{2+}]_i$ increasing phase is $Ca^{2+}$-dependent event. Both quinidine and amiloride suppressed this $Mg^{2+}$ decrease to a similar extent (Figure 3b). Quinidine abolished the small $[Ca^{2+}]_i$ increase evoked by the glutamate stimulus, and amiloride partially attenuated it (Figure 3c). This suggests that both inhibitors affect the glutamate-induced $Ca^{2+}$ mobilization other than $Ca^{2+}$ influx from extracellular medium. This might have some contribution to the attenuation of glutamate-induced $Ca^{2+}$ response by these inhibitors shown in Figure 2. These results also indicate that the decrease in $[Mg^{2+}]_i$ does not require the preceding $[Mg^{2+}]_i$ and $[Ca^{2+}]_i$ increases. This means that the $[Mg^{2+}]_i$-decreasing phase is not a homeostatic cellular response attempting to maintain a normal $[Mg^{2+}]_i$, but rather a glutamate-induced activation of $Mg^{2+}$ extrusion via channels and/or transporters, and that this occurs in response to a neuronal excitation or an intracellular signal other than $Ca^{2+}$. Taken together, these results indicate that glutamate stimulus induces both $Ca^{2+}$-dependent $Mg^{2+}$ release from the mitochondria and $Mg^{2+}$ extrusion from the cytosol, simultaneously.

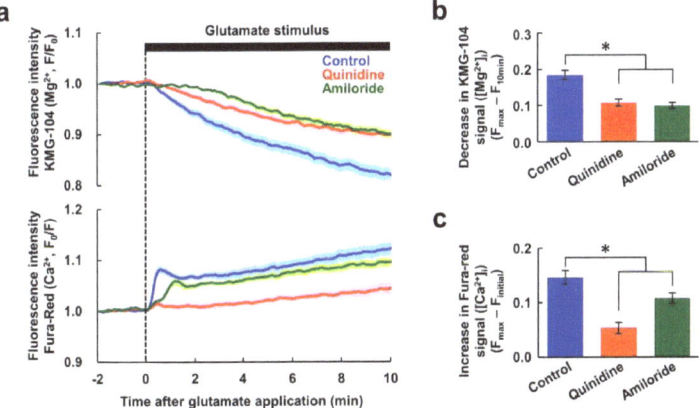

**Figure 3.** The $Mg^{2+}$ and $Ca^{2+}$ increases were not required for $Mg^{2+}$ extrusion. (**a**) Time-courses of $[Mg^{2+}]_i$ (upper) and $[Ca^{2+}]_i$ (bottom) response to glutamate stimulus in nominally $Ca^{2+}$-free conditions in the presence or absence of inhibitor (blue: control, $n = 60$ cells; red: with quinidine at 200 µM, $n = 52$ cells; green: with amiloride at 500 µM, $n = 54$ cells, from 5 replicates each). Data are presented as the average (solid line) ± SEM (shaded area). Comparison of decrease in KMG-104 signal ($[Mg^{2+}]_i$) (**b**) and increase in Fura-red signal ($[Ca^{2+}]_i$) (**c**) at 10 min after application of the glutamate stimulus extracted from the data shown in (**a**). Error bars indicate SEM. * indicates $p < 0.05$ compared with control by Dunnett's test.

## 3.4. Effect of the Inhibition of the $Mg^{2+}$ Extrusion on Cellular Energy Metabolism

Next, we investigated whether inhibitors of $Mg^{2+}$ extrusion had any effect on cellular energy metabolism. Since mitochondria play a central role in neuronal ATP synthesis, changes in mitochondrial membrane potential during glutamate stimulus were measured. In response to the glutamate stimulus, mitochondrial membrane potential decreased gradually (Figure 4a). It was still decreasing at 10 min, indicating the neurons still alive at this point. Quinidine partially attenuated this decrease, but amiloride had no effect on it (Figure 4b). Thus, although these inhibitors attenuated the glutamate-induced $[Ca^{2+}]_i$ increase to the same extent (Figure 2), amiloride did not suppress the depolarization of mitochondrial membrane potential. To further confirm the contribution of $Mg^{2+}$ for maintenance of mitochondrial membrane potential, glutamate stimulus-induced changes in the mitochondrial membrane potential under high $Mg^{2+}$ conditions were compared to those under normal conditions. High $Mg^{2+}$ partially attenuated the decrease in mitochondrial membrane potential (Figure 4c,d), presumably by keeping $[Mg^{2+}]_i$ high. This supports the idea that $Mg^{2+}$ has a protective effect on depolarization of mitochondrial membrane potential.

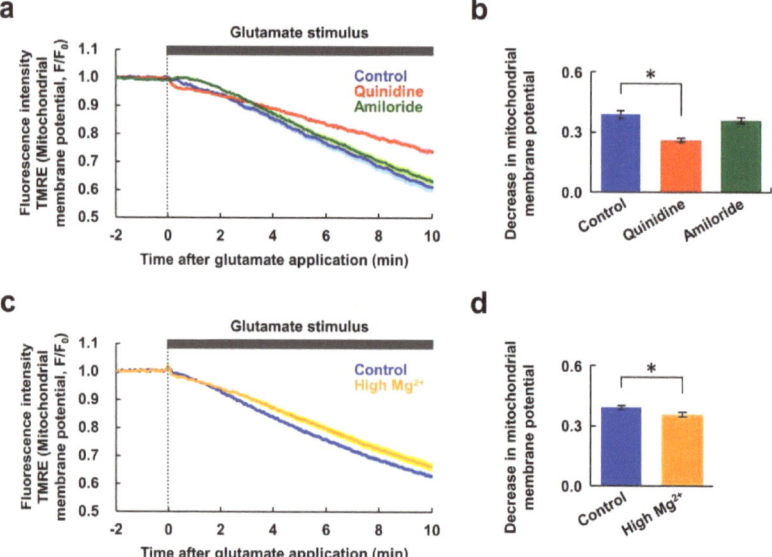

**Figure 4.** Changes in mitochondrial membrane potential in response to glutamate stimulus. (**a**) Time-courses of changes in mitochondrial membrane potential in response to the glutamate stimulus (blue: control, $n = 90$ cells; red: with quinidine at 200 µM, $n = 89$ cells; green: with amiloride at 500 µM, $n = 100$ cells, from 5 replicates each). Data are presented as the average (solid line) ± SEM (shaded area). (**b**) Comparison of decreases in mitochondrial membrane potential after 10 min extracted from the data shown in (**a**). Error bars indicate SEM. * indicates $p < 0.05$ compared with control by Dunnett's test. (**c**) Time-courses of changes in mitochondrial membrane potential in response to the glutamate stimulus (blue: control, $n = 75$ cells; yellow: under high $Mg^{2+}$ (8 mM) condition, $n = 80$ cells, from 5 replicates each). Data are presented as the average (solid line) ± SEM (shaded area). (**d**) Comparison of decreases in mitochondrial membrane potential after 10 min extracted from the data shown in (**c**). Error bars indicate SEM. * indicates $p < 0.05$ by t-test.

We also investigated whether the inhibition of $Mg^{2+}$ extrusion and the resulting high $[Mg^{2+}]_i$ lead to the maintenance of cellular ATP concentrations during the excitotoxicity process. Intracellular ATP concentrations were compared using a genetically encoded ATP sensor, ATeam [31]. The presence of quinidine or amiloride did not affect ATP levels in steady-state neurons, but both inhibitors partially attenuated the decrease in ATP levels evoked by the glutamate stimulus (Figure 5a). High $Mg^{2+}$ also suppressed the glutamate-induced decrease in ATP level (Figure 5b). These results suggest that the high $[Mg^{2+}]_i$ resulting from the inhibition of $Mg^{2+}$ extrusion contributes to the maintenance of intracellular ATP levels.

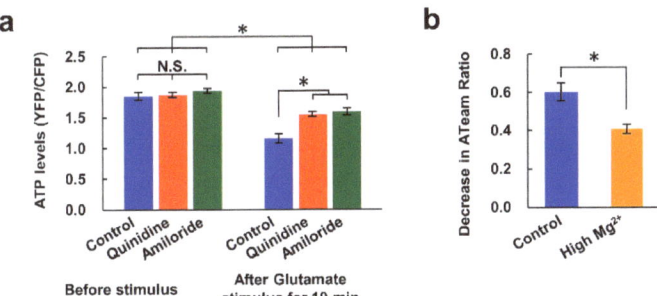

**Figure 5.** Inhibitors of $Mg^{2+}$ extrusion attenuated the glutamate-induced decrease in intracellular ATP level. (**a**) Comparison of intracellular ATP levels (YFP/CFP ratio of ATeam) before and 10 min after application of the glutamate stimulus in the presence or absence of inhibitor ($n$ = 10, 16, 18 cells from 6 replicates for each). Error bars indicate SEM. * indicates $p$ < 0.05 among all possible combinations by Tukey's test. N.S. indicates no statistically significant difference. (**b**) Comparison of decrease in ATeam ratio induced by glutamate stimulus in 10 min under control and high $Mg^{2+}$ (8 mM) conditions (control: $n$ = 32 cells from 9 replicates; High $Mg^{2+}$: $n$ = 34 cells from 11 replicates). Error bars indicate SEM. * indicates $p$ < 0.05 by t-test.

## 3.5. Involvement of mTOR in the Attenuation of Excitotoxicity by $Mg^{2+}$

In the previous study, we demonstrated that increase in $[Mg^{2+}]_i$ activates mTOR in cultured hippocampal neurons [23]. mTOR is an important signal implicated in the regulation of energy metabolism, cell growth and cell death [35,36]. We therefore examined the effect of mTOR inhibitor, Torin1, on the glutamate-induced neuronal cells death and its attenuation by the inhibitors of $Mg^{2+}$ extrusion, according to the procedure shown in Figure 1a. While torin1 did not enhance the toxicity of glutamate stimulus, it abolished the attenuation of the toxicity by quinidine or amiloride (Figure 6). This suggests that the mTOR signal is involved in the attenuation of excitotoxicity by increasing $[Mg^{2+}]_i$.

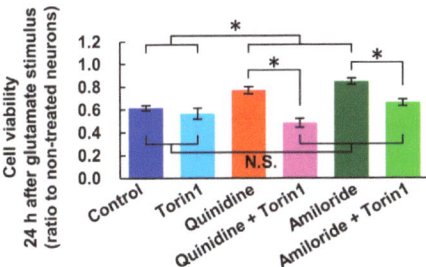

**Figure 6.** Inhibitor of mTOR abolished the attenuation of excitotoxicity by inhibiting $Mg^{2+}$ extrusion. Comparison of cell viability 24 h after ten-minute glutamate stimulus with or without inhibitor of mTOR (Torin1: 2.5 µM) and/or inhibitors of $Mg^{2+}$ extrusion (Quinidine: 200 µM, Amiloride: 500 µM) ($n$ = 12, 9, 12, 12, 12, 12 for each). Error bars indicate SEM. * indicates $p$ < 0.05 among all possible combinations by Tukey's test. N.S. indicates no statistically significant difference.

## 4. Discussion

In this study, we demonstrated that inhibitors of $Mg^{2+}$ extrusion attenuate glutamate excitotoxicity in cultured rat hippocampal neurons (Figure 1). A toxic concentration of glutamate induced $Mg^{2+}$ extrusion from neurons, which was suppressed by both quinidine and amiloride (Figures 2 and 3). These inhibitors also attenuated the glutamate-induced depression of cellular energy metabolism (Figure 5). The protective effect of those inhibitors against excitotoxicity was abolished by inhibition of mTOR (Figure 6). While quinidine and amiloride partially attenuated the glutamate-evoked $[Ca^{2+}]_i$ response (Figure 2e,f), in particular, mobilization of $Ca^{2+}$ from other than extracellular medium (Figure 3), they did not suppress the increasing phase of the $[Mg^{2+}]_i$ transient (Figure 2c), in spite of its $Ca^{2+}$-dependency (Figures 2g and 3). Amiloride also had no effect on the glutamate-induced mitochondrial depolarization (Figure 4), despite attenuating the cytosolic $Ca^{2+}$ increase (Figure 2e,f). Given that the depolarization of mitochondrial membrane potential is a critical event in the process of excitotoxicity [27], the partial attenuation of the $[Ca^{2+}]_i$ increase by the amiloride does not affect glutamate excitotoxicity. Therefore, the neuroprotective effect of the inhibitors is probably due to the maintenance of high $[Mg^{2+}]_i$ during the glutamate stimulus. This idea is also supported by the results that glutamate-induced decrease in mitochondrial membrane potential and cytosolic ATP level were partially attenuated under high $Mg^{2+}$ conditions (Figure 4c,d and Figure 5b), presumably by keeping $[Mg^{2+}]_i$ high. Our data therefore indicate the importance of $Mg^{2+}$ regulation in neuronal cell death and survival.

One of the key roles of $Mg^{2+}$ in ensuring cell survival involves the regulation of mitochondrial functions [37]. It affects enzymatic activities in the tricarboxylic acid (TCA) cycle in mitochondria; thus, a decrease in mitochondrial $Mg^{2+}$ concentration results in a downregulation of mitochondrial membrane potential and ATP synthesis [38–40]. On the other hand, extra-mitochondrial $Mg^{2+}$ suppresses mitochondrial $Ca^{2+}$ uptake [41] which depolarizes mitochondrial membrane potential. Either or both of these processes might contribute to the attenuation of mitochondrial membrane potential depolarization by quinidine (Figure 4) and the suppression of the decrease in ATP levels caused by quinidine and amiloride (Figure 5). The idea that elevated $[Mg^{2+}]_i$ contributes to the maintenance of cellular ATP levels was also supported by our previous work [42]. Although amiloride did not inhibit the depolarization of mitochondrial membrane potential (Figure 4), it did suppress the decrease in ATP levels and cell death (Figures 1 and 5). This suggests that $Mg^{2+}$ affects not only mitochondrial functions but also other mechanisms that are involved in maintaining the cellular ATP concentration and cell viability.

One of the ways in which $Mg^{2+}$ might do this is by regulating intracellular signal transductions [17]. In particular, mTOR is considered an important downstream target of $Mg^{2+}$ [18,19,43]. The mTOR pathway is implicated in the regulation of cellular metabolism and proliferation [36]. In pancreatic cancer cells, overexpression of SLC41A1, a membrane protein for $Mg^{2+}$ extrusion [32], and the resulting low $[Mg^{2+}]_i$, inhibited the Akt/mTOR pathway and cancer proliferation [44]. In cultured hippocampal neurons, $[Mg^{2+}]_i$ within the physiological concentration range is correlated with mTOR activity [23]. These findings strongly suggest that $Mg^{2+}$ is a key regulator of mTOR and mTOR-related signal transductions. Inhibition of the Akt/mTOR pathway in cultured neurons in response to toxic concentrations of glutamate has been reported [45]. This inhibition might result from the glutamate-induced decrease in $[Mg^{2+}]_i$ demonstrated in this study (Figures 2 and 3). The fact that the Akt/mTOR signal is also involved in regulation of both cellular ATP synthesis and also apoptotic signals [35,46,47] would explain why maintaining a high $[Mg^{2+}]_i$ prevent neuronal cell death. In this study, we showed that inhibition of mTOR abolished the attenuation of glutamate excitotoxicity by inhibiting $Mg^{2+}$ extrusion (Figure 6). This suggests that maintaining high $[Mg^{2+}]_i$ protect neurons against excitotoxicity via the mTOR signal.

As discussed above and summarized in Figure 7, $Mg^{2+}$ is a key factor in saving neurons from neurodegenerative diseases. Since inhibition of cellular $Mg^{2+}$ extrusion and resulting maintenance of high $[Mg^{2+}]_i$ protect cells from neurodegenerative disorders, the $Mg^{2+}$ extrusion mechanisms might be a therapeutic target for neuronal diseases. Among $Mg^{2+}$ channels and transporters, SLC41A1 is considered to be the major transporter for cellular $Mg^{2+}$ extrusion [32,48], while it was also suggested that cells has several different mechanisms for extruding cytosolic $Mg^{2+}$ [49]. Preventing cellular $Mg^{2+}$ loss by inhibiting such transporters might delay the progression of neurodegeneration. Because $Mg^{2+}$ is a broad-spectrum regulator of cellular metabolisms and signaling, change in $[Mg^{2+}]_i$ affects multiple aspects of the process of excitotoxicity, even though it is not a main signal for inducing neuronal cell death (Figure 7). In present study, we have demonstrated that keeping $Mg^{2+}$ inside cells is important for neuronal survival under conditions of cellular stress. It may protect neurons from neurodegeneration also in vivo because it has been reported that gain-of-function mutation in SLC41A1, that leads to enhanced $Mg^{2+}$ release from cytosol, and mutation in a $Mg^{2+}$ and $Ca^{2+}$ permeable channel, TRPM7, that attenuates cellular $Mg^{2+}$ uptake, are associated with the pathogenesis of neurodegenerative diseases in human brain [50,51]. To support this, further studies on $Mg^{2+}$ dynamics in vivo are needed. Our results demonstrated in this study suggest the potential of cellular $Mg^{2+}$ homeostasis and the $Mg^{2+}$ transport system for therapy and the prevention of neurodegenerative diseases.

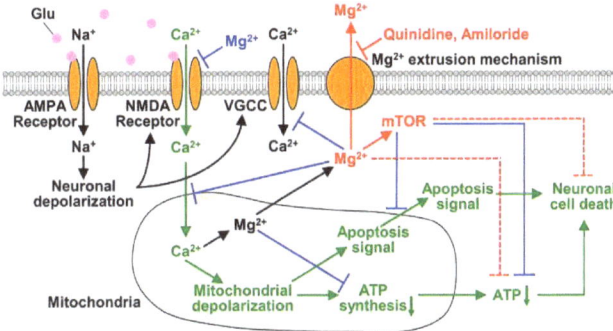

**Figure 7.** Schematic of roles of $Mg^{2+}$ in excitotoxicity. The transport and effects of $Mg^{2+}$ and mTOR demonstrated in this study are shown in red. In this study, we demonstrated that glutamate stimulus induces not only $Mg^{2+}$ released from mitochondria but also $Mg^{2+}$ extrusion from the cytosol, and quinidine and amiloride inhibit it. Maintaining $Mg^{2+}$ in neurons suppressed glutamate-induced decrease in cellular ATP level and also attenuated neuronal cell death via mTOR signaling pathway. Dotted lines indicate pathways that are still obscure whether they are direct or indirect effects. The effects of $Mg^{2+}$ demonstrated previous studies are shown in blue, and the main excitotoxic signals are shown in green.

**Supplementary Materials:** The following are available online at http://www.mdpi.com/2072-6643/12/9/2768/s1, Figure S1: Immunofluorescence images of a synapse marker.

**Author Contributions:** Data curation, Y.S.; Formal analysis, Y.S. and R.Y.; Funding acquisition, K.O.; Project administration, K.O.; Supervision, K.O.; Writing—original draft, Y.S.; Writing—review & editing, R.Y., K.H. and K.O. All authors have read and agreed to the published version of the manuscript.

**Funding:** This research was funded by JSPS KAKENHI Grant Number 16H01751 and Life Innovation Platform (LIP) YOKOHAMA.

**Acknowledgments:** We thank H. Imamura and H. Noji for providing the ATeam plasmid.

**Conflicts of Interest:** The authors declare no conflict of interest.

## References

1. Yamanaka, R.; Shindo, Y.; Oka, K. Magnesium Is a Key Player in Neuronal Maturation and Neuropathology. *Int. J. Mol. Sci.* **2019**, *20*, 3439. [CrossRef]
2. Kirkland, A.; Sarlo, G.; Holton, K. The Role of Magnesium in Neurological Disorders. *Nutrients* **2018**, *10*, 730. [CrossRef]
3. Slutsky, I.; Abumaria, N.; Wu, L.J.; Huang, C.; Zhang, L.; Li, B.; Zhao, X.; Govindarajan, A.; Zhao, M.G.; Zhuo, M.; et al. Enhancement of Learning and Memory by Elevating Brain Magnesium. *Neuron* **2010**, *65*, 165–177. [CrossRef] [PubMed]
4. Jia, S.; Liu, Y.; Shi, Y.; Ma, Y.; Hu, Y.; Wang, M.; Li, X. Elevation of Brain Magnesium Potentiates Neural Stem Cell Proliferation in the Hippocampus of Young and Aged Mice. *J. Cell. Physiol.* **2016**, *231*, 1903–1912. [CrossRef] [PubMed]
5. Andrási, E.; Páli, N.; Molnár, Z.; Kösel, S. Brain aluminum, magnesium and phosphorus contents of control and Alzheimer-diseased patients. *J. Alzheimer's Dis.* **2005**, *7*, 273–284.
6. Yasui, M.; Kihira, T.; Ota, K. Calcium, magnesium and aluminum concentrations in Parkinson's disease. *Neurotoxicology* **1992**, *13*, 593–600. [PubMed]
7. Oyanagi, K.; Kawakami, E.; Kikuchi-Horie, K.; Ohara, K.; Ogata, K.; Takahama, S.; Wada, M.; Kihira, T.; Yasui, M. Magnesium deficiency over generations in rats with special references to the pathogenesis of the parkinsonism-dementia complex and amyotrophic lateral sclerosis of Guam. *Neuropathology* **2006**, *26*, 115–128. [CrossRef]
8. Hashimoto, T.; Nishi, K.; Nagasao, J.; Tsuji, S.; Oyanagi, K. Magnesium exerts both preventive and ameliorating effects in an in vitro rat Parkinson disease model involving 1-methyl-4-phenylpyridinium (MPP$^+$) toxicity in dopaminergic neurons. *Brain Res.* **2008**, *1197*, 143–151. [CrossRef]
9. Shen, Y.; Dai, L.; Tian, H.; Xu, R.; Li, F.; Li, Z.; Zhou, J.; Wang, L.; Dong, J.; Sun, L. Treatment of Magnesium-L-Threonate Elevates the Magnesium Level in the Cerebrospinal Fluid and Attenuates Motor Deficits and Dopamine Neuron Loss in a Mouse Model of Parkinson's disease. *Neuropsychiatr. Dis. Treat.* **2019**, *15*, 3143–3153. [CrossRef]
10. Shindo, Y.; Yamanaka, R.; Suzuki, K.; Hotta, K.; Oka, K. Altered expression of Mg$^{2+}$ transport proteins during Parkinson's disease-like dopaminergic cell degeneration in PC12 cells. *Biochim. Biophys. Acta Mol. Cell Res.* **2016**, *1863*, 1979–1984. [CrossRef]
11. Li, W.; Yu, J.; Liu, Y.; Huang, X.; Abumaria, N.; Zhu, Y.; Huang, X.; Xiong, W.; Ren, C.; Liu, X.G.; et al. Elevation of brain magnesium prevents synaptic loss and reverses cognitive deficits in Alzheimer's disease mouse model. *Mol. Brain* **2014**, *7*, 1–20. [CrossRef] [PubMed]
12. Romani, A.M.P. Cellular magnesium homeostasis. *Arch. Biochem. Biophys.* **2011**, *512*, 1–23. [CrossRef] [PubMed]
13. de Baaij, J.H.F.; Hoenderop, J.G.J.; Bindels, R.J.M. Magnesium in Man: Implications for Health and Disease. *Physiol. Rev.* **2015**, *95*, 1–46. [CrossRef] [PubMed]
14. Kubota, T.; Shindo, Y.; Tokuno, K.; Komatsu, H.; Ogawa, H.; Kudo, S.; Kitamura, Y.; Suzuki, K.; Oka, K. Mitochondria are intracellular magnesium stores: Investigation by simultaneous fluorescent imagings in PC12 cells. *Biochim. Biophys. Acta Mol. Cell Res.* **2005**, *1744*, 19–28. [CrossRef]
15. Quamme, G.A. Molecular identification of ancient and modern mammalian magnesium transporters. *Am. J. Physiol. Physiol.* **2010**, *298*, C407–C429. [CrossRef]
16. Nadler, M.J.S.; Hermosura, M.C.; Inabe, K.; Perraud, A.-L.; Zhu, Q.; Stokes, A.J.; Kurosaki, T.; Kinet, J.-P.; Penner, R.; Scharenberg, A.M.; et al. LTRPC7 is a Mg·ATP-regulated divalent cation channel required for cell viability. *Nature* **2001**, *411*, 590–595. [CrossRef]
17. Li, F.Y.; Chaigne-Delalande, B.; Kanellopoulou, C.; Davis, J.C.; Matthews, H.F.; Douek, D.C.; Cohen, J.I.; Uzel, G.; Su, H.C.; Lenardo, M.J. Second messenger role for Mg$^{2+}$ revealed by human T-cell immunodeficiency. *Nature* **2011**, *475*, 471–476. [CrossRef]
18. Feeney, K.A.; Hansen, L.L.; Putker, M.; Olivares-Yañez, C.; Day, J.; Eades, L.J.; Larrondo, L.F.; Hoyle, N.P.; O'Neill, J.S.; van Ooijen, G. Daily magnesium fluxes regulate cellular timekeeping and energy balance. *Nature* **2016**, *532*, 375–379. [CrossRef]
19. Rubin, H. The logic of the Membrane, Magnesium, Mitosis (MMM) model for the regulation of animal cell proliferation. *Arch. Biochem. Biophys.* **2007**, *458*, 16–23. [CrossRef]

20. Maeshima, K.; Matsuda, T.; Shindo, Y.; Imamura, H.; Tamura, S.; Imai, R.; Kawakami, S.; Nagashima, R.; Soga, T.; Noji, H.; et al. A Transient Rise in Free $Mg^{2+}$ Ions Released from ATP-Mg Hydrolysis Contributes to Mitotic Chromosome Condensation. *Curr. Biol.* **2018**, *28*, 444–451.e6. [CrossRef]
21. Slutsky, I.; Sadeghpour, S.; Li, B.; Liu, G. Enhancement of Synaptic Plasticity through Chronically Reduced $Ca^{2+}$ Flux during Uncorrelated Activity. *Neuron* **2004**, *44*, 835–849. [CrossRef]
22. Nowak, L.; Bregestovski, P.; Ascher, P.; Herbet, A.; Prochiantz, A. Magnesium gates glutamate-activated channels in mouse central neurones. *Nature* **1984**, *307*, 462–465. [CrossRef] [PubMed]
23. Yamanaka, R.; Shindo, Y.; Hotta, K.; Suzuki, K.; Oka, K. GABA-Induced Intracellular $Mg^{2+}$ Mobilization Integrates and Coordinates Cellular Information Processing for the Maturation of Neural Networks. *Curr. Biol.* **2018**, *28*, 3984–3991.e5. [CrossRef] [PubMed]
24. Yamanaka, R.; Shindo, Y.; Hotta, K.; Suzuki, K.; Oka, K. NO/cGMP/PKG signaling pathway induces magnesium release mediated by mitoK ATP channel opening in rat hippocampal neurons. *FEBS Lett.* **2013**, *587*, 2643–2648. [CrossRef] [PubMed]
25. Yamanaka, R.; Shindo, Y.; Karube, T.; Hotta, K.; Suzuki, K.; Oka, K. Neural depolarization triggers $Mg^{2+}$ influx in rat hippocampal neurons. *Neuroscience* **2015**, *310*, 731–741. [CrossRef] [PubMed]
26. Shindo, Y.; Fujimoto, A.; Hotta, K.; Suzuki, K.; Oka, K. Glutamate-induced calcium increase mediates magnesium release from mitochondria in rat hippocampal neurons. *J. Neurosci. Res.* **2010**, *88*, 3125–3132. [CrossRef]
27. Abramov, A.Y.; Duchen, M.R. Mechanisms underlying the loss of mitochondrial membrane potential in glutamate excitotoxicity. *Biochim. Biophys. Acta Bioenerg.* **2008**, *1777*, 953–964. [CrossRef]
28. Rueda, C.B.; Llorente-Folch, I.; Traba, J.; Amigo, I.; Gonzalez-Sanchez, P.; Contreras, L.; Juaristi, I.; Martinez-Valero, P.; Pardo, B.; del Arco, A.; et al. Glutamate excitotoxicity and $Ca^{2+}$-regulation of respiration: Role of the $Ca^{2+}$ activated mitochondrial transporters (CaMCs). *Biochim. Biophys. Acta Bioenerg.* **2016**, *1857*, 1158–1166. [CrossRef]
29. Komatsu, H.; Iwasawa, N.; Citterio, D.; Suzuki, Y.; Kubota, T.; Tokuno, K.; Kitamura, Y.; Oka, K.; Suzuki, K. Design and Synthesis of Highly Sensitive and Selective Fluorescein-Derived Magnesium Fluorescent Probes and Application to Intracellular 3D $Mg^{2+}$ Imaging. *J. Am. Chem. Soc.* **2004**, *126*, 16353–16360. [CrossRef]
30. Trapani, V.; Farruggia, G.; Marraccini, C.; Iotti, S.; Cittadini, A.; Wolf, F.I. Intracellular magnesium detection: Imaging a brighter future. *Analyst* **2010**, *135*, 1855–1866. [CrossRef]
31. Imamura, H.; Huynh Nhat, K.P.; Togawa, H.; Saito, K.; Iino, R.; Kato-Yamada, Y.; Nagai, T.; Noji, H. Visualization of ATP levels inside single living cells with fluorescence resonance energy transfer-based genetically encoded indicators. *Proc. Natl. Acad. Sci. USA* **2009**, *106*, 15651–15656. [CrossRef] [PubMed]
32. Kolisek, M.; Nestler, A.; Vormann, J.; Schweigel-Röntgen, M. Human gene SLC41A1 encodes for the $Na^+/Mg^{2+}$ exchanger. *Am. J. Physiol. Physiol.* **2012**, *302*, C318–C326. [CrossRef] [PubMed]
33. Cefaratti, C.; Romani, A.; Scarpa, A. Characterization of two $Mg^{2+}$ transporters in sealed plasma membrane vesicles from rat liver. *Am. J. Physiol. Physiol.* **1998**, *275*, C995–C1008. [CrossRef] [PubMed]
34. Brunet, S.; Scheuer, T.; Catterall, W.A. Cooperative regulation of Cav12 channels by intracellular $Mg^{2+}$, the proximal C-terminal EF-hand, and the distal C-terminal domain. *J. Gen. Physiol.* **2009**, *134*, 81–94. [CrossRef]
35. Zheng, X.; Boyer, L.; Jin, M.; Kim, Y.; Fan, W.; Bardy, C.; Berggren, T.; Evans, R.M.; Gage, F.H.; Hunter, T. Alleviation of neuronal energy deficiency by mTOR inhibition as a treatment for mitochondria-related neurodegeneration. *eLife* **2016**, *5*, 1–23. [CrossRef]
36. Saxton, R.A.; Sabatini, D.M. mTOR Signaling in Growth, Metabolism, and Disease. *Cell* **2017**, *168*, 960–976. [CrossRef]
37. Pilchova, I.; Klacanova, K.; Tatarkova, Z.; Kaplan, P.; Racay, P. The Involvement of $Mg^{2+}$ in Regulation of Cellular and Mitochondrial Functions. *Oxid. Med. Cell. Longev.* **2017**, *2017*, 6797460. [CrossRef]
38. Yamanaka, R.; Tabata, S.; Shindo, Y.; Hotta, K.; Suzuki, K.; Soga, T.; Oka, K. Mitochondrial $Mg^{2+}$ homeostasis decides cellular energy metabolism and vulnerability to stress. *Sci. Rep.* **2016**, *6*, 30027. [CrossRef]
39. Panov, A.; Scarpa, A. $Mg^{2+}$ Control of Respiration in Isolated Rat Liver Mitochondria. *Biochemistry* **1996**, *35*, 12849–12856. [CrossRef]
40. Rodríguez-Zavala, J.S.; Moreno-Sánchez, R. Modulation of Oxidative Phosphorylation by $Mg^{2+}$ in Rat Heart Mitochondria. *J. Biol. Chem.* **1998**, *273*, 7850–7855. [CrossRef]

41. Boelens, A.D.; Pradhan, R.K.; Blomeyer, C.A.; Camara, A.K.S.; Dash, R.K.; Stowe, D.F. Extra-matrix $Mg^{2+}$ limits $Ca^{2+}$ uptake and modulates $Ca^{2+}$ uptake–independent respiration and redox state in cardiac isolated mitochondria. *J. Bioenerg. Biomembr.* **2013**, *45*, 203–218. [CrossRef] [PubMed]
42. Shindo, Y.; Yamanaka, R.; Suzuki, K.; Hotta, K.; Oka, K. Intracellular magnesium level determines cell viability in the MPP+ model of Parkinson's disease. *Biochim. Biophys. Acta Mol. Cell Res.* **2015**, *1853*, 3182–3191. [CrossRef] [PubMed]
43. Trapani, V.; Wolf, F.I. Dysregulation of $Mg^{2+}$ homeostasis contributes to acquisition of cancer hallmarks. *Cell Calcium* **2019**, *83*, 102078. [CrossRef] [PubMed]
44. Xie, J.; Cheng, C.; Zhu, X.Y.; Shen, Y.H.; Song, L.B.; Chen, H.; Chen, Z.; Liu, L.M.; Meng, Z.Q. Magnesium transporter protein solute carrier family 41 member 1 suppresses human pancreatic ductal adenocarcinoma through magnesium-dependent Akt/mTOR inhibition and bax-associated mitochondrial apoptosis. *Aging (Albany NY)* **2019**, *11*, 2681–2698. [CrossRef]
45. Pomytkin, I.; Krasil'nikova, I.; Bakaeva, Z.; Surin, A.; Pinelis, V. Excitotoxic glutamate causes neuronal insulin resistance by inhibiting insulin receptor/Akt/mTOR pathway. *Mol. Brain* **2019**, *12*, 112. [CrossRef]
46. Sponder, G.; Abdulhanan, N.; Fröhlich, N.; Mastrototaro, L.; Aschenbach, J.R.; Röntgen, M.; Pilchova, I.; Cibulka, M.; Racay, P.; Kolisek, M. Overexpression of $Na^+/Mg^{2+}$ exchanger SLC41A1 attenuates pro-survival signaling. *Oncotarget* **2018**, *9*, 5084–5104. [CrossRef]
47. Hung, Y.P.; Teragawa, C.; Kosaisawe, N.; Gillies, T.E.; Pargett, M.; Minguet, M.; Distor, K.; Rocha-Gregg, B.L.; Coloff, J.L.; Keibler, M.A.; et al. Akt regulation of glycolysis mediates bioenergetic stability in epithelial cells. *eLife* **2017**, *6*, 1–25. [CrossRef]
48. Schäffers, O.J.M.; Hoenderop, J.G.J.; Bindels, R.J.M.; de Baaij, J.H.F. The rise and fall of novel renal magnesium transporters. *Am. J. Physiol. Physiol.* **2018**, *314*, F1027–F1033. [CrossRef]
49. Cefaratti, C.; Romani, A.M.P. Functional characterization of two distinct $Mg^{2+}$ extrusion mechanisms in cardiac sarcolemmal vesicles. *Mol. Cell. Biochem.* **2007**, *303*, 63–72. [CrossRef]
50. Kolisek, M.; Sponder, G.; Mastrototaro, L.; Smorodchenko, A.; Launay, P.; Vormann, J.; Schweigel-Röntgen, M. Substitution pA350V in $Na^+/Mg^{2+}$ Exchanger SLC41A1, Potentially Associated with Parkinson's Disease, Is a Gain-of-Function Mutation. *PLoS ONE* **2013**, *8*, e71096. [CrossRef]
51. Hermosura, M.C.; Nayakanti, H.; Dorovkov, M.V.; Calderon, F.R.; Ryazanov, A.G.; Haymer, D.S.; Garruto, R.M. A TRPM7 variant shows altered sensitivity to magnesium that may contribute to the pathogenesis of two Guamanian neurodegenerative disorders. *Proc. Natl. Acad. Sci. USA* **2005**, *102*, 11510–11515. [CrossRef] [PubMed]

© 2020 by the authors. Licensee MDPI, Basel, Switzerland. This article is an open access article distributed under the terms and conditions of the Creative Commons Attribution (CC BY) license (http://creativecommons.org/licenses/by/4.0/).

*Review*

# Headaches and Magnesium: Mechanisms, Bioavailability, Therapeutic Efficacy and Potential Advantage of Magnesium Pidolate

**Jeanette A. Maier [1],\*, Gisele Pickering [2], Elena Giacomoni [3], Alessandra Cazzaniga [1] and Paolo Pellegrino [3]**

1. Dipartimento di Scienze Biomediche e Cliniche L. Sacco, Università di Milano, 20157 Milano, Italy; alessandra.cazzaniga@unimi.it
2. Department of Clinical Pharmacology, University Hospital and Inserm 1107 Fundamental and Clinical Pharmacology of Pain, Medical Faculty, F-63000 Clermont-Ferrand, France; gisele.pickering@uca.fr
3. Sanofi Consumer Health Care, 20158 Milan, Italy; Elena.Giacomoni@sanofi.com (E.G.); Paolo.Pellegrino@sanofi.com (P.P.)
\* Correspondence: jeanette.maier@unimi.it

Received: 23 June 2020; Accepted: 21 August 2020; Published: 31 August 2020

**Abstract:** Magnesium deficiency may occur for several reasons, such as inadequate intake or increased gastrointestinal or renal loss. A large body of literature suggests a relationship between magnesium deficiency and mild and moderate tension-type headaches and migraines. A number of double-blind randomized placebo-controlled trials have shown that magnesium is efficacious in relieving headaches and have led to the recommendation of oral magnesium for headache relief in several national and international guidelines. Among several magnesium salts available to treat magnesium deficiency, magnesium pidolate may have high bioavailability and good penetration at the intracellular level. Here, we discuss the cellular and molecular effects of magnesium deficiency in the brain and the clinical evidence supporting the use of magnesium for the treatment of headaches and migraines.

**Keywords:** magnesium; pidolate; deficiency; headache; migraine; BBB

## 1. Background

A large body of literature suggests a relationship between magnesium deficiency and mild and moderate tension-type headaches and migraines [1–9]. The International Classification of Headache Disorders (ICHD-3-beta) divides all headache entities into primary and secondary disorders [10] and approximately 90% of headaches seen in general practice are of the primary variety, such as migraine, tension-type headache, or cluster headache [11]. Magnesium for headaches offers an alternative to traditional medication that brings with it issues, such as addiction and side effects. Magnesium, with its relative lack of side effects, is particularly compelling for use in groups in which side effects are less well tolerated, such as children, pregnant women and the elderly population.

Magnesium is the fourth most abundant cation in the human body [12,13] and is involved in several important functions, such as enzyme activity, oxidative phosphorylation, DNA and protein synthesis, neuromuscular excitability and parathyroid hormone secretion [14].

Approximately 99% of total body magnesium is stored intracellularly in soft tissue and muscle (~40%) or resides as a component of bone on the surface of hydroxyapatite crystals (~60%) [15–17]. The absorption of magnesium occurs predominantly in the small intestine (and to a lesser extent in the colon) and depends on two different pathways: a passive paracellular transport, which facilitates bulk magnesium absorption, and an active transcellular pathway responsible for mediating the fine-tuning of magnesium absorption [18]. In the kidney, 80% of total serum magnesium is filtered in the glomeruli, with more than 95% being re-absorbed in the nephron. The renal re-absorption of magnesium contributes to maintaining magnesium homeostasis, as it declines to near zero in the presence of high levels of magnesium and reaches over 99% in the presence of magnesium depletion (Figure 1) [19]. Serum magnesium concentration is strictly regulated by the balance between intestinal absorption, renal excretion and bone buffer (Figure 1).

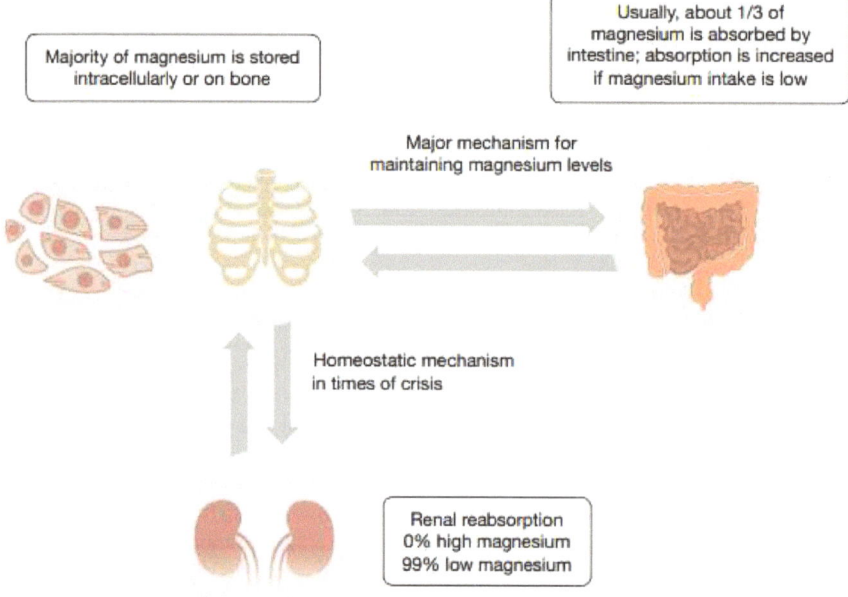

**Figure 1.** Schematic presentation of magnesium homeostasis.

Magnesium is surrounded by two hydration shells. Consequently, the radius of hydrated magnesium is about 400 times larger than its dehydrated radius. This creates steric constraints for magnesium transporters, which need to dehydrate magnesium, an event that is highly energy consuming, before transferring it through the membrane [12]. Over the last 20 years, several putative magnesium channels and transporters have been described [20], but the workings of intracellular magnesium homeostasis remain a conundrum.

Magnesium deficiency may occur for several reasons: inadequate intake, gastrointestinal loss and renal loss, or re-distribution from the extracellular to the intracellular space. Acute magnesium deficiency may be asymptomatic or associated with various disorders, such as nausea, vomiting, lethargy, [19] nervousness/anxiety and stress [21,22]. Chronic deficiency may lead to severe neuromuscular and cardiovascular pathologies [19]. There are multiple studies that suggest a relationship between magnesium deficiency and headaches, and these will be discussed further in the text [1–9]. Challenges exist for measuring magnesium concentration [23], and standardized laboratory tests that accurately evaluate magnesium levels are lacking [24]. Currently, in adults the reference interval for serum magnesium ranges between 0.75–0.95 mmol/L (1.82–2.30 mg/dL), while serum ionized magnesium ranges between 0.50–0.69 mmol/L [19,25]. These values are based on data reported in the 1970s [26]. However, since serum magnesium respond to dietary manipulation [23], and magnesium content in fruits, cereals and vegetables markedly declined over the past 40 years [24], the distribution of serum magnesium in normal population should be updated. In addition, serum magnesium concentration is most often used to assess magnesium status yet only 1% of total body magnesium is present in blood. In some instances, magnesium deficiency may be masked as the large proportion of magnesium residing in bone provides a large exchangeable pool to buffer changes in serum magnesium concentration [16]. For example, in an analysis carried out in women with normal serum values, a significantly greater magnesium retention was shown in osteoporotic patients compared with healthy individuals, thus suggesting the presence of magnesium deficiency despite normal magnesium serum values [27]. The magnesium load test, which analyzes urine samples over 24 h, is currently used to measure whole body magnesium, although it can prove difficult to administer as measurements are taken over 24 h in order to take account of circadian rhythms [28]. The ionized magnesium of erythrocyte cells can also be used as a measure of total body magnesium as, among intracellular magnesium compartments, erythrocytes make up more than 90% of the total blood cells, therefore mainly affect the intracellular blood magnesium content [29].

Magnesium salts used in current clinical practice to treat magnesium deficiency can be organic, such as magnesium pidolate and magnesium lactate, or inorganic, such as magnesium chloride and magnesium carbonate (Table 1). Different salts have been noted to have varying absorption efficiency and soluble properties, leading to a variation in bioavailability.

Table 1. Inorganic and organic salts used for magnesium supplementation [30–34].

| Inorganic Magnesium Salts | Organic Magnesium Salts | Combinations/Different Formulations |
|---|---|---|
| Carbonate | Acetate | Citrate + hydrogen-L-glutamate |
| Chloride | Aspartate | Dicitrate |
| Oxide | Citrate | Glycinate lysinate chelate |
| Sulfate | Gluconate | Oxide + glycerophosphate |
|  | Lactate | Pyrrolidone carboxylic acid |
|  | Pidolate | Trimagnesium dicitrate |
|  |  | U-aspartate-hydrochloride-trihydrate |

Magnesium pidolate may have high bioavailability [35,36] and good penetration at the intracellular level [37]. Furthermore, magnesium pidolate is able to reverse magnesium deficiency responsible for headaches, even after a short administration period [31], and to prevent pediatric tension-type headaches [38]. Taking this into consideration, the unique mechanism of action of magnesium pidolate and the efficacy and safety of magnesium salts for the treatment of headaches is considered.

## 2. Why Should Magnesium Be Used to Treat Headaches?

Multiple studies have suggested a relationship between magnesium deficiency and headaches (Table 2) [8]. In a case-control study of patients suffering from migraine, reduced magnesium levels were found in serum [7], cerebrospinal fluid [1] and the ictal and interictal regions within the brain [2]. Similar results were observed in several other case-control studies [4–6,8,9]. For example, Sarchielli and colleagues have shown that migraine sufferers with and without aura and tension-type headaches have significantly lower levels of serum and salivary magnesium [8]. Importantly, a study by Trauninger and colleagues using the magnesium load test revealed a greater retention of magnesium in patients suffering from migraines compared with healthy controls, suggesting a systemic magnesium deficiency associated with migraine [6]. Furthermore, a 2-week trial revealed that, when 29 migraine patients took mineral water containing 110 mg/L magnesium daily, their total magnesium in erythrocytes significantly increased, compared with 18 healthy controls [4]. A recent observation by Assarzadegan and colleagues [9] indicated that a decrease in magnesium levels in serum increased the odds of acute migraine headaches by a factor of 35 in 40 patients with migraine versus 40 healthy controls, and that magnesium deficiency is an independent risk factor in the incidence of migraines. Studies carried out by Mauskop and colleagues [3,39] estimated the frequency of magnesium deficiency among migraine sufferers by evaluating the efficacy of the intravenous infusion of 1 g of magnesium sulfate for the treatment of patients with headaches. They investigated the correlation of clinical responses and basal serum ionized magnesium level and reported that a 50% reduction in pain was noted after infusion [3,39]. Taken together, these results suggest a correlation between magnesium deficiency and headaches, and of note, they suggest that magnesium deficiency represents an independent risk factor for migraine occurrence.

Table 2. Studies investigating the relationship between magnesium levels and headache.

| Year | Type of Headache | Number of Patients | Outcome | Reference |
|---|---|---|---|---|
| 1985 | Migraine | 57 adults | Reduced magnesium levels in cerebrospinal fluid | [1] |
| 1989 | Migraine | 11 adults | Reduced magnesium levels in the brain | [2] |
| 1995 | Cluster | 22 adults | Up to 50% of migraine patients were found to be magnesium-deficient | [3] |
| 2000 | Migraine | 29 adults plus 18 healthy controls | Total magnesium in erythrocytes significantly increased compared with healthy controls | [4] |
| 2002 | Tension/migraine | 25 adults plus 20 healthy controls | Reduced magnesium levels in serum and saliva | [5] |
| 2002 | Migraine | 20 adults plus 20 healthy controls | Increased systemic retention of magnesium vs. controls | [6] |
| 2011 | Migraine | 140 adults plus 140 healthy controls | Total serum magnesium levels significantly lower vs. controls | [7] |
| 2012 | Migraine | 50 adults plus 50 healthy controls | Total serum magnesium levels significantly lower vs. controls | [8] |
| 2016 | Acute migraine | 40 adults plus 40 healthy controls | Decreased magnesium indicates a 35-fold increased risk of acute migraine | [9] |

The intravenous infusion of magnesium sulfate as a treatment for acute headaches was assessed in a systematic review with varying results [40]. Initial efficacy was demonstrated in adults with low serum magnesium [3,39], and a small study confirmed that the treatment was safe in adolescents [41]. A recent systematic review indicated no benefit immediately after infusion, but potential benefits in pain control beyond the first hour [40]. There is, however, a counterargument that only a proportion of patients with acute headaches have magnesium deficiency [39], which may mask the extent of the therapeutic effectiveness of magnesium infusion in the emergency setting [42].

Magnesium can act as a calcium channel antagonist in neurons, where it is believed to prevent the excessive activation of the excitatory synapses (e.g., N-methyl-D-aspartate [NMDA] receptors); it has also been shown to downregulate inflammation through inhibiting pro-inflammatory intracellular signaling, such as the nuclear factor kappa B pathway [43]. Of interest, magnesium homeostasis in the brain has been found to be dysregulated in various neurological disorders [44]. Lower concentrations of magnesium than in healthy controls were found in the brains of patients with Alzheimer's and Parkinson's diseases [44] and in the occipital lobes of patients with migraine and cluster headaches [45]. In magnesium-deficient individuals, magnesium supplementation attenuates anxiety and stress symptoms [23,46]. Similarly, magnesium-deficient mice exhibit an anxiety-related behavior, which is due, in part, to the increased response of the hypothalamic–pituitary–adrenal axis, the central stress response system [47].

A number of mechanisms have been described to explain the relationship between magnesium deficiency and headaches (Figure 2) [48]. Magnesium deficiency has been associated with cortical spreading depression (CSD), thought to be responsible for the aura associated with migraines [48], imbalanced neurotransmitter release [49], platelet activity [50] and vasoconstriction [51]. In CSD, substance P, a neuropeptide which acts as a neurotransmitter and neuromodulator, is released as a result of magnesium deficiency, possibly acting on sensory fibers and producing headache pain [52]. Magnesium has also been shown to decrease the level of circulating calcitonin gene-related peptide (CGRP), which is involved in migraine pathogenesis through its ability to dilate intracranial blood vessels and produce nociceptive stimuli [48,53]. External magnesium may help to diminish various aspects of neurogenic inflammation as it is involved in the control of NMDA glutamate receptors, which play an important role in pain transmission within the nervous system [54], the regulation of cerebral blood flow [55] and the initiation and spread of CSD. It has been shown that ionized magnesium can block CSD by regulating glutamatergic neurotransmission, closing the NMDA receptor calcium channel and modulating the cyclic adenosine monophosphate (cAMP) response element-binding protein signaling [56,57]. The modulation of the cerebral blood flow by circulating nitric oxide (NO) is one of the mechanisms involved in headaches, and it has been shown to be influenced by magnesium intake [48,58]. Magnesium can also increase vasodilation directly through blocking calcium-sensitive potassium channels on smooth muscle cells [59]. There is some evidence that magnesium may be most beneficial in migraines with aura [60,61].

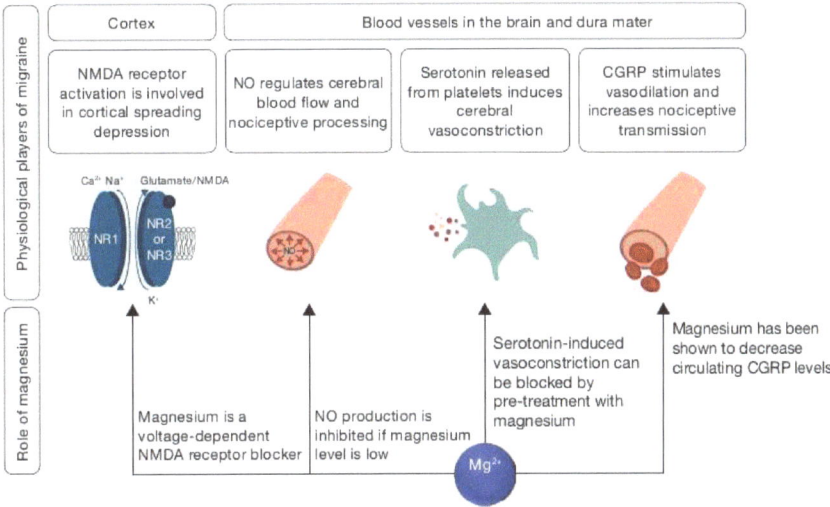

**Figure 2.** Mechanisms involved in migraine and possible role of magnesium. CGRP, circulating calcitonin gene-related peptide; NMDA, N-methyl-D-aspartate; NO, nitric oxide.

Another key molecule in migraine pathogenesis is serotonin, a potent cerebral vasoconstrictor released from platelets during a migraine attack—it also promotes nausea and vomiting [62]. A decrease in serum ionized magnesium level and an elevation of the serum ratio of ionized calcium to ionized magnesium may increase the likelihood for cerebral vascular muscle serotonin receptor sites, potentiate cerebral vasoconstriction induced by serotonin and facilitate serotonin release from neuronal storage sites [62]. Vasoconstriction induced by serotonin can be blocked by pretreatment with ionized magnesium [63].

## 3. Magnesium Supplementation—Therapeutic Efficacy

The therapeutic efficacy of magnesium supplementation in headache patients has been shown in two double-blind, placebo-controlled randomized trials [31,32]. The first study was conducted in 20 women with menstrual migraine. It is known that the magnesium level of erythrocytes and leukocytes of women with premenstrual syndrome is lower than that in the women without the syndrome [64]. For this reason, magnesium supplementation is widely used to treat premenstrual syndrome [31,65,66]. Women received two cycles of 360 mg of magnesium pyrrolidone carboxylic acid or placebo taken daily from ovulation to the first day of their period. Patients receiving active treatment had a significant reduction in the frequency of headaches and total pain index [31]. A larger double-blind, placebo-controlled randomized study of 81 adult patients with migraines, according to the International Headache Society (IHS) criteria, also showed significant improvements in patients on active therapy [32]. The active group received 600 mg of trimagnesium dicitrate in a water-soluble granular powder every morning and had a significant reduction ($p < 0.05$) in the frequency of attacks (41.6%) compared with the placebo group (15.8%). A further randomized controlled trial of 118 children 3–17 years of age receiving 9 mg/kg daily oral magnesium oxide or placebo showed that treatment led to a significant reduction in headache days [67].

One trial, enrolling 69 patients taking 242 mg magnesium-u-aspartate-hydrochloride-trihydrate daily, showed no effect on migraines [33]. Diarrhea occurred in almost half of the 35 patients receiving magnesium compared with a quarter of the 34 patients on placebo indicating that the magnesium salt may be poorly absorbed, which may account for the observed lack of efficacy.

The duration of 1500 mg daily oral magnesium pidolate treatment needed to normalize serum magnesium levels was investigated by Aloisi and colleagues in a study on 40 children designed to evaluate the correlation between magnesium deficiency and the effect on visual evoked potentials. The analysis showed that a treatment lasting 20 days was sufficient to normalize serum magnesium levels in 90% of treated patients [68].

Koseoglu and colleagues evaluated the prophylactic effects of 600 mg daily oral magnesium citrate supplementation in 30 migraine patients without aura compared with 10 patients on placebo treatment. Migraine attack frequency, severity, and P1 amplitude (in visual evoked potential examination) decreased after magnesium treatment compared with pretreatment values and placebo [69].

Karimi and colleagues, in a randomized, double-blind, controlled, crossover trial, gave 63 patients oral daily 500 mg magnesium oxide followed by 800 mg valproate sodium (400 mg every 12 h) or vice versa for 24 weeks. Patients showed a similar number and mean duration of migraine attacks in both groups, indicating that magnesium oxide is as effective as valproate in migraine prophylaxis without significant adverse effects [70].

A recent systematic review of five randomized, double-blind, placebo-controlled trials in adult migraine patients showed possible evidence for the prevention of migraines with 600 mg magnesium dicitrate daily, and that it is a well-tolerated and cost efficient strategy in clinical use [71].

In view of the results of these studies, several national and international guidelines added the recommendation of oral magnesium for headache patients [72–74]. The Italian Headache Society (SISC) guideline mentions magnesium pidolate supplementation for menstrual migraine and pre-menstrual syndrome patients, but a precise administration schedule has not been established [74]. Notably, magnesium pidolate is used at much higher concentrations than other magnesium salts (Table 3).

Side effects were not measured in all studies, but in those that were, diarrhoea and gastric effects were the most common, although mild in all instances, and did not prevent patients from completing treatment [32,33,69,70,75]. See Table 3 for studies describing the efficacy and safety of magnesium in treating headache symptoms.

Table 3. Efficacy and safety of magnesium in treating headache symptoms.

| Type of Study | Author/Year | Study Length | Country | Type of Headache | Number of Patients | Magnesium Salt | Efficacy Outcome | Safety Outcome | Reference |
|---|---|---|---|---|---|---|---|---|---|
| Children | | | | | | | | | |
| Multi-arm | Aloisi, 1997 | 20 days | Italy | Tension, migraine | 60 male and female children 6-13 years | 1500 mg daily oral magnesium pidolate | 20 days treatment sufficiently normalizes serum Magnesium levels in 90% of migraine patients | NR | [68] |
| Double-blind, placebo-controlled randomized trial | Wang, 2003 | 16 weeks | USA | Migraine | 118 male and female children 3-17 years (n = 60, placebo) | 9 mg/kg daily oral magnesium oxide | Significant reduction in headache days | NR | [67] |
| Open label trial | Grazzi, 2007 | 3 months | Italy | Tension | 45 male and female children 8-16 years | 2250 mg x2 daily oral magnesium pidolate | Headache days decreased by 69.9% | No significant side effects | [38] |
| Adults | | | | | | | | | |
| Double-blind, controlled, randomized, crossover trial | Karimi, 2019 | 24 weeks | Iran | Migraine | 63 adult male and females | 500 mg daily oral magnesium oxide (800 mg sodium valproate) | Magnesium oxide appears to be as effective as valproate in migraine prophylaxis without significant adverse effects | No side effects on top of headache symptoms | [70] |
| Systematic review (five clinical trials below) | Von Luckner, 2018 | 2-4 months | Various countries | Migraine | Five clinical trials of adult male and females | Different salts different doses | Possibly effective in preventing migraine. Safe and cost efficient | NA | [71] |
| 1. Double-blind, placebo-controlled randomized trial | Facchinetti, 1991 | 2 months | Italy | Menstrual migraine | 20 females | 360 mg daily oral magnesium pyrrolidone carboxylic acid | Significant reduction in the frequency of headache and total pain index | NR | [31] |
| 2. Double-blind, placebo-controlled randomized trial | Peikert, 1996 | 12 weeks | Germany | Migraine | 81 male and female adults (n = 38, placebo) | 600 mg daily oral trimagnesium dicitrate | Significant improvement in patients on active therapy | Diarrhoea and gastric complaints (mild and tolerable) | [32] |
| 3. Double-blind, placebo-controlled randomized trial | Pfaffenrath, 1996 | 12 weeks | Germany | Migraine | 69 male and female adults (n = 34, placebo) | 242 mg daily oral magnesium-τ-aspartate-hydrochloride-trihydrate | No effect | Soft stool, diarrhoea (mild) | [33] |
| 4. Double-blind, placebo-controlled randomized trial | Koseoglu, 2008 | 3 months | Turkey | Migraine | 40 male and female adults (n = 10, placebo) | 600 mg daily oral magnesium citrate | Migraine attack frequency, severity, and P1 amplitude decreased | Diarrhoea, soft stools, gastric irritation (mild) | [69] |
| 5. Multicenter, crossover trial | Taubert, 1994 | 2 × 2 months | Germany | Migraine | 63 adult male and females | 600 mg daily oral trimagnesium dicitrate or placebo | Statistically significant reduction in the frequency of attacks compared with placebo | Diarrhoea | [75] |

NA, not applicable; NR, not reported.

## 4. Magnesium Salt Bioavailability—Pidolate Versus Other Salts

Magnesium pidolate is an organic salt and, based on animal studies, may have a high bioavailability [35,36]. The bioavailability of magnesium is of high importance in treating headaches as the more magnesium that can be absorbed, the more effective the treatment. In a study by Coudray and colleagues in rats, absorption was 13% higher from organic than inorganic magnesium salts and particularly high urinary excretion with magnesium gluconate and pidolate was observed [35]. Magnesium pidolate exhibited higher bioavailability compared with other organic salts in mice: the post-oral serum magnesium increase was higher in mice receiving magnesium pidolate (100% versus baseline) than in mice treated with magnesium lactate (50% versus baseline) [36]. Other magnesium salts have been studied in a limited number of studies in humans conducted in the early 1990s, with mixed results. In some studies, there was no difference between organic and inorganic magnesium salts [76–79]; others demonstrated slightly higher bioavailability of organic magnesium salts under standardized conditions [18,34,80–83]. Magnesium pidolate is an organic salt and organic salts have been found to be consistently more bioavailable that inorganic salts in many human studies. Despite the lack of studies specifically analyzing magnesium pidolate bioavailability, it could be postulated that magnesium pidolate availability is, in part, due to its organic properties [18,34,80–83].

Magnesium pidolate has good intracellular penetration, which has been shown in vivo. Ten patients with sickle cell disease were treated with daily oral magnesium pidolate (540 mg/70 kg), which resulted in a reduced number of dense erythrocytes and improved erythrocyte membrane transport abnormalities in patients [37]. However, a recent review of the literature conducted by Zhang and colleagues failed to demonstrate any efficacy of the most common oral salts of magnesium [84]. The reason for this discrepancy may be due to differences in the ability of various salts to enter different cell lines. A recent study showed that the bioavailability at the cellular level of magnesium pidolate is different from that of two inorganic salts (magnesium chlorate and sulfate) in cell cultures of osteogenic sarcoma, which could suggest a lower capacity of magnesium pidolate to enter bone cells, the body's main deposit for magnesium. This would explain the greater availability for other tissues and cells, such as lymphocytes and polymorphonuclear cells [85].

## 5. Magnesium Pidolate and Brain Penetration

The dysfunction of the blood–brain barrier (BBB) has been described in several neurological disorders, including ischemic stroke and inherited and neurodegenerative diseases [86,87]. This topic remains controversial: while some studies did not find changes in BBB during a migraine attack [88], there are studies in human subjects and animals that indicate that BBB permeability may be increased with migraine and headaches [89,90]. BBB disruption has been associated with magnesium deficiency in the brain [91,92]. It is therefore interesting to distinguish agents that exert a protective role on BBB and prevent its impairment in response to various challenges. There is evidence that magnesium has a protective role on the BBB in vivo [93,94], and a recent paper has highlighted that 10 mmol/L magnesium sulfate reduces the permeability in an in vitro model of the human BBB [95]. This effect could be the result of the antagonism between calcium and magnesium in the endothelial actin cytoskeleton, which remodels intercellular gap formation, thus inhibiting the paracellular movement of molecules through the tight junctions [96].

Romeo et al. (2019) [95] compared the effect of different magnesium salts at the same concentration (5 mmol/L) in in vitro in models of rat and human BBBs. All salts decreased BBB permeability; among them, magnesium pidolate and magnesium threonate were the most efficient in the rat model, and magnesium pidolate was the most efficient in the human model, suggesting differences in response between humans and rodents.

Another aspect evaluated in Romeo's study [95] was that the transport of magnesium through the BBB is more efficient after magnesium pidolate treatment. Magnesium has been found to cross the intact BBB and enter the central nervous system in rats, to an extent proportional to magnesium serum levels [93,94]. In humans with an intact BBB, a modest but significant increase in magnesium concentration in the cerebrospinal fluid was reported after systemic administration of magnesium sulfate [97]. The use of magnesium pidolate may result in more magnesium crossing the BBB compared with other salts and, therefore, may have special relevance for the treatment of neurological conditions with a known connection to magnesium deficiency [95].

## 6. Magnesium Pidolate and Headache: A Challenge for the Future

Headache is characterized by high lifetime prevalence [98] and, rather than taking preventative medications [98], most patients use non-steroidal anti-inflammatory drugs (NSAIDs), mainly purchased in an over the counter setting without medical advice or prescription.

A literature evidence base suggests that magnesium deficiency increases the risk of headache. As shown in several clinical studies and reported in various national and international guidelines [72–74], the treatment of magnesium deficiency can reduce the frequency of headaches and, as a direct consequence, the use of NSAIDs and other therapies [93,94].

Magnesium pidolate has high bioavailability and good intracellular penetration [82] and it may reverse the magnesium deficiency responsible for headaches, even after a short administration period. Tissue culture and animal model studies indicate that magnesium pidolate may be slightly more effective than other magnesium salts in crossing the BBB [95], and magnesium in general is believed to exert neuroprotective functions. Further studies on the tissue distribution of magnesium pidolate may help to better understand its specific properties.

## 7. Conclusions

Taken together, these results confirm a correlation between magnesium deficiency and headaches. In addition, they suggest magnesium deficiency could be an independent risk factor for migraine occurrence. Some of the trials presented in this review date from the 1990s; however, it is encouraging to see a revitalization of this subject with more recent systematic reviews and clinical trials.

Magnesium deficiency is more often present in postmenopausal women with osteoporosis (84%) [99] and in women aged 18 to 22 (20%) [100]. The use of magnesium with its relatively low side effects is particularly pertinent for these populations who are also particularly susceptible to the side effects of traditional drugs.

When assessing the efficacy of magnesium salt, variations in dosage, study design, methods of assessment and study population, all need to be evaluated, which can make it difficult to interpret which salt is preferable for treating headache. In terms of magnesium pidolate, it may have a lower capacity to enter bone cells, the body's main deposit for magnesium [82], and may cause more magnesium to cross the BBB compared with other salts [95]. Due to its potential high bioavailability, it may have special relevance for the treatment of neurological conditions with a known connection to magnesium deficiency, such as headache. Based on the information in the literature, there is an argument for the use of magnesium pidolate in Italy. However, it needs to be borne in mind that only a limited number of studies have shown the benefits of magnesium pidolate in headaches, and further controlled studies are needed. This is particularly important with regard to elucidating any side effects, as the 1500–4500 mg dose is high compared to the other salts, which range between 242 mg and 600 mg.

Overall, the use of oral magnesium salt represents a well-tolerated and inexpensive addition for the treatment of headache patients, to reduce the frequency of attacks and the costs of treatment both in terms of economic burden and adverse events.

**Author Contributions:** J.A.M., E.G. and P.P., writing—original draft preparation, review and editing; A.C. and G.P., writing—review and editing. All authors have read and agreed to the published version of the manuscript.

**Funding:** Editorial support was provided by Ella Palmer and Olga Ucar, Springer Healthcare Ltd., UK, and the study was funded by Sanofi.

**Acknowledgments:** The authors thank Lionel Noah for providing scientific insight throughout manuscript development.

**Conflicts of Interest:** E.G. and P.P. are employees of Sanofi. A.C. and J.A.M. declare no conflict of interest. G.P. has no conflict of interest related to this publication, but has served on Sanofi advisory boards related to other projects. Sanofi commissioned the review, but had no role in the design of the review or interpretation of data.

**References**

1. Jain, A.; Sethi, N.; Balbar, P. A clinical electroencephalographic and trace element study with special reference to zinc, copper and magnesium in serum and cerebrospinal fluid (CSF) in cases of migraine. *J. Neurol.* **1985**, *232*, S161.
2. Ramadan, N.M.; Halvorson, H.; Vande-Linde, A.; Levine, S.R.; Helpern, J.A.; Welch, K.M. Low brain magnesium in migraine. *Headache* **1989**, *29*, 416–419. [CrossRef]
3. Mauskop, A.; Altura, B.T.; Cracco, R.Q.; Altura, B.M. Intravenous magnesium sulfate relieves cluster headaches in patients with low serum ionized magnesium levels. *Headache* **1995**, *35*, 597–600. [CrossRef]
4. Thomas, J.; Millot, J.M.; Sebille, S.; Delabroise, A.M.; Thomas, E.; Manfait, M.; Arnaud, M.J. Free and total magnesium in lymphocytes of migraine patients—Effect of magnesium-rich mineral water intake. *Clin. Chim. Acta Int. J. Clin. Chem.* **2000**, *295*, 63–75. [CrossRef]
5. Sarchielli, P.; Coata, G.; Firenze, C.; Morucci, P.; Abbritti, G.; Gallai, V. Serum and Salivary Magnesium Levels in Migraine and Tension-Type Headache. Results in a Group of Adult Patients. *Cephalalgia* **1992**, *12*, 21–27. [CrossRef] [PubMed]
6. Trauninger, A.; Pfund, Z.; Koszegi, T.; Czopf, J. Oral magnesium load test in patients with migraine. *Headache* **2002**, *42*, 114–119. [CrossRef]
7. Talebi, M.; Savadi-Oskouei, D.; Farhoudi, M.; Mohammadzade, S.; Ghaemmaghamihezaveh, S.; Hasani, A.; Hamdi, A. Relation between serum magnesium level and migraine attacks. *Neuroscirnce* **2011**, *16*, 320–323.
8. Samaie, A.; Asghari, N.; Ghorbani, R.; Arda, J. Blood Magnesium levels in migraineurs within and between the headache attacks: A case control study. *Pan Afr. Med. J.* **2012**, *11*, 46.
9. Assarzadegan, F.; Asgarzadeh, S.; Hatamabadi, H.R.; Shahrami, A.; Tabatabaey, A.; Asgarzadeh, M. Serum concentration of magnesium as an independent risk factor in migraine attacks: A matched case-control study and review of the literature. *Int. Clin. Psychopharm.* **2016**, *31*, 287–292. [CrossRef]
10. The International Classification of Headache Disorders, 3rd edition (beta version). *Cephalalgia Int. J. Headache* **2013**, *33*, 629–808. [CrossRef]
11. Ravishankar, K. The art of history-taking in a headache patient. *Ann. Indian Acad. Neurol.* **2012**, *15*, S7–S14. [CrossRef] [PubMed]
12. Maguire, M.E.; Cowan, J.A. Magnesium chemistry and biochemistry. *Biometals* **2002**, *15*, 203–210. [CrossRef] [PubMed]
13. Wacker, W. *Magnesium and Man*; Havard University Press: Cambridge, MA, USA, 1980; Volume 1.
14. Whang, R.; Hampton, E.M.; Whang, D.D. Magnesium homeostasis and clinical disorders of magnesium deficiency. *Ann. Pharmacother.* **1994**, *28*, 220–226. [CrossRef]
15. Elin, R.J. Assessment of magnesium status for diagnosis and therapy. *Magnes. Res.* **2010**, *23*, S194–S198. [CrossRef]
16. Alfrey, A.C.; Miller, N.L. Bone magnesium pools in uremia. *J. Clin. Investig.* **1973**, *52*, 3019–3027. [CrossRef] [PubMed]
17. Aikawa, J. *Magnesium: Its Biological Significance*; CRC Press: Boca Raton, FL, USA, 1981.
18. Fine, K.D.; Ana, C.A.S.; Porter, J.L.; Fordtran, J.S. Intestinal absorption of magnesium from food and supplements. *J. Clin. Investig.* **1991**, *88*, 396–402. [CrossRef]
19. Ayuk, J.; Gittoes, N.J. Contemporary view of the clinical relevance of magnesium homeostasis. *Ann. Clin. Biochem.* **2014**, *51*, 179–188. [CrossRef]
20. Kolisek, M.; Sponder, G.; Pilchova, I.; Cibulka, M.; Tatarkova, Z.; Werner, T.; Racay, P. Magnesium Extravaganza: A Critical Compendium of Current Research into Cellular Mg(2+) Transporters Other than TRPM6/7. *Rev. Physiol. Biochem. Pharmacol.* **2019**, *176*, 65–105. [CrossRef]
21. Botturi, A.; Ciappolino, V.; Delvecchio, G.; Boscutti, A.; Viscardi, B.; Brambilla, P. The Role and the Effect of Magnesium in Mental Disorders: A Systematic Review. *Nutrients* **2020**, *12*, 1661. [CrossRef]
22. Boyle, N.B.; Lawton, C.; Dye, L. The Effects of Magnesium Supplementation on Subjective Anxiety and Stress-A Systematic Review. *Nutrients* **2017**, *9*, 429. [CrossRef]

23. Witkowski, M.; Hubert, J.; Mazur, A. Methods of assessment of magnesium status in humans: A systematic review. *Magnes. Res.* **2011**, *24*, 163–180. [CrossRef] [PubMed]
24. Workinger, J.; Doyle, R.; Bortz, J. Challenges in the Diagnosis of Magnesium Status. *Nutrients* **2018**, *10*, 1202. [CrossRef] [PubMed]
25. Greenway, D.C.; Hindmarsh, J.T.; Wang, J.; Khodadeen, J.A.; Hébert, P.C. Reference interval for whole blood ionized magnesium in a healthy population and the stability of ionized magnesium under varied laboratory conditions. *Clin. Biochem.* **1996**, *29*, 515–520. [CrossRef]
26. Lowenstein, F.W.; Stanton, M.F. Serum magnesium levels in the United States, 1971-1974. *J. Am. Coll. Nutr.* **1986**, *5*, 399–414. [CrossRef] [PubMed]
27. Nicar, M.J.; Pak, C.Y. Oral magnesium load test for the assessment of intestinal magnesium absorption. Application in control subjects, absorptive hypercalciuria, primary hyperparathyroidism, and hypoparathyroidism. *Min. Electrolyte Metab.* **1982**, *8*, 44–51.
28. Jahnen-Dechent, W.; Ketteler, M. Magnesium basics. *Clin. Kidney J.* **2012**, *5*, i3–i14. [CrossRef]
29. Xiong, W.; Liang, Y.; Li, X.; Liu, G.; Wang, Z. Erythrocyte intracellular Mg(2+) concentration as an index of recognition and memory. *Sci. Rep.* **2016**, *6*, 26975. [CrossRef]
30. Blancquaert, L.; Vervaet, C.; Derave, W. Predicting and Testing Bioavailability of Magnesium Supplements. *Nutrients* **2019**, *11*, 1663. [CrossRef]
31. Facchinetti, F.; Sances, G.; Borella, P.; Genazzani, A.R.; Nappi, G. Magnesium prophylaxis of menstrual migraine: Effects on intracellular magnesium. *Headache* **1991**, *31*, 298–301. [CrossRef]
32. Peikert, A.; Wilimzig, C.; Kohne-Volland, R. Prophylaxis of migraine with oral magnesium: Results from a prospective, multi-center, placebo-controlled and double-blind randomized study. *Cephalalgia* **1996**, *16*, 257–263. [CrossRef]
33. Pfaffenrath, V.; Wessely, P.; Meyer, C.; Isler, H.R.; Evers, S.; Grotemeyer, K.H.; Taneri, Z.; Soyka, D.; Gobel, H.; Fischer, M. Magnesium in the prophylaxis of migraine—A double-blind placebo-controlled study. *Cephalalgia* **1996**, *16*, 436–440. [CrossRef] [PubMed]
34. Schuchardt, J.P.; Hahn, A. Intestinal Absorption and Factors Influencing Bioavailability of Magnesium—An Update. *Curr. Nutr. Food Sci.* **2017**, *13*, 260–278. [CrossRef] [PubMed]
35. Coudray, C.; Rambeau, M.; Feillet-Coudray, C.; Gueux, E.; Tressol, J.C.; Mazur, A.; Rayssiguier, Y. Study of magnesium bioavailability from ten organic and inorganic Mg salts in Mg-depleted rats using a stable isotope approach. *Magnes. Res.* **2005**, *18*, 215–223. [PubMed]
36. Decollogne, S.; Tomas, A.; Lecerf, C.; Adamowicz, E.; Seman, M. NMDA receptor complex blockade by oral administration of magnesium: Comparison with MK-801. *Pharmacol. Biochem. Behav.* **1997**, *58*, 261–268. [CrossRef]
37. De Franceschi, L.; Bachir, D.; Galacteros, F.; Tchernia, G.; Cynober, T.; Alper, S.; Platt, O.; Beuzard, Y.; Brugnara, C. Oral magnesium supplements reduce erythrocyte dehydration in patients with sickle cell disease. *J. Clin. Investig.* **1997**, *100*, 1847–1852. [CrossRef]
38. Grazzi, L.; Andrasik, F.; Usai, S.; Bussone, G. Magnesium as a preventive treatment for paediatric episodic tension-type headache: Results at 1-year follow-up. *Neurol. Sci.* **2007**, *28*, 148–150. [CrossRef]
39. Mauskop, A.; Altura, B.T.; Cracco, R.Q.; Altura, B.M. Intravenous magnesium sulphate relieves migraine attacks in patients with low serum ionized magnesium levels: A pilot study. *Clin. Sci.* **1995**, *89*, 633–636. [CrossRef] [PubMed]
40. Miller, A.C.; Pfeffer, B.K.; Lawson, M.R.; Sewell, K.A.; King, A.R.; Zehtabchi, S. Intravenous Magnesium Sulfate to Treat Acute Headaches in the Emergency Department: A Systematic Review. *Headache* **2019**, *59*, 1674–1686. [CrossRef]
41. Gertsch, E.; Loharuka, S.; Wolter-Warmerdam, K.; Tong, S.; Kempe, A.; Kedia, S. Intravenous magnesium as acute treatment for headaches: A pediatric case series. *J. Emerg. Med.* **2014**, *46*, 308–312. [CrossRef]
42. Mauskop, A. Intravenous Magnesium Sulfate to Treat Acute Headaches in the Emergency Department: A Systematic Review—A Comment. *Headache* **2020**, *60*, 624. [CrossRef]
43. Lingam, I.; Robertson, N.J. Magnesium as a Neuroprotective Agent: A Review of Its Use in the Fetus, Term Infant with Neonatal Encephalopathy, and the Adult Stroke Patient. *Dev. Neurosci.* **2018**, *40*, 1–12. [CrossRef] [PubMed]
44. Kirkland, A.E.; Sarlo, G.L.; Holton, K.F. The Role of Magnesium in Neurological Disorders. *Nutrients* **2018**, *10*, 730. [CrossRef] [PubMed]

45. Lodi, R.; Iotti, S.; Cortelli, P.; Pierangeli, G.; Cevoli, S.; Clementi, V.; Soriani, S.; Montagna, P.; Barbiroli, B. Deficient energy metabolism is associated with low free magnesium in the brains of patients with migraine and cluster headache. *Brain Res. Bull.* **2001**, *54*, 437–441. [CrossRef]
46. Pouteau, E.; Kabir-Ahmadi, M.; Noah, L.; Mazur, A.; Dye, L.; Hellhammer, J.; Pickering, G.; Dubray, C. Superiority of magnesium and vitamin B6 over magnesium alone on severe stress in healthy adults with low magnesemia: A randomized, single-blind clinical trial. *PLoS ONE* **2018**, *13*, e0208454. [CrossRef] [PubMed]
47. Sartori, S.B.; Whittle, N.; Hetzenauer, A.; Singewald, N. Magnesium deficiency induces anxiety and HPA axis dysregulation: Modulation by therapeutic drug treatment. *Neuropharmacology* **2012**, *62*, 304–312. [CrossRef] [PubMed]
48. Sun-Edelstein, C.; Mauskop, A. Role of magnesium in the pathogenesis and treatment of migraine. *Expert Rev. Neurother.* **2009**, *9*, 369–379. [CrossRef]
49. Coan, E.J.; Collingridge, G.L. Magnesium ions block an N-methyl-D-aspartate receptor-mediated component of synaptic transmission in rat hippocampus. *Neurosci. Lett.* **1985**, *53*, 21–26. [CrossRef]
50. Baudouin-Legros, M.; Dard, B.; Guicheney, P. Hyperreactivity of platelets from spontaneously hypertensive rats. Role of external magnesium. *Hypertension* **1986**, *8*, 694–699. [CrossRef]
51. Altura, B.M.; Altura, B.T. *Factors Affecting Responsiveness of Blood Vessels to Prostaglandins and Other Chemical Mediators of Injury and Shock*; Raven Press: New York, NY, USA, 1982.
52. Innerarity, S. Hypomagnesemia in acute and chronic illness. *Crit. Care Nurs. Q.* **2000**, *23*, 1–19. [CrossRef]
53. Myrdal, U.; Leppert, J.; Edvinsson, L.; Ekman, R.; Hedner, T.; Nilsson, H.; Ringqvist, I. Magnesium sulphate infusion decreases circulating calcitonin gene-related peptide (CGRP) in women with primary Raynaud's phenomenon. *Clin. Physiol.* **1994**, *14*, 539–546. [CrossRef]
54. Foster, A.C.; Fagg, G.E. Neurobiology. Taking apart NMDA receptors. *Nature* **1987**, *329*, 395–396. [CrossRef] [PubMed]
55. Huang, Q.F.; Gebrewold, A.; Zhang, A.; Altura, B.T.; Altura, B.M. Role of excitatory amino acids in regulation of rat pial microvasculature. *Am. J. Physiol. Regul. Integr. Comp. Physiol.* **1994**, *266*, R158–R163. [CrossRef] [PubMed]
56. Mody, I.; Lambert, J.D.; Heinemann, U. Low extracellular magnesium induces epileptiform activity and spreading depression in rat hippocampal slices. *J. Neurophysiol.* **1987**, *57*, 869–888. [CrossRef] [PubMed]
57. Hou, H.; Wang, L.; Fu, T.; Papasergi, M.; Yule, D.I.; Xia, H. Magnesium Acts as a Second Messenger in the Regulation of NMDA Receptor-Mediated CREB Signaling in Neurons. *Mol. Neurobiol.* **2020**. [CrossRef]
58. Chiarello, D.I.; Marin, R.; Proverbio, F.; Coronado, P.; Toledo, F.; Salsoso, R.; Gutierrez, J.; Sobrevia, L. Mechanisms of the effect of magnesium salts in preeclampsia. *Placenta* **2018**, *69*, 134–139. [CrossRef]
59. Murata, T.; Dietrich, H.H.; Horiuchi, T.; Hongo, K.; Dacey, R.G. Mechanisms of magnesium-induced vasodilation in cerebral penetrating arterioles. *Neurosci. Res.* **2016**, *107*, 57–62. [CrossRef]
60. Bigal, M.E.; Bordini, C.A.; Tepper, S.J.; Speciali, J.G. Intravenous magnesium sulphate in the acute treatment of migraine without aura and migraine with aura. A randomized, double-blind, placebo-controlled study. *Cephalalgia* **2002**, *22*, 345–353. [CrossRef]
61. Choi, H.; Parmar, N. The use of intravenous magnesium sulphate for acute migraine: Meta-analysis of randomized controlled trials. *Eur. J. Emerg. Med.* **2014**, *21*, 2–9. [CrossRef]
62. Peters, J.A.; Hales, T.G.; Lambert, J.J. Divalent cations modulate 5-HT3 receptor-induced currents in N1E-115 neuroblastoma cells. *Eur. J. Pharmacol.* **1988**, *151*, 491–495. [CrossRef]
63. Goldstein, S.; Zsotér, T.T. The effect of magnesium on the response of smooth muscle to 5-hydroxytryptamine. *Br. J. Pharmacol.* **1978**, *62*, 507–514. [CrossRef]
64. Salamat, S.; Ismail, K.M.K.; O'Brien, S. Premenstrual syndrome. *Obstet. Gynaecol. Reprod. Med.* **2008**, *18*, 29–32. [CrossRef]
65. Quaranta, S.; Buscaglia, M.A.; Meroni, M.G.; Colombo, E.; Cella, S. Pilot study of the efficacy and safety of a modified-release magnesium 250 mg tablet (Sincromag) for the treatment of premenstrual syndrome. *Clin. Drug Investig.* **2007**, *27*, 51–58. [CrossRef] [PubMed]
66. Walker, A.F.; De Souza, M.C.; Vickers, M.F.; Abeyasekera, S.; Collins, M.L.; Trinca, L.A. Magnesium supplementation alleviates premenstrual symptoms of fluid retention. *J. Women Health* **1998**, *7*, 1157–1165. [CrossRef] [PubMed]

67. Wang, F.; Van Den Eeden, S.K.; Ackerson, L.M.; Salk, S.E.; Reince, R.H.; Elin, R.J. Oral magnesium oxide prophylaxis of frequent migrainous headache in children: A randomized, double-blind, placebo-controlled trial. *Headache* **2003**, *43*, 601–610. [CrossRef] [PubMed]
68. Aloisi, P.; Marrelli, A.; Porto, C.; Tozzi, E.; Cerone, G. Visual evoked potentials and serum magnesium levels in juvenile migraine patients. *Headache* **1997**, *37*, 383–385. [CrossRef] [PubMed]
69. Koseoglu, E.; Talaslioglu, A.; Gonul, A.S.; Kula, M. The effects of magnesium prophylaxis in migraine without aura. *Magnes. Res.* **2008**, *21*, 101–108. [PubMed]
70. Karimi, N.; Razian, A.; Heidari, M. The efficacy of magnesium oxide and sodium valproate in prevention of migraine headache: A randomized, controlled, double-blind, crossover study. *Acta Neurol. Belg.* **2019**. [CrossRef]
71. Von Luckner, A.; Riederer, F. Magnesium in Migraine Prophylaxis-Is There an Evidence-Based Rationale? A Systematic Review. *Headache* **2018**, *58*, 199–209. [CrossRef]
72. Antonaci, F.; Dumitrache, C.; De Cillis, I.; Allena, M. A review of current European treatment guidelines for migraine. *J. Headache Pain* **2010**, *11*, 13–19. [CrossRef]
73. Holland, S.; Silberstein, S.D.; Freitag, F.; Dodick, D.W.; Argoff, C.; Ashman, E. Evidence-based guideline update: NSAIDs and other complementary treatments for episodic migraine prevention in adults: Report of the Quality Standards Subcommittee of the American Academy of Neurology and the American Headache Society. *Neurology* **2012**, *78*, 1346–1353. [CrossRef]
74. Sarchielli, P.; Granella, F.; Prudenzano, M.P.; Pini, L.A.; Guidetti, V.; Bono, G.; Pinessi, L.; Alessandri, M.; Antonaci, F.; Fanciullacci, M.; et al. Italian guidelines for primary headaches: 2012 revised version. *J. Headache Pain* **2012**, *13*, S31–S70. [CrossRef]
75. Taubert, K. Magnesium in migraine. Results of a multicenter pilot study. *Fortschr. Der Med.* **1994**, *112*, 328–330.
76. Bøhmer, T.; Røseth, A.; Holm, H.; Weberg-Teigen, S.; Wahl, L. Bioavailability of oral magnesium supplementation in female students evaluated from elimination of magnesium in 24-h urine. *Magnes. Trace Elem.* **1990**, *9*, 272–278. [PubMed]
77. Koenig, K.; Padalino, P.; Alexandrides, G.; Pak, C.Y. Bioavailability of potassium and magnesium, and citraturic response from potassium-magnesium citrate. *J. Urol.* **1991**, *145*, 330–334. [CrossRef]
78. Altura, B.T.; Wilimzig, C.; Trnovec, T.; Nyulassy, S.; Altura, B.M. Comparative effects of a Mg-enriched diet and different orally administered magnesium oxide preparations on ionized Mg, Mg metabolism and electrolytes in serum of human volunteers. *J. Am. Coll. Nutr.* **1994**, *13*, 447–454. [CrossRef] [PubMed]
79. White, J.; Massey, L.; Gales, S.K.; Dittus, K.; Campbell, K. Blood and urinary magnesium kinetics after oral magnesium supplements. *Clin. Ther.* **1992**, *14*, 678–687. [PubMed]
80. Lindberg, J.S.; Zobitz, M.M.; Poindexter, J.R.; Pak, C.Y. Magnesium bioavailability from magnesium citrate and magnesium oxide. *J. Am. Coll. Nutr.* **1990**, *9*, 48–55. [CrossRef]
81. Mühlbauer, B.; Schwenk, M.; Coram, W.M.; Antonin, K.H.; Etienne, P.; Bieck, P.R.; Douglas, F.L. Magnesium-L-aspartate-HCl and magnesium-oxide: Bioavailability in healthy volunteers. *Eur. J. Clin. Pharmacol.* **1991**, *40*, 437–438. [CrossRef]
82. Firoz, M.; Graber, M. Bioavailability of US commercial magnesium preparations. *Magnes. Res.* **2001**, *14*, 257–262.
83. Walker, A.F.; Marakis, G.; Christie, S.; Byng, M. Mg citrate found more bioavailable than other Mg preparations in a randomised, double-blind study. *Magnes. Res.* **2003**, *16*, 183–191.
84. Zhang, X.; Del Gobbo, L.C.; Hruby, A.; Rosanoff, A.; He, K.; Dai, Q.; Costello, R.B.; Zhang, W.; Song, Y. The Circulating Concentration and 24-h Urine Excretion of Magnesium Dose- and Time-Dependently Respond to Oral Magnesium Supplementation in a Meta-Analysis of Randomized Controlled Trials. *J. Nutr.* **2016**, *146*, 595–602. [CrossRef]
85. Farruggia, G.; Castiglioni, S.; Sargenti, A.; Marraccini, C.; Cazzaniga, A.; Merolle, L.; Iotti, S.; Cappadone, C.; Maier, J.A. Effects of supplementation with different Mg salts in cells: Is there a clue? *Magnes. Res.* **2014**, *27*, 25–34. [CrossRef]
86. Montagne, A.; Barnes, S.R.; Sweeney, M.D.; Halliday, M.R.; Sagare, A.P.; Zhao, Z.; Toga, A.W.; Jacobs, R.E.; Liu, C.Y.; Amezcua, L.; et al. Blood-brain barrier breakdown in the aging human hippocampus. *Neuron* **2015**, *85*, 296–302. [CrossRef] [PubMed]

87. Sandoval, K.E.; Witt, K.A. Blood-brain barrier tight junction permeability and ischemic stroke. *Neurobiol. Dis.* **2008**, *32*, 200–219. [CrossRef] [PubMed]
88. Amin, F.M.; Hougaard, A.; Cramer, S.P.; Christensen, C.E.; Wolfram, F.; Larsson, H.B.W.; Ashina, M. Intact blood-brain barrier during spontaneous attacks of migraine without aura: A 3T DCE-MRI study. *Eur. J. Neurol.* **2017**, *24*, 1116–1124. [CrossRef] [PubMed]
89. Kim, Y.S.; Kim, M.; Choi, S.H.; You, S.H.; Yoo, R.E.; Kang, K.M.; Yun, T.J.; Lee, S.T.; Moon, J.; Shin, Y.W. Altered Vascular Permeability in Migraine-associated Brain Regions: Evaluation with Dynamic Contrast-enhanced MRI. *Radiology* **2019**, *292*, 713–720. [CrossRef]
90. Mi, X.; Ran, L.; Chen, L.; Qin, G. Recurrent Headache Increases Blood-Brain Barrier Permeability and VEGF Expression in Rats. *Pain Physician* **2018**, *21*, E633–E642.
91. Kaya, M.; Ahishali, B. The role of magnesium in edema and blood brain barrier disruption. In *Magnesium in the Central Nervous System*; Vink, R., Nechifor, M., Eds.; University of Adelaide Press: Adelaide, Australia, 2011.
92. Maier, J.A.; Bernardini, D.; Rayssiguier, Y.; Mazur, A. High concentrations of magnesium modulate vascular endothelial cell behaviour in vitro. *Biochim. Biophys. Acta* **2004**, *1689*, 6–12. [CrossRef]
93. Esen, F.; Erdem, T.; Aktan, D.; Orhan, M.; Kaya, M.; Eraksoy, H.; Cakar, N.; Telci, L. Effect of magnesium sulfate administration on blood-brain barrier in a rat model of intraperitoneal sepsis: A randomized controlled experimental study. *Crit. Care* **2005**, *9*, R18–R23. [CrossRef]
94. Johnson, A.C.; Tremble, S.M.; Chan, S.L.; Moseley, J.; LaMarca, B.; Nagle, K.J.; Cipolla, M.J. Magnesium sulfate treatment reverses seizure susceptibility and decreases neuroinflammation in a rat model of severe preeclampsia. *PLoS ONE* **2014**, *9*, e113670. [CrossRef]
95. Romeo, V.; Cazzaniga, A.; Maier, J.A.M. Magnesium and the blood-brain barrier in vitro: Effects on permeability and magnesium transport. *Magnes. Res.* **2019**, *32*, 16–24. [CrossRef] [PubMed]
96. Zhu, D.; You, J.; Zhao, N.; Xu, H. Magnesium Regulates Endothelial Barrier Functions through TRPM7, MagT1, and S1P1. *Adv. Sci.* **2019**, *6*, 1901166. [CrossRef]
97. Euser, A.G.; Cipolla, M.J. Magnesium sulfate for the treatment of eclampsia: A brief review. *Stroke* **2009**, *40*, 1169–1175. [CrossRef] [PubMed]
98. Allena, M.; Steiner, T.J.; Sances, G.; Carugno, B.; Balsamo, F.; Nappi, G.; Andree, C.; Tassorelli, C. Impact of headache disorders in Italy and the public-health and policy implications: A population-based study within the Eurolight Project. *J. Headache Pain* **2015**, *16*, 100. [CrossRef] [PubMed]
99. Cohen, L.; Kitzes, R. Infrared spectroscopy and magnesium content of bone mineral in osteoporotic women. *Isr. J. Med. Sci.* **1981**, *17*, 1123–1125.
100. Vormann, J. Magnesium: Nutrition and metabolism. *Mol. Asp. Med.* **2003**, *24*, 27–37. [CrossRef]

© 2020 by the authors. Licensee MDPI, Basel, Switzerland. This article is an open access article distributed under the terms and conditions of the Creative Commons Attribution (CC BY) license (http://creativecommons.org/licenses/by/4.0/).

Review

# Magnesium Status and Stress: The Vicious Circle Concept Revisited

Gisèle Pickering [1], André Mazur [2], Marion Trousselard [3], Przemyslaw Bienkowski [4], Natalia Yaltsewa [5], Mohamed Amessou [6], Lionel Noah [6,*] and Etienne Pouteau [6]

1. Plateforme d'Investigation Clinique/CIC Inserm 1405, University Hospital CHU, 63000 Clermont-Ferrand, France; gisele.pickering@uca.fr
2. INRAE, UNH, Unité de Nutrition Humaine, Université Clermont Auvergne, 63001 Clermont-Ferrand, France; andre.mazur@inrae.fr
3. Independent Regulatory Board for Auditors (IRBA), 91220 Bretigny-sur-Orge, France; marion.trousselard@gmail.com
4. Department of Psychiatry, Warsaw Medical University, 02-091 Warsaw, Poland; pbienko@yahoo.com
5. Department of General Medicine, The Yaroslavl State Medical University Institute of Postgraduate Education, 150000 Yaroslavl, Russia; yaltzewa@yandex.ru
6. Medical Affairs Department, Consumer HealthCare, Sanofi, Gentilly, 94250 Paris, France; Mohamed.Amessou@sanofi.com (M.A.); Etienne.Pouteau@sanofi.com (E.P.)
* Correspondence: Lionel.Noah@sanofi.com; Tel.: +33-06-0213-6165

Received: 11 November 2020; Accepted: 25 November 2020; Published: 28 November 2020

**Abstract:** Magnesium deficiency and stress are both common conditions among the general population, which, over time, can increase the risk of health consequences. Numerous studies, both in pre-clinical and clinical settings, have investigated the interaction of magnesium with key mediators of the physiological stress response, and demonstrated that magnesium plays an inhibitory key role in the regulation and neurotransmission of the normal stress response. Furthermore, low magnesium status has been reported in several studies assessing nutritional aspects in subjects suffering from psychological stress or associated symptoms. This overlap in the results suggests that stress could increase magnesium loss, causing a deficiency; and in turn, magnesium deficiency could enhance the body's susceptibility to stress, resulting in a magnesium and stress vicious circle. This review revisits the magnesium and stress vicious circle concept, first introduced in the early 1990s, in light of recent available data.

**Keywords:** stress; magnesium; hypomagnesemia; magnesium deficiency; vicious circle; dietary intake; magnesium supplementation

## 1. Introduction

Stress, often intended as a psychological response to external stressors, has become a common issue of modern life [1]. From a neurobiology perspective, stress is an adaptive system that continuously assesses and interacts physically, physiologically, or psychosocially, with the environment. When this stress system is overloaded, negative health outcomes could result [2]. Magnesium is a fundamental nutrient, the role of which in human health is widely recognized [3]. Today, magnesium deficiency is also a common condition among the general population [4], and given its importance in the functioning of many reactions of the human body, this deficiency can increase the risk of physical and mental health illness over time. Of note, symptoms of magnesium deficiency and stress are very similar, the most common being fatigue, irritability, and mild anxiety [5–7]; further symptoms are shown in Table 1.

Table 1. Symptoms of magnesium deficiency and symptoms of stress.

| Most Frequently Reported Symptoms of Stress [6,7] | Symptoms of Magnesium Deficiency [5,8] |
|---|---|
| **Fatigue** | **Tiredness** |
| **Irritability or anger** | **Irritability** |
| **Feeling nervous** | **Mild anxiety/nervousness** |
| Lack of energy | Muscle weakness |
| Upset stomach | Gastrointestinal spasms |
| Muscle tension | Muscle cramps |
| **Headache** | **Headache** |
| Sadness/depression | Mild sleep disorders |
| Chest pain/hyperventilation | Nausea/vomiting |

Note: Similar symptoms are highlighted in bold.

The idea of a bidirectional relationship between magnesium and stress was first introduced by Galland and Seelig, in the early 1990s [9,10] and then referred to as the vicious circle. This vicious circle implies that stress can increase magnesium loss, causing a deficiency; in turn, magnesium deficiency can enhance the body's susceptibility to stress [10].

Taking into account the increasing prevalence of stress in modern societies [11], and its related consequences to health, this review revisits the magnesium and stress vicious circle concept, with a focus on the role of magnesium on the body's response to stress and the pathways that regulate such a response. In particular, the scope of this article was to assess the evidence available on the need of an adequate intake of magnesium, and strengthen the hypothesis that a revision of the current recommended intake of magnesium is needed for the general population when exposed to stress, in order to reduce associated health risks.

## 2. Magnesium: Biological Role and Dietary Needs

### 2.1. Biological Role of Magnesium and Homeostasis

Magnesium is an essential mineral for humans [12]. Being the second most abundant intracellular cation [5], magnesium is involved in almost all major metabolic and biochemical processes [13]. It acts as a cofactor in hundreds of enzymatic reactions [13], its primary functions including protein and nucleic acid synthesis, regulation of metabolic pathways, neuronal transmission, neuromuscular function, and regulation of cardiac rhythm [8,14,15]. In addition, magnesium is a naturally occurring calcium channel blocker, is involved in the maintenance of electrolyte balance (e.g., regulation of sodium–potassium ATPase activity), and plays a key role in membrane excitability [5,12].

It is estimated that an adult human body contains around 21–28 g of magnesium, 50–60% of which is stored in the bones, with the remainder distributed in soft tissues such as muscles [14,16]. Magnesium is also an essential component of the extracellular fluid (ECF) and the cerebrospinal fluid (CSF) in the central nervous system [17,18]. Magnesium enters the brain through the blood–brain barrier which maintains the passage of nutrients and electrolytes for the ECF homeostasis, and is actively transported by choroidal epithelial cells into the CSF [17,18]. Although little has been revealed about the exact mechanisms of magnesium transport into the brain, it is known that magnesium concentration is higher in the CSF than in plasma [19]. Under conditions of deficiency, magnesium levels still decline in the CSF, but slower when compared to the changes observed in plasma magnesium levels [19]. Experimental studies have shown that in magnesium-deficient animals, the brain uptake of magnesium is almost doubled compared to normal-fed controls [20] and CFS magnesium concentration was readily repleted, showing that magnesium is an essential mineral for the brain homeostasis [19,20].

Only 1% of the total magnesium is extracellular and 0.3% of this circulates in serum in three different forms [21]: Free (unbound; 60%), which represents the biologically active form; albumin-bound (30%); or in a complex with other ions (10%) [13].

Magnesium homeostasis is tightly regulated and relies on the dynamic balance between intestinal absorption, kidney excretion, and storage in bones (Figure 1) [22]. Magnesium is mainly absorbed in the distal parts of the small intestine [22], and mostly stored in bones [22], where it serves as a reservoir to maintain the equilibrium with its extracellular concentration [22]. The kidneys play a critical part in magnesium homeostasis by eliminating its excess [22].

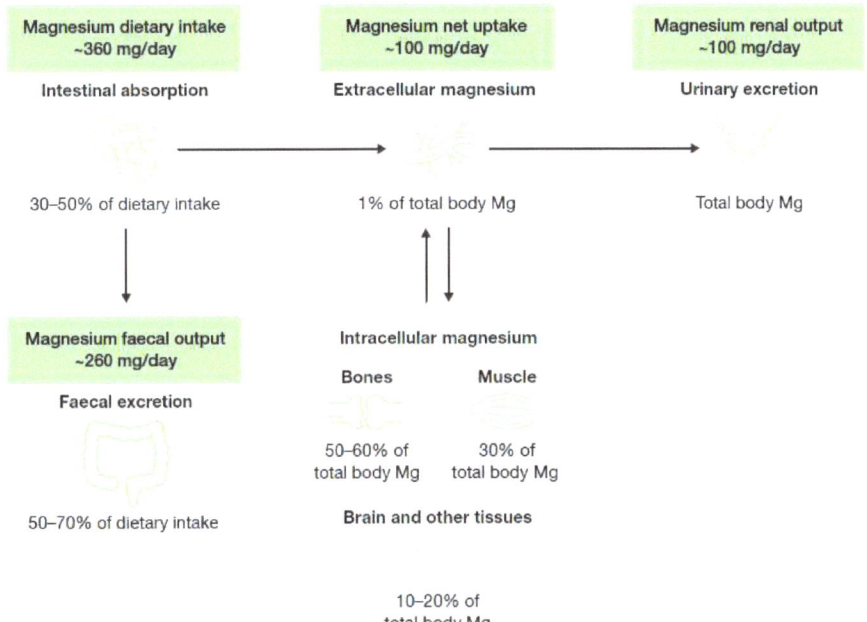

**Figure 1.** Magnesium homeostasis. Figure adapted from Jahnen–Dechent, 2012 [17]. Data from Elin, 1988 [19] and deBaaij, 2015 [8]. Mg, magnesium.

Many factors can affect magnesium balance: A diet high in sodium, calcium, and protein [5,23–25], the consumption of caffeine and alcohol [5,13,26], and the use of certain medicines such as diuretics, proton-pump, and inhibitors or antibiotics [13,26,27], which can all cause lower magnesium retention. In healthy individuals, some physiological conditions such as pregnancy [28,29], menopause [30], or ageing [31,32] are associated with changes in the need for magnesium. Pathological conditions, particularly those affecting the absorption and the elimination of nutrients (e.g., diabetes, renal function impairment, and physiological stress), may also result in significant magnesium loss or malabsorption [4,5,26,33,34]. Studies on hereditary forms of magnesium deficiency have contributed to the identification of both recessive and dominant genetic disorders directly affecting the transport of magnesium at a cellular level [35]. Although mutated transporting-proteins mainly contribute to renal wasting or intestinal malabsorption of magnesium, the mechanisms at the molecular level remain to be elucidated [8,36]. Notably, several studies showed that lower magnesium levels are involved in the course of several mental disorders, especially depression [37]. A summary of the factors affecting magnesium homeostasis is presented in Table 2.

**Table 2.** Factors contributing to magnesium deficiency.

| Diet related |
| --- |
| Inadequate magnesium intake [5,13,26] |
| High protein diet [5,25] |
| High sodium diet [5,25] |
| High calcium diet [5,23–25] |
| High caffeine intake [5] |
| Alcohol dependence [5,13,26] |
| **Lifestyle** |
| Sports [25,38–40] |
| Sleep quality and quantity [41,42] |
| Chronic stress [43,44] |
| **Pharmacological related** |
| Diuretics, e.g., furosemide [13,26,27] |
| Proton-pump inhibitors, e.g., omeprazole [13,27] |
| Cisplatin [13,26,27] |
| Antibiotics, e.g., gentamicin [13,27] |
| **Physiological conditions** |
| Pregnancy [28,29] |
| Ageing [31,32] |
| Menopause [30] |
| **Pathological conditions** |
| Genetic disorders [8,35,36] |
| Type 2 diabetes mellitus [4,5] |
| Gastrointestinal disorders [5,26] |
| Kidney failure [5,33] |
| Cardiovascular diseases [5,45] |
| Metabolic syndrome [5,34] |
| Osteoporosis [15,25,33] |

*2.2. Food Sources, Current Recommended Intakes and Safety*

Nuts, legumes, whole cereals, and fruits have the highest magnesium content of all foods [16]. Coffee or cocoa-based products may also contain significant amounts of magnesium, while fish, meats, and milk have an intermediate amount [46,47]. Drinking water, especially harder water, can also be rich in magnesium salts [48]. The source of dietary magnesium varies widely according to gender, age, and dietary habits. For example, French adults in 2016 obtained more than 21% of their magnesium from hot beverages including coffee, 9% from bread, and 6% from vegetables [47], whereas in a sample of American adults, the main sources of magnesium were vegetables (13%), milk (7%) and meat (7%) [49]. In a sample of Polish adults, dietary magnesium requirements were mostly maintained by the consumption of cereal products (11.8–15.3%) [50], and milk or dairy products (10.9%) [51]. A study investigating the Italian diet found that cereals (27%) are the primary source of magnesium in adults [52].

Over time, public health agencies have reviewed and established recommendations for dietary intake of magnesium (and other nutrients). These include the estimated average requirement (EAR), which represents the average daily intake that satisfies the nutritional requirement of 50% of the population considered; and the recommended dietary allowance (RDA), which is the daily intake that meets the requirement of 97.5% of the same population [15,53]. Values are set on the basis of dietary balance experiments and/or results from clinical studies and meta-analyses [54–57].

Dietary balance studies performed in the 1980s in the USA concluded that the EAR for magnesium was 310–330 in men and 255–265 mg/day in women [55–57]. As a consequence, in 1997, the Standing Committee on the Scientific Evaluation of Dietary Reference Intakes (for the USA and Canada) set an RDA of 400–420 and 310–320 mg/day for men and women, respectively [58]. Nowadays, nutrient requirements and dietary guidelines are available in every country. For instance, in Poland, the RDA for magnesium is 400–420 for men and 310–320 mg/day for women [59], and in Russia, it is 300 mg/day for both men and women [60]. The European Food Safety Agency (EFSA) did not consider the available scientific evidence strong enough to determine RDAs, and has suggested an "adequate intake" of 350 and 300 mg/day for men and women, respectively [16]. Within the EU, national governmental bodies have set local RDAs. In 2015, the Japanese Ministry of Health, Labor and Welfare updated the dietary reference intake guidelines and set the RDA at 320–340 and 220–230 mg/day for adult men and women, respectively [61]. Recommended intakes are summarized by country in Table 3.

**Table 3.** Current magnesium recommended dietary allowances (RDAs) across countries.

| Country | Magnesium, mg/day | |
|---|---|---|
| | Men | Women |
| Italy [62] | 240 | 240 |
| Russia [60] | 300 | 300 |
| Japan [61] | 320–340 | 220–230 |
| Poland [59] | 400–420 | 310–320 |
| USA and Canada [58] | 400–420 | 310–320 |
| France [63] | 420 | 360 |

Values shown refer to adult population only (≥19 years).

Studies have consistently shown that the dietary magnesium intake is often inadequate across different countries [5,64]. In 2005, King et al. reported that approximately 60% of Americans do not reach the recommended daily intake of magnesium through their diet [65]. In the USA, between 2003–2006, the average intakes of magnesium from food were 268 for men and 234 mg for women, which meant that 63% of men and 69% of women did not meet the EAR [66]. These results were confirmed and supported by the Dietary Guidelines Advisory Committee, which, in 2015, concluded that magnesium is an under-consumed nutrient for many Americans [46].

In Europe, the situation is similar. The National Diet and Nutrition Survey conducted in the UK between 2014–2016 showed that men's mean dietary intake in magnesium was 302 and women's was 238 mg/day [67]. In France in 2007, the mean daily dietary intake was 323 in men and 263 mg in women and more than two-thirds of the French adult population (67.4% of men and 76.7% of women, aged 18 to 54) had an inadequate magnesium intake [68]. Furthermore, in Spain, the Anthropometry, Intake and Energy Balance in Spain (ANIBES) study revealed that the mean consumption of magnesium in the population was 222 mg/day, indicating that 79% of the population had an intake below 80% of the national RDA [69]. The Mediterranean Healthy Eating, Ageing, and Lifestyle (MEAL) observational study conducted in Italy found that the dietary intake of magnesium was adequate both in men (397) and women (390 mg/day), with cereals, dairy products, and legumes being the main food sources [70]. Lastly, an analysis based on national surveys conducted across European countries showed that the mean magnesium intake among adults (18–60 years old) in Poland was 396 in men and 264 mg/day in women; [71] whereas German adults had the highest mean intake magnesium, 522 in men and 418 mg/day in women [71].

When there is a need for optimizing the magnesium status, a variety of oral supplements are available. Magnesium supplementation is considered well-tolerated, with diarrhea typically being the main manifestation of an excessive intake [72]. The upper limit for magnesium supplementation in healthy adults is 350 mg/day [73]. Normally, an increased renal filtration can reverse a wide range of serum magnesium concentration to normal levels. However, serious adverse effects have been reported for serum magnesium concentration exceeding 1.74–2.61 mmol/L. Symptoms of magnesium toxicity include hypotension, nausea, flushing of the face, retention of urine, and lethargy, and may progress to difficulty breathing, extreme hypotension, irregular heartbeat, and cardiac arrest [72].

## 2.3. Magnesium Deficiency: Causes and Health Consequences

In clinical practice, measurement of serum magnesium levels is the most common means of assessing nutrient status [13], with normal values considered to be within the 0.7–1.0 mmol/L range [74,75]. Hypomagnesemia is clinically defined when serum concentrations drop below 0.7 mmol/L [8]. Severe hypomagnesemia (<0.4 mmol/L) is rare and occurs mostly in serious pathological conditions [76]. Symptoms may include neuromuscular dysfunction (muscular weakness, tremors, seizures or tetany); cardiovascular signs (electrocardiographic abnormalities and arrhythmias); and hypokalemia and hypocalcemia [13]. However, mild hypomagnesemia (0.5–0.7 mmol/L) is common and estimated to affect around 2.5–15% of the population [4,26]. In the majority of cases, magnesium deficiency is not identified, as low serum levels are compensated by the release of magnesium from the bone reservoir [48]. In addition, mild deficiency can remain undetected as it often occurs with nonspecific symptoms such as irritability, nervousness, mild anxiety, muscle contractions, weakness, fatigue, and digestive troubles [26]. In addition, it has been suggested that chronic latent magnesium deficiency could start developing below 0.85 mmol/L with a potential impact on human health [12,77]. A recent study by Noah et al. found that nearly half (~44%) of the subjects screened for stress had chronic latent magnesium deficiency (defined as serum magnesium <0.85 mmol/L) [78]. Moreover, subclinical, chronic magnesium depletion may contribute to various dysfunctions and diseases and the scientific literature is rich in studies highlighting the association between low dietary magnesium intake and a higher risk of type 2 diabetes, cardiovascular diseases, osteoporosis, and metabolic syndrome [8,79,80].

Several factors contribute to magnesium deficiency (Table 2). Dietary surveys point to an inadequate magnesium intake from food. Surveys conducted in different countries have consistently showed a substantial inadequate intake of magnesium from food in the general population, particularly in young adults, those over 70 years of age [81], and in women [71]. Of note, over the past 60 years, intensive farming practices have caused a significant depletion of the mineral content of the soil [82–84], including a decrease in magnesium of up to 30% [85,86]. Additionally, western diets typically have a greater proportion of processed food, where several products are mostly refined, with magnesium being depleted by up to 80–90% in the process [5,8]. Factors and behaviors associated with the western lifestyle, including intense sport and physical activity [38], poor sleep quality and quantity [41], and psychological stress [43,44], can also induce magnesium loss. Magnesium deficiency is linked to many health conditions, from those affecting its metabolism, such as gastrointestinal diseases, type 2 diabetes, alcohol dependence, or kidney failure [5,26], to genetic disorders [35]. A growing body of evidence also suggests that chronic stress may cause magnesium loss/deficiency [43]. Numerous studies have shown lower magnesium levels associated with different neurological and psychiatric disorders, particularly depression and post-traumatic stress disorder [87,88] but also anxiety disorders, attention deficit hyperactivity disorder, and bipolar disorder [37,89]. Although evidence on a causal factor between mental disorders and magnesium deficiency has yet to be confirmed, stress appears as a key component in the relationship between mental health illness and magnesium deficiency.

## 3. Stress

Stress is commonly described as a trigger that evokes a physiological and psychological response of the body [90]. Over the past decades, the understanding of stress biology has largely evolved. Stress is no longer considered as a temporary response to occasional threats, but rather an ongoing and adaptive system that enables an individual to assess, cope, and predict constantly changing conditions. However, the capacity of this stress system is limited and can be overloaded, resulting in poor health outcomes, particularly those related to mental illness like depression or cognitive deficits [2].

Stress not only affects the mental health status of an individual, but it is also characterized by a physical response of the body that, depending on the type and length of exposure, may lead to short-term effects (e.g., increased blood pressure, increased heart and respiration rates, increased alertness) [91], or long-term effects (e.g., impaired hippocampal neurogenesis, cognitive and memory disorders) [92].

The following sections summarize widely recognized theoretical models of stress and describe possible physiological roles of magnesium in the stress response. Here, the term "stress model" refers to a theoretical framework used to predict outcomes and to explain specific processes.

*3.1. Neurobiological Stress and Allostatic Load Model*

In the 1950s, Selye proposed the general adaptation syndrome model to describe stress as the reaction of the body to emergency situations [93]. This theory divides the response to stressful stimuli into three phases: (1) Alarm: Upon perceiving a stressor, the body reacts with a "fight-or-flight" response, the sympathetic nervous system (SNS) is stimulated, and the body's resources are mobilized to meet the threat. (2) Resistance: The body resists and compensates as the parasympathetic nervous system (PNS) attempts to return many physiological functions to normal levels, while the body remains on alert and focuses resources against the stressor. (3) Exhaustion: If the stressor continues beyond the body's capacity, the resources are depleted and the body becomes susceptible to disease (distress) [91,93].

A more contemporary concept, which better describes the cumulative impact of stressor exposure on health outcomes, is that of allostasis. Allostasis is the process by which constant changes allow an organism to achieve and maintain normal functions, thus reflecting the ability of the body to adapt to daily situations like exercising or hunger, effectively [94]. However, this continual maintenance costs the body energy and resources, and over time, may lead to symptoms of allostatic load—the functional and structural damage caused by "the wear and tear" of the body's resources in response to stress [95]. Therefore, the response to a new stressor depends on the body's resources available following the previous stress response [96]. The allostatic load is characterized by a cumulative effect, which becomes greatest when stress is chronic or intense [96].

The hypothalamic–pituitary–adrenal (HPA) axis and the autonomic nervous system (comprising SNS and PNS) have been identified as the mediators of this neurobiological stress model [90,91,95]. First, corticotrophin-releasing factor (CRF) is secreted from the paraventricular nucleus in the hypothalamus; the subsequent secretion of adrenocorticotropic hormone (ACTH) from the anterior pituitary stimulates the release of glucocorticoids (mainly cortisol) from the adrenal cortex [97]. Noradrenaline (NA) and adrenaline are also released from the sympathetic nerves and the adrenal medulla, and together with the glucocorticoids regulate the stress response [90,91]. Cortisol also interacts with the serotonergic pathway, adjusting the release of serotonin (5-hydroxytryptamine or 5-HT) neurotransmitter in response to acute or chronic stressors [98]. Serotoninergic neurons modulate the stress response either via direct neurotransmission to the hypothalamus, or by stimulation of noradrenergic neurons [97]. In addition to the regulation through feedback mechanisms, the HPA axis is also modulated by other central systems, particularly by the inhibitory action of the γ-aminobutyric acid (GABA), and the excitatory effect of glutamate [99].

In this neurobiological model, cortisol is a well-known mediator of the stress response. The nocturnal cortisol urinary excretion in apparently healthy subjects reflects the basal tone of the HPA axis [100]; conversely, the blood cortisol concentration measured in a challenging environment is a sign of stress activity [101]. It has been shown that cortisol coordinates the central response to stress at several levels [102], and indirectly influences mechanisms of neuroprotection [103]. Neurotrophic factor production, represented by the brain-derived neurotrophic factor (BNDF), intervenes in allostasis through protecting neurons [104]. Normally, BNDF promotes neuronal survival and plasticity [104]; however, changes in BNDF expression have been reported following exposure to stressful stimuli. An increase of BNDF has been observed in response to moderate stress [105], whereas a decrease has been associated with high levels of stress [106]. Furthermore, increasing evidence shows a link between cortisol responses and oxidant elevation [107]. The accumulation of free radicals and other reactive oxygen species is also a sign of allostatic load, resulting from the imbalance between cellular metabolic activities and antioxidant defense mechanisms [108,109].

Noteworthy, magnesium interacts with all these stress mediators [17,110–112], overall serving an inhibitory function in the regulation and central neurotransmission of the stress response (details of these interactions are summarized in chapter 6).

### 3.2. Generalized Unsafety Theory of Stress (GUTS) Model

Conventional theories of stress have historically focused on the assumption that stress is a response to an actual environmental threat (either internal or external to the body), making it difficult to explain the relation between stress and disease. In contrast, GUTS is a new psychological and cognitive theoretical model proposed by Brosschot in 2016 [113] that revises and expands the stress theory by focusing on safety instead of threat, and by including risk factors that have hitherto not been attributed to stress [113]. Based on neurobiological and evolutionary evidence, GUTS hypothesizes that stressors are not necessary for a chronic stress response to occur but the perception of an unsafe state is enough. In GUTS, PNS is the key system controlling the stress response (particularly the vagus nerve and the prefrontal cortex activity) [113]. Of note, preclinical data suggest that magnesium may be important for the functionality of these central systems. An excess of magnesium or magnesium deficiency have been shown to modulate the autonomic nervous system, but further research is still needed [114–116].

The GUTS model suggests that the default stress response can be chronically activated in various situations, three of which are particularly susceptible to health risks. (1) Reduced body capacity: In compromised physical conditions, e.g., obesity or aging, the brain perceives the body as inadequate to be able to "fight-or-flight", and therefore maintains a state of general alarm, or unsafety [117]. (2) Compromised social network: Being part of a group is a fundamental aspect of survival for social animals and humans, and isolation is one of the main conditions in which safety is lacking [117]. Interestingly, there is evidence that patients suffering from metabolic syndrome [118] or congestive heart failure [5] (both conditions of reduced body capacity as described in GUTS) exhibit lower serum magnesium concentration. (3) Perceived aversive environment: In cases of specific stressors (e.g., work stressors), a neutral daily environment (e.g., an office working environment) can be perceived as unsafe [117]. The GUTS model suggests that repetitive negative thinking may result in the impairment of key systems controlling the stress response [119]; however, the relationship between general unsafety and magnesium status is to be elucidated yet.

## 4. Evidence of the Impact of Stress on Magnesium Homeostasis

Initially, the shift from intracellular to extracellular magnesium following a stressor exposure plays a protective and regulatory role [90]. Normally, magnesium inhibits the glutamatergic transmission while promoting GABA activity, resulting in a mostly inhibitory effect at the central level [42]. Magnesium also tends to diminish the stress response mediated by catecholamines and glucocorticoids. However, a chronic stressor exposure may result in a depletion of various resources as described by Selye, including magnesium [42,93]. The progressive loss of magnesium from the reservoir in bone can eventually compromise its physiological inhibitory action and lead to an over-activation of the HPA axis and neuronal hyperactivity [120]. The impact of stress on magnesium status has been extensively investigated in both animal and human studies [42,121].

*Pre-clinical evidence.* Animal studies have shown a transient hypermagnesemia in the short-term period after the exposure to acute stress stimuli [122–125]. A series of experiments conducted on cats by Classen et al. showed that stimuli such as withdrawal of blood, infusion of catecholamines, or potassium poisoning all caused an increase of blood magnesium concentration [122]. This increase was not influenced by pre-treatment with an adrenergic blocking agent (e.g., reserpine), suggesting that other mechanisms rather than catecholamines are responsible for the change in magnesium levels [122]. Similarly, a shift of magnesium from erythrocytes to serum was reported in different studies investigating the effect of acute noise on magnesium-deficient guinea pigs [123] and rats. As a consequence, a net renal excretion of magnesium occurs, leaving the animal magnesium deficient.

irritability, fatigue, and sleep disorders, and 60% presented magnesium deficiency [144]. In other studies, subclinical chronic magnesium deficiency was found in up to 45% of the stressed subjects enrolled [78,145,146]. Nielsen et al. found that out of 96 American adults complaining of sleep disorders, a potential source of stress, 58% were consuming less than the EAR for magnesium and had higher levels of C-reactive protein (CRP), an indicator of inflammatory stress [145]. Lastly, it is generally recognized that chronic stress and magnesium deficiency may influence an individual's susceptibility to depressive disorders [147]. In a group of Australian patients experiencing depression and/or anxiety, the analysis of their nutrition status showed that 22% of participants did not meet magnesium EAR. Furthermore, magnesium intake (expressed as % EAR) was negatively correlated with stress, depression, and total Depression Anxiety Stress Scale (DASS) scores [148].

A summary of the pre-clinical and clinical evidence supporting a relation between magnesium status and stress is shown in Table 5.

**Table 5.** Summary of the pre-clinical and clinical evidence supporting magnesium status on stress susceptibility. [a] Only symptoms shown in ≥70% of women at baseline are reported.

| | | | | |
|---|---|---|---|---|
| **Evidence of the Impact of Magnesium Status on Stress Susceptibility** | | | | |
| | **Population Tested** | **Mg Status** | **Stress Stimulus** | **Impact on Stress Mediator/Stress** |
| Pre-clinical | Rats (N = 84) | Mg-deficient | Noise | ↑Catecholamines (NA, adrenaline, dopamine) [136] |
| | Mice (N = 120) | Mg-deficient | Genetic selection | ↑NA [137] |
| | Mice (N = 80) | Mg-deficient | Genetic selection; forced swimming test; four-plate test | ↑NA [138] |
| | Mice (N = 100) | Mg-deficient | Genetic selection; immobilization test | ↑Gastric ulcers [139] |
| | Mice (N = 20/test) | Dietary Mg restriction | Hyperthermia; open field test; light/dark test; hyponeophagia test | ↑CRH; ↑ACHT [140] |
| | Mice | Mg-deficient | Light/dark test | Depression-like behavior [42,140] |
| | Rats | Dietary Mg restriction | Forced swimming test | Depression-like behavior [142,143] |
| | Rats | Dietary Mg restriction | Open field test | Stress/anxiety [142,143] |
| Clinical | Women (N = 100) | Mg-deficient | - | Chronic emotional stress; irritability; fatigue; sleep disturbance; headache [a] [144] |
| | Adults (N = 264) | Mg-deficient | - | Severe stress [78,145,146] |
| | Adults (N = 100) | Mg-deficient | Poor sleep quality | ↑CRP [145] |
| | Adults (N = 109) | Mg-deficient | - | Depression/anxiety [148] |

ACHTH, adrenocorticotropic hormone; CRH, corticotrophin-releasing hormone; CRP, C-reactive protein; Mg, magnesium; NA, noradrenaline; ↑, increase.

## 6. Proposed Model for the Vicious Circle of Stress and Magnesium Deficiency

Over the years, a growing body of evidence has consistently shown that magnesium acts on several key physiological steps involved in the response to stressful stimuli.

- Magnesium and HPA. *5-HT transmission:* Magnesium directly enhances the interaction between 5-HT and its membrane receptor, and it promotes the cellular transmission of the serotoninergic signal (Figure 2A) [90]. Additionally, magnesium is a cofactor of tryptophan hydroxylase, the enzyme involved in 5-HT synthesis [90]. Glutamatergic transmission: Magnesium inhibits the glutamate directly and indirectly by blocking the glutamate N-methyl-D-aspartate (NMDA) receptor and by enhancing its reuptake in the synaptic vesicles through stimulation of the sodium–potassium ATPase, respectively (Figure 2B) [42]. GABA transmission: A GABA-agonistic activity of magnesium has been observed, although the mechanism has not yet been elucidated,

(Figure 2B) [42]. *Cortisol:* Magnesium indirectly reduces the release of ACTH by modulating the neurotransmission pathways, and therefore decreases cortisol levels in the body [42];
- Magnesium and neuroprotection. Studies on the antidepressant effects of magnesium have shown the positive impact of this mineral on the expression of BNDF in the brain [149,150];
- Magnesium and oxidative stress. Magnesium may be involved in suppressing the production of free radicals in various tissues including the brain [17], and several laboratory studies have shown that magnesium-deficient animals are more at risk of oxidative stress [112,151].

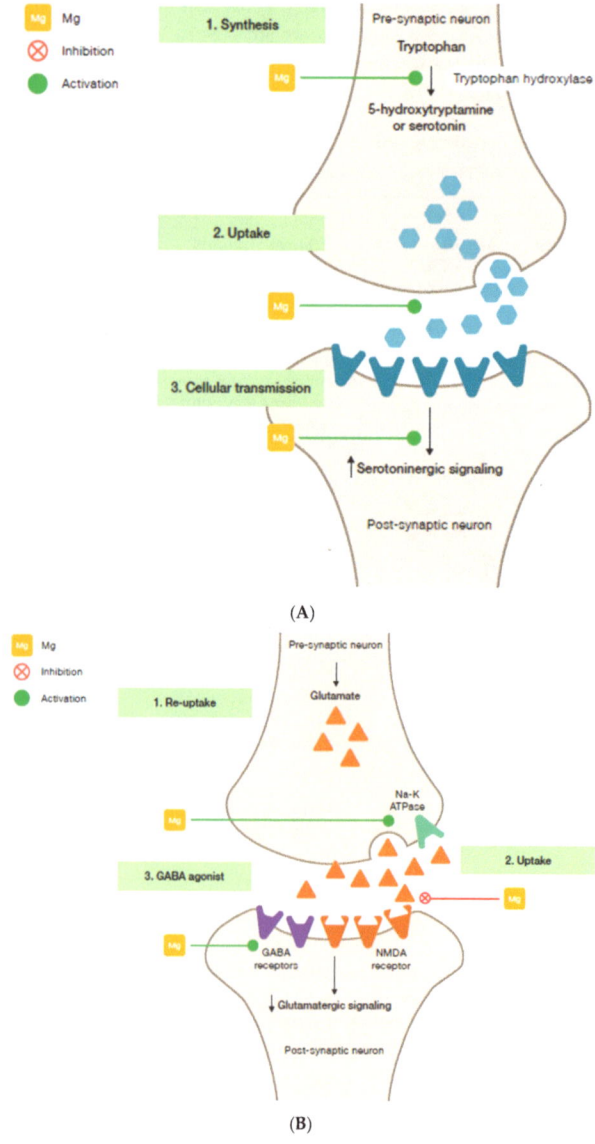

**Figure 2.** Interaction of magnesium and neurological stress mediators. (**A**) Serotoninergic transmission. (**B**) Glutamatergic and GABAergic transmission. GABA, γ-aminobutyric acid; Mg, magnesium; NMDA, N-methyl-D-aspartate.

Mild hypomagnesemia can be observed in response to mid- or long-term exposure to stress. A study conducted on guide dog candidates at different levels of a training program (elementary, intermediate, and advanced) showed the effects of temperature and physical stress on serum magnesium levels. First, it was demonstrated that serum magnesium levels were significantly lower in winter than in summer (average temperature was 6 and 29 °C, respectively), suggesting an impact of seasonality on magnesium homeostasis. Thereafter, it was noticed that physical exercise had a greater impact on serum magnesium levels of dog candidates in the elementary class compared to more trained ones. These results were lastly confirmed by a third experiment, assessing both the impact of physical stress and temperature on serum magnesium levels and finding that serum magnesium levels after exercise were significantly lower in winter than in summer. [126]. The impact of physical stressors was assessed also in another study conducted on rats. A greater serum magnesium reduction was observed in those administrated with ethanol and then exposed to restraint stress, compared to control rats facing the same restraint test but receiving water [127]. An additional study conducted by Heroux et al. on rats fed with a magnesium-deficient diet and kept at low temperature (6 °C) for about 17 months found that the studied animals were capable of adapting to cold stress despite suboptimal magnesium intake; initial signs of magnesium deficiency (including skin sores, reduced growth rate, lower levels of magnesium in most organs) gradually disappeared after two months. However, regardless this adaptation, the long-term stress resistance (measured as cold resistance at −20 °C) of magnesium-deficient rats was reported to decline over time when compared to controls [128]. Lastly, exposure to cold (2–5 °C) and a deficient dietary intake of magnesium significantly reduced plasma magnesium in sheep, whereas no effect was observed in normally fed sheep [129].

*Clinical evidence.* To help elucidate the stress hormone-induced magnesium deficiency and its clinical relevance, Whyte et al. investigated the effect produced by the infusion of adrenaline on plasma magnesium concentrations [130]. They found that magnesium levels were significantly reduced not only during the infusion time but also an hour after test cessation, without any sign of recovery [130]. A variety of tests have demonstrated that magnesium levels, both in serum and urine, are affected by the exposure to stress stimuli. Significant reductions in plasma and total magnesium concentrations were reported in a 3-month analysis on young adults exposed to either chronic or sub-chronic stressful conditions (e.g., acts of intolerance or fear of military actions) [43]. A similar effect was also seen in a study investigating the effect of temporary (one day) and chronic (one month) sleep deprivation on magnesium levels; in a group of otherwise healthy men, chronic sleep restriction was associated with greater reductions in erythrocyte magnesium concentrations [131]. University students during an exam period reported an increase in anxiety that was also associated with an increased urinary excretion of magnesium [44]. In a similar study conducted on college students during the 4 weeks following an examination period, erythrocyte magnesium content was found to be significantly depleted [132]. Interestingly, the variations in blood and urine magnesium levels were confirmed by Mocci et al. who studied the effect of noise on catecholamines and magnesium serum and urinary excretions on healthy men [133]. Mocci and his study group also noted how the timing for the change to occur was very different between the two variables, with serum magnesium increasing a few hours after the exposure to the noise (probably reflecting extracellular flux immediately after the stress), and urine excretion reaching a peak in a few hours but lasting up two days [133]. A similar result was reported by Ising et al. on a study investigating the effect of traffic noise on workers' performance. Under noise stress (7 h), a decrease in erythrocyte magnesium levels was observed, followed by an increase of serum levels and urine excretion of magnesium [134]. The impact of acute stress on transient hypermagnesemia was noted also under physical stress. Short- and long-term exercise (20 min versus 1 h, respectively) had a different influence on the plasma magnesium levels: an increase of plasma magnesium was reported after short-term exercise but not after long-term exercise. However, after both physical tests, magnesium levels dropped below the pre-exercise values [135]. A summary of the pre-clinical and clinical evidence is shown in Table 4.

**Table 4.** Summary of the pre-clinical and clinical evidence supporting the impact of stress on magnesium homeostasis.

| Evidence of the Impact of Stress on Magnesium Homeostasis | | | |
|---|---|---|---|
| | **Population Tested** | **Stress Stimulus** | **Impact on Magnesium** |
| Pre-clinical | Cats (N = 30) | Withdrawal of blood; infusion of catecholamines; potassium poisoning | ↑Blood Mg [122] |
| | Guinea pigs (41) | Noise | ↑Serum Mg, ↓Erythrocytes Mg [123] |
| | Rats (88) | Noise | ↑Serum Mg, ↓Erythrocytes Mg [124] |
| | Rats | Noise | ↓Serum Mg, ↓Erythrocytes Mg |
| | Dogs | Physical exercise, temperature | ↓Serum Mg [126] |
| | Rats | Ethanol/Restraint stress | ↓Serum Mg [127] |
| | Rats | Cold | ↓Tissue content of Mg [129] |
| | Sheep | Dietary Mg restriction, cold | ↓Plasma Mg [129] |
| Clinical | Adults (N = 8) | Adrenaline infusion | ↓Plasma Mg [130] |
| | Young adults (N = 35) | Chronic or sub-chronic psychological stress | ↓Plasma Mg [43] |
| | Healthy men (N = 16) | Chronic sleep deprivation | ↓Erythrocyte Mg [131] |
| | Young adults (N = 35) | University exams | ↑Urinary Mg [44] |
| | Young adults (N = 30) | University exams | ↓Erythrocyte Mg [132] |
| | Young adults (N = 25) | Noise | ↑Urinary Mg ↑Serum Mg [133] |
| | Healthy men (56) | Noise | ↑Serum Mg, ↓Erythrocytes Mg; ↑Urinary Mg [134] |
| | Healthy men | Short- and long-term physical exercise | ↑Plasma Mg [135] |

Mg, magnesium; ↑, increase; ↓, decrease.

## 5. Evidence of the Impact of Magnesium Status on Stress Susceptibility

*Pre-clinical evidence.* The relationship between magnesium deficiency and stress-related behavior is well documented. In 1986, Caddell et al. reported an increase of circulating catecholamines following the exposure of magnesium-deficient rats to a noise stress test [136]. The relationship between low serum magnesium concentrations and the increased release of catecholamines in the central nervous system was then confirmed in studies conducted on mice selected for low (MGL) and high (MGH) blood magnesium. The simple selection of genetic traits inducing low blood magnesium was found to significantly affect the metabolism of NA but not that of other neurotransmitters [137]. MGL mice not only showed higher NA levels (17% in the brain; 200% in urine) but also a more restless behavior and higher rectal temperature, all signs of an exaggerated stress response [138]. In a different study also conducted on MGL and MGH mice, both fed with the same magnesium-rich diet, the number of stress-induced gastric ulcers through the immobilization test was higher in the magnesium-deficient mice [139]. Besides the noradrenergic hyperactivity in basal magnesium deficiency conditions, dietary magnesium restrictions have also been associated with an upregulation of the stress system via increases in CRH and ACTH levels [140]. Experimental data indicate that magnesium-deficient rats exhibit more anxiety- and depression-like behavior compared with controls [42,138]. For example, in the light–dark test (often used to screen for anxiolytic and antidepressant drugs) [141], mice with magnesium deficiency showed a net preference for the darker compartment [42,140,141]. Similarly, in the forced swimming test, magnesium-deficient rats spent more time immobile compared with their controls [142,143]. Dietary magnesium deficiency in laboratory animals was also associated with stress-like behavior in the open field test. Rats with a reduced dietary magnesium content tended to visit the bright and central area less frequently [142,143], even when motivated by the presence of food, showing a psychological stress caused by the open space [140].

*Clinical evidence.* Results in human studies are consistent with animal findings and show low magnesium status in stressed/depressed populations. In a study investigating the potential benefit of magnesium supplementation in Russian women who suffered from chronic emotional stress, Akarachkova et al. found that at baseline the majority of women were suffering from symptoms like

In response to a stressful stimulus, stress hormones are released, causing an increase of magnesium extracellular levels [90]. As a consequence, higher magnesium concentrations are excreted through the kidneys [133]. When the stressor persists over time, this mechanism may contribute to magnesium cation depletion and deficiency [42,130], and trigger the stress and magnesium vicious circle as illustrated in Figure 3.

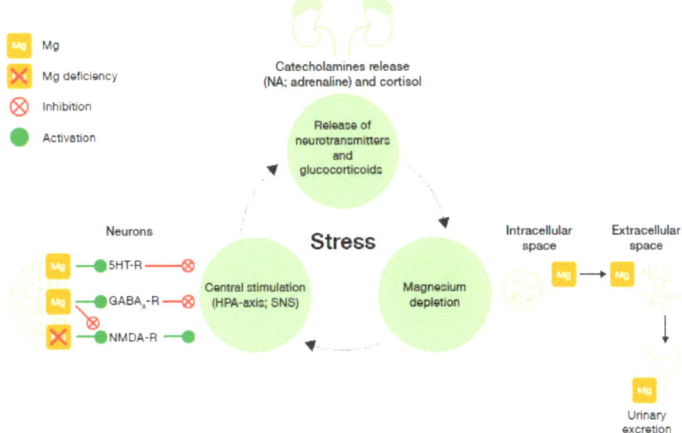

**Figure 3.** The vicious circle of stress and magnesium. GABAA-R, γ-aminobutyric acid-A receptor; Mg, magnesium; NMDA-R, N-methyl-D-aspartate receptor; NA, noradrenaline; SNS, sympathetic nervous system, 5HT-R, 5-hydroxytryptamine receptor.

Comprehensively, both pre-clinical and clinical studies' results point to the bi-directional relationship between magnesium levels and stress: Magnesium deficiency can induce symptoms and increase susceptibility to stress, and acute and chronic stress can precipitate magnesium deficiency [10,104,106].

## 7. Magnesium Supplementation

Magnesium supplementation has proven benefits for the treatment of symptoms of psychological daily stress (fatigue, irritability, sleep) [144]. It has been shown that subjects with mental and physical stress can benefit from a daily intake of magnesium. Male students experiencing common stress factors such as sleep deprivation, malnutrition, and a lack of physical activity, and receiving magnesium 250 mg/day for four weeks not only presented an increase in erythrocyte magnesium content but also a reduction of serum cortisol [152]. Magnesium supplementation of 400 mg/day was associated with a clear improvement of the heart rate variability, measured as an indicator of the parasympathetic and vagal systems' response to stress, in subjects who were asked to complete moderate muscle endurance training once weekly [153]. The daily supplementation with 300 mg (combined or not with vitamin B6, 30 mg) provided positive results on stress relief [154], particularly on subjects who reported severe stress levels at baseline, with a reduction in Depression Anxiety Stress Scale scores of up to 45% from baseline [154]. It is interesting to note that several studies investigating the potential benefit of magnesium supplementation in populations with symptoms of stress reported a subclinical chronic magnesium deficiency or a low magnesium status at baseline in the majority of the subjects enrolled [37,78,144–146]. Nevertheless, despite several studies reporting an association between magnesium deficiency and stress, the effect of magnesium supplementation on stress has been less documented than its effects on depression [37,155] and anxiety disorders [156]; therefore, further investigation is still needed on stress symptoms. A possible limiting factor to the performance of such studies could be the difficulty of setting up optimal experimental conditions for studying the effect of stress; however, this challenge may be overcome in future analyses by focusing on well-defined conditions (e.g., psychological stress), and by using robust and validated tools to assess stress (such as DASS scores).

## 8. Conclusions: Implications in Terms of Dietary Magnesium Needs

Over the past decades, increasing evidence, as shown in the present narrative review, has investigated and supported the link between magnesium deficiency and increased susceptibility to stress disorders, and further suggested that stress itself can lead to magnesium depletion. Magnesium is an essential element involved in reactions regulating the body's stress response at several levels. Severe magnesium deficiency is rare, but chronic latent deficiency appears to be common among the general population and even more among those suffering from a number of chronic diseases or stress [5]. Although the current intake of magnesium through our diet seems sufficient to avoid overt signs of magnesium deficiency in the majority of the population, it might not be adequate to provide optimal health and risk reduction of chronic diseases [5]. Stress is also an increasing condition worldwide and its effects can negatively impact health outcomes. Noteworthy, magnesium intake has been found negatively correlated with subjective stress in some populations [148], and magnesium supplementation has shown benefits in stressed but otherwise healthy subjects [153,154]. Additionally, magnesium intake is safe with limited side-effects in cases of chronic overconsumption [72].

To conclude, while there is good evidence from animal and human studies of the bi-directional link between magnesium and stress, further research is needed to better understand the impact of this correlation and the benefit of magnesium supplementation on general health. Additional studies should apply standard methodologies (e.g., magnesium load test) to evaluate the magnesium status in well-characterized stressed population. These studies would help to demonstrate the increased need of magnesium supplementation during stress periods, and further strengthen our initial hypothesis. Further, in line with the GUTS model, repetitive negative thinking could be considered as a cognitive indicator of stress and evaluated in relation to blood magnesium levels in a cohort of subjects exposed to chronic stress. Given the strong association of stress with mental and physical diseases, these studies are fundamental to further support adequate magnesium dietary needs.

**Author Contributions:** Conceptualization, M.A., L.N.; writing—original draft preparation, and writing—review and editing, G.P., A.M., M.T., P.B., N.Y., M.A., L.N., E.P.; supervision, project administration, funding acquisition, A.M., L.N. All authors have read and agreed to the published version of the manuscript.

**Funding:** This review was funded by Sanofi.

**Acknowledgments:** Editorial support was provided by Martina Klinger Sikora and Mark Davies of inScience Communications, Springer Healthcare Ltd., UK.

**Conflicts of Interest:** A.M. reports consultancy fees from Sanofi, unrelated to this publication. E.P. is an employee at Sanofi. G.P. has no conflicts of interest. L.N. is an employee at Sanofi. M.A. is an employee at Sanofi. M.T. has no conflicts of interest. P.B. reports speaker's honoraria from: Abbott, Adamed, Angelini, Apotex, Astellas, Asteriamed, Bausch Health, BGP Products, BMS, Bioton, Boehringer Ingelheim, Bonnier Business, Chiesi, CROS, Fundacja Syntonia, Fundacja Zdrowie i Opieka, Gedeon Richter, + pharma, G-Pharma, Ipsen, Item Publishing, Janssen, Kimze, Krka, Lilly, Lundbeck, Medforum, Mediadore, Medical Education, Medical Experts, Medycyna Praktyczna, Mylan, Neoart, Novo Nordisk, P2P, Pfizer, Pierre Fabre, Polfa Tarchomin, Polpharma, Promed, QAH, Sandoz, Sanofi-Aventis, Servier, S & P Partners, Takeda, Termedia, Teva, Warsaw Voice, White Solutions, Valeant, Via Medica, VM Media, Zentiva. Y.N. has no conflicts of interest.

## References

1. Kessler, R.C.; Aguilar-Gaxiola, S.; Alonso, J.; Chatterji, S.; Lee, S.; Ormel, J.; Üstün, T.B.; Wang, P.S. The global burden of mental disorders: An update from the WHO World Mental Health (WMH) Surveys. *Epidemiol. Psichiatr. Soc.* **2009**, *18*, 23–33. [CrossRef] [PubMed]
2. McEwen, B.S.; Akil, H. Revisiting the Stress Concept: Implications for Affective Disorders. *J. Neurosci.* **2020**, *40*, 12–21. [CrossRef] [PubMed]

3. Konrad, M.; Schlingmann, K.P.; Gudermann, T. Insights into the molecular nature of magnesium homeostasis. *Am. J. Physiol. Renal. Physiol.* **2004**, *286*, F599–F605. [CrossRef] [PubMed]
4. Pham, P.C.; Pham, P.M.; Pham, S.V.; Miller, J.M.; Pham, P.T. Hypomagnesemia in patients with type 2 diabetes. *Clin. J. Am. Soc. Nephrol.* **2007**, *2*, 366–373. [CrossRef]
5. DiNicolantonio, J.J.; O'Keefe, J.H.; Wilson, W. Subclinical magnesium deficiency: A principal driver of cardiovascular disease and a public health crisis. *Open Heart* **2018**, *5*, e000668. [CrossRef]
6. American Psychological Association. Stress Effects on the Body. Available online: https://www.apa.org/helpcenter/stress (accessed on 30 June 2020).
7. American Psychological Association. Understanding Chronic Stress. Available online: https://www.apa.org/helpcenter/understanding-chronic-stress.aspx (accessed on 30 June 2020).
8. De Baaij, J.H.; Hoenderop, J.G.; Bindels, R.J. Magnesium in man: Implications for health and disease. *Physiol. Rev.* **2015**, *95*, 1–46. [CrossRef]
9. Galland, L. Magnesium, stress and neuropsychiatric disorders. *Magnes. Trace Elem.* **1991**, *10*, 287–301.
10. Seelig, M.S. Consequences of magnesium deficiency on the enhancement of stress reactions; preventive and therapeutic implications (a review). *J. Am. Coll. Nutr.* **1994**, *13*, 429–446. [CrossRef]
11. American Psychological Association. 2015 Stress in America. Available online: https://www.apa.org/news/press/releases/stress/2015/snapshot (accessed on 10 August 2020).
12. Costello, R.; Wallace, T.C.; Rosanoff, A. Magnesium. *Adv. Nutr.* **2016**, *7*, 199–201. [CrossRef]
13. Reddy, S.T.; Soman, S.S.; Yee, J. Magnesium Balance and Measurement. *Adv. Chronic Kidney Dis.* **2018**, *25*, 224–229. [CrossRef] [PubMed]
14. Glasdam, S.M.; Glasdam, S.; Peters, G.H. The Importance of Magnesium in the Human Body: A Systematic Literature Review. *Adv. Clin. Chem.* **2016**, *73*, 169–193. [CrossRef] [PubMed]
15. Bergman, C.; Gray-Scott, D.; Chen, J.J.; Meacham, S. What is next for the Dietary Reference Intakes for bone metabolism related nutrients beyond calcium: Phosphorus, magnesium, vitamin D, and fluoride? *Crit Rev Food Sci. Nutr.* **2009**, *49*, 136–144. [CrossRef] [PubMed]
16. EFSA Panel on Dietetic Products Nutrition and Allergies (NDA). Scientific Opinion on Dietary Reference Values for magnesium. *EFSA J.* **2015**, *13*, 4186. [CrossRef]
17. Yamanaka, R.; Shindo, Y.; Oka, K. Magnesium Is a Key Player in Neuronal Maturation and Neuropathology. *Int. J. Mol. Sci.* **2019**, *20*, 3439. [CrossRef] [PubMed]
18. Ghabriel, M.N.; Vink, R. Magnesium transport across the blood-brain barriers. In *Magnesium in the Central Nervous System*; Nechifor, M., Vink, R., Eds.; The University of Adelaide Press: Adelaide, Australia, 2011; pp. 59–74. [CrossRef]
19. Morris, M.E. Brain and CSF magnesium concentrations during magnesium deficit in animals and humans: Neurological symptoms. *Magnes. Res.* **1992**, *5*, 303–313. [PubMed]
20. Chutkow, J.G. Uptake of magnesium into the brain of the rat. *Exp. Neurol.* **1978**, *60*, 592–602. [CrossRef]
21. Jahnen-Dechent, W.; Ketteler, M. Magnesium basics. *Clin. Kidney J.* **2012**, *5*, i3–i14. [CrossRef]
22. Ross, C.A.; Caballero, B.; Cousins, R.J.; Tucker, K.L.; Ziegler, T.R. *Modern Nutrition in Health and Disease*, 11th ed.; Lippincott Williams & Wilkins: Baltimore, MD, USA, 2014; pp. 159–175.
23. Clarkson, E.M.; Warren, R.L.; McDonald, S.J.; de Wardener, H.E. The effect of a high intake of calcium on magnesium metabolism in normal subjects and patients with chronic renal failure. *Clin. Sci.* **1967**, *32*, 11–18.
24. Norman, D.A.; Fordtran, J.S.; Brinkley, L.J.; Zerwekh, J.E.; Nicar, M.J.; Strowig, S.M.; Pak, C.Y. Jejunal and ileal adaptation to alterations in dietary calcium: Changes in calcium and magnesium absorption and pathogenetic role of parathyroid hormone and 1,25-dihydroxyvitamin D. *J. Clin. Investig.* **1981**, *67*, 1599–1603. [CrossRef]
25. Johnson, S. The multifaceted and widespread pathology of magnesium deficiency. *Med. Hypotheses* **2001**, *56*, 163–170. [CrossRef]
26. Ayuk, J.; Gittoes, N.J. Contemporary view of the clinical relevance of magnesium homeostasis. *Ann. Clin. Biochem.* **2014**, *51*, 179–188. [CrossRef] [PubMed]
27. Gröber, U. Magnesium and Drugs. *Int. J. Mol. Sci.* **2019**, *20*, 2094. [CrossRef] [PubMed]
28. Dalton, L.M.; DM, N.F.; Gaydadzhieva, G.T.; Mazurkiewicz, O.M.; Leeson, H.; Wright, C.P. Magnesium in pregnancy. *Nutr. Rev.* **2016**, *74*, 549–557. [CrossRef] [PubMed]
29. Morton, A. Hypomagnesaemia and pregnancy. *Obstet. Med.* **2018**, *11*, 67–72. [CrossRef] [PubMed]

30. Parazzini, F.; Di Martino, M.; Pellegrino, P. Magnesium in the gynecological practice: A literature review. *Magnes. Res.* **2017**, *30*, 1–7. [CrossRef]
31. Seelig, M.S.; Preuss, H.G. Magnesium metabolism and perturbations in the elderly. *Geriatr. Nephrol. Urol.* **1994**, *4*, 101–111. [CrossRef]
32. Lo Piano, F.; Corsonello, A.; Corica, F. Magnesium and elderly patient: The explored paths and the ones to be explored: A review. *Magnes. Res.* **2019**, *32*, 1–15. [CrossRef]
33. Swaminathan, R. Magnesium metabolism and its disorders. *Clin. Biochem. Rev.* **2003**, *24*, 47–66.
34. Guerrero-Romero, F.; Rodríguez-Morán, M. Low serum magnesium levels and metabolic syndrome. *Acta Diabetol.* **2002**, *39*, 209–213. [CrossRef]
35. Viering, D.H.H.M.; de Baaij, J.H.F.; Walsh, S.B.; Kleta, R.; Bockenhauer, D. Genetic causes of hypomagnesemia, a clinical overview. *Pediatr. Nephrol.* **2017**, *32*, 1123–1135. [CrossRef]
36. Naderi, A.S.A.; Reilly, R.F. Hereditary etiologies of hypomagnesemia. *Nat. Clin. Pract. Nephrol.* **2008**, *4*, 80–89. [CrossRef] [PubMed]
37. Botturi, A.; Ciappolino, V.; Delvecchio, G.; Boscutti, A.; Viscardi, B.; Brambilla, P. The Role and the Effect of Magnesium in Mental Disorders: A Systematic Review. *Nutrients* **2020**, *12*, 1661. [CrossRef] [PubMed]
38. Nielsen, F.H.; Lukaski, H.C. Update on the relationship between magnesium and exercise. *Magnes. Res.* **2006**, *19*, 180–189. [PubMed]
39. Laires, M.J.; Monteiro, C. Exercise and Magnesium. In *New Perspectives in Magnesium Research: Nutrition and Health*; Nishizawa, Y., Morii, H., Durlach, J., Eds.; Springer: London, UK, 2007; pp. 173–185. [CrossRef]
40. Warburton, D.E.R.; Welsh, R.C.; Haykowsky, M.J.; Taylor, D.A.; Humen, D.P. Biochemical changes as a result of prolonged strenuous exercise. *Br. J. Sports Med.* **2002**, *36*, 301. [CrossRef]
41. Ikonte, C.J.; Mun, J.G.; Reider, C.A.; Grant, R.W.; Mitmesser, S.H. Micronutrient Inadequacy in Short Sleep: Analysis of the NHANES 2005–2016. *Nutrients* **2019**, *11*, 2335. [CrossRef]
42. Murck, H. Magnesium and affective disorders. *Nutr. Neurosci.* **2002**, *5*, 375–389. [CrossRef]
43. Cernak, I.; Savic, V.; Kotur, J.; Prokic, V.; Kuljic, B.; Grbovic, D.; Veljovic, M. Alterations in magnesium and oxidative status during chronic emotional stress. *Magnes. Res.* **2000**, *13*, 29–36.
44. Grases, G.; Pérez-Castelló, J.A.; Sanchis, P.; Casero, A.; Perelló, J.; Isern, B.; Rigo, E.; Grases, F. Anxiety and stress among science students. Study of calcium and magnesium alterations. *Magnes. Res.* **2006**, *19*, 102–106.
45. Classen, H.G. Systemic stress, magnesium status and cardiovascular damage. *Magnesium* **1986**, *5*, 105–110.
46. United States Department of Health and Human Services. *Scientific Report of the 2015 Dietary Guidelines Advisory Committee—Advisory Report to the Secretary of Health and Human Services and the Secretary of Agriculture*; USDA Agricoltural Research Service: Washngton, DC, USA, 2015.
47. Agence Nationale de Sécurité Sanitaire de L'alimentation de L'environnement et du Travail (ANSES). *TANSES-CIQUAL French Food Composition Table, Version 2017*; ANSES: Maisons-Alfort, France, 2017.
48. Costello, R.B.; Elin, R.J.; Rosanoff, A.; Wallace, T.C.; Guerrero-Romero, F.; Hruby, A.; Lutsey, P.L.; Nielsen, F.H.; Rodriguez-Moran, M.; Song, Y.; et al. Perspective: The Case for an Evidence-Based Reference Interval for Serum Magnesium: The Time Has Come. *Adv. Nutr.* **2016**, *7*, 977–993. [CrossRef]
49. Ford, E.S.; Mokdad, A.H. Dietary Magnesium Intake in a National Sample of U.S. Adults. *J. Nutr.* **2003**, *133*, 2879–2882. [CrossRef] [PubMed]
50. Winiarska-Mieczan, A.; Zaricka, E.; Kwiecień, M.; Kwiatkowska, K.; Baranowska-Wójcik, E.; Danek-Majewska, A. Can Cereal Products Be an Essential Source of Ca, Mg and K in the Deficient Diets of Poles? *Biol. Trace Elem. Res.* **2020**, *195*, 317–322. [CrossRef] [PubMed]
51. Górska-Warsewicz, H.; Rejman, K.; Laskowski, W.; Czeczotko, M. Milk and Dairy Products and Their Nutritional Contribution to the Average Polish Diet. *Nutrients* **2019**, *11*, 1771. [CrossRef] [PubMed]
52. Lombardi-Boccia, G.; Aguzzi, A.; Cappelloni, M.; Di Lullo, G.; Lucarini, M. Total-diet study: Dietary intakes of macro elements and trace elements in Italy. *Br. J. Nutr.* **2003**, *90*, 1117–1121. [CrossRef] [PubMed]
53. Nielsen, F.H. The Problematic Use of Dietary Reference Intakes to Assess Magnesium Status and Clinical Importance. *Biol. Trace Elem. Res.* **2019**, *188*, 52–59. [CrossRef] [PubMed]
54. Dong, J.Y.; Xun, P.; He, K.; Qin, L.Q. Magnesium intake and risk of type 2 diabetes: Meta-analysis of prospective cohort studies. *Diabetes Care* **2011**, *34*, 2116–2122. [CrossRef] [PubMed]
55. Schwartz, R.; Apgar, B.J.; Wien, E.M. Apparent absorption and retention of Ca, Cu, Mg, Mn, and Zn from a diet containing bran. *Am. J. Clin. Nutr.* **1986**, *43*, 444–455. [CrossRef]

56. Lakshmanan, F.L.; Rao, R.B.; Kim, W.W.; Kelsay, J.L. Magnesium intakes, balances, and blood levels of adults consuming self-selected diets. *Am. J. Clin. Nutr.* **1984**, *40*, 1380–1389. [CrossRef]
57. Greger, J.L.; Baier, M.J. Effect of dietary aluminum on mineral metabolism of adult males. *Dev. Med. Child Neurol.* **1983**, *38*, 411–419. [CrossRef]
58. Standing Commitee on the Scientific Evaluation of Dietary Reference Intakes Institute of Medicine. Dietary reference intakes for calcium, phosphorus, magnesium, vitamin D, and fluoride. In *Dietary Reference Intakes for Calcium, Phosphorus, Magnesium, Vitamin D, and Fluoride*; National Acedemies Press, Ed.; The National Academies Press: Washington, DC, USA, 1997; pp. 190–249.
59. Jarosz, M. *Nutritional Guidelines for the Polish Population*; National Food and Nutrition Institute, Ed.; National Food and Nutrition Institute: Warsaw, Poland, 2017.
60. Tutel'ian, V.A. Norms of physiological requirements in energy and nutrients in various groups of population in Russian Federation. *Vopr. Pitan.* **2009**, *78*, 4–15.
61. Ministry of Health Labour and Welfare. *Dietary Reference Intakes for Japanese*; Japan Government Printing Office, Ed.; Ministry of Health Labour and Welfare: Tokyo, Japan, 2015.
62. Società Italiana di Nutrizione Umana-SINU. LARN—Livelli di Assunzione di Riferimento per la Popolazione Italiana: MINERALI. Assunzione Raccomandata per la Popolazione (PRI in Grassetto) e Assunzione Adeguata (AI in Corsivo): Valori su Base Giornaliera. Available online: https://sinu.it/2019/07/09/minerali-assunzione-raccomandata-per-la-popolazione-pri-e-assunzione-adeguataai/ (accessed on 30 June 2020).
63. Agence Nationale de Sécurité Sanitaire de L'alimentation de L'environnement et du Travail (ANSES). Avis de l'ANSES. In *Actualisation des Repères du PNNS: Élaborationdes Références Nutritionnelles*; ANSES, Ed.; ANSES: Maisons-Alfort, France, 2016.
64. Vormann, J. Magnesium: Nutrition and metabolism. *Mol. Asp. Med.* **2003**, *24*, 27–37. [CrossRef]
65. King, D.E.; Mainous, A.G., 3rd; Geesey, M.E.; Woolson, R.F. Dietary magnesium and C-reactive protein levels. *J. Am. Coll. Nutr.* **2005**, *24*, 166–171. [CrossRef] [PubMed]
66. Bailey, R.L.; Fulgoni, V.L., 3rd; Keast, D.R.; Dwyer, J.T. Dietary supplement use is associated with higher intakes of minerals from food sources. *Am. J. Clin. Nutr.* **2011**, *94*, 1376–1381. [CrossRef] [PubMed]
67. Public Health England. *Official Statistics NDNS: Results from Years 7 and 8 (Combined). Results of the National Diet and Nutrition Survey (NDNS) Rolling Programme for 2014 to 2015 and 2015 to 2016*; Public Health England: England, UK, 2018.
68. Agence Nationale de Sécurité Sanitaire de L'alimentation de L'environnement et du Travail (ANSES). *Avis de l'ANSES relatif à l'évaluation des apports en vitamines et minéraux issus de l'alimentation non enrichie, de l'alimentation enrichie et des compléments alimentaires dans la population Française: estimation des aApports usuels, des prévalences d'inadéquation et des risques de dépassement des limites de sécurité (Saisine n° 2012-SA-0142)*; ANSES, Ed.; ANSES: Maisons-Alfort, France, 2015.
69. Olza, J.; Aranceta-Bartrina, J.; González-Gross, M.; Ortega, R.M.; Serra-Majem, L.; Varela-Moreiras, G.; Gil, Á. Reported Dietary Intake, Disparity between the Reported Consumption and the Level Needed for Adequacy and Food Sources of Calcium, Phosphorus, Magnesium and Vitamin D in the Spanish Population: Findings from the ANIBES Study. *Nutrients* **2017**, *9*, 168. [CrossRef]
70. Castiglione, D.; Platania, A.; Conti, A.; Falla, M.; D'Urso, M.; Marranzano, M. Dietary Micronutrient and Mineral Intake in the Mediterranean Healthy Eating, Ageing, and Lifestyle (MEAL) Study. *Antioxidants* **2018**, *7*, 79. [CrossRef]
71. Mensink, G.B.; Fletcher, R.; Gurinovic, M.; Huybrechts, I.; Lafay, L.; Serra-Majem, L.; Szponar, L.; Tetens, I.; Verkaik-Kloosterman, J.; Baka, A.; et al. Mapping low intake of micronutrients across Europe. *Br. J. Nutr.* **2013**, *110*, 755–773. [CrossRef]
72. National Institute of Health, Office of Dietary Supplements. Magnesium. Available online: https://ods.od.nih.gov/factsheets/Magnesium-HealthProfessional/#en1 (accessed on 10 August 2020).
73. Institute of Medicine. *Dietary Reference Intakes for Calcium, Phosphorus, Magnesium, Vitamin D, and Fluoride*; The National Academies Press: Washington, DC, USA, 1997; p. 448. [CrossRef]
74. Spätling, L.; Classen, H.G.; Külpmann, W.R.; Manz, F.; Rob, P.M.; Schimatschek, H.F.; Vierling, W.; Vormann, J.; Weigert, A.; Wink, K. Diagnosing magnesium deficiency. Current recommendations of the Society for Magnesium Research. *Fortschr. Med. Orig.* **2000**, *118* (Suppl. 2), 49–53.
75. Lowenstein, F.W.; Stanton, M.F. Serum magnesium levels in the United States, 1971–1974. *J. Am. Coll. Nutr.* **1986**, *5*, 399–414. [CrossRef]

76. Topf, J.M.; Murray, P.T. Hypomagnesemia and hypermagnesemia. *Rev. Endocr. Metab. Disord.* **2003**, *4*, 195–206. [CrossRef]
77. Elin, R.J. Assessment of magnesium status for diagnosis and therapy. *Magnes. Res.* **2010**, *23*, S194–S198. [CrossRef]
78. Noah, L.; Pickering, G.; Mazur, A.; Dubray, C.; Hitier, S.; Dualé, C.; Pouteau, E. Impact of magnesium supplementation, in combination with vitamin B6, on stress and magnesium status: Secondary data from a randomised controlled trial. *Magnes. Res. J.* **2020**, *33*, 45–57.
79. Rosanoff, A.; Weaver, C.M.; Rude, R.K. Suboptimal magnesium status in the United States: Are the health consequences underestimated? *Nutr. Rev.* **2012**, *70*, 153–164. [CrossRef] [PubMed]
80. Meyer, T.E.; Verwoert, G.C.; Hwang, S.J.; Glazer, N.L.; Smith, A.V.; van Rooij, F.J.; Ehret, G.B.; Boerwinkle, E.; Felix, J.F.; Leak, T.S.; et al. Genome-wide association studies of serum magnesium, potassium, and sodium concentrations identify six Loci influencing serum magnesium levels. *PLoS Genet.* **2010**, *6*, e1001045. [CrossRef] [PubMed]
81. Moshfegh, A.; Goldman, J.; Ahuja, J.; Rhodes, D.; LaComb, R. *What We Eat in America, NHANES 2005–2006. Usual Nutrient Intakes from Food and Water Compared to 1997 Dietary Reference Intakes for Vitamid D, Calcium, Phosphorus, and Magnesium*; United States Department of Agriculture, Agricultural Research Service, Eds.; 2009. Available online: https://www.ars.usda.gov/ARSUserFiles/80400530/pdf/0506/usual_nutrient_intake_vitD_ca_phos_mg_2005-06.pdf (accessed on 30 June 2020).
82. Guo, W.; Nazim, H.; Liang, Z.; Yang, D. Magnesium deficiency in plants: An urgent problem. *Crop J.* **2016**, *4*, 83–91. [CrossRef]
83. Cakmak, I. Magnesium in crop production, food quality and human health. *Plant Soil* **2013**, *368*, 1–4. [CrossRef]
84. Workinger, J.L.; Doyle, R.P.; Bortz, J. Challenges in the Diagnosis of Magnesium Status. *Nutrients* **2018**, *10*, 1202. [CrossRef]
85. Worthington, V. Nutritional quality of organic versus conventional fruits, vegetables, and grains. *J. Altern. Complement. Med.* **2001**, *7*, 161–173. [CrossRef]
86. Thomas, D. The mineral depletion of foods available to us as a nation (1940–2002)—A Review of the 6th Edition of McCance and Widdowson. *Nutr. Health* **2007**, *19*, 21–55. [CrossRef]
87. Du, J.; Zhu, M.; Bao, H.; Li, B.; Dong, Y.; Xiao, C.; Zhang, G.Y.; Henter, I.; Rudorfer, M.; Vitiello, B. The Role of Nutrients in Protecting Mitochondrial Function and Neurotransmitter Signaling: Implications for the Treatment of Depression, PTSD, and Suicidal Behaviors. *Crit. Rev. Food Sci. Nutr.* **2016**, *56*, 2560–2578. [CrossRef]
88. Fromm, L.; Heath, D.L.; Vink, R.; Nimmo, A.J. Magnesium Attenuates Post-Traumatic Depression/Anxiety Following Diffuse Traumatic Brain Injury in Rats. *J. Am. Coll. Nutr.* **2004**, *23*, 529S–533S. [CrossRef]
89. Veronese, N.; Solmi, M. Impaired Magnesium Status and Depression. In *Handbook of Famine, Starvation, and Nutrient Deprivation: From Biology to Policy*; Preedy, V.R., Patel, V.B., Eds.; Springer: Berlin/Heidelberg, Germany, 2019; pp. 1861–1872. [CrossRef]
90. Cuciureanu, M.; Vink, R. Magnesium and stress. In *Magnesium in the Central Nervous System*; Vink, R., Nechifor, M., Eds.; University of Adelaide Press: Adelaide, Australia, 2011.
91. Chrousos, G.P. Stress and disorders of the stress system. *Nat. Rev. Endocrinol.* **2009**, *5*, 374–381. [CrossRef] [PubMed]
92. Yaribeygi, H.; Panahi, Y.; Sahraei, H.; Johnston, T.P.; Sahebkar, A. The impact of stress on body function: A review. *EXCLI J.* **2017**, *16*, 1057–1072. [CrossRef] [PubMed]
93. Selye, H. Stress and the general adaptation syndrome. *Br. Med. J.* **1950**, *1*, 1383–1392. [CrossRef] [PubMed]
94. Clark, M.S.; Bond, M.J.; Hecker, J.R. Environmental stress, psychological stress and allostatic load. *Psychol. Health Med.* **2007**, *12*, 18–30. [CrossRef] [PubMed]
95. McEwen, B.S. Protective and damaging effects of stress mediators. *N. Engl. J. Med.* **1998**, *338*, 171–179. [CrossRef]
96. McEwen, B.S. Protection and damage from acute and chronic stress: Allostasis and allostatic overload and relevance to the pathophysiology of psychiatric disorders. *Ann. N. Y. Acad Sci* **2004**, *1032*, 1–7. [CrossRef]
97. Carrasco, G.A.; Van de Kar, L.D. Neuroendocrine pharmacology of stress. *Eur. J. Pharm.* **2003**, *463*, 235–272. [CrossRef]

98. Lanfumey, L.; Mongeau, R.; Cohen-Salmon, C.; Hamon, M. Corticosteroid-serotonin interactions in the neurobiological mechanisms of stress-related disorders. *Neurosci. Biobehav. Rev.* **2008**, *32*, 1174–1184. [CrossRef]
99. Herman, J.P.; Mueller, N.K.; Figueiredo, H. Role of GABA and glutamate circuitry in hypothalamo-pituitary-adrenocortical stress integration. *Ann. N. Y. Acad. Sci.* **2004**, *1018*, 35–45. [CrossRef]
100. Nakamura, J.; Yakata, M. Two-cycle liquid-chromatographic quantitation of cortisol in urine. *Clin. Chem.* **1982**, *28*, 1497–1500. [CrossRef]
101. Singh, A.; Petrides, J.S.; Gold, P.W.; Chrousos, G.P.; Deuster, P.A. Differential hypothalamic-pituitary-adrenal axis reactivity to psychological and physical stress. *J. Clin. Endocrinol. Metab.* **1999**, *84*, 1944–1948. [CrossRef] [PubMed]
102. Ullmann, E.; Perry, S.W.; Licinio, J.; Wong, M.-L.; Dremencov, E.; Zavjalov, E.L.; Shevelev, O.B.; Khotskin, N.V.; Koncevaya, G.V.; Khotshkina, A.S.; et al. From Allostatic Load to Allostatic State—An Endogenous Sympathetic Strategy to Deal With Chronic Anxiety and Stress? *Front. Behav. Neurosci.* **2019**, *13*. [CrossRef] [PubMed]
103. de Assis, G.G.; Gasanov, E.V. BDNF and Cortisol integrative system—Plasticity vs. degeneration: Implications of the Val66Met polymorphism. *Front. Neuroendocrinol.* **2019**, *55*. [CrossRef] [PubMed]
104. Karatsoreos, I.N.; McEwen, B.S. Psychobiological allostasis: Resistance, resilience and vulnerability. *Trends Cogn. Sci.* **2011**, *15*, 576–584. [CrossRef]
105. Wallingford, J.K.; Deurveilher, S.; Currie, R.W.; Fawcett, J.P.; Semba, K. Increases in mature brain-derived neurotrophic factor protein in the frontal cortex and basal forebrain during chronic sleep restriction in rats: Possible role in initiating allostatic adaptation. *Neuroscience* **2014**, *277*, 174–183. [CrossRef]
106. Yuluğ, B.; Ozan, E.; Gönül, A.S.; Kilic, E. Brain-derived neurotrophic factor, stress and depression: A minireview. *Brain Res. Bull.* **2009**, *78*, 267–269. [CrossRef]
107. Colaianna, M.; Schiavone, S.; Zotti, M.; Tucci, P.; Morgese, M.G.; Bäckdahl, L.; Holmdahl, R.; Krause, K.-H.; Cuomo, V.; Trabace, L. Neuroendocrine profile in a rat model of psychosocial stress: Relation to oxidative stress. *Antioxid. Redox. Signal* **2013**, *18*, 1385–1399. [CrossRef]
108. Kapczinski, F.; Vieta, E.; Andreazza, A.C.; Frey, B.N.; Gomes, F.A.; Tramontina, J.; Kauer-Sant'anna, M.; Grassi-Oliveira, R.; Post, R.M. Allostatic load in bipolar disorder: Implications for pathophysiology and treatment. *Neurosci. Biobehav. Rev.* **2008**, *32*, 675–692. [CrossRef]
109. Danhof-Pont, M.B.; van Veen, T.; Zitman, F.G. Biomarkers in burnout: A systematic review. *J. Psychosom. Res.* **2011**, *70*, 505–524. [CrossRef]
110. Pochwat, B.; Szewczyk, B.; Sowa-Kucma, M.; Siwek, A.; Doboszewska, U.; Piekoszewski, W.; Gruca, P.; Papp, M.; Nowak, G. Antidepressant-like activity of magnesium in the chronic mild stress model in rats: Alterations in the NMDA receptor subunits. *Int. J. Neuropsychopharmacol.* **2014**, *17*, 393–405. [CrossRef]
111. Vink, R. Magnesium in the CNS: Recent advances and developments. *Magnes. Res.* **2016**, *29*, 95–101. [CrossRef] [PubMed]
112. Zheltova, A.A.; Kharitonova, M.V.; Iezhitsa, I.N.; Spasov, A.A. Magnesium deficiency and oxidative stress: An update. *Biomed. Taipei* **2016**, *6*, 20. [CrossRef] [PubMed]
113. Brosschot, J.F.; Verkuil, B.; Thayer, J.F. The default response to uncertainty and the importance of perceived safety in anxiety and stress: An evolution-theoretical perspective. *J. Anxiety. Disord.* **2016**, *41*, 22–34. [CrossRef] [PubMed]
114. Somjen, G.G.; Baskerville, E.N. Effect of Excess Magnesium on Vagal Inhibition and Acetylcholine Sensitivity of the Mammalian Heart in situ and in vitro. *Nature* **1968**, *217*, 679–680. [CrossRef]
115. Toda, N.; West, T. Interaction between Na, Ca, Mg, and vagal stimulation in the S-A node of the rabbit. *Am. J. Physiol. Leg. Content* **1967**, *212*, 424–430. [CrossRef]
116. Murasato, Y.; Harada, Y.; Ikeda, M.; Nakashima, Y.; Hayashida, Y. Effect of Magnesium Deficiency on Autonomic Circulatory Regulation in Conscious Rats. *Hypertension* **1999**, *34*, 247–252. [CrossRef]
117. Brosschot, J.F.; Verkuil, B.; Thayer, J.F. Generalized Unsafety Theory of Stress: Unsafe Environments and Conditions, and the Default Stress Response. *Int. J. Environ. Res. Public Health* **2018**, *15*, 464. [CrossRef]
118. Stoian, M.; Stoica, V. The role of distubances of phosphate metabolism in metabolic syndrome. *Maedica Buchar.* **2014**, *9*, 255–260. [CrossRef]
119. Brosschot, J.F. Ever at the ready for events that never happen. *Eur. J. Psychotraumatol.* **2017**, *8*, 1309934. [CrossRef]

120. Kirkland, A.E.; Sarlo, G.L.; Holton, K.F. The Role of Magnesium in Neurological Disorders. *Nutrients* **2018**, *10*, 730. [CrossRef]
121. Lopresti, A.L. The Effects of Psychological and Environmental Stress on Micronutrient Concentrations in the Body: A Review of the Evidence. *Adv. Nutr.* **2019**, *11*, 103–112. [CrossRef] [PubMed]
122. Classen, H.G.; Marquardt, P.; Späth, M.; Schumacher, K.A. Hypermagnesemia Following Exposure to Acute Stress. *Pharmacology* **1971**, *5*, 287–294. [CrossRef] [PubMed]
123. Ising, H.; Handrock, M.; Günther, T.; Fischer, R.; Dombrowski, M. Increased noise trauma in guinea pigs through magnesium deficiency. *Arch. Oto Rhino Laryngol.* **1982**, *236*, 139–146. [CrossRef] [PubMed]
124. Joachims, Z.; Babisch, W.; Ising, H.; Günther, T.; Handrock, M. Dependence of noise-induced hearing loss upon perilymph magnesium concentration. *J. Acoust. Soc. Am.* **1983**, *74*, 104–108. [CrossRef] [PubMed]
125. Ising, H. Interaction of noise-induced stress and Mg decrease. *Artery* **1981**, *9*, 205–211.
126. Ando, I.; Karasawa, K.; Yokota, S.; Shioya, T.; Matsuda, H.; Tanaka, A. Analysis of serum magnesium ions in dogs exposed to external stress: A pilot study. *Open Vet. J.* **2017**, *7*, 367–374. [CrossRef]
127. Yasmin, F.; Haleem, D.J.; Haleem, M.A. Effects of repeated restraint stress on serum electrolytes in ethanol-treated and water-treated rats. *Pak. J. Pharm. Sci.* **2007**, *20*, 51–55.
128. Heroux, O.; Peter, D.; Heggtveit, A. Long-term Effect of Suboptimal Dietary Magnesium on Magnesium and Calcium Contents of Organs, on Cold Tolerance and on Lifespan, and its Pathological Consequences in Rats. *J. Nutr.* **1977**, *107*, 1640–1652. [CrossRef]
129. Terashima, Y.; Tucker, R.E.; Deetz, L.E.; Degregorio, R.M.; Muntifering, R.B.; Mitchell, G.E., Jr. Plasma Magnesium Levels as Influenced by Cold Exposure in Fed or Fasted Sheep. *J. Nutr.* **1982**, *112*, 1914–1920. [CrossRef]
130. Whyte, K.F.; Addis, G.J.; Whitesmith, R.; Reid, J.L. Adrenergic control of plasma magnesium in man. *Clin. Sci. Lond.* **1987**, *72*, 135–138. [CrossRef]
131. Tanabe, K.; Osada, N.; Suzuki, N.; Nakayama, M.; Yokoyama, Y.; Yamamoto, A.; Oya, M.; Murabayashi, T.; Yamamoto, M.; Omiya, K.; et al. Erythrocyte magnesium and prostaglandin dynamics in chronic sleep deprivation. *Clin. Cardiol.* **1997**, *20*, 265–268. [CrossRef] [PubMed]
132. Takase, B.; Akima, T.; Uehata, A.; Ohsuzu, F.; Kurita, A. Effect of chronic stress and sleep deprivation on both flow-mediated dilation in the brachial artery and the intracellular magnesium level in humans. *Clin. Cardiol.* **2004**, *27*, 223–227. [CrossRef] [PubMed]
133. Mocci, F.; Canalis, P.; Tomasi, P.A.; Casu, F.; Pettinato, S. The effect of noise on serum and urinary magnesium and catecholamines in humans. *Occup. Med.* **2001**, *51*, 56–61. [CrossRef] [PubMed]
134. Ising, H.; Dienel, D.; Günther, T.; Markert, B. Health effects of traffic noise. *Int. Arch. Occup. Environ. Health* **1980**, *47*, 179–190. [CrossRef]
135. Joborn, H.; Akerström, G.; Ljunghall, S. Effects of exogenous catecholamines and exercise on plasma magnesium concentrations. *Clin. Endocrinol.* **1985**, *23*, 219–226. [CrossRef]
136. Caddell, J.; Kupiecki, R.; Proxmire, D.; Satoh, P.; Hutchinson, B. Plasma Catecholamines in Acute Magnesium Deficiency in Weanling Rats. *JN* **1986**, *116*, 1896–1901. [CrossRef]
137. Amyard, N.; Leyris, A.; Monier, C.; Francès, H.; Boulu, R.G.; Henrotte, J.G. Brain catecholamines, serotonin and their metabolites in mice selected for low (MGL) and high (MGH) blood magnesium levels. *Magnes. Res.* **1995**, *8*, 5–9.
138. Henrotte, J.G.; Franck, G.; Santarromana, M.; Frances, H.; Mouton, D.; Motta, R. Mice selected for low and high blood magnesium levels: A new model for stress studies. *Physiol. Behav.* **1997**, *61*, 653–658. [CrossRef]
139. Henrotte, J.G.; Aymard, N.; Allix, M.; Boulu, R.G. Effect of pyridoxine and magnesium on stress-induced gastric ulcers in mice selected for low or high blood magnesium levels. *Ann. Nutr. Metab.* **1995**, *39*, 285–290. [CrossRef]
140. Sartori, S.B.; Whittle, N.; Hetzenauer, A.; Singewald, N. Magnesium deficiency induces anxiety and HPA axis dysregulation: Modulation by therapeutic drug treatment. *Neuropharmacology* **2012**, *62*, 304–312. [CrossRef]
141. Bourin, M.; Hascoët, M. The mouse light/dark box test. *Eur. J. Pharmacol.* **2003**, *463*, 55–65. [CrossRef]
142. Spasov, A.A.; Iezhitsa, I.N.; Kharitonova, M.V.; Kravchenko, M.S. Depression-like and anxiety-related behaviour of rats fed with magnesium-deficient diet. *Zhurnal Vysshei Nervnoi Deiatelnosti Imeni IP Pavlova* **2008**, *58*, 476–485.

143. Iezhitsa, I.N.; Spasov, A.A.; Kharitonova, M.V.; Kravchenko, M.S. Effect of magnesium chloride on psychomotor activity, emotional status, and acute behavioural responses to clonidine, d-amphetamine, arecoline, nicotine, apomorphine, and L-5-hydroxytryptophan. *Nutr. Neurosci.* **2011**, *14*, 10–24. [CrossRef] [PubMed]
144. Akarachkova, E. The role of magnesium deficiency in the formation of clinical manifestation of stress in women. *Probl. Women Health* **2013**, *8*, 57.
145. Nielsen, F.H.; Johnson, L.K.; Zeng, H. Magnesium supplementation improves indicators of low magnesium status and inflammatory stress in adults older than 51 years with poor quality sleep. *Magnes. Res.* **2010**, *23*, 158–168. [CrossRef]
146. Hermes Sales, C.; Azevedo Nascimento, D.; Queiroz Medeiros, A.C.; Costa Lima, K.; Campos Pedrosa, L.F.; Colli, C. There is chronic latent magnesium deficiency in apparently healthy university students. *Nutr. Hosp.* **2014**, *30*, 200–204. [CrossRef]
147. Eby, G.A.; Eby, K.L. Magnesium for treatment-resistant depression: A review and hypothesis. *Med. Hypotheses* **2010**, *74*, 649–660. [CrossRef]
148. Forsyth, A.K.; Williams, P.G.; Deane, F.P. Nutrition status of primary care patients with depression and anxiety. *Aust. J. Prim. Health* **2012**, *18*, 172–176. [CrossRef]
149. Abumaria, N.; Yin, B.; Zhang, L.; Li, X.Y.; Chen, T.; Descalzi, G.; Zhao, L.; Ahn, M.; Luo, L.; Ran, C.; et al. Effects of elevation of brain magnesium on fear conditioning, fear extinction, and synaptic plasticity in the infralimbic prefrontal cortex and lateral amygdala. *J. Neurosci.* **2011**, *31*, 14871–14881. [CrossRef]
150. Pochwat, B.; Sowa-Kucma, M.; Kotarska, K.; Misztak, P.; Nowak, G.; Szewczyk, B. Antidepressant-like activity of magnesium in the olfactory bulbectomy model is associated with the AMPA/BDNF pathway. *Psychopharmacology* **2015**, *232*, 355–367. [CrossRef]
151. Pilchova, I.; Klacanova, K.; Tatarkova, Z.; Kaplan, P.; Racay, P. The Involvement of Mg(2+) in Regulation of Cellular and Mitochondrial Functions. *Oxid Med. Cell Longev.* **2017**, *2017*, 6797460. [CrossRef] [PubMed]
152. Zogović, D.; Pesić, V.; Dmitrasinović, G.; Dajak, M.; Plećas, B.; Batinić, B.; Popović, D.; Ignjatović, S. Pituitary-gonadal, pituitary-adrenocortical hormonal and IL-6 levels following long-term magnesium supplementation in male students. *J. Med. Biochem.* **2014**, *33*, 291–298. [CrossRef]
153. Wienecke, E.; Nolden, C. Long-term HRV analysis shows stress reduction by magnesium intake. *MMW Fortschr. Med.* **2016**, *158*, 12–16. [CrossRef] [PubMed]
154. Pouteau, E.; Kabir-Ahmadi, M.; Noah, L.; Mazur, A.; Dye, L.; Hellhammer, J.; Pickering, G.; Dubray, C. Superiority of magnesium and vitamin B6 over magnesium alone on severe stress in healthy adults with low magnesemia: A randomized, single-blind clinical trial. *PLoS ONE* **2018**, *13*, e0208454. [CrossRef] [PubMed]
155. Eby, G.A.; Eby, K.L. Rapid recovery from major depression using magnesium treatment. *Med. Hypotheses* **2006**, *67*, 362–370. [CrossRef]
156. Boyle, N.B.; Lawton, C.; Dye, L. The Effects of Magnesium Supplementation on Subjective Anxiety and Stress-A Systematic Review. *Nutrients* **2017**, *9*, 429. [CrossRef]

**Publisher's Note:** MDPI stays neutral with regard to jurisdictional claims in published maps and institutional affiliations.

© 2020 by the authors. Licensee MDPI, Basel, Switzerland. This article is an open access article distributed under the terms and conditions of the Creative Commons Attribution (CC BY) license (http://creativecommons.org/licenses/by/4.0/).

Review

# Effectively Prescribing Oral Magnesium Therapy for Hypertension: A Categorized Systematic Review of 49 Clinical Trials

Andrea Rosanoff [1,*], Rebecca B. Costello [1] and Guy H. Johnson [2]

1 CMER Center for Magnesium Education &Research, Pahoa, HI 96778, USA; RBCostello@earthlink.net
2 Johnson Nutrition Solutions LLC, Minneapolis, MN 55416, USA; guy@NutritionSolutions.net
* Correspondence: ARosanoff@gmail.com

**Abstract:** Trials and meta-analyses of oral magnesium for hypertension show promising but conflicting results. An inclusive collection of 49 oral magnesium for blood pressure (BP) trials were categorized into four groups: (1) Untreated Hypertensives; (2) Uncontrolled Hypertensives; (3) Controlled Hypertensives; (4) Normotensive subjects. Each group was tabulated by ascending magnesium dose. Studies reporting statistically significant ($p < 0.05$) decreases in both systolic BP (SBP) and diastolic BP (DBP) from both baseline and placebo (if reported) were labeled "Decrease"; all others were deemed "No Change." Results: Studies of Untreated Hypertensives (20 studies) showed BP "Decrease" only when Mg dose was >600 mg/day; <50% of the studies at 120–486 mg Mg/day showed SBP or DBP decreases but not both while others at this Mg dosage showed no change in either BP measure. In contrast, all magnesium doses (240–607 mg/day) showed "Decrease" in 10 studies on Uncontrolled Hypertensives. Controlled Hypertensives, Normotensives and "magnesium-replete" studies showed "No Change" even at high magnesium doses (>600 mg/day). Where magnesium did not lower BP, other cardiovascular risk factors showed improvement. Conclusion: Controlled Hypertensives and Normotensives do not show a BP-lowering effect with oral Mg therapy, but oral magnesium (≥240 mg/day) safely lowers BP in Uncontrolled Hypertensive patients taking antihypertensive medications, while >600 mg/day magnesium is required to safely lower BP in Untreated Hypertensives; <600 mg/day for non-medicated hypertensives may not lower both SBP and DBP but may safely achieve other risk factor improvements without antihypertensive medication side effects.

**Keywords:** magnesium; oral magnesium therapy; hypertension; blood pressure; anti-hypertensive medications

**Citation:** Rosanoff, A.; Costello, R.B.; Johnson, G.H. Effectively Prescribing Oral Magnesium Therapy for Hypertension: A Categorized Systematic Review of 49 Clinical Trials. *Nutrients* **2021**, *13*, 195. https://doi.org/10.3390/nu13010195

Received: 27 November 2020
Accepted: 7 January 2021
Published: 10 January 2021

**Publisher's Note:** MDPI stays neutral with regard to jurisdictional claims in published maps and institutional affiliations.

**Copyright:** © 2021 by the authors. Licensee MDPI, Basel, Switzerland. This article is an open access article distributed under the terms and conditions of the Creative Commons Attribution (CC BY) license (https://creativecommons.org/licenses/by/4.0/).

## 1. Introduction

More than any other modifiable risk factor, hypertension is responsible for cardiovascular disease deaths both globally [1,2] and in the United States [3]. Given the potential harms and costs of hypertension and the need for safe lowering of this important risk factor [4–8], we probed the large trove of research on oral magnesium therapy for hypertension and BP hoping to discern any prescription guidance.

Oral magnesium therapy for the treatment of hypertension has been well studied over the last 35 years but results are highly mixed. The trials differ not only in oral magnesium dose and form of magnesium but also in normotensive vs. hypertensive status at baseline as well as use or non-use of antihypertensive medications. Fourteen clinical trials have shown that oral magnesium therapy significantly lowers both systolic blood pressure (SBP) and diastolic blood pressure (DBP), whereas > twice that number of studies have shown no statistically significant lowering of either SBP, DBP, or both with oral magnesium therapy. Of six meta-analyses on this topic conducted to date [9–14], one shows no effect of oral magnesium on BP, one shows lowering of DBP but not SBP, four show that oral magnesium therapy lowers both SBP and DBP but only one of these suggests that the BP reductions are clinically relevant. Such results do not lend confidence in prescribing oral magnesium therapy to control or prevent high blood pressure. However, magnesium's low cost, safety, positive research in cardiovascular risks [15–17] plus its partial beneficial BP results encourages this inclusive analytical categorization of all of these studies. We are looking for information on when and at what dose oral magnesium therapy is beneficial in the treatment of hypertension. This is an inclusive but neither quantitative nor precise analysis, but we hope it will provide guidance for both future meta-analyses and the prescribing of oral Mg therapy for high BP.

## 2. Materials and Methods

### 2.1. Data Sources and Searches

For this analysis, all articles from six meta-analyses on this subject [9–14] and articles used in CMER's 2016 "Petition for the Authorization of a Qualified Health Claim for Magnesium and Reduced Risk of High Blood Pressure (Hypertension)" [18] submitted to the US Food and Drug Administration (FDA) were added to the CMER collection which began in 1997 (see below and Figure 1). We categorized each of these studies into four groups—Untreated Hypertensive, Uncontrolled Hypertensive, Controlled Hypertensive and Normotensive—and then tabulated each category by ascending oral magnesium dose in milligrams per day. Studies reporting a statistically significant decrease in both SBP and DBP from baseline as well as placebo (if reported) were labeled "Decrease"; all others were deemed "No change" in BP. Our goal was to determine whether these groups might influence the effect of oral magnesium therapy on BP.

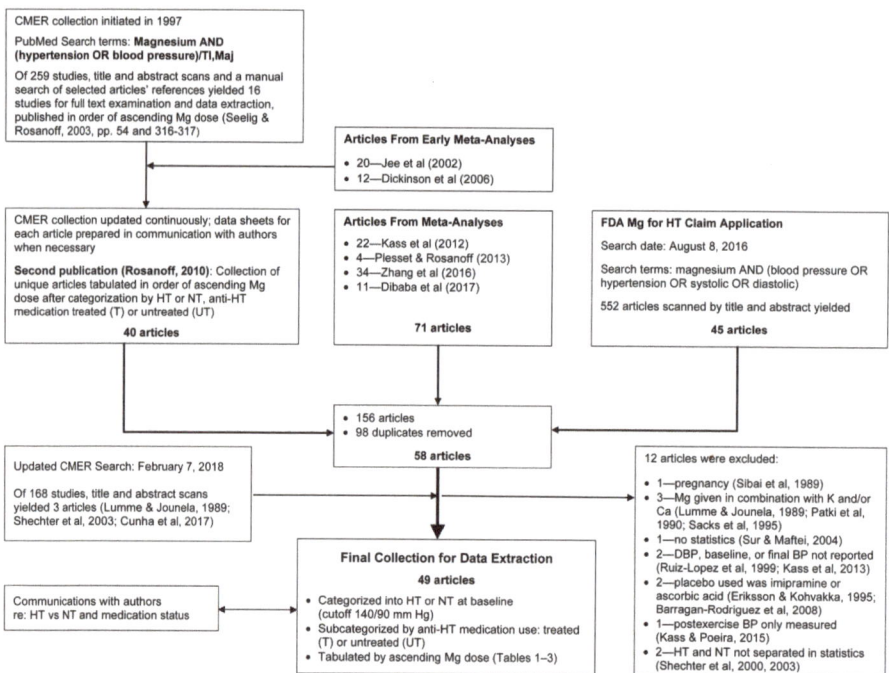

**Figure 1.** Flow chart of the article acquisition, examination, categorization, and tabulation process. Abbreviations: BP, blood pressure; Ca, calcium; CMER, Center for Magnesium Education & Research; DBP, diastolic blood pressure; HT, hypertensive or hypertension; K, potassium; Maj, Major Topic for National Library of Medicine Pubmed search term; Mg, magnesium; NT, normotensive; T, treated with anti-hypertensive medications; TI, title field for Pubmed search term; UT, untreated, i.e. non-use of anti-hypertensive medications.

## 2.2. Study Selection

The initial 1997 CMER search was performed in PubMed using the following search terms: magnesium AND (blood pressure OR hypertension) [each term limited to Title AND/OR Medical Subject Heading fields]. This initial search yielded 259 studies. A title/abstract scan plus additions from manual searching resulted in 16 studies of oral magnesium therapy for BP for full-text examination and data extraction. These studies were listed by ascending magnesium dose and published in 2003 by Seelig and Rosanoff [19] (non-peer reviewed). CMER continuously updated this preliminary 1997 PubMed search, creating data extraction sheets with data tables for each article and adding appropriate articles cited in published meta-analyses [11,12]. Data sheets for each newly added article were created and updated with correspondence with authors when appropriate. By 2010, the collection had 40 relevant articles that had undergone full-text examination and data extraction. This collection of studies was published, again listed by ascending magnesium dose, after being first categorized as to normotensive (NT) or hypertensive (HT) status as well as medication usage [20] (Figure 1).

In 2017, the 71 articles in four additional published meta-analyses of oral magnesium therapy for BP [9,10,13,14] were added to the CMER collection of 40 articles along with the 45 articles collected for CMER's 2016 "Petition for a Qualified Health Claim for Magnesium for Hypertension" [18]. This gathering resulted in a total of 156 articles. Duplicates were removed, yielding 58 articles appropriate for an analysis of oral magnesium therapy for BP. CMER then updated its original PubMed search on February 7, 2018, yielding 168 further studies, three of which were appropriate for addition to the collection. All 61 full-text articles plus existing data sheets were gathered for final analysis.

Twelve articles were excluded [21–32], yielding 49 articles (see Figure 1). Exclusion criteria were as follows: pregnant subjects [22], use of oral magnesium supplements in combination with any other mineral nutrient [21,23,24], no statistical analysis [25], baseline and/or final SBP or DBP not reported [26,30], a control group using ascorbic acid or imipramine rather than placebo [27,28], only pre- and post-exercise BP values were measured over 2 days [29], and Hypertensive subjects were not separated from Normotensive subjects in the statistical analysis [31,32].

## 2.3. Data Extraction and Quality Assessment

Each article in the final collection of 49 articles was examined for starting SBP/DBP for both magnesium test and placebo control groups to determine BP status at baseline. Subjects with average baseline BP $\geq$ 140/90 mm Hg or mean blood pressure (MBP) $\geq$106 mm Hg were deemed hypertensive; all others were deemed normotensive. Studies were furthered examined for antihypertensive medication usage. Antihypertensive medications known to be used in these studies included diuretics (thiazide, spironolactone), calcium channel blockers, beta-blockers, angiotensin-converting enzyme inhibitors, and alpha-blockers.

## 2.4. Data Synthesis and Analysis

Studies were separated into four categories:

- Untreated Hypertensive, i.e., subjects were treatment naive or not using antihypertensive medications before or during the study and were hypertensive at baseline.
- Uncontrolled Hypertensive, i.e., subjects were using antihypertensive medications during and previous to the study but were still hypertensive at baseline.
- Controlled Hypertensive, i.e., subjects were using antihypertensive medications during and previous to the study and were normotensive at baseline.
- Normotensive subjects, untreated with antihypertensive medications plus normotensive at baseline.

SBP/DBP results with statistical findings and any other cardiovascular-related relevant information were extracted from each article. Studies showing a statistically significant decrease in both SBP and DBP from baseline as well as placebo (if reported) were labeled

"Decrease"; all others were deemed "No Change" in BP. Correspondence with authors aided proper categorization as to BP status as well as antihypertensive medication usage.

Each article was examined for oral magnesium form and dose, and each of the four categories was tabulated in order of ascending oral magnesium dose within that category.

*2.5. Role of the Funding Source*

CMER is a group of independent scholars long interested in oral magnesium therapy's effect on BP [9,10,14,18,20,33–39]. Funding to sponsor the Food & Drug Administration (FDA)-qualified health claim petition was provided by the Almond Board of California, PepsiCo, Inc., Council for Responsible Nutrition, Pfizer Consumer Healthcare, Premier Magnesia, and Adobe Springs. These sponsors' donations funded a consultant (Johnson Nutrition Solutions LLC, G.H.J.) to work with CMER scholars to search the literature, evaluate studies by FDA criteria, extract data, and write the FDA petition. The sponsors provided no input into any aspect of this study, data collection, data extraction, writing, or decision to publish.

## 3. Results

Results for Untreated Hypertensive, Uncontrolled Hypertensive, Controlled Hypertensive and Normotensive subjects are shown in Tables 1–4, respectively.

*3.1. Blood Pressure Outcomes with Oral Mg Therapy*

In studies on Untreated Hypertensives (Table 1), decrease in BP was highly influenced by magnesium dose: of the 20 studies, only 4 showed "Decrease" in BP by the strict criteria of this analysis, and all of these occurred at daily magnesium doses >600 mg/day. Daily magnesium doses between 120 and 486 mg/day sometimes showed a decrease in either SBP or DBP but not both. The influence of magnesium dose in Untreated Hypertensives is confirmed in the only study using three different magnesium doses in the same subjects (Widman), which showed "No change" at 365 mg Mg/day but "Decrease" at magnesium doses >600 mg/day. It is tempting to predict that Untreated Hypertensive subjects need high doses of oral magnesium ($\geq$600 mg/day) to consistently lower both SBP and DBP, but 2 studies, Walker and Zemel, were both at doses >600 mg/day but showed "No change". Both of these studies' authors observed that these subjects were "magnesium replete" either by high measured dietary Mg intake or healthier magnesium-dependent lipoprotein values at baseline, suggesting that the oral Mg therapy >600 mg/day lowers both SBP and DBP in hypertensives but only when subjects are in states of Mg deficit.

In contrast to studies on Untreated Hypertensives, those on Uncontrolled Hypertensive subjects (Table 2) showed that oral magnesium therapy doses as low as 240 mg/day to as high as 607 mg/day consistently and significantly lowered both SBP and DBP (Table 2).

For Controlled Hypertensive subjects (Table 3), Mg doses of 304 to 583 mg/day showed no change in BP in the only two studies of this category.

For Normotensive subjects (Table 4), several studies consistently showed "No change" in BP from Mg doses as low as 250 mg/day and as high as 632 mg/day.

Table 1. Summary of magnesium supplementation studies for blood pressure in Untreated Hypertensives (subjects untreated with antihypertensive medications, hypertensive at baseline).

| Study Citation | Mg Dose, mg/day | Form of Mg | BP Status at Baseline, NT or HT | Medical Status at Baseline, T or UT | BP Outcome [1] | Notes |
|---|---|---|---|---|---|---|
| Borrello et al. (1996) [40] | 120 | MgO | HT | UT | No change [2] | Decrease in SBP only |
| Nowson and Morgan (1989) [41] | 240 | Aspartate | HT | UT | No change | |
| Ferrara et al. (1992) [42] | 365 | Pidolate | HT | UT | No change | |
| Lind et al. (1991) [43] | 365 | Lactate and citrate | HT | UT | No change [2,3] | |
| de Valk et al. (1998) [44] | 365 | Aspartate HCl | HT | UT | No change [2] | |
| Plum-Wirell et al. (1994) [45] | 365 | Aspartate | HT | UT | No change [2] | |
| Wirell et al. (1993) [46] | 365 | Aspartate | HT | UT | No change [2] | Decrease in DBP only; medication interrupted 2–3 months pre-study |
| Cappuccio et al. (1985) [47] | 365 | Aspartate | HT | UT | No change [2] | Baseline BP: 150/82 mm Hg; some perhaps taking medications |
| Barbagallo et al. (2010) [48] | 368 | Pidolate | HT | UT | No change [2,4] | Decrease in DBP only; medication interrupted 4 weeks pre-study |
| Reyes et al. (1984) [49] | 384 | MgCl$_2$ | HT | UT | No change | |
| Olhaberry et al. (1987) [50] | 384 | MgCl$_2$ | HT | UT | No change | Decrease in SBP only |
| Purvis et al. (1994) [51] | 389 | MgCl$_2$ | HT | UT | No change | Decrease in SBP only |
| Cohen et al. (1984) [52] | 450 | MgO | HT | UT | No change [2,5,6] | |
| Witteman et al. (1994) [53] | 486 | Aspartate HCl | HT | UT | No change | Decrease in DBP only |
| Walker et al. (2002) [54] | 607 | Amino acid chelate | HT | UT | No change | Mg replete [7] |
| Haga (1992) [55] | 607 | MgO | HT | UT | Decrease [6] | MBP measured in HT vs. NT "control" [8] |
| Motoyama et al. (1989) [56] | 607 | MgO | HT | UT | Decrease [6] | No medications during study or 1 month pre-study, at least |
| Sanjuliani et al. (1996) [57] | 607 | MgO | HT | UT | Decrease | No medications 2 weeks pre-study or during study |
| Zemel et al. (1990) [58] | 972 | Aspartate | HT | UT | No change | Mg replete [9]; no medications 3 mo pre-study at least |
| Widman et al. (1993) [59] | 365 | Mg(OH)$_2$ | HT | UT | No change | Only titrated Mg dose study |
|  | 729 | Mg(OH)$_2$ | HT | UT | Decrease [10] |  |
|  | 972 | Mg(OH)$_2$ | HT | UT | Decrease |  |

Abbreviations: BP, blood pressure; DBP, diastolic blood pressure; HT, hypertensive at baseline; MBP, mean blood pressure; Mg, magnesium; NT, normotensive at baseline; SBP, systolic blood pressure; T, most or all subjects treated with antihypertensive medications including diuretics; UT, most or all subjects treatment naive or not taking any antihypertensive medications during or before the study. [1] Studies showing a statistically significant decrease in both SBP and DBP from baseline as well as placebo (if reported) were labeled "Decrease"; all others were deemed "No change". [2] Rise in serum or plasma Mg in Mg test group. [3] Improved sodium excretion in Mg test group. [4] Improved endothelial function in Mg test group. [5] Reversal of retinal vasospasm. [6] Study not included in most meta-analyses due to no true placebo control group. [7] Walker et al. [54] showed a very large placebo effect. In addition, this study found dietary Mg especially high (485 mg/day) in the Mg group compared to placebo (346 mg/day) leading authors to suggest those subjects had been "magnesium replete.". [8] Haga [55] gave 600 mg Mg/day to 17 HT and 8 NT "control" subjects. Only HT subjects showed a decrease in MBP. [9] Zemel et al. [58] noted that the placebo group at baseline had higher cholesterol, triglycerides, and low-density lipoprotein cholesterol and lower high-density lipoprotein than the Mg group at baseline (i.e., the Mg group was the "healthier" of the two). These differences persisted throughout the study. The authors suggest that oral Mg therapy only lowers BP in "states of Mg deficiency". [10] At the 30 mmol Mg/day dose (729 mg/day), Widman et al. [59] reported both SBP and DBP decreases that were not significant in this crossover design, uncorrected for carryover effects. Statistical analyses were performed on the 30 mmol Mg period against a middle placebo period, which included subjects (50%) who had previously spent 12 weeks taking a 15, 30, and then 40 mmol Mg/day supplement. These statistical analyses did not separate middle placebo group subjects as to pre— or post—Mg arms of the crossover, and neither tested nor corrected for any carryover effect. When a t-test is performed on 30 mmol Mg period BP values against Mg test arm baseline, placebo baseline, and pre-Mg placebo arm final values, both SBP and DBP show significant decreases in all three tests ($p < 0.01$).

**Table 2.** Summary of magnesium supplementation studies for blood pressure in Uncontrolled Hypertensives (subjects treated with antihypertensive medications, hypertensive at baseline).

| Study Citation | Mg Dose, mg/day | Form of Mg | BP Status at Baseline, NT or HT | Medical Status at Baseline, T or UT | BP Outcome [1] | Notes |
|---|---|---|---|---|---|---|
| Shafique et al. (1993) [60] | 240 | MgCl$_2$ | HT | T | Decrease [2] | Diuretics >1 year |
| Sebekova et al. (1992) [61] | 255 | Aspartate HCl | HT | T | Decrease [2] | Interrupted medications |
| Michon (2002) [62] | 323 | Slow-mag/B$_6$ | HT | T | Decrease [2] | Beta-blockers, ACE inhibitors, calcium channel blockers, diuretics |
| Wirell et al. (1994) [63] | 365 | Aspartate | HT | T | Decrease | Beta-blockers |
| Dyckner and Wester (1983) [64] | 365 | Aspartate HCl | HT | T | Decrease | Beta-blockers |
| Paolisso et al. (1992) [65] | 384 | Pidolate | HT | T | Decrease [3] | Thiazide diuretics—long term |
| Guerrero-Romero and Rodriguez-Moran (2009) [66] | 450 | MgCl$_2$ | HT | T | Decrease | All taking medications ≥6 months pre-study, type not specified |
| Kawano et al. (1998) [67] | 486 | MgO | HT | T | Decrease | 33% untreated; 30% monotherapy; 37% combination therapy; therapy included calcium channel blockers, beta-blockers, ACE inhibitors, thiazides, spironolactone, alpha-blockers |
| Cunha et al. (2017) [68] | 600 | Mg chelate | HT | T | Decrease | Hydrochlorothiazide |
| Hattori et al. (1988) [69] | 607 | MgO | HT | T | Decrease [4] | Thiazide diuretics—long term |
| | | | NT | T | No change [4] | Thiazide diuretics—long term |

Abbreviations: ACE, angiotensin-converting enzyme; BP, blood pressure; DBP, diastolic blood pressure; HT, hypertensive at baseline; MBP, mean blood pressure; Mg, magnesium; NT, normotensive at baseline; SBP, systolic blood pressure; T, most or all subjects treated with antihypertensive medications including diuretics, ACE inhibitors, calcium channel blockers, beta-blockers, or alpha-blockers. [1] Studies showing a statistically significant decrease in both SBP and DBP from baseline as well as placebo (if reported) were labeled "Decrease"; all others were deemed "No change" in BP. [2] Study not included in most meta-analyses due to no true placebo control group. [3] Mg test group showed lower sodium in red blood cells. [4] Hattori et al. [69] showed significant decreases in both SBP and DBP from baseline and placebo in these 20 thiazide-treated subjects. However, a baseline BP of 134/80 mm Hg would categorize them as NT. However, the authors separated the 9 HT subjects (baseline MBP = 104.8 mm Hg) from the 11 NT subjects (baseline MBP = 93) and found that the former showed a decrease in MBP (−11 ± 2.0 mm Hg, $p < 0.05$) and the latter showed no change in MBP (+0.1 ± 0.46 mm Hg).

**Table 3.** Summary of magnesium supplementation studies for blood pressure in Controlled Hypertensives (subjects treated with antihypertensive medications, normotensive at baseline).

| Study Citation | Mg Dose, mg/day | Form of Mg | BP Status at Baseline, NT or HT | Medical Status at Baseline, T or UT | BP Outcome [1] | Notes |
|---|---|---|---|---|---|---|
| Henderson et al. (1986) [70] | 304 | MgO | NT | T | No change | Potassium depleting diuretics ≥ 6 months |
| Itoh et al. (1997) [71] | 413–583 | Mg(OH)$_2$ | NT | T and UT | No change [2,3] | Some subjects were borderline HT; medications kept constant "when necessary" (medications not specified) |

[1] Studies showing a statistically significant decrease in both SBP and DBP from baseline as well as placebo (if reported) were labeled "Decrease"; all others were deemed "No change" in BP. [2] Rise in Na excretion; decrease in serum Na. [3] Faulty baseline statistics; final SBP and DBP significantly lower than baseline but change in SBP and DBP not significantly different from those of placebo, thus the "No change" categorization. Only as a percentage of run-in, pre-baseline value was final Mg SBP significantly lower than placebo's percentage of run-in SBP. Abbreviations: BP, blood pressure; NT, normotensive at baseline; HT, hypertensive at baseline; UT, most or all subjects treatment naïve or not taking any antihypertensive medications during or before the study; T, most or all subjects treated with antihypertensive medications including diuretics.

Table 4. Summary of magnesium supplementation studies for blood pressure in Normotensives (subjects untreated with antihypertensive medications, normotensive at baseline).

| Study Citation | Mg Dose, mg/day | Form of Mg | BP Status at Baseline, NT or HT | Medical Status at Baseline, T or UT | BP Outcome [1] | Notes |
|---|---|---|---|---|---|---|
| Doyle et al. (1999) [72] | 250 | Mg(OH)$_2$ | NT | UT | No change | |
| Lee et al. (2009) [73] | 300 | MgO | NT | Unknown | No change | |
| Guerrero-Romero et al. (2004) [74] | 304 | MgCl$_2$ | NT | UT | No change | |
| Sacks et al. (1998) [75] | 340 | Lactate | NT | UT | No change | |
| Joris et al. (2016) [76] | 350 | Citrate | NT | UT | No change | Overweight, healthy |
| TOHP Study Group (1992) [77] | 365 | Diglycine | NT | UT | No change | |
| Mooren et al. (2011) [78] | 365 | Aspartate HCl | NT | Not reported | No change | |
| Simental-Mendia et al. (2014) [79] | 382 | MgCl$_2$ | NT | UT | No change | |
| Simental-Mendia et al. (2012) [80] | 382 | MgCl$_2$ | NT | UT | No change | Hyperglycemic, insulin resistant, hypertriglyceridemic, hypomagnesemic, normal weight |
| Rodriguez-Moran and Guerrero-Romero (2014) [81] | 381 | | NT | UT | No change [2] | |
| Cosaro et al. (2014) [82] | 394 | Pidolate | NT | UT | No change | |
| Rodriguez-Moran and Guerrero-Romero (2003) [83] | 450 | MgCl$_2$ | Borderline HT/NT | UT | No change | |
| Rodriguez-Hernandez et al. (2010) [84] | 450 | MgCl$_2$ | NT | UT | No change | |
| Daly et al. (1990) [85] | 500 | MgO | NT | UT | No change | |
| Kisters et al. (1993) [86] | 505 | Aspartate | NT | UT | No change | |
| Wary et al. (1999) [87] | 600 | Lactate + B$_6$ | NT | UT | No change | |
| Guerrero-Romero and Rodriguez-Moran (2011) [88] | 632 | MgCl$_2$ | NT | UT | No change [3] | Subjects had low serum Mg that normalized with Mg therapy |

Abbreviations: BP, blood pressure; DBP, diastolic blood pressure; HOMA-IR, homeostatic model assessment–insulin resistance; HT, hypertensive at baseline; MBP, mean blood pressure; Mg, magnesium; NT, normotensive at baseline; SBP, systolic blood pressure; T, most or all subjects treated with antihypertensive medications including diuretics; UT, most or all subjects treatment naive or not taking any antihypertensive medications during or before the study. [1] Studies showing a statistically significant decrease in both SBP and DBP from baseline as well as placebo (if reported) were labeled "Decrease"; all others were deemed "No change" in BP. [2] SBP and DBP in the Mg group showed no statistically significant change from baseline, but both decreased significantly compared with the placebo group (which showed slight increases in both SBP and DBP), thus the "no change" categorization. Nonetheless, subjects taking Mg significantly improved fasting glucose, HOMA-IR index, triglycerides, and serum Mg when statistically compared with both baseline and placebo. [3] Author statistics showed baseline Mg group vs. placebo SBP and DBP to be not significant and final Mg group vs. placebo SBP and DBP to be significantly different ($p < 0.05$). They did not calculate $p$ for final vs. baseline. In our calculations, the placebo group showed no change in both DBP and SBP from baseline; Mg test group final vs. baseline for SBP was highly significant ($p = 0.0003$) but borderline for DBP ($p = 0.0642$) which technically requires a "No change" by criteria of this analysis. Any reasonable person would deem this a "Decrease" which a quantitative meta-analysis would incorporate.

### 3.2. Other Cardiovascular Risk Factors

Even when oral magnesium therapy did not lower BP in Untreated Hypertensives, Controlled Hypertensives or Normotensive subjects, their studies often showed improved parameters linked to cardiovascular health, such as serum or plasma magnesium [40,43,44,46–48,74,77,79–81,83,84], improved endothelial function [48], reversal of retinal vasospasm [52], improved sodium excretion [43,71], lower sodium in red blood cells [65], higher serum potassium [63], lower C-reactive protein (CRP) [79], improved fasting glucose and insulin resistance [74,78,81,83,88], and lower triglycerides and total cholesterol as well as higher high-density lipoprotein (HDL) cholesterol [74].

### 3.3. Form of Magnesium

Several forms of magnesium, both organic and inorganic, were used in these studies, and it is interesting to note that the only effective doses of >600 mg/day in Untreated Hypertensives were in studies using MgO, often noted in advertising as being poorly absorbed. Magnesium given as aspartate, chloride, oxide, pidolate, lactate, citrate, amino-acid chelate at doses below 600 mg/day were not effective in this category. It was Mg dose, not form of Mg, that made the difference. Likewise, the several forms of magnesium showing BP-lowering effects in Uncontrolled Hypertensives (Table 2) included six forms of magnesium, both organic and inorganic, including MgO, and all were effective in lowering BP by the criteria of this analysis.

### 3.4. Magnesium-Replete Subjects

Two studies on Untreated Hypertensive subjects given 607 and 972 mg Mg/day showed "no change" in BP [54,58]. Walker et al. [54] noted that these subjects were "replete" in magnesium status because their dietary magnesium was high in the magnesium test group (485 vs. 346 mg/day in the placebo group). This author's term, "magnesium replete" may relate to the intake being >RDA for their Mg test group subjects but <RDA in their placebo group. Zemel et al. [58] noted that their magnesium-treated group was "healthier" than their placebo group at baseline in risk factors affected by magnesium status (see Table 1, footnote 9), and they suggested that oral magnesium "lowers BP only in states of Mg deficiency" (i.e., not magnesium replete). Other studies did not specifically report on the general magnesium status of their subjects, so the question of subjects' magnesium status playing a role in in magnesium's effect on BP is not resolved with this analysis even though it is suggested by these two studies.

### 3.5. Treated or Untreated Normotensive Subjects

None of the studies on subjects normotensive at baseline, be they Controlled Hypertensives or Normotensives (Tables 3 and 4), showed a decrease in SBP and/or DBP with magnesium doses ranging from 250 to 600 mg/day. In contrast, several but not all studies on subjects hypertensive at baseline, be they Untreated Hypertensives or Uncontrolled Hypertensives (Tables 1 and 2), showed oral Mg therapy to have a BP-lowering effect.

Tending to confirm this difference between response to oral magnesium therapy between NT and HT subjects, Haga [55] administered 600 mg Mg/day to 17 HT and 8 NT subjects. Only the HT subjects showed a significant decrease in BP. The NT subjects showed no change even at this high level of oral magnesium therapy (see Table 1, footnote 8). In addition, Hattori et al. [69] added a separate analysis of HT versus NT subjects in their magnesium-treated group and found a decrease with the high magnesium dose (607 mg/day) in HT subjects but "no change" in NT subjects (see Table 2, footnote 4).

### 3.6. Side Effects of Oral Magnesium Therapy in These Studies

The trials included in this analysis observed no serious adverse reactions to magnesium supplementation reported among participants receiving up to 972 mg Mg per day. The adverse effects that were reported were minor, transient and were often reported in both experimental and control groups. A full analysis of side effects reported in these trials is available in the Petition to FDA for a Health Claim for Magnesium and Reduced Risk of Hypertension [18] on page 130–132.

*3.7. Safety of Magnesium Doses in Effective Range*

Effective magnesium doses in this analysis ranged from 240 to 972 mg/day. For Untreated Hypertensives, the minimum effective dose was ≥600 mg/day. The tolerable upper intake level (UL) of magnesium for non-food sources is 350 mg/day for adults [15,89]. However, this UL was based on limited data and "although a few studies have noted mild diarrhea and other mild gastrointestinal complaints in a small percentage of patients at levels of 360 to 380 mg per day, it is noteworthy that many other individuals have not encountered such effects even when receiving substantially more than this UL of supplementary magnesium" [18]. Very high intakes of magnesium supplements can be dangerous, even to people without renal or intestinal disease, but such concentrations of magnesium supplement intake are in the range of ≥5000 mg magnesium/d, i.e., ≥10-fold higher than the additional amounts discussed in this article [15,90].

*3.8. How Does This Analysis Build Upon Existing Meta-analyses?*

As noted in the Introduction, six meta-analyses on this topic have been published so far [9–14] with mixed results and mostly high heterogeneity. All included randomized trials. This qualitative categorization builds on their quantitative work by suggesting possible origins of that heterogeneity.

One meta-analysis shows no effect of oral magnesium and "high" heterogeneity using 20 trials from all four categories of this collection [12]. A second shows lowering of DBP but not SBP, again with high heterogeneity ($I^2 = 62\%/47\%$), using 12 of the trials from three categories of this collection [11]. Another meta-analysis [10] using 11 trials of only unhealthy subjects drew from three categories of this analysis to show low heterogeneity ($I^2 = 2.1\%$) in the lowering of both SBP and DBP. Only one published meta-analysis [9] showed zero heterogeneity ($I^2 = 0\%$) with a clinically relevant lowering of both SBP and DBP ($-18.7/-10.9$ mm Hg) but using only four trials, all from Group 2. (However, this study's conclusions are limited by its use of some trials without a placebo control group, as most unbiased data come from subtracting the placebo response from the magnesium BP response so the magnesium effect will not be overestimated in persons with elevated blood pressure at baseline because of regression to the mean.) The two largest and probably most reliable meta-analyses show that oral magnesium therapy lowers both SBP and DBP [13,14] using 33 and 34 trials from all four categories of this analysis, but again with high heterogeneity ($I^2 = 80\% +$ and $62\% +$).

Combined with our qualitative findings, these results suggest that use of studies on both HT and NT subjects as well as the different effective Mg dose in Untreated Hypertensives vs. Uncontrolled Hypertensive subjects are sources of heterogeneity in Mg for BP meta-analyses.

## 4. Discussion

This categorization clearly shows that NT study subjects, both Controlled Hypertensives and Normotensive (i.e., those with an untreated healthy BP), will not show lower BP with oral magnesium therapy, even at high doses. However, several studies in these normotensive categories reported significant improvement in blood magnesium, lipoproteins, C-reactive protein, fasting glucose and insulin resistance, reversal of retinal vasospasm and increased sodium excretion, all of cardiovascular risk factor benefit. Oral magnesium therapy in NT patients, treated with antihypertensive medications or not, may not show improved BP readings, but these individuals may benefit from improved cardiovascular risk factors.

Among subjects who are hypertensive (≥140/90 mm Hg; MBP ≥ 106 mm Hg) at baseline, both low and high doses of oral magnesium therapy show significant decreases in both SBP and DBP only if the subjects are concurrently taking antihypertensive medications, i.e., partially or Uncontrolled Hypertensives. In the studies of Untreated Hypertensive subjects taking no antihypertensive medications, only the studies with Mg supplement doses >600 mg/day demonstrated statistically significant improvements in blood pressure by the criteria of this analysis. Subjects on lower magnesium doses showed other improvements in measures important to cardiovascular health such as serum magnesium, endothelial function and sodium excretion.

Magnesium-replete subjects, even those who are hypertensive, did not show a decrease in BP with oral magnesium therapy, even at doses as high as 972 mg/day [58]. This finding indicates that a person can have adequate magnesium status and still have high BP. Other essential electrolytes besides magnesium can impact BP. For these patients, potassium could be low, especially when concurrent with a high sodium and/or low calcium intake.

The main limitation to this study is the lack of quantification of the BP changes, instead using the statistics and conclusions from each individual study, which varied widely. This study is not a precise meta-analysis and makes no attempt to fully quantify the impact of the categories derived from this analysis. This, rather, is the job of future meta-analyses, and we see this categorization as a preliminary study to guide future meta-analysis that may provide enhanced information about oral Mg therapy for BP while hopefully achieving lower heterogeneity than existing meta-analyses without losing precision. Nonetheless, this categorization of studies by hypertensive as well as medication status plus magnesium dose yields an informative framework for the prescription of oral magnesium therapy for high BP. It well accommodates large and small studies ($n$ = 7–227 receiving magnesium therapy), short-term and long-term studies (2–26 weeks), 11 different forms of magnesium preparations (four inorganic and seven organic), parallel as well as crossover study designs, and placebo control or not (see Michon et al. [62], Sebekova et al. [61], Shafique et al. [60], Motoyama et al. [56], Cohen et al. [52], and Haga [55], which are studies not included in most meta-analyses due to no true placebo group).

Over 30 years ago, magnesium was shown to alter vascular constriction [91] and several studies have since shown that the physiology and cellular biochemistry of magnesium is important to the functionality of endothelial and smooth muscle cells and regulation of vascular tone [92]. Decreased magnesium concentrations have been implicated in altered vascular reactivity, endothelial dysfunction, vascular inflammation, and structural remodeling [93]. Low dietary magnesium has been associated with a higher risk of hypertension [94]. In the United States, 67% of the population aged $\geq$51 years is low in dietary magnesium [95] and 55% of adults aged 19 to 50 years, 60% aged 51 to 70 years, and 78% aged >71 years do not consume their estimated average requirement for magnesium [96]. Therefore, it is not surprising that prescribing oral magnesium therapy can lower a high BP. However, this categorized review of clinical trials shows that medication status, hypertensive status, and magnesium dose all must be considered in the use of this inexpensive, non-invasive, safe, readily available, "lifestyle" therapy to prevent and treat high BP as well as other conditions for which high BP is a risk factor. Pervasive low dietary magnesium status affects the health and health care systems of national and global populations [39,97]. Chronic low dietary magnesium quite likely constitutes one of the "lifestyle" components in the high risk of cardiovascular disease of our time [39,98,99].

## 5. Conclusions

This categorization study shows that oral magnesium therapy added to treatment regimens of patients with partially controlled hypertension holds promise as a way of safely achieving lower BP without increasing antihypertensive medications. Prescribing magnesium supplements to hypertensive but untreated patients may not lower BP unless the daily magnesium dose meets or exceeds 600 mg/day, which can be safely and economically accomplished, but magnesium doses below this level can achieve other cardiovascular risk factor improvements without the side effects of antihypertensive medications [99].

**Author Contributions:** A.R.: creation and maintenance of CMER collection of oral magnesium for BP studies, initial and follow-up searches, scanning and article selection, data extraction, data analysis, correspondence with study researchers, design of categorization, tabulation, and writing. R.B.C.: article evaluation, data extraction, expert consultation, and writing. G.H.J.: performed independent search for the FDA-qualified health claim petition, article scanning and selection, data extraction, evaluation of studies by FDA criteria, and writing. All authors have read and agreed to the published version of the manuscript.

**Funding:** Support for this study was achieved via an internal grant from CMER Center for Magnesium Education & Research plus the use of, in part, insights gained from the preparation of "Petition for the Authorization of a Qualified Health Claim for Magnesium and Reduced Risk of High Blood Pressure (Hypertension)," which was funded by the Almond Board of California, PepsiCo, Inc., Council for Responsible Nutrition, Pfizer Consumer Healthcare, Premier Magnesia, and Adobe Springs.

**Institutional Review Board Statement:** Not applicable.

**Informed Consent Statement:** Not applicable.

**Data Availability Statement:** Data sharing not applicable.

**Conflicts of Interest:** A.R. and R.B.C. were co-initiators of the 2016 "Petition for the Authorization of a Qualified Health Claim for Magnesium and Reduced Risk of High Blood Pressure (Hypertension)" submitted to the US Food and Drug Administration. A.R. has received speaker honoraria and travel expenses from Pharmavite and book royalties for The Magnesium Factor (published in 2003). R.B.C. is an ad hoc consultant to RMJ Holdings biotechnology firm. G.H.J is Executive Director of the McCormick Science Institute and has provided consulting services to a variety of food and beverage companies over the years. G.H.J also authored the aforementioned health claim petition.

## References

1. Lim, S.S.; Vos, T.; Flaxman, A.D.; Danaei, G.; Shibuya, K.; Adair-Rohani, H.; AlMazroa, M.A.; Amann, M.; Anderson, H.R.; Andrews, K.G.; et al. A comparative risk assessment of burden of disease and injury attributable to 67 risk factors and risk factor clusters in 21 regions, 1990–2010: A systematic analysis for the Global Burden of Disease Study 2010. *Lancet* **2012**, *380*, 2224–2260. [CrossRef]
2. Forouzanfar, M.H.; Liu, P.; Roth, G.A.; Ng, M.; Biryukov, S.; Marczak, L.; Alexander, L.; Estep, K.; Abate, K.H.; Akinyemiju, T.F.; et al. Global Burden of Hypertension and Systolic Blood Pressure of at Least 110 to 115 mm Hg, 1990–2015. *JAMA* **2017**, *317*, 165–182. [CrossRef] [PubMed]
3. Danaei, G.; Ding, E.L.; Mozaffarian, D.; Taylor, B.; Rehm, J.; Murray, C.J.L.; Ezzati, M. The Preventable Causes of Death in the United States: Comparative Risk Assessment of Dietary, Lifestyle, and Metabolic Risk Factors. *PLoS Med.* **2009**, *6*, e1000058. [CrossRef] [PubMed]
4. Wilt, T.J.; Kansagara, D.; Qaseem, A.; Clinical Guidelines Committee of the American College of Physicians. Hypertension Limbo: Balancing Benefits, Harms, and Patient Preferences Before We Lower the Bar on Blood Pressure. *Ann. Intern. Med.* **2018**, *168*, 369–370. [CrossRef] [PubMed]
5. Whelton, P.K.; Carey, R.M.; Aronow, W.S.; Casey, D.E., Jr.; Collins, K.J.; Dennison Himmelfarb, C.; DePalma, S.M.; Gidding, S.; Jamerson, K.A.; Jones, D.W.; et al. 2017 ACC/AHA/AAPA/ABC/ACPM/AGS/APhA/ASH/ASPC/NMA/PCNA Guideline for the Prevention, Detection, Evaluation, and Management of High Blood Pressure in Adults: Executive Summary: A Report of the American College of Cardiology/American Heart Association Task Force on Clinical Practice Guidelines. *Hypertension* **2018**, *71*, 1269–1324. [CrossRef] [PubMed]
6. Ioannidis, J.P.A. Diagnosis and Treatment of Hypertension in the 2017 ACC/AHA Guidelines and in the Real World. *JAMA* **2018**, *319*, 115. [CrossRef] [PubMed]
7. Muntner, P.; Carey, R.M.; Gidding, S.; Jones, D.W.; Taler, S.J.; Wright, J.T.; Whelton, P.K. Potential US Population Impact of the 2017 ACC/AHA High Blood Pressure Guideline. *Circulation* **2018**, *137*, 109–118. [CrossRef] [PubMed]
8. Carey, R.M.; Whelton, P.K. Prevention, Detection, Evaluation, and Management of High Blood Pressure in Adults: Synopsis of the 2017 American College of Cardiology/American Heart Association Hypertension Guideline. *Ann. Intern. Med.* **2018**, *168*, 351–358. [CrossRef]
9. Rosanoff, A.; Plesset, M.R. Oral magnesium supplements decrease high blood pressure (SBP > 155mmHg) in hypertensive subjects on anti-hypertensive medications: A targeted meta-analysis. *Magnes. Res.* **2013**, *26*, 93–99. [CrossRef]
10. Dibaba, D.T.; Xun, P.; Song, Y.; Rosanoff, A.; Shechter, M.; He, K. The effect of magnesium supplementation on blood pressure in individuals with insulin resistance, prediabetes, or noncommunicable chronic diseases: A meta-analysis of randomized controlled trials. *Am. J. Clin. Nutr.* **2017**, *106*, 921–929. [CrossRef]

11. Dickinson, H.O.; Nicolson, D.; Campbell, F.; Cook, J.V.; Beyer, F.R.; A Ford, G.; Mason, J. Magnesium supplementation for the management of primary hypertension in adults. *Cochrane Database Syst. Rev.* **2006**, *3*, CD004640. [CrossRef] [PubMed]
12. Jee, S.H.; Miller, E.R.; Guallar, E.; Singh, V.K.; Appel, L.J.; Klag, M.J. The effect of magnesium supplementation on blood pressure: A meta-analysis of randomized clinical trials. *Am. J. Hypertens.* **2002**, *15*, 691–696. [CrossRef]
13. Kass, L.S.; Weekes, J.; Carpenter, L.W. Effect of magnesium supplementation on blood pressure: A meta-analysis. *Eur. J. Clin. Nutr.* **2012**, *66*, 411–418. [CrossRef] [PubMed]
14. Zhang, X.; Li, Y.; Del Gobbo, L.C.; Rosanoff, A.; Wang, J.; Zhang, W.; Song, Y. Effects of Magnesium Supplementation on Blood Pressure. *Hypertension* **2016**, *68*, 324–333. [CrossRef] [PubMed]
15. Costello, R.B.; Rosanoff, A. Magnesium. In *Present Knowledge in Nutrition*, 11th ed.; Basic Nutrition and Metabolism; Marriott, B., Birt, D.F., Stalling, V., Yates, A., Eds.; ILSI-Academic Press: Washington, DC, USA, 2020; pp. 349–373.
16. Rosique-Esteban, N.; Guasch-Ferré, M.; Hernandez-Alonso, P.; Salas-Salvadó, J. Dietary Magnesium and Cardiovascular Disease: A Review with Emphasis in Epidemiological Studies. *Nutrients* **2018**, *10*, 168. [CrossRef]
17. Champagne, C. Magnesium in Hypertension, Cardiovascular Disease, Metabolic Syndrome, and Other Conditions: A Review. *Nutr. Clin. Pract.* **2008**, *23*, 142–151. [CrossRef]
18. Center for Magnesium Education and Research LLC. Petition for the Authorization of a Qualified Health Claim for Magnesium and Reduced Risk of High Blood Pressure (Hypertension) (FDA Docket ID FDA-2016-Q-3770). Available online: https://www.noticeandcomment.com/FDA-2016-Q-3770-fdt-138630.aspx (accessed on 30 March 2018).
19. Seelig, M.S.; Rosanoff, A. *The Magnesium Factor*; Avery Penguin Group: New York, NY, USA, 2003.
20. Rosanoff, A. Magnesium supplements may enhance the effect of antihypertensive medications in stage 1 hypertensive subjects. *Magnes. Res.* **2010**, *23*, 27–40. [CrossRef]
21. Patki, P.S.; Singh, J.; Gokhale, S.V.; Bulakh, P.M.; Shrotri, D.S.; Patwardhan, B. Efficacy of potassium and magnesium in essential hypertension: A double-blind, placebo controlled, crossover study. *BMJ* **1990**, *301*, 521–523. [CrossRef]
22. Sibai, B.M.; A Villar, M.; Bray, E.; L, M.A.V. Magnesium supplementation during pregnancy: A double-blind randomized controlled clinical trial. *Am. J. Obstet. Gynecol.* **1989**, *161*, 115–119. [CrossRef]
23. Sacks, F.M.; Brown, L.E.; Appel, L.; Borhani, N.O.; Evans, D.; Whelton, P. Combinations of Potassium, Calcium, and Magnesium Supplements in Hypertension. *Hypertension* **1995**, *26*, 950–956. [CrossRef]
24. Lumme, J.A.; Jounela, A.J. The effect of potassium and potassium plus magnesium supplementation on ventricular extrasystoles in mild hypertensives treated with hydrochlorothiazide. *Int. J. Cardiol.* **1989**, *25*, 93–97. [CrossRef]
25. Sur, G.; Maftei, O. Role of magnesium in essential hypertension in teenagers. In *Advances in Magnesium Research—New Data*; Porr, P.J., Nechifor, M., Durlack, J., Eds.; John Libbey Eurotext: Montrouge, France, 2006; pp. 55–60.
26. Ruiz-López, M.; Gil-Extremera, B.; Maldonado-Martín, A.; Huertas-Hernández, F.; Ceballos-Atienza, R.; Muñoz-Parra, F.; Cruz-Benayas, M.; León-Espinosa-Monteros, M.; Cobo-Martínez, F.; Soto-Mas, J. Blood pressure and metabolic syndrome in essential hypertensive patients treated with losartan or verapamil after oral magnesium supplement. *Am. J. Hypertens.* **1999**, *4*, 129. [CrossRef]
27. Eriksson, J.; Kohvakka, A. Magnesium and Ascorbic Acid Supplementation in Diabetes mellitus. *Ann. Nutr. Metab.* **1995**, *39*, 217–223. [CrossRef] [PubMed]
28. Barragán-Rodríguez, L.; Rodríguez-Morán, M.; Guerrero-Romero, F. Efficacy and safety of oral magnesium supplementation in the treatment of depression in the elderly with type 2 diabetes: A randomized, equivalent trial. *Magnes. Res.* **2008**, *21*, 218–223.
29. Kass, L.S.; Poeira, F. The effect of acute vs chronic magnesium supplementation on exercise and recovery on resistance exercise, blood pressure and total peripheral resistance on normotensive adults. *J. Int. Soc. Sports Nutr.* **2015**, *12*, 1–8. [CrossRef]
30. Kass, L.S.; Skinner, P.; Poeira, F. A Pilot Study on the Effects of Magnesium Supplementation with High and Low Habitual Dietary Magnesium Intake on Resting and Recovery from Aerobic and Resistance Exercise and Systolic Blood Pressure. *J. Sports Sci. Med.* **2013**, *12*, 144–150.
31. Shechter, M.; Sharir, M.; Labrador, M.J.P.; Forrester, J.; Silver, B.; Merz, C.N.B. Oral Magnesium Therapy Improves Endothelial Function in Patients with Coronary Artery Disease. *Circulation* **2000**, *102*, 2353–2358. [CrossRef]
32. Shechter, M.; Merz, C.B.; Stuehlingen, H.-G.; Slany, J.; Pachinger, O.; Rabinowitz, B. Effects of Oral Magnesium Therapy on Exercise Tolerance, Exercise-Induced Chest Pain, and Quality of Life in Patients with Coronary Artery Disease. *Am. J. Cardiol.* **2003**, *91*, 517–521. [CrossRef]
33. Zhang, X.; Del Gobbo, L.C.; Hruby, A.; Rosanoff, A.; He, K.; Dai, Q.; Costello, R.B.; Zhang, W.; Song, Y. The Circulating Concentration and 24-h Urine Excretion of Magnesium Dose- and Time-Dependently Respond to Oral Magnesium Supplementation in a Meta-Analysis of Randomized Controlled Trials. *J. Nutr.* **2016**, *146*, 595–602. [CrossRef]
34. Rosanoff, A. Importance of magnesium dose in the treatment of hypertension. In *Advances in Magnesium Research: New Data*; Porr, P., Nechifor, M., Durlach, J., Eds.; Libbey Eurotext: Montrouge, France, 2006; pp. 97–104.
35. Rosanoff, A. Magnesium and hypertension. *Clin. Calcium* **2005**, *15*, 255–260.
36. Rosanoff, A. Importance of dosage and experimental design in trials testing the effect of magnesium supplementation on hypertension. In Proceedings of the European Magnesium Congress, Cluj-Napoca, Romania, 25–28 May 2004.
37. Seelig, M.S.; Rosanoff, A. High blood pressure, salt and magnesium. In *The Magnesium Factor*; Avery Penguin Group: New York, NY, USA, 2003; pp. 50–84, 315–324.
38. Rosanoff, A. Changing crop magnesium concentrations: Impact on human health. *Plant Soil* **2013**, *368*, 139–153. [CrossRef]

39. Rosanoff, A.; Weaver, C.M.; Rude, R.K. Suboptimal magnesium status in the United States: Are the health consequences underestimated? *Nutr. Rev.* **2012**, *70*, 153–164. [CrossRef] [PubMed]
40. Borrello, G.; Mastroroberto, P.; Curcio, F.; Chello, M.; Zofrea, S.; Mazza, M.L. The effects of magnesium oxide on mild essential hypertension and quality of life. *Curr. Ther. Res.* **1996**, *57*, 767–774. [CrossRef]
41. Nowson, C.; Morgan, T.O. Magnesium supplementation in mild hypertensive patients on a moderately low sodium diet. *Clin. Exp. Pharmacol. Physiol.* **1989**, *16*, 299–302. [CrossRef]
42. Ferrara, L.A.; Iannuzzi, R.; Castaldo, A.; Iannuzzi, A.; Russo, A.D.; Mancini, M. Long-Term Magnesium Supplementation in Essential Hypertension. *Cardiology* **1992**, *81*, 25–33. [CrossRef]
43. Lind, L.; Lithell, H.; Pollare, T.; Ljunghall, S. Blood Pressure Response During Long-Term Treatment With Magnesium Is Dependent on Magnesium Status. *Am. J. Hypertens.* **1991**, *4*, 674–679. [CrossRef]
44. De Valk, H.W.; Verkaaik, R.; van Rijn, H.J.; Geerdink, R.A.; Struyvenberg, A. Oral magnesium supplementation in insu-lin-requiring type 2 diabetic patients. *Diabet. Med.* **1998**, *15*, 503–507. [CrossRef]
45. Plum-Wirell, M.; Stegmayr, B.G.; O Wester, P. Nutritional magnesium supplementation does not change blood pressure nor serum or muscle potassium and magnesium in untreated hypertension. A double-blind crossover study. *Magnes. Res.* **1994**, *7*, 277–283.
46. Wirell, M.M.; Wester, P.O.; Stegmayr, B.G. Nutritional dose of magnesium given to short-term thiazide treated hypertensive patients does not alter the blood pressure or the magnesium and potassium in muscle-A double blind cross-over study. *Magnes. Bull.* **1993**, *15*, 50–54.
47. Cappuccio, F.P.; Markandu, N.D.; Beynon, G.W.; Shore, A.C.; Sampson, B.; A MacGregor, G. Lack of effect of oral magnesium on high blood pressure: A double-blind study. *BMJ* **1985**, *291*, 235–238. [CrossRef]
48. Barbagallo, M.; Dominguez, L.; Galioto, A.; Pineo, A.; Belvedere, M. Oral magnesium supplementation improves vascular function in elderly diabetic patients. *Magnes. Res.* **2010**, *23*, 131–137. [PubMed]
49. Reyes, A.J.; Leary, W.P.; Acosta-Barrios, T.N.; Davis, W.H. Magnesium supplementation in hypertension treated with hy-drochlorothiazide. *Curr. Ther. Res.* **1984**, *36*, 332–340.
50. Olhaberry, J.; Reyes, A.J.; Acosta-Barrios, T.N.; Leary, W.P.; Queiruga, G. Pilot evaluation of the putative antihypertensive effect of magnesium. *Magnes. Bull.* **1987**, *9*, 181–184.
51. Purvis, J.R. Effect of oral magnesium supplementation on selected cardiovascular risk factors in non-insulin-dependent diabetics. *Arch. Fam. Med.* **1994**, *3*, 503–508. [CrossRef] [PubMed]
52. Cohen, L.; Laor, A.; Kitzes, R. Reversible retinal vasospasm in magnesium-treated hypertension despite no significant change in blood pressure. *Magnesium* **1984**, *3*, 159–163. [PubMed]
53. Witteman, J.C.; E Grobbee, D.; Derkx, F.H.; Bouillon, R.; De Bruijn, A.M.; Hofman, A. Reduction of blood pressure with oral magnesium supplementation in women with mild to moderate hypertension. *Am. J. Clin. Nutr.* **1994**, *60*, 129–135. [CrossRef]
54. Walker, A.F.; Marakis, G.; Morris, A.P.; Robinson, P.A. Promising hypotensive effect of hawthorn extract: A randomized double-blind pilot study of mild, essential hypertension. *Phytother. Res.* **2002**, *16*, 48–54. [CrossRef]
55. Haga, H. Effects of Dietary Magnesium Supplementation on Diurnal Variations of Blood Pressure and Plasma Na+, K+-ATPase Activity in Essential Hypertension. *Jpn. Heart J.* **1992**, *33*, 785–800. [CrossRef]
56. Motoyama, T.; Sano, H.; Fukuzaki, H. Oral magnesium supplementation in patients with essential hypertension. *Hypertension* **1989**, *13*, 227–232. [CrossRef]
57. Sanjuliani, A.F.; Fagundes, V.G.D.A.; Francischetti, E.A. Effects of magnesium on blood pressure and intracellular ion levels of Brazilian hypertensive patients. *Int. J. Cardiol.* **1996**, *56*, 177–183. [CrossRef]
58. Zemel, P.C.; Zemel, M.; Urberg, M.; Douglas, F.L.; Geiser, R.; Sowers, J.R. Metabolic and hemodynamic effects of magnesium supplementation in patients with essential hypertension. *Am. J. Clin. Nutr.* **1990**, *51*, 665–669. [CrossRef] [PubMed]
59. Widman, L.; Wester, P.; Stegmayr, B.; Wirell, M. The Dose-Dependent Reduction in Blood Pressure Through Administration of Magnesium A Double Blind Placebo Controlled Cross-Over Study. *Am. J. Hypertens.* **1993**, *6*, 41–45. [CrossRef] [PubMed]
60. Shafique, M.; Misbah ul, A.; Ashraf, M. Role of magnesium in the management of hypertension. *J. Pak. Med. Assoc.* **1993**, *43*, 77–78. [PubMed]
61. Sebeková, K.; Revúsová, V.; Polakovicová, D.; Drahosová, J.; Zverková, D.; Dzúrik, R. Anti-hypertensive treatment with magnesium-aspartate-dichloride and its influence on peripheral serotonin metabolism in man: A subacute study. *Cor et Vasa* **1992**, *34*, 390–401.
62. Michoń, P. Level of total and ionized magnesium fraction based on biochemical analysis of blood and hair and effect of supplemented magnesium (Slow Mag B6) on selected parameters in hypertension of patients treated with various groups of drugs. *Ann. Acad. Med. Stetin.* **2002**, *48*, 85–97.
63. Wirell, M.P.; Wester, P.O.; Stegmayr, B. Nutritional dose of magnesium in hypertensive patients on beta blockers lowers systolic blood pressure: A double-blind, cross-over study. *J. Intern. Med.* **1994**, *236*, 189–195. [CrossRef]
64. Dyckner, T.; O Wester, P. Effect of magnesium on blood pressure. *BMJ* **1983**, *286*, 1847–1849. [CrossRef]
65. Paolisso, G.; Di Maro, G.; Cozzolino, D.; Salvatore, T.; D'Amore, A.; Lama, D.; Varricchio, M.; D'Onofrio, F. Chronic Maenesium Administration Enhances Oxidative Glucose Metabolism in Thiazide Treated Hypertensive Patients. *Am. J. Hypertens.* **1992**, *5*, 681–686. [CrossRef]

66. Guerrero-Romero, F.; Rodriguez-Moran, M. The effect of lowering blood pressure by magnesium supplementation in diabetic hy-pertensive adults with low serum magnesium levels: A randomized, double-blind, placebo-controlled clinical trial. *J. Hum. Hypertens.* **2009**, *23*, 245–251. [CrossRef]
67. Kawano, Y.; Matsuoka, H.; Takishita, S.; Omae, T. Effects of Magnesium Supplementation in Hypertensive Patients. *Hypertension* **1998**, *32*, 260–265. [CrossRef]
68. Cunha, A.R.; D'El-Rei, J.; Medeiros, F.; Umbelino, B.; Oigman, W.; Touyz, R.M.; Neves, M.F. Oral magnesium supplementation improves endothelial function and attenuates subclinical atherosclerosis in thiazide-treated hypertensive women. *J. Hypertens.* **2017**, *35*, 89–97. [CrossRef] [PubMed]
69. Hattori, K.; Saito, K.; Sano, H.; Fukuzaki, H. Intracellular magnesium deficiency and effect of oral magnesium on blood pressure and red cell sodium transport in diuretic-treated hypertensive patients. *Jpn. Circ. J.* **1988**, *52*, 1249–1256. [CrossRef] [PubMed]
70. Henderson, D.G.; Schierup, J.; Schodt, T. Effect of magnesium supplementation on blood pressure and electrolyte concentrations in hypertensive patients receiving long term diuretic treatment. *BMJ* **1986**, *293*, 664–665. [CrossRef] [PubMed]
71. Itoh, K.; Kawasaki, T.; Nakamura, M. The effects of high oral magnesium supplementation on blood pressure, serum lipids and related variables in apparently healthy Japanese subjects. *Br. J. Nutr.* **1997**, *78*, 737–750. [CrossRef] [PubMed]
72. Doyle, L.; Flynn, A.; Cashman, K. The effect of magnesium supplementation on biochemical markers of bone metabolism or blood pressure in healthy young adult females. *Eur. J. Clin. Nutr.* **1999**, *53*, 255–261. [CrossRef] [PubMed]
73. Lee, S.; Park, H.; Son, S.; Lee, C.W.; Kim, I.; Kim, H. Effects of oral magnesium supplementation on insulin sensitivity and blood pressure in normo-magnesemic nondiabetic overweight Korean adults. *Nutr. Metab. Cardiovasc. Dis.* **2009**, *19*, 781–788. [CrossRef] [PubMed]
74. Guerrero-Romero, F.; E Tamez-Perez, H.; González-González, G.; Salinas-Martínez, A.M.; Montes-Villarreal, J.; Treviño-Ortiz, J.H.; Rodríguez-Morán, M. Oral Magnesium supplementation improves insulin sensitivity in non-diabetic subjects with insulin resistance. A double-blind placebo-controlled randomized trial. *Diabetes Metab.* **2004**, *30*, 253–258. [CrossRef]
75. Sacks, F.M.; Willett, W.C.; Smith, A.; Brown, L.E.; Rosner, B.; Moore, T.J. Effect on blood pressure of potassium, calcium, and magnesium in women with low habitual intake. *Hypertension* **1998**, *31*, 131–138. [CrossRef]
76. Joris, P.J.; Plat, J.; Bakker, S.J.L.; Mensink, R.P. Long-term magnesium supplementation improves arterial stiffness in overweight and obese adults: Results of a randomized, double-blind, placebo-controlled intervention trial. *Am. J. Clin. Nutr.* **2016**, *103*, 1260–1266. [CrossRef]
77. Whelton, P.K.; Appel, L.; Charleston, J.; Dalcin, A.T.; Ewart, C.; Fried, L.; Kaidy, D.; Klag, M.J.; Kumanyika, S.; Steffen, L.; et al. The Effects of Nonpharmacologic Interventions on Blood Pressure of Persons with High Normal Levels. *JAMA* **1992**, *267*, 1213–1220. [CrossRef]
78. Mooren, F.C.; Krüger, K.; Völker, K.; Golf, S.W.; Wadepuhl, M.; Kraus, A. Oral magnesium supplementation reduces insulin resistance in non-diabetic subjects - a double-blind, placebo-controlled, randomized trial. *Diabetes Obes. Metab.* **2011**, *13*, 281–284. [CrossRef] [PubMed]
79. Simental-Mendía, L.E.; Rodríguez-Morán, M.; Guerrero-Romero, F. Oral Magnesium Supplementation Decreases C-reactive Protein Levels in Subjects with Prediabetes and Hypomagnesemia: A Clinical Randomized Double-blind Placebo-controlled Trial. *Arch. Med. Res.* **2014**, *45*, 325–330. [CrossRef] [PubMed]
80. Simental-Mendía, L.E.; Rodríguez-Morán, M.; Reyes-Romero, M.A.; Guerrero-Romero, F. No positive effect of oral magnesium supplementation in the decreases of inflammation in subjects with prediabetes: A pilot study. *Magnes. Res.* **2012**, *25*, 140–146. [CrossRef] [PubMed]
81. Rodríguez-Moran, M.; Guerrero-Romero, F. Oral Magnesium Supplementation Improves the Metabolic Profile of Metabolically Obese, Normal-weight Individuals: A Randomized Double-blind Placebo-controlled Trial. *Arch. Med. Res.* **2014**, *45*, 388–393. [CrossRef] [PubMed]
82. Cosaro, E.; Bonafini, S.; Montagnana, M.; Danese, E.; Trettene, M.; Minuz, P.; Delva, P.; Fava, C. Effects of magnesium supplements on blood pressure, endothelial function and metabolic parameters in healthy young men with a family history of metabolic syndrome. *Nutr. Metab. Cardiovasc. Dis.* **2014**, *24*, 1213–1220. [CrossRef] [PubMed]
83. Rodríguez-Morán, M.; Guerrero-Romero, F. Oral Magnesium Supplementation Improves Insulin Sensitivity and Metabolic Control in Type 2 Diabetic Subjects: A randomized double-blind controlled trial. *Diabetes Care* **2003**, *26*, 1147–1152. [CrossRef]
84. Rodriguez-Hernandez, H.; Cervantes-Huerta, M.; Rodríguez-Morán, M.; Guerrero-Romero, F. Oral magnesium supplementation decreases alanine aminotransferase levels in obese women. *Magnes. Res.* **2010**, *23*, 90–96.
85. Daly, N.M.; Allen, K.G.D.; Harris, M. Magnesium supplementation and blood pressure in borderline hypertensive subjects: A double blind study. *Magnes. Bull.* **1990**, *12*, 149–154.
86. Kisters, K.; Spieker, C.; Tepel, M.; Zidek, W. New data about the effects of oral physiological magnesium supplementation on several cardiovascular risk factors (lipids and blood pressure). *Magnes. Res.* **1993**, *6*, 355–360.
87. Wary, C.; Brillault-Salvat, C.; Bloch, G.; Leroy-Willig, A.; Roumenov, D.; Grognet, J.M.; Leclerc, J.H.; Carlier, P.G. Effect of chronic magnesium supplementation on magnesium distribution in healthy volunteers evaluated by 31 P-NMRS and ion selective electrodes. *Br. J. Clin. Pharmacol.* **1999**, *48*, 655–662. [CrossRef]
88. Guerrero-Romero, F.; Rodríguez-Morán, M. Magnesium improves the beta-cell function to compensate variation of insulin sensitivity: Double-blind, randomized clinical trial. *Eur. J. Clin. Investig.* **2011**, *41*, 405–410. [CrossRef] [PubMed]

89. Institute of Medicine (US) Standing Committee on the Scientific Evaluation of Dietary Reference Intakes. *Dietary Reference Intakes for Calcium, Phosphorus, Magnesium, Vitamin D, and Fluoride*; The National Academies Press: Washington, DC, USA, 1997.
90. Rosanoff, A. Perspective: US adult magnesium requirements need updating: Impacts of rising body weights and data-derived variance. *Adv. Nutr.* **2020**. [CrossRef] [PubMed]
91. Turlapaty, P.; Altura, B. Magnesium deficiency produces spasms of coronary arteries: Relationship to etiology of sudden death ischemic heart disease. *Science* **1980**, *208*, 198–200. [CrossRef] [PubMed]
92. Kolte, D.; Vijayaraghavan, K.; Khera, S.; Sica, D.A.; Frishman, W.H. Role of Magnesium in Cardiovascular Diseases. *Cardiol. Rev.* **2014**, *22*, 182–192. [CrossRef] [PubMed]
93. Touyz, R.M. Transient receptor potential melastatin 6 and 7 channels, magnesium transport, and vascular biology: Implications in hypertension. *Am. J. Physiol. Circ. Physiol.* **2008**, *294*, H1103–H1118. [CrossRef] [PubMed]
94. Kass, L.; Sullivan, K.R. Low Dietary Magnesium Intake and Hypertension. *World J. Cardiovasc. Dis.* **2016**, *6*, 447–457. [CrossRef]
95. Institute of Medicine (US) Food Forum. *Providing Healthy and Safe Foods as We Age*; The National Academies Press: Washington, DC, USA, 2010. Available online: https://pubmed.ncbi.nlm.nih.gov/21391340/ (accessed on 15 May 2020).
96. Blumberg, J.B.; Frei, B.; Fulgoni, V.L.; Weaver, C.; Zeisel, S.H. Contribution of Dietary Supplements to Nutritional Adequacy in Various Adult Age Groups. *Nutrients* **2017**, *9*, 1325. [CrossRef]
97. US Department of Agriculture; US Department of Health and Human Services. *Dietary Guidelines for Americans, 2015–2020*, 8th ed. Available online: https://health.gov/dietaryguidelines/2015/resources/2015-2020_Dietary_Guidelines.pdf (accessed on 7 March 2018).
98. Greenland, P. Cardiovascular Guideline Skepticism vs Lifestyle Realism? *JAMA* **2018**, *319*, 117. [CrossRef]
99. Dominguez, L.J.; Veronese, N.; Barbagallo, M. Magnesium and Hypertension in Old Age. *Nutrients* **2020**, *13*, 139. [CrossRef]

Article

# Circulating Ionized Magnesium as a Measure of Supplement Bioavailability: Results from a Pilot Study for Randomized Clinical Trial

Jiada Zhan [1], Taylor C. Wallace [2,3,4,*], Sarah J. Butts [5], Sisi Cao [5], Velarie Ansu [6], Lisa A. Spence [6], Connie M. Weaver [7] and Nana Gletsu-Miller [6]

1. Public Health Nutrition, Case Western Reserve University, 10900 Euclid Avenue, Cleveland, OH 44106, USA; jxz1119@case.edu
2. Department of Nutrition and Food Studies, George Mason University, MS1F7, 4400 University Drive, Fairfax, VA 22030, USA
3. Think Healthy Group, Inc., 1301 20th Street NW, Washington, DC 20036, USA
4. Center for Magnesium Education & Research, 13-1255 Malama Street, Pahoa, HI 96778, USA
5. Department of Nutrition Science, Purdue University, 700 West State Street, West Lafayette, IN 47907, USA; sjb1602@gmail.com (S.J.B.); caosisi0703@gmail.com (S.C.)
6. Department of Applied Health Science, School of Public Health, Indiana University Bloomington, Bloomington, IN 47405, USA; vansu@iu.edu (V.A.); lisspenc@iu.edu (L.A.S.); ngletsum@indiana.edu (N.G.-M.)
7. Weaver and Associates Consulting, LLC, West Lafayette, IN 47906, USA; weaverconnie1995@gmail.com
* Correspondence: twallac9@gmu.edu; Tel.: +1-812-855-7643

Received: 24 March 2020; Accepted: 22 April 2020; Published: 28 April 2020

**Abstract:** Oral supplementation may improve the dietary intake of magnesium, which has been identified as a shortfall nutrient. We conducted a pilot study to evaluate appropriate methods for assessing responses to the ingestion of oral magnesium supplements, including ionized magnesium in whole blood ($iMg^{2+}$) concentration, serum total magnesium concentration, and total urinary magnesium content. In a single-blinded crossover study, 17 healthy adults were randomly assigned to consume 300 mg of magnesium from $MgCl_2$ (ReMag®, a picosized magnesium formulation) or placebo, while having a low-magnesium breakfast. Blood and urine samples were obtained for the measurement of $iMg^{2+}$, serum total magnesium, and total urine magnesium, during 24 h following the magnesium supplement or placebo dosing. Bioavailability was assessed using area-under-the-curve (AUC) as well as maximum ($C_{max}$) and time-to-maximum ($T_{max}$) concentration. Depending on normality, data were expressed as the mean ± standard deviation or median (range), and differences between responses to $MgCl_2$ or placebo were measured using the paired *t*-test or Wilcoxon signed-rank test. Following $MgCl_2$ administration versus placebo administration, we observed significantly greater increases in $iMg^{2+}$ concentrations (AUC = 1.51 ± 0.96 vs. 0.84 ± 0.82 mg/dL•24h; $C_{max}$ = 1.38 ± 0.13 vs. 1.32 ± 0.07 mg/dL, respectively; both $p < 0.05$) but not in serum total magnesium (AUC = 27.00 [0, 172.93] vs. 14.55 [0, 91.18] mg/dL•24h; $C_{max}$ = 2.38 [1.97, 4.01] vs. 2.24 [1.98, 4.31] mg/dL) or in urinary magnesium (AUC = 201.74 ± 161.63 vs. 139.30 ± 92.84 mg•24h; $C_{max}$ = 26.12 [12.91, 88.63] vs. 24.38 [13.51, 81.51] mg/dL; $p > 0.05$). Whole blood $iMg^{2+}$ may be a more sensitive measure of acute oral intake of magnesium compared to serum and urinary magnesium and may be preferred for assessing supplement bioavailability.

**Keywords:** magnesium; iMg; biomarkers; nutritional status; diet

## 1. Introduction

Magnesium is an element with an atomic number of 12 and a mass of 24.32 Da. It is the fourth most abundant mineral in the human body, with > 99% residing in the bone, muscle, and nonmuscular soft tissue and < 1% residing in the serum and red blood cells [1,2]. Magnesium is also the second most abundant intracellular cation [3]. Enzymatic databases list over 600 enzymes for which magnesium serves as a cofactor, and an additional 200 in which it may act as an activator [4–6]. The intracellular level of free magnesium ions regulates intermediary metabolism, DNA and RNA synthesis and structure, cell growth, reproduction, and membrane structure. Thus, the cation exerts numerous physiological functions, including control of neuronal activity, cardiac excitability, neuromuscular transmission, muscular contraction, vasomotor tone, blood pressure, and peripheral blood flow [3]. Magnesium also plays a role in the movement of sodium and potassium across membranes [7]. The antiarrhythmic potency of magnesium has been described repeatedly since 1935, both as a factor in human disease and animal experiments [8,9].

Substantial portions of the U.S. population fail to meet dietary recommendations for magnesium [10,11]. Approximately 48% of the U.S. population consumes less than the estimated average requirement [12]. Furthermore, racial or ethnic differences in magnesium intake exist and may contribute to some health disparities [10]. The top 10 contributors to dietary magnesium intake in the United States are plant-based protein foods (9.52%); breads, rolls, and tortillas (7.42%); coffee and tea (7.03%); vegetables, excluding potatoes (6.12%); plain water (4.56%); milk (4.23%); fruits (4.20%); ready-to-eat cereals (3.68%); mixed dishes (meat, poultry, fish) (3.56%); and grain-based mixed dishes (3.25%) [13]. Decreased intake can result from inadequate dietary consumption, starvation, and alcohol dependence. Hypomagnesemia is an electrolyte disturbance that is caused by magnesium deficiency and is clinically defined as a serum total magnesium concentration more than 2 standard deviations below the mean of the general population [14,15]. Causes include low dietary intake, alcoholism, diarrhea, increased urinary loss, poor absorption from the gut, and diabetes mellitus [4,16]. Some medications, including proton pump inhibitors and furosemide [17], also cause low magnesium. The prevalence of hypomagnesemia is higher (11%–48%) in those with diabetes mellitus [15,18]. Hypomagnesemia is common in hospitalized patients (20%) [19] and is even more frequent in patients with other coexisting electrolyte abnormalities [20–22] and in critically ill patients (20%–65%) [23,24].

Many studies have focused on the measurement of serum total magnesium concentration because of ease of measurement rather than its free bioactive $iMg^{2+}$ form, making it difficult to correlate to disease states [25] or to truly assess status. Estimation of the $iMg^{2+}$ levels in serum or plasma by analysis of ultrafiltrates (complexed magnesium + $iMg^{2+}$) is better than total magnesium measures, but it does not distinguish the truly ionized form from that which is bound to organic and inorganic anions [25]. Technological advances have improved the reliability for measuring $iMg^{2+}$ in complex matrices such as whole blood, plasma, and serum by improving the durability of the ion-selective electrodes and by reducing the interference on the electrode from other cations such $Ca^{2+}$ [26]. Given the capacity for rapid, accurate, and reliable analysis of electrolytes, commercially available analyzers are now routinely used in the clinical setting [26,27]. However, these instruments have rarely been used for research purposes. An early bioavailability study used ion-selective electrode technology to measure $iMg^{2+}$ responses in serum following administration of magnesium oxide formulations [28]. However, the study participants ingested high magnesium diets beforehand, creating magnesium-saturated individuals who are not a representative sample of the population. Our research group is currently evaluating the bioavailability of ReMag®, a formulation of $MgCl_2$ against a commonly marketed Mg supplement, MgO, which is often used for bioavailability studies [28–30]. We are utilizing a randomized clinical study design and newer technology with analytical advances in the assessment of $iMg^{2+}$ to test acute oral doses of these two supplements against a placebo. In the pilot study reported here, our objective was to determine the utility of whole blood concentrations of $iMg^{2+}$, compared to concentrations of serum total magnesium and total urinary magnesium, as a biomarker of response to an oral challenge consisting of a single dose of magnesium chloride, among healthy individuals.

By evaluating the acute bioavailability of a one-time dose of ReMag®, a novel formulation of MgCl$_2$, our results may be extrapolated to assess the feasibility of this formulation to increase iMg$^{2+}$ status. This pilot study provided critical learnings for the current study and both studies will be useful for any future study assessing chronic administration of magnesium supplements. Future work will investigate effects of longer-term dosage of magnesium supplementation on magnesium status and markers of disease risk.

## 2. Materials and Methods

The main study will be a randomized, single-blind, placebo-controlled crossover trial that aims to compare acute pharmacokinetics following ingestion of 300 mg (12.34 mmol) of magnesium from ReMag® (a formulation of MgCl$_2$) against MgO. We describe the similarly designed pilot study here, which aimed to evaluate the methodology for assessing magnesium and to generate data to inform sample size calculations for the main study. For the pilot study, we compared a formulation of magnesium chloride (MgCl$_2$) called ReMag® (New Capstone, Inc., Mooresville, NC, USA) to a vehicle placebo of water (and lemon juice to mask the taste of the MgCl$_2$ formulation). Using a molecular size analyzer (Malvern Zetasizer Nano ZS, Malvern, UK), we confirmed that the majority of the particle size of ReMag® formulation was in the picometer range (52% was 800 pm and 48% was 2.5 μm); we speculated that the small particle size would promote rapid absorption. Using atomic absorption spectrometry (5100 PC; Perkin-Elmer, Waltham, MA, USA), we determined that the concentration of magnesium in the ReMag® formulation was 44.73 mg/g of sample and this concentration was used to deliver a dose of 300 mg of magnesium from the ReMag® formulation. The study protocol was approved by the Purdue University Institutional Review Board (protocol number 1802020279) and is registered at ClinicalTrials.gov (NCT04139928).

### 2.1. Study Population

Recruitment was conducted via flyers, emails, and word-of-mouth. Prior to the screening visit at the clinic, interested volunteers were contacted by study staff and prescreened using a medical questionnaire. The medical questionnaire included questions related to participants' medical history (occurrence of any cardiac, digestive, renal, and other diseases, i.e., thyroid disorder, diabetes, and cancer). Other questions included recent blood donation, intake of Mg supplements, pregnancy or intention to become pregnant (females only), and medications that participants were currently using. Potential participants were then mailed a consent form and asked to complete a 3-d food log to capture habitual dietary intake before their screening visit. Participants fasted overnight prior to the screening visit. Patients provided written informed consent before entering the study.

Healthy adult men and women of all races/ethnicities, aged 18–65 years, with a body mass index (BMI) of 18–35 kg/m$^2$ could enter the study after the screening visit. Individuals were excluded if they met any of the following criteria: (1) had a diagnosis of hypertension, prehypertension, diabetes, cardiovascular disease, or other chronic diseases (e.g., cancer); (2) had been diagnosed with hypermagnesemia (> 2.28 mg/dL); (3) had been diagnosed with gastrointestinal disease, hepatitis, anemia, or hepatic enzyme abnormalities or were currently taking magnesium supplements or medications that interfere with magnesium absorption or metabolism within 2 wk of screening; (4) were currently pregnant or trying to become pregnant; (5) had a history of hospitalization for acute illness within 1 month prior to screening; (6) were unable to speak English or were unable able to comprehend the informed consent; (7) had a habitual diet which contained an excess of high magnesium foods (by reviewing dietary intake from 3-day food logs prior to the screening visit); or (8) were unable or failed to complete the full medical questionnaire. Participants were financially compensated for their participation in the study.

## 2.2. Magnesium Dosing

Enrolled participants (Males = 7, Females = 10, Total $N = 17$) were randomly assigned to a single-dose treatment of $MgCl_2$ (ReMag®; in a solution of lemon juice and water) or placebo (lemon juice and water only) with a low-magnesium (~50 mg Mg) breakfast after 8 h of fasting. Participants were blinded to their treatment assignment, but the research team was not blinded. Participants partook in two clinic visits with a minimum of a 7-d washout period between treatments. A low-magnesium lunch, dinner, and evening snack, designed by a research dietitian to contain < 160 mg of magnesium, and low-mineral water (Aquafina; PepsiCo Inc., Purchase, Harrison, NY, USA) were also provided on the days of the clinical visits. Participants were asked to refrain from consuming foods and beverages, aside from those provided, on the days they visited the clinic. Examples of foods and beverages served during participants clinic visit days were: (1) breakfast—omelet with eggs and vegetables (peppers and zucchini) and water or apple juice; (2) lunch—stir fry with rice and vegetables (cabbage, zucchini, carrots); (3) dinner—rice noodle stir fry with vegetables (carrots, celery, peppers) and (4) evening snack—vanilla ice cream. Lunch and dinner were provided to coincide with the 4- and 8-h sample collection periods. After the 8-h sample collection, participants were given their evening snack and water, allowed to leave the clinic, and asked to return the next morning for the 24-h sample collections.

## 2.3. Specimen Collection and Measurements

Blood samples were obtained beginning with a fasting sample (at 15 min before the dosing) and at 0, 0.5, 1, 2, 4, 6, 8, and 24 h following the dosing. The timepoints chosen are standard for bioavailability tests to assess serum concentrations [28,31]. Specimens were collected within ± 15 min of the hourly time points and within ± 30 min of the 24-h timepoint. Venous blood samples were collected in lithium heparinized tubes and in serum separator tubes for measurement of $iMg^{2+}$ and serum total magnesium concentrations, respectively.

The whole blood concentration of $iMg^{2+}$ was determined using previously described standardized methods [32] with a Nova 8 Electrolyte Analyzer (Nova Biomedical, Waltham, MA, USA). The instrument is designed as a point-of-care analyzer for the critical care setting, and it is rapid and easy to use along with automated quality control. Previous studies reported that $iMg^{2+}$ was fairly stable for at least 6 h when stored in capped lithium heparinized tubes at either room temperature or 4 °C [26,33]. However, our in-house testing suggested that the whole blood $iMg^{2+}$ level was relatively stable when stored at 4 °C for over 4 h, but there was a mean decrease of 7.14% after storing at room temperature for 2 h. Thus, to ensure accuracy and consistency, blood samples were measured within 10 min of collection. The intra- and inter-day coefficient of variation (CV) values for $iMg^{2+}$ were less than 3%, as reported previously [26].

Serum was separated from whole blood samples and frozen before analysis. Urine specimens were collected from participants 15 min before dosing and the total urine collected at 2, 4, 6, 8, 8–21, and 24 h post dosing were pooled in batches. For each urine sample, the specific gravity was measured to determine the urine concentration and the hydration status, and the sample was frozen before analysis. Concentration of serum total magnesium and total urinary magnesium content were determined by atomic absorption mass spectrometry using previously described standardized methods [34]. All CVs for each serum and urine magnesium measurement were below 4%.

## 2.4. Statistical Analyses

We did not conduct power analysis for the pilot study because we intended for the pilot data to inform the sample size of the main study, but the size of our sample compares to a similar bioavailability study by Altura et al. [28]. We analyzed the 24-h area under the curve (AUC) to obtain the maximum concentration ($C_{max}$) and the time to maximum concentrations ($T_{max}$) for each of the $iMg^{2+}$ and serum total magnesium concentrations and the total urinary magnesium content responses using OriginPro software (2019 version; OriginLab, Northampton, MA, USA). Data were analyzed in terms of descriptive statistics and tested for normality (Z-skewness cutoff = 1.96). The mean ± standard deviation (SD) was used to report data with a normal distribution. When the data were not normally distributed, we reported the median and range. Differences between MgCl$_2$ and placebo responses were compared using paired $t$-tests, or the nonparametric equivalent, Wilcoxon signed-rank test, based on the normality of the groups. To access the impact of confounding variables such as age, baseline serum total magnesium, or BMI on iMg$^{2+}$, serum total magnesium, or urine magnesium measures, we conducted partial correlations as well as linear regression. A $p$ value of 0.05 was used to determine statistical significance. IBM SPSS software (version 26; IBM, Armonk, NY, USA) was used for all statistical analyses.

## 3. Results

A total of 17 participants were enrolled in the pilot study and their baseline characteristics are presented in Table 1. These participants were young adults (median age 25 y [18–44 y]), with normal weight (mean BMI, 24.7 ± 3.7 kg/m$^2$), slightly more female (58.8%) than male (41.2%) participants, and representation from various racial and ethnic groups. We excluded from the data analysis participants who had data missing for the iMg$^{2+}$, serum, or urinary magnesium assays after administration of MgCl$_2$ or placebo; thus, we collected complete matched pairs data with 14 participants for iMg$^{2+}$, 17 participants for serum total magnesium, and 12 participants for urinary magnesium. Graphs of the absorption curves and total AUC of iMg$^{2+}$, serum, and urinary magnesium measures over 24-h post-treatment of MgCL$_2$ vs. placebo can be found in Figure 1.

**Table 1.** Baseline characteristics of the study participants ($N$ = 17).

| Characteristic | Value [1] | Reference Range |
|---|---|---|
| Age (y) | 25.0 (18–44) | |
| Male (%) | 41.2 | |
| Female (%) | 58.8 | |
| Race/ethnicity (%) | | |
| White | 58.8 | |
| Black | 11.8 | |
| Hispanic | 5.9 | |
| Asian | 23.5 | |
| Body mass index (kg/m$^2$) | 24.7 ± 3.7 | |
| iMg$^{2+}$ (mmol/L) | 0.52 ± 0.03 | 0.44–0.59 |
| Serum total magnesium (mmol/L) | 0.84 ± 0.01 | 0.75–0.95 |
| Serum creatinine (mmol/L) | | |
| Female (mmol/L) | 0.30 ± 0.07 | 0.23–0.42 |
| Male (mmol/L) | 0.37 ± 0.05 | 0.30–0.49 |

[1] Values are given as the mean ± SD or the median (range) depending on the normality of the data. The following citations provided reference ranges: iMg$^{2+}$ [33]; serum total magnesium [35]; serum creatinine [36].

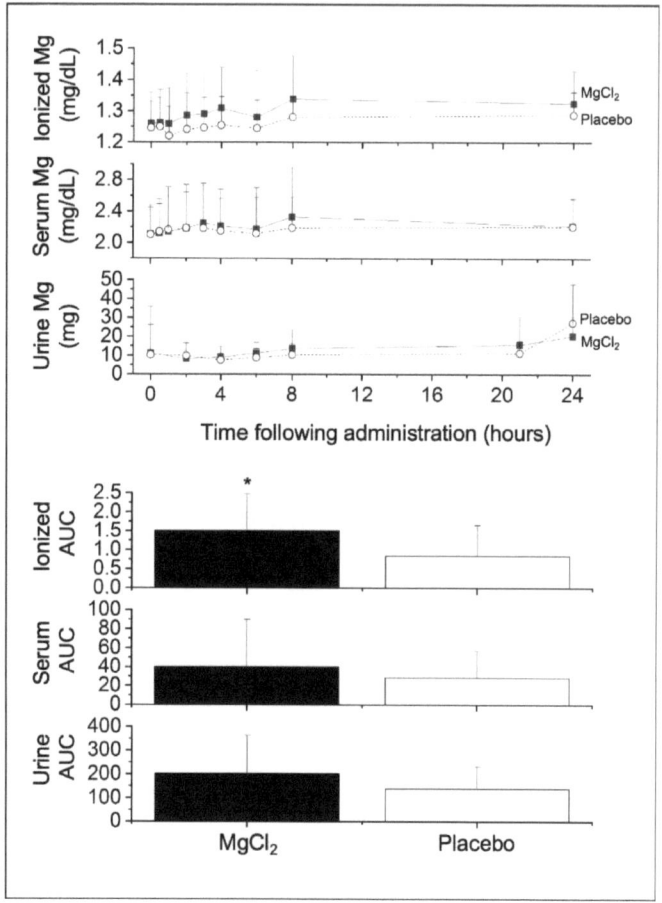

**Figure 1.** Average whole blood concentration of ionized magnesium (i$Mg^{2+}$) and serum total magnesium concentrations, as well as urinary magnesium content (top); total area under the curve (AUC) in mg/dL•24h for i$Mg^{2+}$ and serum total Mg or mg•24h for urine Mg (bottom) over 24 h post-treatment for $MgCl_2$ treatment vs. placebo. * Differences between $MgCl_2$ versus placebo, $p < 0.05$. To convert between mg/dL and mmol/L, divide by 2.43.

Following administration of $MgCl_2$, we found a significant greater increase in the 24-h AUC of whole blood concentration of i$Mg^{2+}$ (1.51 ± 0.96 mg/dL•24h) compared to that increase after administration of placebo (0.84 ± 0.82 mg/dL•24h) ($p = 0.029$, Table 2). We also observed a greater increase in $C_{max}$ for i$Mg^{2+}$ between $MgCl_2$ (1.38 ± 0.13 mg/dL) vs. placebo (1.32 ± 0.07 mg/dL) ($p = 0.034$). There was no difference in $T_{max}$ for i$Mg^{2+}$ between $MgCl_2$ vs. placebo. We found no difference in AUC between $MgCl_2$ vs. placebo for serum total magnesium ($p = 0.097$) or for urinary magnesium ($p = 0.118$). Similarly, no differences in $C_{max}$ and $T_{max}$ were found between $MgCl_2$ vs. placebo for serum or urinary magnesium. In linear regression analyses, with AUC of i$Mg^{2+}$, serum total magnesium, and urine magnesium as the dependent measures, and age, serum total magnesium at screening, or BMI as independent measures in the model, we did not find any significant coefficients, suggesting that none of these factors were confounders.

**Table 2.** Absorption kinetics for whole blood $iMg^{2+}$ and serum total magnesium concentrations and urinary magnesium content.

| Parameter | Placebo | $MgCl_2$ | $p$ Value |
| --- | --- | --- | --- |
| $iMg^{2+}$ ($n = 14$) | | | |
| Baseline (mg/dL) | 1.25 ± 0.08 | 1.26 ± 0.01 | 0.242 |
| AUC (mg/dL•24h) | 0.84 ± 0.82 | 1.51 ± 0.96 | 0.029 |
| $C_{max}$ (mg/dL) | 1.32 ± 0.07 | 1.38 ± 0.13 | 0.034 |
| $T_{max}$ (h) | 9.54 ± 9.85 | 10.36 ± 9.30 | 0.396 |
| Serum magnesium ($n = 17$) | | | |
| Baseline (mg/dL) | 1.95 (1.78–3.30) | 2.08 (1.76–3.19) | 0.491 |
| AUC (mg/dL•24h) | 14.55 (0–91.18) | 27.00 (0–172.93) | 0.097 |
| $C_{max}$ (mg/dL) | 2.24 (1.98–4.31) | 2.38 (1.97–4.01) | 0.221 |
| $T_{max}$ (h) | 11.38 ± 9.93 | 10.82 ± 9.11 | 0.396 |
| Urine magnesium ($n = 12$) | | | |
| Baseline (mg) | 2.90 (0.81–47.15) | 2.95 (0.35–88.63) | 0.267 |
| AUC (mg•24h) | 139.30 ± 92.84 | 201.74 ± 161.63 | 0.118 |
| $C_{max}$ (mg) | 24.38 (13.51–81.51) | 26.12 (12.91–88.63) | 0.469 |
| $T_{max}$ (h) | 24.00 (0–24.00) | 24.00 (0–24.00) | 0.363 |

Whole blood concentration of $iMg^{2+}$, concentration of total magnesium in serum, and total content of magnesium in urine collected during 24 h of $MgCl_2$ versus placebo administration are shown. Values are given as the mean ± SD or the median (range) depending on the normality of the data. AUC = the area under the curve, $C_{max}$ = maximum (or peak) concentration, and $T_{max}$ = time (in hours) at which $C_{max}$ is observed. One-sided $p$ values are shown.

## 4. Discussion

In this pilot study, we demonstrated the superiority of concentrations of $iMg^{2+}$ in blood, compared to concentrations of total magnesium in serum and total urine magnesium content, as a rapid and sensitive measure of dietary intake of magnesium in healthy humans. The finding that a single dose of 300 mg of magnesium can alter $iMg^{2+}$, but not total serum total magnesium, suggests that the $iMg^{2+}$ method is more sensitive and reflects a subject's dietary intake.

Our findings of a more robust $iMg^{2+}$ response, when compared to other magnesium biomarkers, are similar to those reported by Altura et al. [28], when measurement of $iMg^{2+}$ using ion-selective electrodes was new. In their study, they similarly administered a 300-mg dose of magnesium, but they did not have a placebo comparator; instead, participants were randomized to three different formulations of magnesium oxide. The MgO used in the previous study is known to have low bioavailability compared to the $MgCl_2$ formulation used in the current study [37]. In the current study, we also show the importance of having a placebo, which enabled us to observe circadian fluctuations of $iMg^{2+}$ as well as of serum or urinary magnesium. Another difference between the previous study by Altura et al. and the current study was that we measured $iMg^{2+}$ in whole blood that was freshly obtained, whereas the previous study measured $iMg^{2+}$ in frozen plasma. Accurate and precise measurement of $iMg^{2+}$ in blood has only recently been available in clinical practice and for research purposes. The use of more precise ion-selective electrodes allows one to make measurements of whole blood $iMg^{2+}$ within minutes. Given that $iMg^{2+}$ readings change over time of storage, measures taken from freshly obtained whole blood are likely more accurate and reliable [33]. Differences notwithstanding, our current study and the previous study by Altura et al. support the utility of $iMg^{2+}$ in bioavailability studies for measuring the impact of acute oral administration of magnesium. To determine whether a more rapid response, due to a potentially enhanced bioavailability of the $MgCl_2$ formulation (ReMag®), will be observed compared with another magnesium formulation (MgO) is an objective for the main study.

Other studies have examined the pharmacokinetics of the various Mg formulations. Blancquaert et al. conducted an in vivo study that aimed at evaluating methods that could be used in assessing Mg absorption after a single acute dosing of supplement (Ultractive Mg with 392 mg of elementary Mg vs two placebos) over six hours using blood and urine for the analysis. Results showed a significant difference for the absorption of serum total magnesium after four hours of supplement versus placebo ingestion as well as significant differences in total AUC for serum Mg. In a study by Rooney et al. [29], of chronic magnesium supplementation for 10 weeks at 400 mg magnesium per day via MgO, the findings were greater increases in both iMg$^{2+}$ concentrations in whole blood, and serum total magnesium concentrations after Mg supplementation as compared to placebo.

In the current study, because we only assessed iMg$^{2+}$ after an acute small oral dose of magnesium, we did not evaluate whether iMg$^{2+}$ is an appropriate measure of magnesium status. In the crossover, a randomized clinical trial by Rooney in which healthy participants received 10 weeks of chronic supplementation of MgO versus placebo, both fresh blood concentration of iMg$^{2+}$ and total circulating magnesium concentrations increased [29]. Although serum magnesium is used more often to assess status, the relationship between serum total magnesium and iMg$^{2+}$ in health and chronic disease has not been extensively studied. As subclinical magnesium deficiency can occur when serum total magnesium levels are within the lower reference range [38], considerations based solely on serum total magnesium levels may not be sufficient for diagnosis of deficiency. Thus, typical symptoms and data from patient history are critical when evaluating magnesium status and the diagnosis of magnesium deficiency has therefore been based on three criteria: 1) clinical symptoms, 2) patient history, and 3) laboratory analysis of serum samples [39]. Since free ionized magnesium is considered as metabolically active, assessment of whole blood iMg$^{2+}$ makes sense from a biological perspective. It may be important to screen for magnesium status in various populations who may be at risk due to low dietary intake, underlying diseases, or genetic susceptibility. Clinical investigations have demonstrated that iMg$^{2+}$, but not serum total magnesium, is depressed in a number of clinical conditions such as in patients with migraine, individuals with noninsulin-dependent diabetes, patients with asthma, and women with high-risk pregnancies [28]. Magnesium homeostasis is likely to be affected by common genetic variants, similarly to most nutrients. A recent investigation found associations between genetic variations in magnesium-related ion channel genes and type 2 diabetes risk in both African American and Hispanic American women [40]. Prior to the revision of the 1998 Dietary Reference Intakes, we strongly advocate for increased government funding to assess whether whole blood iMg$^{2+}$ better reflects the magnesium requirement in humans.

Our pilot study has some limitations. We did not exclude participants who were smokers, which may impact the assessment of iMg$^{2+}$, as thiocyanate, a product of smoking, interferes with the Nova 8 magnesium sensor [41]. We expect this issue to be minimal since our university campus, like many others, participates in a smoke-free policy [42]. We monitored participants' typical dietary intake before the start of the study but did not control the participants' dietary intake of magnesium during the study, except during testing of the acute dose of magnesium. However, even if participants consumed high intakes of magnesium prior to either the placebo or MgCl$_2$ on study days, we do not expect the background diet to impact response to the Mg supplement, given the lack of relationship between baseline serum total Mg and the iMg$^{2+}$, serum Mg, and urine Mg responses. Additionally, the study conducted by Altura et al. [28] was done in individuals who were Mg loaded, yet they were able to observe an increase in iMg$^{2+}$ after administration of the Mg dose. A larger sample size may have resulted in differences in serum total magnesium concentration and total urinary magnesium after an acute dose of magnesium compared to placebo. There is a possibility of acidosis following MgCl$_2$ administration which would promote urinary Mg loss. We did not measure blood pH in this study. However, to address potential issue of acidosis we will obtain information on blood and urine pH from the ionized selective electrode during the main study. Thus, each of the limitations of the pilot study will be overcome in the main study.

## 5. Conclusions

Whole blood concentration of iMg$^{2+}$ may be a more sensitive indicator of an acute response to a magnesium load compared to concentrations of serum total magnesium and total urinary magnesium. To use iMg$^{2+}$ as a valid nutritional biomarker or a biomarker of magnesium status, it is essential to establish a reliable reference range in a healthy population with a large sample size. This work is currently underway.

**Author Contributions:** Conceptualization, N.G.-M., S.J.B., T.C.W. and C.M.W.; methodology, N.G.-M., T.C.W. and C.M.W.; software, N.G.-M.; validation, N.G.-M., J.Z., S.C., V.A. and L.A.S.; formal analysis, N.G.-M., J.Z., S.C. and V.A.; investigation, J.Z., T.C.W., S.J.B., S.C., C.M.W. and N.G.-M.; resources, T.C.W., C.M.W. and N.G.-M.; data curation, J.Z., S.J.B., T.C.W., S.C., V.A., L.A.S., C.M.W. and N.G.-M.; writing—original draft preparation, J.Z., T.C.W., N.G.-M.; writing—review and editing, J.Z., T.C.W., S.J.B., S.C., V.A., L.A.S., C.M.W. and N.G.-M.; visualization, N.G.-M. and T.C.W.; supervision, N.G.-M. and T.C.W.; project administration, N.G.-M. and T.C.W.; funding acquisition, T.C.W. All authors have read and agreed to the published version of the manuscript.

**Funding:** Funding for this study was provided through an unrestricted educational grant from New Capstone Inc., the manufacturer of ReMag®, to Think Healthy Group Inc. The funding body had no role in the design, analysis, interpretation, or presentation of the data and results. The authors and sponsor strictly adhered to the American Society for Nutrition guiding principles for private funding for food science and nutrition research [43].

**Acknowledgments:** We are grateful for the support of Arthur D. Rosen, Professor of Clinical Neurology at Indiana University, who served as a study physician and assisted with determining participant eligibility by reviewing the laboratories of study participants from screening visits.

**Conflicts of Interest:** TCW serves on the Scientific Advisory Board for The Vitamin Shoppe and has received research grants from Pfizer Consumer Healthcare. All his conflicts are listed at www.drtaylorwallace.com. CMW is on the scientific advisory boards of the Yogurt in Nutrition Initiative (YINI) and the U.S. Food and Drug Administration and serves on the Board of Trustees of the International Life Sciences Institute (ILSI). JZ, SJB, SC, VA, LAS, and NG-M have no conflicts of interest to disclose.

## References

1. Newhouse, I.J.; Finstad, E.W. The effects of magnesium supplementation on exercise performance. *Clin. J. Sport Med.* **2000**, *10*, 195–200. [CrossRef] [PubMed]
2. Fawcett, W.J.; Haxby, E.J.; Male, D.A. Magnesium: Physiology and pharmacology. *Br. J. Anaesth.* **1999**, *83*, 302–320. [CrossRef] [PubMed]
3. Ahmed, F.; Mohammed, A. Magnesium: The forgotten electrolyte—A review on hypomagnesemia. *Med. Sci.* **2019**, *7*, E56. [CrossRef] [PubMed]
4. De Baaij, J.H.; Hoenderop, J.G.; Bindels, R.J. Magnesium in man: Implications for health and disease. *Physiol. Rev.* **2015**, *95*, 1–46. [CrossRef] [PubMed]
5. Bairoch, A. The ENZYME database in 2000. *Nucleic Acids Res.* **2000**, *28*, 304–305. [CrossRef]
6. Caspi, R.; Altman, T.; Dreher, K.; Fulcher, C.A.; Subhraveti, P.; Keseler, I.M.; Kothari, A.; Krummenacker, M.; Latendresse, M.; Mueller, L.A.; et al. The MetaCyc database of metabolic pathways and enzymes and the BioCyc collection of pathway/genome databases. *Nucleic Acids Res.* **2012**, *40*, D742–D753. [CrossRef]
7. Bara, M.; Guiet-Bara, A.; Durlach, J. Regulation of sodium and potassium pathways by magnesium in cell membranes. *Magnes. Res.* **1993**, *6*, 167–177.
8. Laban, E.; Charbon, G.A. Magnesium and cardiac arrhythmias: Nutrient or drug? *J. Am. Coll. Nutr.* **1986**, *5*, 521–532. [CrossRef]
9. Zwillinger, L. Uber die magnesiumwirking auf das. *Herz Klin Wochenschr.* **1935**, *14*, 1429–1433. [CrossRef]
10. Ford, E.S.; Mokdad, A.H. Dietary magnesium intake in a national sample of US adults. *J. Nutr.* **2003**, *133*, 2879–2882. [CrossRef]
11. Fulgoni, V.L., III; Keast, D.R.; Bailey, R.L.; Dwyer, J. Foods, fortificants, and supplements: Where do Americans get their nutrients? *J. Nutr.* **2011**, *141*, 1847–1854. [CrossRef] [PubMed]
12. Moshfegh, A.J.; Rhodes, D.G.; Baer, D.J.; Murayi, T.; Clemens, J.C.; Rumpler, W.V.; Paul, D.R.; Sebastian, R.S.; Kuczynski, K.J.; Ingwersen, L.A.; et al. The US Department of Agriculture Automated Multiple-Pass Method reduces bias in the collection of energy intakes. *Am. J. Clin. Nutr.* **2008**, *88*, 324–332. [CrossRef] [PubMed]
13. Papanikolaou, Y.; Fulgoni, V.L. Grains contribute shortfall nutrients and nutrient density to older US adults: Data from the National Health and Nutrition Examination Survey, 2011–2014. *Nutrients* **2018**, *10*, E534. [CrossRef] [PubMed]
14. McNair, P.; Christensen, M.S.; Christiansen, C.; Madsbad, S.; Transbol, I. Renal hypomagnesaemia in human diabetes mellitus: Its relation to glucose homeostasis. *Eur. J. Clin. Invest.* **1982**, *12*, 81–85. [CrossRef] [PubMed]

15. Pham, P.C.; Pham, P.M.; Pham, P.A.; Pham, S.V.; Pham, H.V.; Miller, J.M.; Yanagawa, N.; Pham, P.T. Lower serum magnesium levels are associated with more rapid decline of renal function in patients with diabetes mellitus type 2. *Clin. Nephrol.* **2005**, *63*, 429–436. [CrossRef] [PubMed]
16. Gommers, L.M.; Hoenderop, J.G.; Bindels, R.J.; de Baaij, J.H. Hypomagnesemia in type 2 diabetes: A vicious circle? *Diabetes* **2016**, *65*, 3–13. [CrossRef]
17. Goldman, L.; Schafer, A.I. *Goldman-Cecil Medicine*, 26th ed.; Elsevier Health Sciences: Philadelphia, PA, USA, 2015.
18. Dasgupta, A.; Sarma, D.; Saikia, U.K. Hypomagnesemia in type 2 diabetes mellitus. *Indian J. Endocrinol. Metab.* **2012**, *16*, 1000–1003. [CrossRef]
19. Cheungpasitporn, W.; Thongprayoon, C.; Qian, Q. Dysmagnesemia in hospitalized patients: Prevalence and prognostic importance. *Mayo Clin. Proc.* **2015**, *90*, 1001–1010. [CrossRef]
20. Hayes, J.P.; Ryan, M.F.; Brazil, N.; Riordan, T.O.; Walsh, J.B.; Coakley, D. Serum hypomagnesaemia in an elderly day-hospital population. *Ir. Med. J.* **1989**, *82*, 117–119.
21. Wong, E.T.; Rude, R.K.; Singer, F.R.; Shaw, S.T., Jr. A high prevalence of hypomagnesemia and hypermagnesemia in hospitalized patients. *Am. J. Clin. Pathol.* **1983**, *79*, 348–352. [CrossRef]
22. Whang, R.; Oei, T.O.; Aikawa, J.K.; Watanabe, A.; Vannatta, J.; Fryer, A.; Markanich, M. Predictors of clinical hypomagnesemia. Hypokalemia, hypophosphatemia, hyponatremia, and hypocalcemia. *Arch. Intern. Med.* **1984**, *144*, 1794–1796. [CrossRef] [PubMed]
23. Reinhart, R.A.; Desbiens, N.A. Hypomagnesemia in patients entering the ICU. *Crit. Care Med.* **1985**, *13*, 506–507. [CrossRef] [PubMed]
24. Ryzen, E.; Wagers, P.W.; Singer, F.R.; Rude, R.K. Magnesium deficiency in a medical ICU population. *Crit. Care Med.* **1985**, *13*, 19–21. [CrossRef] [PubMed]
25. Altura, B.M.; Altura, B.T. Role of magnesium in patho-physiological processes and the clinical utility of magnesium ion selective electrodes. *Scand. J. Clin. Lab. Invest.* **1996**, *224*, 211–234. [CrossRef] [PubMed]
26. Thode, J.; Juul-Jorgensen, B.; Seibaek, M.; Elming, H.; Borresen, E.; Jordal, R. Evaluation of an ionized magnesium-pH analyzer—NOVA 8. *Scand. J. Clin. Lab. Invest.* **1998**, *58*, 127–133. [CrossRef] [PubMed]
27. Yeh, D.D.; Chokengarmwong, N.; Chang, Y.; Yu, L.; Arsenault, C.; Rudolf, J.; Lee-Lewandrowski, E.; Lewandrowski, K. Total and ionized magnesium testing in the surgical intensive care unit-Opportunities for improved laboratory and pharmacy utilization. *J. Crit. Care* **2017**, *42*, 147–151. [CrossRef] [PubMed]
28. Altura, B.T.; Wilimzig, C.; Trnovec, T.; Nyulassy, S.; Altura, B.M. Comparative effects of a Mg-enriched diet and different orally administered magnesium oxide preparations on ionized Mg, Mg metabolism and electrolytes in serum of human volunteers. *J. Am. Coll. Nutr.* **1994**, *13*, 447–454. [CrossRef]
29. Rooney, M.R.; Rudser, K.D.; Alonso, A.; Harnack, L.; Saenger, A.K.; Lutsey, P.L. Circulating ionized magnesium: Comparisons with circulating total magnesium and the response to magnesium supplementation in a randomized controlled trial. *Nutrients* **2020**, *12*, E263. [CrossRef]
30. Blancquaert, L.; Vervaet, C.; Derave, W. Predicting and Testing Bioavailability of Magnesium Supplements. *Nutrients* **2019**, *11*, 1663. [CrossRef]
31. Wilimzig, C.; Latz, R.; Vierling, W.; Mutschler, E.; Trnovec, T.; Nyulassy, S. Increase in magnesium plasma level after orally administered trimagnesium dicitrate. *European J. Clin. Pharmacol.* **1996**, *49*, 317–323. [CrossRef]
32. Malon, A.; Brockmann, C.; Fijalkowska-Morawska, J.; Rob, P.; Maj-Zurawska, M. Ionized magnesium in erythrocytes–the best magnesium parameter to observe hypo- or hypermagnesemia. *Clin. Chim. Acta* **2004**, *349*, 67–73. [CrossRef] [PubMed]
33. Greenway, D.C.; Hindmarsh, J.T.; Wang, J.; Khodadeen, J.A.; Hebert, P.C. Reference interval for whole blood ionized magnesium in a healthy population and the stability of ionized magnesium under varied laboratory conditions. *Clin. Biochem.* **1996**, *29*, 515–520. [CrossRef]
34. Palacios, C.; Wigertz, K.; Braun, M.; Martin, B.R.; McCabe, G.P.; McCabe, L.; Pratt, J.H.; Peacock, M.; Weaver, C.M. Magnesium retention from metabolic-balance studies in female adolescents: Impact of race, dietary salt, and calcium. *Am. J. Clin. Nutr.* **2013**, *97*, 1014–1019. [CrossRef] [PubMed]
35. Lowenstein, F.W.; Stanton, M.F. Serum magnesium levels in the United States, 1971–1974. *J. Am. Coll. Nutr.* **1986**, *5*, 399–414. [CrossRef]
36. Norbert Wolfgang Tietz. *Clinical Guide to Laboratory Tests*; W.B. Saunders Company: Philadelphia, PA, USA, 1990.

37. Firoz, M.; Graber, M. Bioavailability of US commercial magnesium preparations. *Magnes. Res.* **2001**, *14*, 257–262.
38. Von Ehrlich, B.; Barbagallo, M.; Classen, H.G.; Guerrero-Romero, F.; Mooren, F.C.; Rodriguez-Moran, M.; Vierling, W.; Vormann, J.; Kisters, K. Significance of magnesium in insulin resistance, metabolic syndrome, and diabetes–recommendations of the Association of Magnesium Research e.V. *Trace Elem. Electrolytes* **2017**, *34*, 124. [CrossRef]
39. Costello, R.B.; Elin, R.J.; Rosanoff, A.; Wallace, T.C.; Guerrero-Romero, F.; Hruby, A.; Lutsey, P.L.; Nielsen, F.H.; Rodriguez-Moran, M.; Song, Y.; et al. Perspective: The case for an evidence-based reference interval for serum magnesium: The time has come. *Adv. Nutr.* **2016**, *7*, 977–993. [CrossRef]
40. Chan, K.H.; Chacko, S.A.; Song, Y.; Cho, M.; Eaton, C.B.; Wu, W.C.; Liu, S. Genetic variations in magnesium-related ion channels may affect diabetes risk among African American and Hispanic American women. *J. Nutr.* **2015**, *145*, 418–424. [CrossRef]
41. McHale, J. Thiocyanate interference with Nova's ionized magnesium electrode. *Clin. Chem.* **1997**, *43*, 1672. [CrossRef]
42. Wong, S.L.; Epperson, A.E.; Rogers, J.; Castro, R.J.; Jackler, R.K.; Prochaska, J.J. A multimodal assessment of tobacco use on a university campus and support for adopting a comprehensive tobacco-free policy. *Prev. Med.* **2020**, *133*, 106008. [CrossRef]
43. Rowe, S.; Alexander, N.; Clydesdale, F.M.; Applebaum, R.S.; Atkinson, S.; Black, R.M.; Dwyer, J.T.; Hentges, E.; Higley, N.A.; Lefevre, M.; et al. Funding food science and nutrition research: Financial conflicts and scientific integrity. *J. Nutr.* **2009**, *139*, 1051–1053. [CrossRef] [PubMed]

© 2020 by the authors. Licensee MDPI, Basel, Switzerland. This article is an open access article distributed under the terms and conditions of the Creative Commons Attribution (CC BY) license (http://creativecommons.org/licenses/by/4.0/).

Article

# Magnesium Deficiency Questionnaire: A New Non-Invasive Magnesium Deficiency Screening Tool Developed Using Real-World Data from Four Observational Studies

Svetlana Orlova [1], Galina Dikke [2], Gisele Pickering [3], Sofya Konchits [4], Kirill Starostin [4,*] and Alina Bevz [4]

1. Department of Dietetics and Clinical Nutritiology of Continuing Medical Education, Medical Institute, RUDN University, 119571 Moscow, Russia; rudn_nutr@mail.ru
2. Department of Obstetrics and Gynecology with a Course of Reproductive Medicine, The Academy of Medical Education F.I. Inozemtsev, 190013 Saint-Petersburg, Russia; galadikke@yandex.ru
3. Department of Clinical Pharmacology, University Hospital and Inserm 1107 Fundamental and Clinical Pharmacology of Pain, Medical Faculty, F-63000 Clermont-Ferrand, France; gisele.pickering@uca.fr
4. Medical Affairs, Sanofi, 125009 Moscow, Russia; sofya.konchits@sanofi.com (S.K.); alina.bevz@sanofi.com (A.B.)
* Correspondence: kirill.starostin@sanofi.com; Tel.: +7-495-7211400

Received: 18 June 2020; Accepted: 8 July 2020; Published: 11 July 2020

**Abstract:** Due to the high estimated prevalence of magnesium deficiency, there is a need for a rapid, non-invasive assessment tool that could be used by patients and clinicians to confirm suspected hypomagnesemia and substantiate laboratory testing. This study analyzed data from four large observational studies of hypomagnesemia in pregnant women and women with hormone-related conditions across Russia. Hypomagnesemia was assessed using a 62-item magnesium deficiency questionnaire (MDQ-62) and a serum test. The diagnostic utility (sensitivity/specificity) of MDQ-62 was analyzed using area under the receiver operating characteristic curve (AUROC). A logistic regression model was applied to develop a shorter, optimized version of MDQ-62. A total of 765 pregnant women and 8836 women with hormone-related conditions were included in the analysis. The diagnostic performance of MDQ-62 was "fair" (AUROC = 0.7–0.8) for women with hormone-related conditions and "poor" for pregnant women (AUROC = 0.6–0.7). The optimized MDQ-23 (23 questions) and MDQ-10 (10 questions) had similar AUROC values; for all versions of the questionnaire, there was a significant negative correlation between score and changes in total serum magnesium levels ($p < 0.0001$ for all comparisons; correlation coefficients ranged from $-0.1667$ to $-0.2716$). This analysis confirmed the value of MDQ in identifying women at risk of hypomagnesemia.

**Keywords:** hypomagnesemia; magnesium deficiency questionnaire; pregnancy; hormone-related conditions

## 1. Introduction

Magnesium deficiency and low magnesium intake are associated with altered levels of other electrolytes, cardiovascular events, various metabolic and neuromuscular conditions, type II diabetes mellitus, and depression [1]. Studies of various populations showed that 15–42% of apparently healthy adults have subnormal serum magnesium levels; magnesium deficiency is more frequent in women than in men, and the proportions are much higher in post-menopausal women and in individuals with obesity or type 2 diabetes [1,2].

Despite its implications in clinical practice, diagnosing magnesium deficiency still presents a challenge. To date, there is no gold standard for how to determine magnesium levels. Most of the magnesium tests used in research, such as 24-hour urine magnesium load test, muscle and bone concentration measurements, nuclear magnetic resonance imaging, and isotope studies, are considered impractical in the clinic [3]. The measurement of total magnesium concentration in serum using spectrophotometry with titan yellow or xylidyl blue is the most practical and reliable approach; however, there is an argument that the results can only serve as an approximation of magnesium concentrations in tissues [1–3]. The optimal cut-off serum concentration that indicates magnesium deficiency is also a matter of ongoing debate. In Russia and other countries, the most commonly used lower reference limit is 0.66 mmol/L [4]. However, the largest historical study in the US identified 0.75 mmol/L as the reference limit, and recent publications suggest that the cut-off should be set at 0.8 mmol/L [2,3,5].

Only about 0.8% of total body magnesium is present in blood, with the rest stored in soft tissue (19%), muscle (27%), and bone (53%), meaning that magnesium deficiency can be present even when serum magnesium levels are in the normal range [1,3,6–8]. Conversely, a low serum magnesium level is a clear indication of overall magnesium deficiency. However, due to the lack of specific symptoms, magnesium deficiency is rarely suspected in the clinic, and furthermore, serum levels or other tests (e.g., magnesium level in red blood cells) may not be reimbursed [3]. For these reasons, in many countries including Russia, magnesium tests are rarely reimbursed by insurance companies, emphasizing the need for a simple, reliable, and affordable screening tool to help identify magnesium deficiency.

Magnesium deficiency can be suspected and diagnosed with the help of the 'magnesium deficiency questionnaire' (MDQ-62), consisting of 62 questions that can be grouped into 5 general categories: wellbeing, lifestyle, pregnancy, disease, and medication [9]. Although MDQ-62 may help identify non-specific symptoms frequently accompanying magnesium deficiency, the questionnaire is cumbersome and time-consuming, and several questions are very similar. It is currently unclear to what extent MDQ-62 scores correlate with total magnesium serum concentration and whether it can be a reliable surrogate for laboratory values.

Four large observational studies conducted between 2012 and 2016 across multiple regions and cities in the Russian Federation assessed the prevalence and clinical management of magnesium deficiency in pregnant women (MAGIC, MAGIC2) and in women with hormone-related conditions (MAGYN, MAGYN2) using MDQ-62 and laboratory tests [10–13]. Here, we report the results of a secondary analysis using pooled data from these studies, designed to describe the prevalence of magnesium deficiency in these populations and to identify associated risk factors and comorbidities. Another key objective of this analysis was to develop shorter, optimized versions of the questionnaire that would offer the same level of accuracy in identifying suspected hypomagnesemia.

## 2. Materials and Methods

This manuscript summarizes a part of the secondary analysis of pooled data collected in four observational studies of magnesium deficiency in pregnant women and women with hormone-related conditions: MAGIC (DIREGL06157), MAGIC2 (DIREGL06468), MAGYN (MAGNEL06863), and MAGYN2 (MAGNEL07741) [10–13].

MAGIC and MAGIC2 enrolled pregnant women ($N = 1130$ and $N = 2117$, respectively) during routine visits to maternity welfare centers. Women were included in the studies if they were >18 years of age, were pregnant, and had suspected magnesium deficiency (fatigue, muscle cramps, etc.). The study excluded women who reported other known or obvious reasons for magnesium deficiency beside pregnancy [11,13]. MAGYN and MAGYN2 studies enrolled women with hormone-related conditions ($N = 9168$ and $N = 11,424$, respectively) attending outpatient clinics. Women were included if they were 18–60 years of age and used hormonal contraception or hormone replacement therapy (HRT) or had one of the following conditions: premenstrual syndrome (PMS), climacteric syndrome without HRT, osteoporosis, or other hormonal conditions (including endometriosis, polycystic ovarian disease, uterine leiomyoma, algodysmenorrhea, endometrial hyperplastic processes). Women were excluded if they had severe conditions potentially hindering their participation in the study or were receiving magnesium supplementation at baseline [10,12].

The present analysis included all patients who fulfilled the inclusion/exclusion criteria in the studies (Figure 1). Patients with missing data, contradictory/inconsistent data, or outlier data were excluded from the analysis (exclusion was performed separately for each variable of interest). Pooled databases were created for patient populations of "pregnant women" (MAGIC and MAGIC2) and 'women with hormone-related conditions' (MAGYN and MAGYN2).

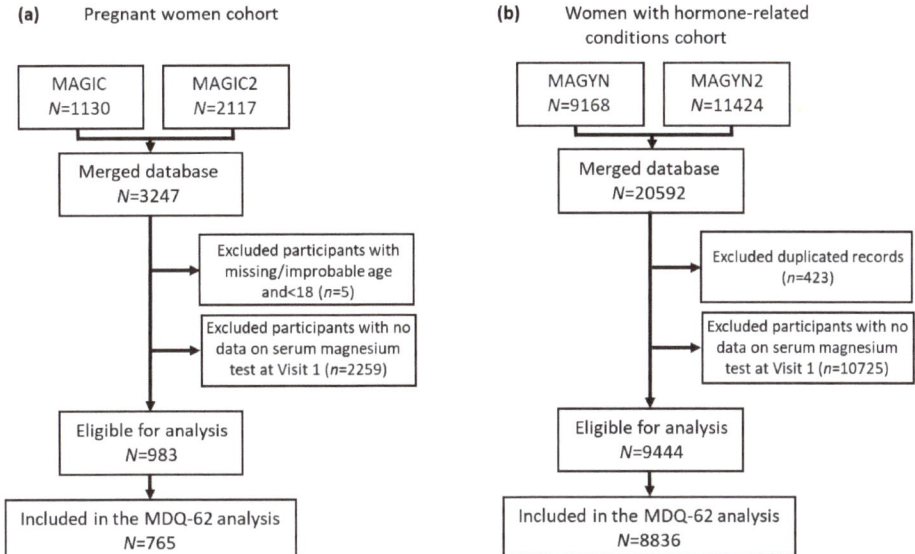

**Figure 1.** Study population: (**a**) Pregnant women cohort; (**b**) Women with hormone-related conditions cohort.

### 2.1. Study Visits and Treatment

MAGIC, MAGIC2, MAGYN, and MAGYN2 were observational studies; during these studies, all treatment decisions were made by the treating physicians.

Epidemiological data were collected at baseline (Visit 1) for all participants. Patients with low serum magnesium at this visit or with suspected deficiency based on MDQ-62 attended Visit 2 and underwent a second assessment of serum magnesium. Visit 2 was scheduled by the treating physicians according to their standard practice; in MAGYN and MAGYN2, it occurred approximately 4 weeks after Visit 1.

The analysis of the effectiveness of magnesium supplementation included participants who had mild hypomagnesemia at Visit 1 (serum levels above 0.5 mmol/L but below 0.66 mmol/L or 0.8 mmol/L, depending on the specific cut-off used) and who were prescribed a combination of magnesium and vitamin B6, Magne B6® or Magne B6 Forte® (Sanofi), for approximately 4 weeks. Patients who received other types of magnesium supplementation were excluded from the analysis.

### 2.2. Study Objectives and Endpoints

The objectives of this secondary analysis were (1) to assess the sensitivity and specificity of MDQ-62; (2) to develop a shortened version of MDQ using regression modelling; (3) to assess the sensitivity and specificity of this new shortened version of MDQ, and (4) to analyze the ability of MDQ and shortened MDQ to reflect dynamic changes in serum magnesium level.

The following research questions were assessed.

Question 1: What are the sensitivity and specificity of MDQ-62 when detecting hypomagnesemia in women with hormone-related conditions, using serum magnesium level cut-offs of 0.66 mmol/L or 0.8 mmol/L?

Question 2: What are the sensitivity and specificity of MDQ-62 when detecting hypomagnesemia in pregnant women, using serum magnesium level cut-offs of 0.66 mmol/L or 0.8 mmol/L?

Question 3: Can the number of questions in MDQ be reduced without a loss of diagnostic quality?

Question 4: Is it possible to detect changes in the serum magnesium level using MDQ-62 or shortened MDQ?

*2.3. Questionnaires*

The MDQ-62, consisting of 62 questions, was developed previously and adapted for use in the Russian Federation [9,11]. Each question contributed 2–5 points to the overall score. A score of 51 points or more was used as an indication that magnesium deficiency was 'highly probable', whereas a score of 30–50 points was interpreted as 'likely' magnesium deficiency [9,11].

To optimize the questionnaire, each of 62 questions was tested for contribution to the total MDQ score as well as for association with the magnesium level using the population of women with hormone-related conditions as the training sample (see Statistical analysis for further details). Questions independently associated with hypomagnesemia were selected and assembled in the modified MDQ. The modified shortened MDQ was tested on the pregnant women cohort (testing sample).

The diagnostic performance of MDQ-62 and modified MDQ was analyzed in women who had serum magnesium level data and questionnaire data at Visit 1.

*2.4. Statistical Analysis*

The sensitivity and specificity of MDQ-62 were analyzed using area under the receiver operating characteristic curve (AUROC), MDQ cut-off, positive and negative predictive values, and the likelihood ratio. Statistical significance of AUROC was analyzed using a Mann–Whitney test.

For the development of the modified MDQ, stepwise multiple regression was applied to obtain a shortened version of the questionnaire. All 62 questions of MDQ-62 were included and excluded into the testing empty model step by step, using a bidirectional stepwise selection approach and assessing explanatory capabilities each time after inclusion or exclusion of a variable with the following parameters or variable to select and to keep: stepwise slEntry = 0.00001, slStay = 0.1, accordingly (software SAS version 9.4, SAS Institute Inc., Cary, NC, USA). The model was developed for both cut-offs, <0.66 mmol/L and <0.8 mmol/L. The quality of the final model was also analyzed and compared with the quality of the initial model.

Optimization was considered to be achieved for the modified MDQ if it consisted of a smaller number of questions than MDQ-62, if its sensitivity was equal or higher than that of MDQ-62, and if the difference in specificity between MDQ-62 and the modified MDQ was equal to or lower than 20%. Alternatively, optimization was considered to be achieved if the modified MDQ consisted of 15 questions, and if the difference in specificity between MDQ-62 and the modified MDQ was equal to or lower than 10%.

The training sample consisted of the cohort of women with hormone-related conditions, and the test sample consisted of the pregnant women cohort.

The sensitivity and specificity of the modified MDQs were analyzed using the same approach as for MDQ-62.

Correlation coefficients (r) were calculated based on changes from baseline to week 4 in terms of MDQ-62/modified MDQ scores and serum magnesium concentration and analyzed as continuous or dichotomous variables (hypomagnesemia: yes/no). Sensitivity and specificity were estimated using standard formulae [14]. The absolute magnesium concentration was defined as reference.

The study group characteristics (such as demographics) were analyzed using descriptive statistics, and differences were analyzed using the chi square test, unpaired t-test, and non-parametric tests. Statistical significance threshold was set at $p < 0.05$.

## 3. Results

### 3.1. Study Population

A flow chart of the study population is presented in Figure 1 (duplicated records, subjects without serum magnesium test, and subjects <18 years old were excluded from the analysis). In total, 983 participants in the "pregnant women" cohort and 9444 participants in the "women with hormone-related conditions" cohort were eligible for analysis. Participants in the "pregnant women" cohort had a median age of 28.0 years (range 18–52 years) and a mean (SD) serum magnesium level of 0.714 (0.125) mmol/L (range 0.12–1.92 mmol/L). Participants in the "women with hormone-related conditions" cohort had a median age of 44.0 years (range 18–60 years) and a mean (SD) serum magnesium level of 0.776 (0.198) mmol/L (range 0.08–4.08 mmol/L).

The prevalence of magnesium deficiency assessed by serum levels in pregnant women was 34.0%/78.9% when using 0.66/0.8 mmol/L, respectively, as the cut-off. In women with hormone-related conditions, the prevalence was 24.1%/54.8% when using 0.66/0.8 mmol/L, respectively, as the cut-off (Figure 1). After taking magnesium supplements for four weeks, a large proportion of women in both cohorts was able to achieve a magnesium level above the target cut-offs (Figure 2).

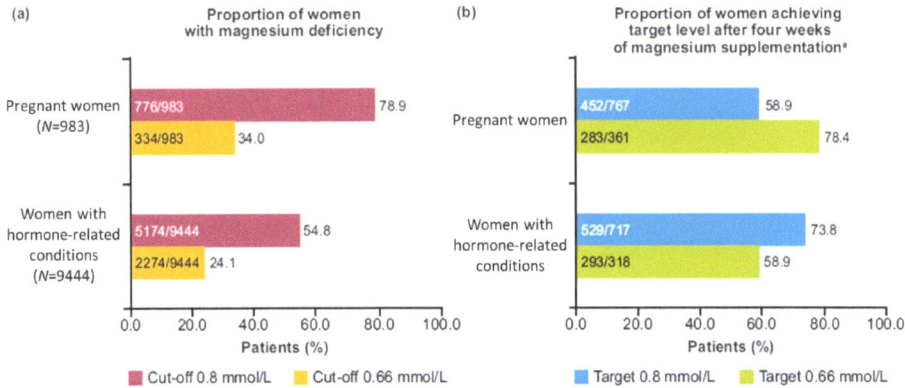

**Figure 2.** Prevalence and clinical management of magnesium deficiency: (**a**) Proportion of women with magnesium deficiency at Visit 1; (**b**) Proportion of women achieving the target level after four weeks of magnesium supplementation. [a] Includes women who had a magnesium serum level below the corresponding target at baseline.

### 3.2. Diagnostic Performance of MDQ-62

The diagnostic performance of MDQ was examined separately in pregnant women and in women with hormone-related conditions using two different cut-offs of magnesium serum levels: 0.8 mmol/L and 0.66 mmol/L. The analysis included 765 pregnant women and 8836 women with hormone-related conditions who both had the results of total serum magnesium test and had filled the MDQ-62 at Visit 1.

In pregnant women, using the cut-off of 0.8 mmol/L, AUROC for MDQ-62 was 0.6301 (standard error [SE] = 0.0251; 95% confidence interval [CI]: 0.5810–0.6792; $p < 0.0001$); using the cut-off of 0.66 mmol/L, AUROC for MDQ-62 was 0.6446 (SE = 0.0210; 95% CI: 0.6036–0.6857; $p < 0.0001$).

The positive predictive value of MDQ-62 was slightly better with the cut-off 0.66 mmol/L than that obtained with the cut-off of 0.8 mmol/L (Supplementary Tables S1 and S2).

In women with hormone-related conditions, using the cut-off of 0.8 mmol/L, AUROC for MDQ-62 was 0.7893 (SE = 0.0049; 95% CI: 0.7797–0.7990; $p < 0.0001$); using the cut-off of 0.66 mmol/L, AUROC for MDQ-62 was 0.7412 (SE = 0.0057; 95% CI: 0.7300–0.7524; $p < 0.0001$) (Figure 3). For this group, the positive predictive value was slightly worse with the cut-off of 0.66 mmol/L compared to that with the cut-off of 0.8 mmol/L (Supplementary Tables S1 and S2).

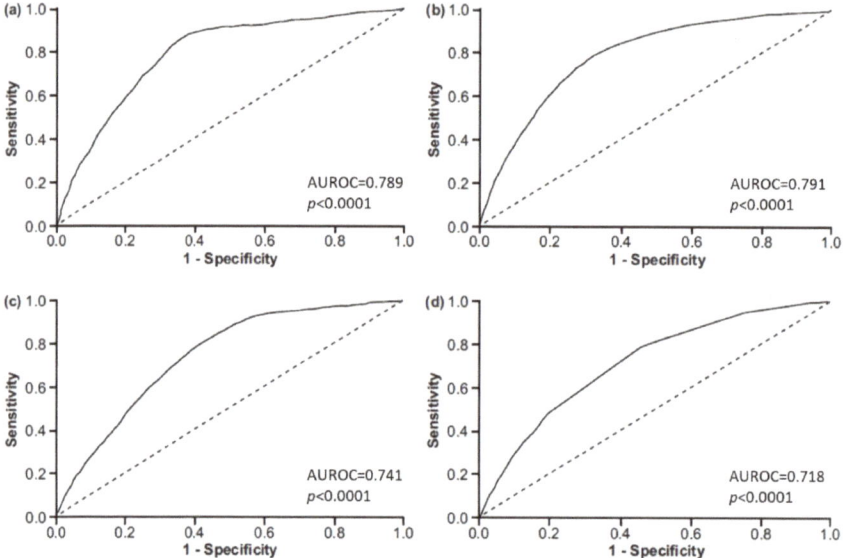

**Figure 3.** AUROC curves showing the questionnaires' ability to predict hypomagnesemia in women with hormone-related conditions for: (**a**) MDQ-62 using the cut-off <0.8 mmol/L; (**b**) MDQ-62 using the cut-off <0.66 mmol/L; (**c**) MDQ-23using the cut-off <0.8 mmol/L; (**d**) MDQ-10 using the cut-off <0.66 mmol/L. AUROC, area under the receiver operating characteristic; MDQ, magnesium deficiency questionnaire; The presented $p$ values are related to ROC curves implying the H0 hypothesis: AUC = 0.5.

In all analyses, an MDQ-62 cut-off ≥51 points provided better specificity, worse sensitivity, better positive likelihood ratio, and worse negative likelihood ratio compared with an MDQ-62 cut-off ≥30. Formally, based on the previously published rough classification system, MDQ-62 had a "poor" (AUROC 0.6–0.7) diagnostic value in pregnant women and a "fair" (AUROC 0.7–0.8) diagnostic value in women with hormone-related conditions [15].

### 3.3. Development of Modified MDQs

Each of the 62 MDQ questions was tested for contribution to the total MDQ score as well as for association with the serum magnesium level in women with hormone-related conditions (training sample), using both cut-offs of <0.8 mmol/L and <0.66 mmol/L. For each of the cut-offs, a modified MDQ was developed that contained questions that showed statistically significant correlation with hypomagnesemia defined as the corresponding cut-off (Supplementary Table S3).

MDQ-23 contained 23 questions that showed a significant correlation with hypomagnesemia defined as <0.8 mmol/L (Supplementary Table S3); its total score range was 0–41, and the optimal cut-off value was >9. MDQ-10 contained 10 questions that showed a significant correlation with hypomagnesemia defined as <0.66 mmol/L (Supplementary Table S4); its total score range was 0–31, and the optimal cut-off value was >5.

The diagnostic performance of the modified MDQs to detect hypomagnesemia at the corresponding cut-offs was tested on the data from pregnant women with symptoms of magnesium deficiency (testing sample).

3.4. Diagnostic Performance of MDQ-23 and MDQ-10

The diagnostic performance of MDQ-23 was compared with that of MDQ-62 using the total serum magnesium level cut-off of 0.8 mmol/L (Table 1; Figure 3). MDQ-23 had similar AUROC, higher sensitivity, and lower specificity than MDQ-62 at cut-off ≥30; the 95% CIs overlapped, suggesting that the slight differences were not statistically significant.

Table 1. Diagnostic performance of MDQ-23 compared with that of MDQ-62 using a serum magnesium level cut-off of 0.8 mmol/L.

|  | MDQ-62 Score ≥ 30 |  | MDQ-62 Score ≥ 51 |  | MDQ-23 Score > 9 |  |
|---|---|---|---|---|---|---|
|  | Value | 95% CI | Value | 95% CI | Value | 95% CI |
| Training Sample: Women with Hormone-Related Conditions | | | | | | |
| Sensitivity | 0.884 | 0.874–0.893 | 0.248 | 0.236–0.260 | 0.888 | 0.879–0.897 |
| Specificity | 0.616 | 0.601–0.631 | 0.938 | 0.930–0.945 | 0.524 | 0.508–0.539 |
| AUROC | 0.789 | 0.780–0.799 | 0.789 | 0.780–0.799 | 0.791 | 0.781–0.800 |
| $p$ value |  | <0.001 |  |  |  | <0.001 |
| Testing Sample: Pregnant Women with Symptoms of Magnesium Deficiency | | | | | | |
| Sensitivity | 0.845 | 0.815–0.872 | 0.283 | 0.248–0.320 | 0.864 | 0.835–0.890 |
| Specificity | 0.235 | 0.165–0.316 | 0.864 | 0.793–0.917 | 0.197 | 0.133–0.275 |
| AUROC | 0.630 | 0.581–0.679 | 0.630 | 0.581–0.679 | 0.610 | 0.560–0.659 |

CI, confidence interval; the presented $p$ values are related to ROC curves implying the H0 hypothesis: AUC = 0.5.

The diagnostic performance of MDQ-10 was compared with that of MDQ-62 using the total serum magnesium level cut-off of 0.66 mmol/L (Table 2; Figure 3). MDQ-10 had a similar AUROC to that of MDQ-62 at score cut-off ≥30 and a higher sensitivity than MDQ-62 in the training sample, but a slightly lower sensitivity in the testing sample and a lower specificity in both samples; the 95% CIs overlapped, suggesting that the detected differences were not statistically significant.

Table 2. Diagnostic performance of MDQ-10 compared with that of MDQ-62 using a serum magnesium level cut-off of 0.66 mmol/L.

|  | MDQ-62 Score ≥ 30 |  | MDQ-62 Score ≥ 51 |  | MDQ-10 Score > 5 |  |
|---|---|---|---|---|---|---|
|  | Value | 95% CI | Value | 95% CI | Value | 95% CI |
| Training Sample: Women with Hormone-Related Conditions | | | | | | |
| Sensitivity | 0.927 | 0.915–0.938 | 0.310 | 0.290–0.331 | 0.940 | 0.929–0.950 |
| Specificity | 0.427 | 0.415–0.439 | 0.881 | 0.873–0.889 | 0.260 | 0.250–0.271 |
| AUROC | 0.741 | 0.730–0.752 | 0.741 | 0.730–0.752 | 0.718 | 0.706–0.730 |
| $p$ value |  | <0.001 |  |  |  | <0.001 |
| Testing Sample: Pregnant Women with Symptoms of Magnesium Deficiency | | | | | | |
| Sensitivity | 0.887 | 0.843–0.922 | 0.358 | 0.301–0.418 | 0.869 | 0.823–0.906 |
| Specificity | 0.200 | 0.165–0.238 | 0.798 | 0.760–0.833 | 0.171 | 0.139–0.207 |
| AUROC | 0.645 | 0.604–0.686 | 0.645 | 0.604–0.686 | 0.610 | 0.568–0.652 |

The presented $p$ values are related to ROC curves implying the H0 hypothesis: AUC = 0.5.

The ability of MDQ-23 and MDQ-10 to detect changes in total serum magnesium level was determined by estimating the correlation between serum magnesium level change and questionnaire score change from Visit 1 to Visit 2. In total, 765 pregnant women and 933 women with hormone-related conditions were included in the analysis (Table 3). In both datasets, statistically significant negative correlations were observed ($p < 0.0001$); however, the correlation coefficients were low (ranging from −0.1667 to −0.2716). For all tested questionnaires, the correlation coefficients were higher for women with hormone-related conditions than for pregnant women.

Table 3. Correlation between changes from Visit 1 to Visit 2 for total serum magnesium level and for initial (MDQ-62) and modified (MDQ-23 and MDQ-10) questionnaire scores.

|  | Women with Hormone-Related Conditions ($N$ = 933) | | Pregnant Women with Symptoms of Magnesium Deficiency ($N$ = 765) | |
| --- | --- | --- | --- | --- |
|  | r | p Value | r | p Value |
| MDQ-62 | −0.2716 | <0.0001 | −0.2253 | <0.0001 |
| MDQ-23 | −0.2890 | <0.0001 | −0.1729 | <0.0001 |
| MDQ-10 | −0.2072 | <0.0001 | −0.1667 | <0.0001 |

r, Pearson correlation coefficient. The presented $p$ values are related to the Pearson correlation coefficient implying the H0 hypothesis: r = 0.

## 4. Discussion

This study is one of the largest and the most comprehensive real-world studies of magnesium deficiency in women. The study population consisted of pregnant women and women with hormone-related conditions from multiple cities and regions of the Russian Federation, providing a wide geographical coverage and a large sample size (a total of 10,427 women).

Among various magnesium tests, total serum concentration is considered to be the most practical for use in the clinic; however, the prevalence of magnesium deficiency may be underestimated if only blood levels are used [1–3,8]. Because of these considerations and the lack of clinical symptoms that specifically indicate magnesium deficiency, serum magnesium tests are rarely ordered and often not reimbursed by insurance companies [3]. There is an unmet need for a non-invasive assessment method that could be used as a surrogate for a clinical diagnosis of magnesium deficiency and at the very least serve as the basis for more detailed laboratory investigations.

This study measured the predictive ability of an existing non-invasive magnesium deficiency screening tool (MDQ-62). It was classified as a "fairly predictive" diagnostic tool for women with hormone-related conditions based on the rough classification of AUROC values [16]; nevertheless, it provided a clinically useful estimate of the magnesium status that may help physicians to identify a possible magnesium deficiency and provide a basis for laboratory testing. This questionnaire may also be used by the general populations to raise or dispel a possible suspicion of magnesium deficiency. The study used the existing MDQ-62 [9] to develop two new short questionnaires (MDQ-23 at cut-off <0.8 mmol/L and MDQ-10 at cut-off <0.66 mmol/L) that have nearly the same diagnostic performance as the original MDQ-62 when assessing likely magnesium deficiency but contain fewer questions and are therefore less time-consuming and easier to administer in routine clinical practice (the initial MDQ-62 and the modified MDQ-23 and MDQ-10 questionnaires with individual questions' scores are presented below (Table 4).

Having a quick non-invasive cost-free method that could provide further evidence in cases of suspected magnesium deficiency would be very valuable for healthcare providers, especially where there are very few symptoms, and the need for serum testing is otherwise unclear. In this study, the initial long and cumbersome questionnaire was shortened by almost three-fold without losing sensitivity and with a minimal loss of specificity (less than 20%). Both MDQ-23 and MDQ-10 scores showed a statistically significant negative correlation with changes in serum magnesium levels, demonstrating their potential clinical utility in everyday practice as a tool to identify suspected cases of magnesium deficiency.

**Table 4.** Initial (MDQ-62) and modified (MDQ-23, MDQ-10) questionnaires with individual questions' scores.

| Question Number in MDQ-62 | Question | Question Score in MDQ-62 | Question Score in MDQ-23 | Question Score in MDQ-10 |
|---|---|---|---|---|
| Q1 | Excessive emotional stress | 2 | 2 | 3 |
| Q2 | Irritable, or easily provoked to anger | 3 | 2 | 3 |
| Q3 | Restless, or hyperactive | 2 | - | - |
| Q4 | Easily startled by sound or light | 4 | - | - |
| Q5 | Insomnia | 2 | 1 | 3 |
| Q6 | Chronic headache or migraine | 3 | - | - |
| Q7 | Convulsions | 2 | 2 | - |
| Q8 | Tremor or shakiness in the hands | 3 | - | - |
| Q9 | Fine, barely noticeable muscle twitching around your eyes, facial muscles, or other muscles of your body | 3 | 1 | - |
| Q10 | Muscle spasms | 3 | - | - |
| Q11 | Muscle spasms in hands or feet | 3 | 2 | - |
| Q12 | Gag or choke from spasms in your esophagus (food tube) | 4 | 2 | - |
| Q13 | Asthma, short breathing, rales | 3 | - | - |
| Q14 | Emphysema, chronic bronchitis, or high respiratory rate (tachypnea) | 2 | - | - |
| Q15 | Osteoporosis | 5 | - | - |
| Q16 | Kidney stone disease (urolithiasis) | 3 | - | 3 |
| Q17 | Chronic kidney disease | 2 | 2 | - |
| Q18 | Diabetes | 4 | - | - |
| Q19 | Hyperfunction of the thyroid or parathyroid gland | 3 | - | - |
| Q20 | High blood pressure | 3 | - | - |
| Q21 | Mitral valve prolapse ("floppy heart valve") | 4 | 4 | 4 |
| Q22 | Tachycardia, irregular heartbeat, or arrhythmia | 3 | - | - |
| Q23 | Chronic bowel disease, ulcerative colitis, Crohn's disease or irritable bowel syndrome | 3 | - | - |
| Q24 | Frequent diarrhea or constipation | 3 | 2 | 3 |
| Q25 | Suffer from premenstrual syndrome or menstrual cramps | 3 | - | - |
| Q26 | Pregnant or recently pregnant | 2 | - | - |
| Q27 | Take Digitalis (Digoxin) | 3 | - | - |
| Q28 | Take any kind of diuretic | 5 | 1 | - |
| Q29 | Recent radiation therapy or other type of radiation exposure | 5 | - | - |
| Q30 | Have more than seven alcohol drinks weekly | 4 | 1 | - |
| Q31 | Problems with excessive alcohol intake | 3 | - | - |
| Q32 | Take more than three portions of caffeine-containing drinks daily | 2 | - | - |
| Q33 | Consumption of sugar-containing products | 2 | - | - |
| Q34 | Crave carbohydrates and/or chocolate | 2 | 2 | - |
| Q35 | Crave salt and/or salt products | 2 | - | - |
| Q36 | Eat a high-processed food/fast food diet | 2 | - | - |
| Q37 | Eat a diet low in greens, leafy vegetables, seeds, and fresh fruit | 2 | 1 | - |
| Q38 | Eat a low-protein diet | 2 | - | - |
| Q39 | Presence of undigested food or fat in feces | 2 | - | - |
| Q40 | High blood pressure or pre-eclampsia in previous pregnancy | 4 | - | - |
| Q41 | Chronic fatigue | 2 | 1 | - |
| Q42 | Muscle weakness | 2 | 2 | - |
| Q43 | Feeling of cold hands and/or feet | 2 | - | - |
| Q44 | Numbness in face, hands, or feet | 2 | - | - |
| Q45 | Persistent tingling in the body | 2 | 2 | 3 |
| Q46 | Feeling of chronic indifference or apathy | 2 | - | 4 |
| Q47 | Poor memory | 2 | - | - |
| Q48 | Loss of concentration | 2 | 1 | - |
| Q49 | Anxiety | 3 | 2 | - |
| Q50 | Chronic depression for no apparent reason | 2 | 2 | - |
| Q51 | Feelings of disorientation as to time or place | 2 | - | - |
| Q52 | Feeling depressed, lack of personal identity | 2 | - | - |
| Q53 | Hallucinations | 2 | - | - |
| Q54 | Feeling of persecution and hostility from others | 2 | - | - |
| Q55 | Pale and puffy face or poor, bad complexion | 2 | - | - |
| Q56 | Loss of considerable sexual energy or vitality | 2 | 2 | - |

Table 4. Cont.

| Question Number in MDQ-62 | Question | Question Score in MDQ-62 | Question Score in MDQ-23 | Question Score in MDQ-10 |
|---|---|---|---|---|
| Q57 | Been told by your attending doctor that your blood calcium is low | 2 | - | 3 |
| Q58 | Been told by your attending doctor that your blood potassium is low | 3 | - | - |
| Q59 | Take calcium supplements regularly without magnesium | 2 | 3 | 2 |
| Q60 | Take iron or zinc supplements regularly without magnesium | 2 | - | - |
| Q61 | Chronic fluoride intake | 2 | - | - |
| Q62 | Frequently use antibiotics, steroids, oral contraceptives, indomethacin, cisplatin, amphotericin B, cholestyramine, synthetic estrogens | 3 | 1 | - |

The symptoms potentially related to magnesium deficiency and included into the MDQ-62 questionnaire are non-specific and may be related to a number of various medical disorders, for instance, B-vitamins deficiencies, low blood calcium, alcohol abuse [16]. Moreover, some of the medical conditions mentioned in the initial MDQ-62 questionnaire (for instance, diabetes) may contribute to some of these symptoms independently of the Mg status [9,17]. Knowing that we did not expect from the beginning a high specificity from the questionnaire based on subjective self-assessment and while establishing research questions, we consciously concentrated our efforts on keeping maximal sensitivity, setting it as a priority in questionnaire optimization. We chose this particular approach to develop a fast non-invasive screening tool to catch most true positive cases implying further laboratory diagnosis verification. The price in this case is usually false positive cases with symptoms/complaints related to other medical conditions and, therefore, a lower specificity of the questionnaire.

MDQ-23 and MDQ-10 showed a slightly worse performance in pregnant women than in women with hormone-related conditions, possibly because some symptoms or signs assessed in the questionnaire may be caused by pregnancy itself rather than by an underlying magnesium deficiency, for example, emotional stress, irritability, or frequent constipation. Another potential reason is that this study used the same questionnaire and cut-offs for pregnant women and women with hormone-related conditions; however, pregnancy may require a different questionnaire design and validation. Finally, pregnancy-related magnesium deficiency may be associated with specific enhanced needs of mother and child, and the natural course of hypomagnesemia may differ in this group of patients. Future studies in pregnant women would be useful to determine whether the questionnaire could be modified further to increase its sensitivity and specificity for this specific population.

The four observational studies (MAGIC, MAGIC2, MAGYN, and MAGYN2) and the analysis presented here were not designed to validate MDQ-62; therefore, it may not be possible to directly extrapolate the results obtained here to the general population without additional validation procedures. One should also bear in mind that the serum levels of magnesium may be misleading, as they do not always accurately reflect the levels in soft tissues and bones; on the other hand, magnesium deficiency can be asymptomatic. Consequently, MDQ and laboratory testing cannot fully replace each other in the assessment of the magnesium status, and the complementary use of both should be considered. It may be advisable to use them step by step, for example to use the questionnaire first to corroborate a suspected deficiency and then follow up with a serum magnesium test.

To our knowledge, MDQ-62 and the two shortened versions presented here are currently the only tools that can be used to assess magnesium deficiency in a non-invasive manner and at no cost. The main advantage of MDQ-23 and MDQ-10 is their relatively high sensitivity (0.8–0.9) in predicting magnesium serum levels <0.8 mmol/L (MDQ-23 score >9) or <0.66 mmol/L (MDQ-10 score >5). Using the modified questionnaires allows clinicians to identify those at a high risk of magnesium deficiency and to verify it further with blood test. The relatively low specificity and therefore potentially high false positive rate are not a major limitation for screening tests, as their main goal is not to miss true positive cases. We believe that MDQs could be very useful in the clinical practice as magnesium deficiency screening tools.

## 5. Conclusions

This analysis determined the sensitivity and specificity of MDQ-62 in identifying patients with suspected magnesium deficiency. Furthermore, we developed two shortened questionnaires (MDQ-23 and MDQ-10), which were non-inferior to MDQ-62. All versions of MDQ showed a better performance in women with hormone-related conditions than in pregnant women.

**Supplementary Materials:** The following are available online at http://www.mdpi.com/2072-6643/12/7/2062/s1, Table S1: Diagnostic utility of MDQ using a serum magnesium level cut-off of 0.8 mmol/L, Table S2: Diagnostic utility of MDQ using a serum magnesium level cut-off of 0.66 mmol/L, Table S3: Development of MDQ-23, Table S4: Development of MDQ-10.

**Author Contributions:** Conceptualization, K.S., A.B, and S.O.; methodology, K.S., S.K., G.D., G.P.; validation, S.O., G.D., K.S.; formal analysis, S.K.; investigation, K.S.; resources, A.B.; data curation, S.K.; writing—original draft preparation, K.S.; writing—review and editing, all authors.; supervision, G.P.; project administration, K.S.; funding acquisition, K.S., A.B. All authors have read and agreed to the published version of the manuscript.

**Funding:** MAGIC, MAGIC2, MAGYN, MAGYN2, and the secondary analysis of these studies presented here were sponsored by Sanofi.

**Acknowledgments:** The authors are grateful to the participants included in the studies, the investigators, coordinators, and the study teams. The authors thank Lionel Noah and Beatrice Bois de Fer of Consumer Healthcare, Sanofi, Paris, France, for critical review of the manuscript. The authors thank Atlant Clinical Ltd. Clinical Research Organization for their contribution to protocol development and conducting the analysis. Editorial support was provided by Olga Ucar and Mark Davies of inScience Communications, Springer Healthcare Ltd., UK, and was funded by Sanofi.

**Conflicts of Interest:** S.K., K.S. and A.B. are employees of Sanofi. Other authors declare no conflict of interest.

## References

1. Workinger, J.L.; Doyle, R.P.; Bortz, J. Challenges in the diagnosis of magnesium status. *Nutrients* **2018**, *10*, 1201. [CrossRef] [PubMed]
2. Costello, R.B.; Elin, R.J.; Rosanoff, A.; Wallace, T.C.; Guerrero-Romero, F.; Hruby, A.; Lutsey, P.L.; Nielsen, F.H.; Rodriguez-Moran, M.; Song, Y.; et al. Perspective: The case for an evidence-based reference interval for serum magnesium: The time has come. *Adv. Nutr.* **2016**, *7*, 977–993. [CrossRef]
3. Elin, R.J. Assessment of magnesium status for diagnosis and therapy. *Magnes. Res.* **2010**, *23*, S194–S198. [CrossRef]
4. Bell, C. *Clinical Guide to Laboratory Tests*, 3rd ed.; Norbert, W.T., Ed.; Saunders, Elsevier: St. Louis, MO, USA, 2009.
5. Lowenstein, F.W.; Stanton, M.F. Serum magnesium levels in the United States, 1971–1974. *J. Am. Coll. Nutr.* **1986**, *5*, 399–414. [CrossRef]
6. Abbott, L.G.; Rude, R.K. Clinical manifestations of magnesium deficiency. *Miner. Electrolyte Metab.* **1993**, *19*, 314–322. [PubMed]
7. DiNicolantonio, J.J.; O'Keefe, J.H.; Wilson, W. Subclinical magnesium deficiency: A principal driver of cardiovascular disease and a public health crisis. *Open Heart* **2018**, *5*, e000668. [CrossRef]
8. Ismail, A.A.A.; Ismail, N.A. Magnesium: A mineral essential for health yet generally underestimated or even ignored. *J. Nutr. Food Sci.* **2016**, *6*. [CrossRef]
9. Slagle, P. Magnificent magnesium. In *The Way Up Newsletter*; The Way Up Ltd.: Rancho Mirage, CA, USA, 2001; Volume 30.
10. Makatsariya, A.D.; Dadak, C.; Bitsadze, V.O.; Solopova, A.G.; Khamani, N.M. Clinical features of patients with hormone-dependent conditions and magnesium deficiency. *Akusherstvo Ginekol.* **2017**, *5*, 124–131. [CrossRef]
11. Makatsariya, A.D.; Bitsadze, V.O.; Khizroeva, D.K.; Dzhobava, E.M. Prevalence of magnesium deficiency in pregnant women. *Vopr. Ginekol. Akusherstva Perinatol.* **2012**, *11*, 25–34.

12. Serov, V.N.; Baranov, I.I.; Blinov, D.V.; Zimovina, U.V.; Sandakova, E.A.; Ushakova, T.I. Results of evaluating Mg deficiency among female patients with hormone-related conditions. *Akusherstvo Ginekol.* **2015**, *6*, 91–97.
13. Serov, V.N.; Blinov, D.V.; Zimovina, U.V.; Dzhobava, E.M. Results of an investigation of the prevalence of magnesium deficiency in pregnant women. *Akusherstvo Ginekol.* **2014**, *6*, 33–40.
14. Trevethan, R. Sensitivity, specificity, and predictive values: Foundations, pliabilities, and pitfalls in research and practice. *Front. Public Health* **2017**, *5*, 307. [CrossRef] [PubMed]
15. Safari, S.; Baratloo, A.; Elfil, M.; Negida, A. Evidence based emergency medicine; Part 5 receiver operating curve and area under the curve. *Emergy (Tehran)* **2016**, *4*, 111–113.
16. Shenkin, A. Micronutrients in health and disease. *Postgrad Med. J.* **2006**, *82*, 559–567. [CrossRef] [PubMed]
17. Ramachandran, A. Know the signs and symptoms of diabetes. *Indian J. Med. Res.* **2014**, *5*, 579–581.

© 2020 by the authors. Licensee MDPI, Basel, Switzerland. This article is an open access article distributed under the terms and conditions of the Creative Commons Attribution (CC BY) license (http://creativecommons.org/licenses/by/4.0/).

*Review*

# Magnesium Oxide in Constipation

Hideki Mori [1], Jan Tack [1] and Hidekazu Suzuki [2,*]

[1] Translational Research Center for Gastrointestinal Diseases (TARGID), University of Leuven, 3000 Leuven, Belgium; koyamaru2002@yahoo.co.jp (H.M.); jan.tack@kuleuven.be (J.T.)
[2] Division of Gastroenterology and Hepatology, Department of Internal Medicine, Tokai University School of Medicine, Isehara 259-1193, Japan
* Correspondence: hsuzuki@tokai.ac.jp; Tel.: +81-463-93-1121 (ext. 2251)

**Abstract:** Magnesium oxide has been widely used as a laxative for many years in East Asia, yet its prescription has largely been based on empirical knowledge. In recent years, several new laxatives have been developed, which has led to a resurgence in interest and increased scientific evidence surrounding the use of magnesium oxide, which is convenient to administer, of low cost, and safe. Despite these advantages, emerging clinical evidence indicates that the use of magnesium oxide should take account of the most appropriate dose, the serum concentration, drug–drug interactions, and the potential for side effects, especially in the elderly and in patients with renal impairment. The aim of this review is to evaluate the evidence base for the clinical use of magnesium oxide for treating constipation and provide a pragmatic guide to its advantages and disadvantages.

**Keywords:** magnesium oxide; constipation; laxative; hypermagnesemia

Citation: Mori, H.; Tack, J.; Suzuki, H. Magnesium Oxide in Constipation. *Nutrients* **2021**, *13*, 421. https://doi.org/10.3390/nu13020421

Academic Editor: Sara Castiglioni
Received: 30 December 2020
Accepted: 26 January 2021
Published: 28 January 2021

**Publisher's Note:** MDPI stays neutral with regard to jurisdictional claims in published maps and institutional affiliations.

**Copyright:** © 2021 by the authors. Licensee MDPI, Basel, Switzerland. This article is an open access article distributed under the terms and conditions of the Creative Commons Attribution (CC BY) license (https://creativecommons.org/licenses/by/4.0/).

## 1. Introduction

Magnesium began to be used medicinally in Western countries after the 1618 discovery by a farmer in Epsom, England that well water had a healing effect on cattle skin diseases [1]. The substance, magnesium sulfate, $MgSO_4$, became known as Epsom salts and was used as a treatment for constipation for the next 350 years. However, the history of the medicinal use of magnesium in Eastern countries is even older. Magnesium nitrate, $Mg(NO_3)_2$, was already being used in Chinese herbal medicine when magnesium sulfate was discovered in the West, and magnesium nitrate was introduced to Japan in the 8th century [2]. The expected laxative actions of magnesium nitrate meant it was used to treat constipation alongside rhubarb, which contains an anthraquinone glycoside compound that acts as a stimulant laxative in Chinese and Japanese herbal medicine [3]. Magnesium oxide was introduced from the West to the East in the 19th century when German physician P.F.B. Siebold brought magnesium oxide to Japan [4,5]. Following this, magnesium oxide took a central position as a laxative of choice in East Asian countries such as Japan, China, and Taiwan [6,7]. On the other hand, the magnesium preparation most commonly used in South Korea and the United States is magnesium hydroxide [8,9]. In European countries, magnesium hydroxide, magnesium citrate, magnesium sulfate, and magnesium oxide are used as saline laxatives, but there are only a limited number of studies that compare the different salt forms and few actual cases of use [10,11]. Therefore, at present, the decision of which magnesium salt to use as a laxative is dependent upon the country in which it is prescribed. Polyethylene glycol is the laxative of first choice in the United States [12] and is also widely used in Europe [10]. In recent years, new drugs such as a type-2 chloride channel activator, a guanylate cyclase 2C receptor agonist, and an inhibitor of the ileal bile acid transporter have been developed to treat constipation and can be given to a patient sequentially [13–15]. While the range of laxatives available has expanded, there is still no consensus on how to use them correctly. Currently, the safety, convenience, and low cost of magnesium oxide, which has been used for many years, mean that it has once again attracted attention (Table 1).

Table 1. Comparison of laxative drugs by cost.

|  | Daily Dose | Cost Per Day (JP) | Cost Per Day (NL) |
|---|---|---|---|
| Magnesium oxide | 2 g | €0.18 | NA |
| Magnesium hydroxide | 2 g | €0.36 | €0.16 |
| Senna | 24 mg | €0.09 | €0.25 |
| Polyethylene glycol | 10 g | €1.02 | €0.58 |
| Psyllium seed | 7 g | NA | €0.45 |
| Lubiprostone | 48 μg | €1.93 | NA |
| Linaclotide | 0.29 mg | €1.44 | €1.87 |
| Elobixibat hydrate | 10 mg | €1.68 | NA |

Abbreviations: JP = Japan, NL = the Netherlands. NA = not applicable. Data was obtained from the following sites. https://www.kegg.jp/kegg/medicus.html (JP), https://www.farmacotherapeutischkompas.nl/ (NL), 1 euro = 125 yen (December, 2020).

In this article, we address the evidence for using magnesium oxide to treat constipation, and highlight the advantages and limitations that should guide clinical decisions about the use of this therapy.

## 2. Mechanism of Action

Magnesium oxide (MgO) is converted into magnesium chloride ($MgCl_2$) under acidic conditions in the stomach. Thereafter, $MgCl_2$ is converted to magnesium bicarbonate $Mg(HCO_3)_2$ by sodium hydrogen carbonate ($NaHCO_3$) from pancreatic secretion in the duodenum, and finally becomes magnesium carbonate ($MgCO_3$) (Figure 1). $Mg(HCO_3)_2$ and $MgCO_3$ increase the osmotic pressure of the intestinal lumen fluid, thereby promoting the transfer of water to the intestinal lumen and increasing the water content and volume of the stool. In addition, the swollen stool stimulates the intestinal wall and intestinal propulsive motor activity.

Anthraquinone-based drugs, which act as stimulant laxatives and include rhubarb and senna, cause tolerance with continuous use, whereas patients do not develop tolerance to magnesium oxide with continuous use.

Although magnesium oxide use can lead to hypermagnesemia, this is rare. This, combined with a lack of pharmacokinetic information on magnesium oxide, means that in clinical practice it is often prescribed with little regard for the optimal dose. Siener et al. administered 404 mg/day of magnesium oxide to healthy volunteers and analyzed the magnesium concentration in blood and urine. There was no significant change in blood concentration of magnesium, but urinary magnesium excretion increased by 40% after administration of magnesium oxide [16]. Yoshimura et al. performed a pharmacokinetic study of orally administered magnesium oxide in rats [17]. They showed that 85% of magnesium is excreted in feces, while 15% of magnesium is absorbed from the intestinal tract and excreted in urine [17]. Furthermore, they showed that the plasma magnesium concentration in rats was maintained at a high level for a relatively long period due to the slow absorption of magnesium [17]. Such data have not yet been shown in humans. The kidney plays a central role in maintaining magnesium homeostasis through active reabsorption, which is influenced by the sodium load in the tubules and possibly by acid–base balance. The kidneys of individuals with a normal glomerular filtration rate (GFR) filter approximately 2000–2400 mg of magnesium per day [18]. Based on this information, the dose of magnesium oxide usually used for laxative purposes is considered to have a low propensity to cause hypermagnesemia, while magnesium oxide should be used carefully in patients with renal dysfunction.

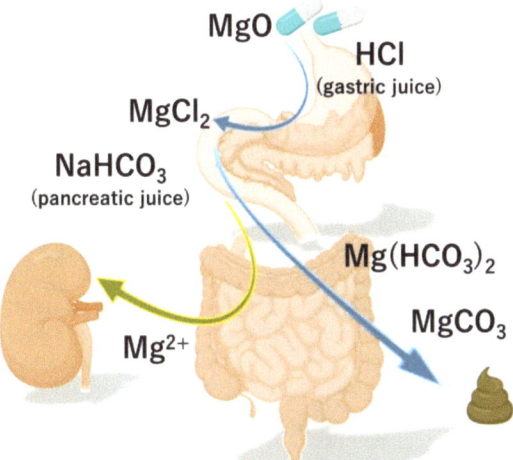

**Figure 1.** The pharmacokinetics of magnesium oxide. Magnesium oxide (MgO) is converted to sodium hydrogen carbonate (NaHCO$_3$) and magnesium carbonate (MgCO$_3$) by gastric and pancreatic juice, and exerts its effect as a salt laxative.

## 3. Evidence Underpinning the Use of Magnesium Oxide

### 3.1. Functional Constipation in Adults

Magnesium oxide has long been used to treat constipation, and an estimated 10 million patients in Japan are treated with this agent annually [19]. Although the prescription of magnesium oxide has been based on empirical evidence for many years, two randomized controlled trials (RCTs) showing the effectiveness of magnesium oxide for treating chronic constipation in adults have recently been reported. Mori S et al. conducted a randomized, double-blind placebo-controlled study comparing magnesium oxide with placebo [20]. In this study, magnesium oxide administration led to superior overall improvement of symptoms, and improved spontaneous bowel movement, stool form, colonic transit time, abdominal symptom, and quality of life. Patients treated with magnesium oxide had a response rate of 70.6% for overall symptom improvement, which was significantly higher than the response rate of 25.0% observed in the placebo group [20]. Morishita et al. conducted a randomized, double-blind placebo-controlled study comparing magnesium oxide with placebo and senna [21]. The response rate for overall improvement was 11.7% in the placebo group, 69.2% in the senna group, and 68.3% in the magnesium oxide group. Patients receiving senna or magnesium oxide had significant improvements in spontaneous bowel movement and constipation-related quality of life, compared with patients in the placebo group. Moreover, no severe treatment-related adverse events were observed in either treatment group. The time to the first spontaneous bowel movement in this study was significantly shortened for the senna group (18.8 h) and magnesium oxide group (17.9 h) compared with the placebo group (22.0 h). However, the authors discussed that senna and magnesium oxide were less effective in reducing time to first spontaneous bowel movement than newer drugs: lubiprostone (3.5 h versus 48.0 h for placebo), linaclotide (6.7 h versus 24.7 h for placebo), and elobixibat (5.1 h versus 25.5 h for placebo) [21]. These differences may serve as a reference for the correct use of these drugs for treating constipation in the future.

### 3.2. Opioid-Induced Constipation

Opioid-induced constipation is the most common and problematic complication of opioid therapy [22–24]. Standard laxatives, such as osmotic agents (macrogol) and stimulants (bisacodyl, picosulfate, and senna), are good first-line choices in the management of opioid-induced constipation [22]. Second-line agents that block μ-opioid receptors in the gastrointestinal tract but do not enter the central nervous system, so called PAMORAs (peripherally acting mu opioid receptor antagonists), can be used to treat opioid-induced constipation without diminishing central analgesic actions [22]. The mu-opioid receptor antagonists methylnaltrexone, naloxegol, and naldemedine are safe and effective treatments for opioid-induced constipation [22,25–29]. Evidence is not yet available on the use of magnesium oxide for treating opioid-induced constipation; however a single-institution, open-label, randomized controlled trial comparing the effectiveness of magnesium oxide with naldemedine for preventing opioid-induced constipation is ongoing [30]. Magnesium oxide could be a promising drug candidate for opioid-induced constipation, yet patients taking opioids often take gastric antisecretory drugs such as proton pump inhibitors and histamine $H_2$ receptor antagonists, which can diminish the efficacy of magnesium oxide [31,32]. Moreover, retrospective studies have shown that 94% of opioid-induced constipation patients did not achieve an inadequate response to one laxative agent with one laxative treatment including magnesium hydrochloride [33]. On the other hand, PAMORAs may be advantageous, as opioids act on receptors in both the small and large intestine and, second, dampen neuronal activity also in the gut [22].

### 3.3. Functional Constipation in Children

Magnesium salts are also used for treatment of functional constipation in children [34]. Two RCTs showing the effectiveness of magnesium oxide for treating chronic constipation in children have been reported. Bu et al. conducted a double-blind placebo-controlled, randomized study to compare *Lactobacillus casei rhamnosus Lcr35* with magnesium oxide and placebo [35]. The patients who received magnesium oxide or the probiotic had a higher defecation frequency, higher percentage of treatment success, less use of glycerin enema, and softer stools than the placebo group. There were no significant differences in the aforementioned comparisons between the magnesium oxide and probiotic groups; however, the onset of effects occurred slightly earlier in patients treated with magnesium oxide than those treated with probiotics [35]. Kubota et al. conducted a double-blind, placebo-controlled, randomized trial to compare the probiotic *Lactobacillus reuteri* DSM 17938 and magnesium oxide for relieving chronic functional constipation in children [36]. They divided the subjects into three groups: the first group received *L. reuteri* DSM 17938, the second group received *L. reuteri* DSM 17938 and magnesium oxide, and the third group received magnesium oxide. There was a significant improvement in the defecation frequency in all groups at the fourth week after treatment compared to baseline. In this study, the authors also investigated the relationship between gut microbiome composition, magnesium oxide, and defecation frequency. They showed that defecation frequency was higher in magnesium oxide-treated patients than in patients whose gut microbiome contained bacteria of the genus *Dialister*, and that defecation frequency negatively correlated with the frequency of bacteria belonging to the genus *Clostridiales* in patients' gut microbiomes [36]. This result suggests that magnesium oxide treatment alters the gut microbiome. Although this is a noteworthy finding, the long-term health effects of an altered gut microbiome induced by magnesium oxide are unclear, and further research is likely necessary.

### 3.4. Guidelines on the Use of Magnesium Precautions for Functional Constipation

Guidelines for using magnesium preparations in various countries and regions are shown in Table 2. Magnesium oxide is mentioned only in Japanese guidelines, in which the recommendation level is "strong" [37]. Although sufficient international evidence of the use of magnesium oxide in adults only recently became available, the experience gained from prescribing magnesium oxide to more than 10 million patients annually was only acknowledged in the Japanese guidelines. Magnesium salts, including magnesium hydroxide, are recommended in the guidelines of the American Gastroenterological Association, the Asian Neurogastroenterology and Motility Association, the Korean Society of Neurogastroenterology and Motility, the Mexican Association of Gastroenterology, the

Italian Association of Hospital Gastroenterologists and the Italian Society of Colorectal Surgery, and the French Society of Gastroenterology [38–43]. However, due to low levels of evidence, the recommendation level in these guidelines is weak. On the other hand, the German Society for Digestive and Metabolic Diseases, the German Society for Neurogastroenterology and Motility, and the German Society for Internal Medicine and the American College of Gastroenterology recommend against the use of magnesium hydroxide because of possible adverse effects [44,45]. The European Society of Neurogastroenterology and Motility, the UK's National Institute for Health and Clinical Excellence, and an expert panel in Hong Kong do not mention magnesium preparations [46–48].

**Table 2.** Guidelines on the use of magnesium precautions for treating functional constipation.

| Institution | Recommendation | Class | Level of Evidence |
|---|---|---|---|
| European Society of Neurogastroenterology and Motility [46] | None | NA | NA |
| American Gastroenterological Association [38] | Magnesium hydroxide and other salts improve stool frequency and consistency. Absorption of magnesium is limited, and these agents are generally safe. However, there are a few case reports of severe hypermagnesemia after use of magnesium-based cathartics in patients with renal impairment. | NA | NA |
| American College of Gastroenterology [45] | There is insufficient data to make a recommendation about the effectiveness of magnesium hydroxide in patients with chronic constipation. | No recommendation | Moderate |
| Asian Neurogastroenterology and Motility Association [39] | Milk of magnesia (magnesium hydroxide) is an osmotic laxative by which the poorly absorbable magnesium ions cause water to be retained in the intestinal lumen. The evidence for its efficacy from randomized control trials is limited. | NA | NA |
| The Japanese Society of Gastroenterology [37] | Osmotic laxatives are useful and recommended for use in chronic constipation. However, regular magnesium measurement is recommended when using salt laxatives containing magnesium. | Strong | High |
| The Korean Society of Neurogastroenterology and Motility [40] | Magnesium salts improve stool frequency and consistency in patients with normal renal function. | Strong | Low |
| An expert panel in Hong Kong [47] | None | NA | NA |
| Mexican Association of Gastroenterology [41] | Magnesium salts are useful in patients with acute constipation associated with immobilization and should not be used chronically because they produce hypermagnesemia, especially in patients with kidney failure. | Weak | Low |
| The UK's National Institute for Health and Clinical Excellence [48] | Substitute a stimulant laxative singly or in combination with an osmotic laxative such as lactulose if polyethylene glycol 3350 plus electrolytes are not tolerated. | NA | NA |

Table 2. *Cont.*

| Institution | Recommendation | Class | Level of Evidence |
|---|---|---|---|
| The German Society for Digestive and Metabolic Diseases, the German Society for Neurogastroenterology and Motility and the German Society for Internal Medicine [44] | Saline laxatives, such as magnesium hydroxide, are not recommended for chronic constipation because of possible adverse effects. | No recommendation | NA |
| The Italian Association of Hospital Gastroenterologists and the Italian Society of Colorectal Surgery [42] | The use of magnesium hydroxide is supported by case-series level of evidence. | Weak | Low |
| French Society of Gastroenterology [43] | The first-line therapeutic interventions recommended by the guidelines are osmotic laxatives (macrogol, lactulose, or milk of magnesia) and bulk-forming laxatives. | Moderate | Moderate |

Abbreviation: NA = not applicable.

## 4. Practical Use of Magnesium Oxide

### 4.1. Dosage and Administration

Magnesium oxide is an osmotic laxative, and its key effect is a softening of hard stools; therefore, it is important to first ask the patient about the hardness of stools and the frequency of bowel movement. In real-life clinical practice, evaluation using the Bristol scale is useful as an objective index [49]. To choose a proper drug, a diagnostic process, including several clinical/psychiatric parameters, is also important [50,51].

The package insert of magnesium oxide advises: "In general, for adults, take 2 g of the active ingredient in 3 divided doses a day before or after meals, or once before bedtime" [52]. However, in practice, 2 g per day can result in hypermagnesemia; therefore, we recommend that a starting dose of approximately 1 g taken as two or three divided doses a day is used, and adjusted appropriately according to symptoms [6]. While there are cases in which 250 mg a day is sufficiently effective, there are rare cases in which sufficient improvements cannot be obtained even with a dose of 2 g a day [6].

If magnesium oxide alone is not effective, other laxatives such as stimulant laxatives, polyethylene glycol, lubiprostone, linaclotide, and elobixibat can also be used as adjunct drugs, in which case magnesium oxide should not be used in excess to avoid hypermagnesemia.

### 4.2. Drug Interactions

Magnesium oxide has an adsorptive action and an antacid action, and so it affects the absorption and excretion of other drugs. Tetracycline, new quinolones, and bisphosphonates may form chelates with magnesium, which diminishes the effects of these drugs. Therefore, there should be a sufficient time interval between dosing if these drugs are prescribed together. Considering the digestion time in the stomach, an interval of at least 2 h is recommended [53]. The effects of iron supplements, digitalis, polycarbophil calcium, and fexofenadine may be diminished by the adsorption action of magnesium or the magnesium oxide-induced increase the intragastric pH. The effects of cation-exchange resins may be decreased because magnesium ions exchange with the cations of these drugs. The effects of some cephem antibiotics, mycophenolate mofetil, delavirdine, zalcitabine, penicillamine, azithromycin, celecoxib, rosuvastatin, rabeprazole, and gabapentin, may be diminished by magnesium; the reasons for this reduced efficacy is not understood. Activated vitamin D supplements may cause hypermagnesemia because they can promote gastrointestinal absorption and reabsorption of magnesium from renal tubules. Consumption of large amounts of milk and calcium supplements may cause hypercalcemia and alkalosis due to increased renal reabsorption of calcium (known as milk–alkali syndrome). Magnesium oxide, an absorbable alkaline preparation, exacerbates calcium retention in the kidneys. The laxative effect of magnesium oxide is decreased in patients receiving a $H_2$ receptor antagonist or a proton pump inhibitor due to the low solubility of magnesium oxide at the higher gastric pH and lower generation of $MgCl_2$ and $Mg(HCO_3)_2$ [31,32].

*4.3. Side Effect Profile and Toxicity*

4.3.1. Hypermagnesemia

The poor bioavailability of magnesium oxide makes it relatively safe, but prolonged treatment may induce hypermagnesemia [54]. In recent years, cases of magnesium oxide-induced hypermagnesemia resulting in serious outcomes have been reported [55–58]. Blood magnesium levels are usually tightly controlled by the kidneys; the normal range is 1.8–2.4 mg/dL, and levels 3.0 mg/dL and above are defined as hypermagnesemia. Serum magnesium concentrations >5.0 mg/dL have been associated with nausea, headache, light-headedness, and cutaneous flushing, and levels above 12 mg/dL have been associated with respiratory failure, complete heart blockage, and cardiac arrest [59]. Recently, the Japanese Ministry of Health, Labour and Welfare recommended that serum magnesium concentrations be measured periodically in geriatric patients and in patients administered magnesium oxide for prolonged periods [60].

Due to a lack of convincing evidence of the degree of risk of using magnesium salts in clinical practice, the authors of this review conducted a retrospective study on the occurrence of hypermagnesemia in patients receiving oral magnesium oxide for treating constipation [6]. Among the patients evaluated for serum magnesium concentration, 5.2% had hypermagnesemia and 16.6% had high serum magnesium concentration (>2.5 mg/dL). Factors associated with hypermagnesemia were impaired renal function and higher magnesium oxide dosage. Patients with a renal function classification of G1 (GFR ≥ 90 mL/min/1.73 m$^2$) and G2 (GFR 60–89 mL/min/1.73 m$^2$) had a serum magnesium concentration of 2.06 ± 0.23 and 2.11 ± 030 mg/dL, respectively, which is within the normal range. Patients classified as G4 (GFR 15–29 mL/min/1.73 m$^2$) and G5 (GFR < 15 mL/min/1.73 m$^2$) had a serum magnesium concentration of 2.46 ± 0.58 and 2.60 ± 0.99 mg/dL respectively; these averages exceeded the upper limit of normal levels. Analysis showed a significant positive correlation between the daily dose of magnesium oxide and blood magnesium concentration. By contrast, age and duration of administration were not correlated with serum magnesium concentration. From these results, we clarified that individuals with decreased renal function and individuals receiving a large daily dose are at high-risk of developing the hypermagnesemia. Wakai et al. also identified risk factors for developing hypermagnesemia in patients prescribed magnesium oxide via a retrospective cohort study [61]. They showed that 23% developed high serum magnesium concentration (>2.5 mg/dL). Renal function, daily dose, and duration of administration were indicated to be independent risk factors. Horibata et al. examined the relationship between renal function and serum magnesium concentration in elderly patients treated with magnesium oxide [62], and found that renal function also significantly correlated with serum magnesium levels. Tatsuki et al. investigated whether children with functional constipation taking daily magnesium oxide develop hypermagnesemia [63]. They showed that the serum magnesium concentration was 2.4 (2.3–2.5 median and interquartile range) mg/dL in children with functional constipation taking daily magnesium oxide, which was significantly higher than in the age- and sex-matched control group (2.2; 2.0–2.2 mg/dL). However, none of the patients had side effects associated with hypermagnesemia. These reports indicated the importance of monitoring serum magnesium levels in patients being treated with magnesium, especially in patients with chronic kidney failure and in patients treated with high dosages of magnesium oxide (Figure 2a).

Magnesium oxide has relatively poor bioavailability (a fractional absorption of 4%) compared to other magnesium salts; the fractional absorption of magnesium chloride, magnesium lactate, and magnesium aspartate are all between 9 and 11% [54]. Another report showed that the bioavailability of magnesium oxide was significantly lower than that of magnesium citrate [64]. The low bioavailability of magnesium oxide may be related to its low solubility in water [64]. These results suggest that magnesium oxide may have a less propensity to cause hypermagnesemia than other magnesium preparations.

4.3.2. Milk–Alkali Syndrome

Milk–alkali syndrome is characterized by the triad of hypercalcemia, metabolic alkalosis, and decreased kidney function, and is caused by excessive intake of calcium and alkali [65]. Magnesium oxide is an absorbable alkaline preparation. Since serum calcium levels are tightly controlled by parathyroid hormone and active vitamin D, hypercalcemia does not easily occur even with excessive calcium intake [66]. However, when the serum calcium concentration rises (as a result of several possible causes) and the calcium concentration in the renal tubule rises, hypercalcemia has a well-known natriuretic and diuretic effect by activating the calcium-sensing receptor, leading to intravascular depletion of calcium. The resulting reduction in GFR further limits the excretion of bicarbonate and calcium, and absorbable alkaline preparations including magnesium oxide exacerbate calcium retention in the kidneys [67] (Figure 2b).

In recent years, common diseases in the aging population include both constipation and osteoporosis, which are often treated at the same time. In other words, individuals are often simultaneously treated with magnesium oxide for constipation and vitamin D preparations for osteoporosis, leading to milk–alkali syndrome [56,68,69]. Considering this scenario, it is necessary to pay attention to the possibility of milk–alkali syndrome and drug–drug interactions in the elderly who copresent with constipation and osteoporosis.

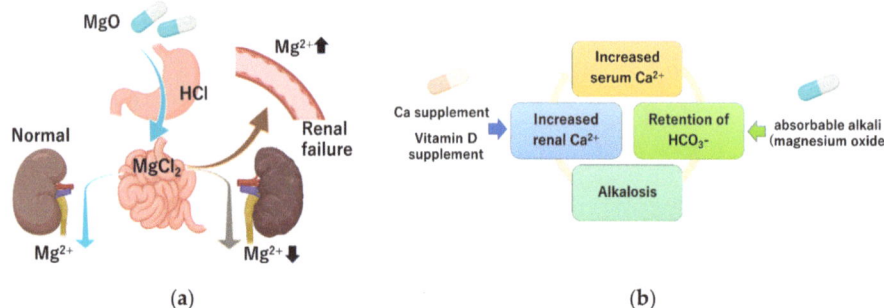

**Figure 2.** Mechanism of hypermagnesemia and milk–alkali syndrome. (**a**) In patients with impaired renal function, poor magnesium excretion increases the risk of hypermagnesemia. (**b**) Simultaneous intake of excessive calcium and magnesium oxide, which is a non-absorbable alkali, can increase the serum calcium concentration.

## 5. Summary and Future Perspectives

Magnesium oxide has been clinically used as a laxative for many years. Due to a lack of alternative treatment options, it was prescribed based on empirical experience. The increasing availability of newer drugs for treating constipation has led to the emergence of scientific evidence surrounding the use of magnesium oxide, which is convenient to administer, of low cost, and safe. Although RCTs have recently shown that magnesium oxide is safe and efficacious for treating constipation, evidence of efficacy for treating symptoms of irritable bowel syndrome, especially the constipation-predominant subgroup, needs to be urgently established. Risk factors for developing hypermagnesemia have been clarified and evidence suggests that appropriate monitoring for this potential side effect is necessary. To be specific, patients with renal impairment of CKD grade G4 or higher and patients who take 1000 mg of magnesium oxide or more daily should be monitored monthly at the time of drug introduction, and monitoring of serum magnesium is recommended in parallel with renal function even during the stable period [6]. However, there is still insufficient evidence to enable comparisons to be made between various laxative drugs and to enable correct prescribing decisions to be made. Drugs such as magnesium oxide are still prescribed based on empirical knowledge, and an accumulation of systematic evidence is still needed (Figure 3).

Chronic constipation is normally treated by general practitioners rather than gastroenterologists. The establishment of systematic, scientifically presented guidelines for treating constipation, which clearly define the position of general practitioners and gastroenterologists and are based on sufficient evidence, are highly desirable.

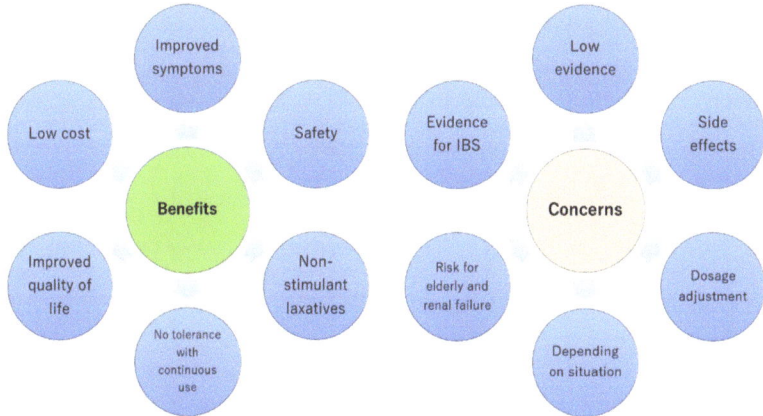

**Figure 3.** The advantages and disadvantages of magnesium oxide therapy. IBS, irritable bowel syndrome.

**Author Contributions:** All authors helped to perform the research; H.M. wrote the manuscript and summarized the references; J.T. and H.S. contributed to the manuscript writing, and study conception and design. All authors have read and agreed to the published version of the manuscript.

**Funding:** This work was partially supported by a Grant from AMED (to Suzuki H.) and by a Grant-in-Aid for Scientific Research B (20H3667, to Suzuki H.).

**Institutional Review Board Statement:** Not applicable.

**Informed Consent Statement:** Not applicable.

**Data Availability Statement:** Data sharing not applicable.

**Acknowledgments:** Figures were created with BioRender.com.

**Conflicts of Interest:** The author Suzuki H. received scholarship funds for research from Daiichi-Sankyo, Otsuka Pharmaceutical Co, Ltd., MSD Co., Mylan EPD, Tanabe Pharm. Co., and Takeda Pharmaceutical Co., and received service honoraria from Astellas Pharma, AstraZeneca K.K., EA Pharma Co., Ltd., Mylan EPD, Otsuka Pharm Co., Takeda Pharm Co., and Tsumura Co. Jan Tack has given Scientific advice to Adare, AlfaWassermann, Allergan, Arena, Bayer, Christian Hansen, Clasado, Danone, Devintec, Falk, Grünenthal, Ironwood, Janssen, Kiowa Kirin, Menarini, Mylan, Neurogastrx, Neutec, Novartis, Noventure, Nutricia, Shionogi, Shire, Takeda, Theravance, Tramedico, Truvion, Tsumura, Zealand and Zeria pharmaceuticals, has received research support from Shire, Sofar and Tsumura, and has served on the Speaker bureau for Abbott, Allergan, AstraZeneca, Janssen, Kyowa Kirin, Menarini, Mylan, Novartis, Shire, Takeda, Truvion and Zeria.

## References

1. Rudolf, R.D. The use of Epsom salts, historically considered. *Can. Med. Assoc. J.* **1917**, *7*, 1069–1071. [PubMed]
2. Watanabe, T. BOSHO and BOKUSHO. *J. Jpn. Soc. Orient. Med.* **1956**, *7*, 27–33.
3. Liu, C.; Zheng, Y.; Xu, W.; Wang, H.; Lin, N. Rhubarb Tannins Extract Inhibits the Expression of Aquaporins 2 and 3 in Magnesium Sulphate-Induced Diarrhoea Model. *BioMed Res. Int.* **2014**, *2014*, 1–14. [CrossRef] [PubMed]
4. Veith, I. Physician Travelers in Japan (Engelbert Kaempfer, Karl Peter Thunberg, Philipp Franz Von Siebold). *JAMA* **1965**, *192*, 137–140. [CrossRef] [PubMed]

5. Miyazaki, M. Von Siebold's Prescription. *J. Jpn. Hist. Pharm.* **1991**, *26*, 12–23.
6. Mori, H.; Suzuki, H.; Hirai, Y.; Okuzawa, A.; Kayashima, A.; Kubosawa, Y.; Kinoshita, S.; Fujimoto, A.; Nakazato, Y.; Nishizawa, T.; et al. Clinical features of hypermagnesemia in patients with functional constipation taking daily magnesium oxide. *J. Clin. Biochem. Nutr.* **2019**, *65*, 76–81. [CrossRef]
7. Chen, I.-C.; Huang, H.J.; Yang, S.F.; Chen, C.C.; Chou, Y.C.; Kuo, T.M. Prevalence and Effectiveness of Laxative Use Among Elderly Residents in a Regional Hospital Affiliated Nursing Home in Hsinchu County. *Nurs. Midwifery Stud.* **2014**, *3*. [CrossRef]
8. Park, K.S.; Choi, S.-C.; Park, M.-I.; Shin, J.-E.; Jung, K.-W.; Kim, S.-E.; Lee, T.-H.; Koo, H.-S. Practical treatments for constipation in Korea. *Korean J. Intern. Med.* **2012**, *27*, 262–270. [CrossRef]
9. Borowitz, S.M.; Cox, D.J.; Kovatchev, B.; Ritterband, L.M.; Sheen, J.; Sutphen, J. Treatment of Childhood Constipation by Primary Care Physicians: Efficacy and Predictors of Outcome. *Pediatrics* **2005**, *115*, 873–877. [CrossRef]
10. Müller-Lissner, S.; Tack, J.; Feng, Y.; Schenck, F.; Gryp, R.S. Levels of satisfaction with current chronic constipation treatment options in Europe—An internet survey. *Aliment. Pharmacol. Ther.* **2013**, *37*, 137–145. [CrossRef]
11. Tack, J.; Müller-Lissner, S.; Stanghellini, V.; Boeckxstaens, G.; Kamm, M.A.; Simren, M.; Galmiche, J.-P.; Fried, M. Diagnosis and treatment of chronic constipation—A European perspective. *Neurogastroenterol. Motil.* **2011**, *23*, 697–710. [CrossRef] [PubMed]
12. Mounsey, A.; Raleigh, M.; Wilson, A. Management of Constipation in Older Adults. *Am. Fam. Physician* **2015**, *92*, 500–504. [PubMed]
13. Simrén, M.; Tack, J. New treatments and therapeutic targets for IBS and other functional bowel disorders. *Nat. Rev. Gastroenterol. Hepatol.* **2018**, *15*, 589–605. [CrossRef] [PubMed]
14. Tack, J.; Vanuytsel, T.; Corsetti, M. Modern Management of Irritable Bowel Syndrome: More than Motility. *Dig. Dis.* **2016**, *34*, 566–573. [CrossRef] [PubMed]
15. Corsetti, M.; Tack, J. New pharmacological treatment options for chronic constipation. *Expert Opin. Pharmacother.* **2014**, *15*, 927–941. [CrossRef]
16. Siener, R.; Jahnen, A.; Hesse, A. Bioavailability of magnesium from different pharmaceutical formulations. *Urol. Res.* **2011**, *39*, 123–127. [CrossRef] [PubMed]
17. Yoshimura, Y.; Fujisaki, K.; Yamamoto, T.; Shinohara, Y. Pharmacokinetic Studies of Orally Administered Magnesium Oxide in Rats. *J. Pharm. Soc. Jpn.* **2017**, *137*, 581–587. [CrossRef] [PubMed]
18. Blaine, J.; Chonchol, M.; Levi, M. Renal Control of Calcium, Phosphate, and Magnesium Homeostasis. *Clin. J. Am. Soc. Nephrol.* **2015**, *10*, 1257–1272. [CrossRef]
19. The Japanese Ministry of Health, Labour and Welfare. Pharmaceuticals and Medical Devices Safety Information (Japanese). 2015. Available online: https://www.pmda.go.jp/files/000208517.pdf (accessed on 30 December 2020).
20. Mori, S.; Tomita, T.; Fujimura, K.; Asano, H.; Ogawa, T.; Yamasaki, T.; Kondo, T.; Kono, T.; Tozawa, K.; Oshima, T.; et al. A Randomized Double-blind Placebo-controlled Trial on the Effect of Magnesium Oxide in Patients With Chronic Constipation. *J. Neurogastroenterol. Motil.* **2019**, *25*, 563–575. [CrossRef]
21. Morishita, D.; Tomita, T.; Mori, S.; Kimura, T.; Oshima, T.; Fukui, H.; Miwa, H. Senna Versus Magnesium Oxide for the Treatment of Chronic Constipation: A Randomized, Placebo-Controlled Trial. *Am. J. Gastroenterol.* **2020**, *116*, 152–161. [CrossRef]
22. Farmer, A.D.; Drewes, A.M.; Chiarioni, G.; De Giorgio, R.; O'Brien, T.; Morlion, B.; Tack, J. Pathophysiology and management of opioid-induced constipation: European expert consensus statement. *United Eur. Gastroenterol. J.* **2019**, *7*, 7–20. [CrossRef] [PubMed]
23. Locasale, R.J.; Datto, C.J.; Margolis, M.K.; Tack, J.; Coyne, K.S. The impact of opioid-induced constipation among chronic pain patients with sufficient laxative use. *Int. J. Clin. Pract.* **2015**, *69*, 1448–1456. [CrossRef] [PubMed]
24. Tack, J. Current and future therapies for chronic constipation. *Best Pract. Res. Clin. Gastroenterol.* **2011**, *25*, 151–158. [CrossRef]
25. Pannemans, J.; Vanuytsel, T.; Tack, J. New developments in the treatment of opioid-induced gastrointestinal symptoms. *United Eur. Gastroenterol. J.* **2018**, *6*, 1126–1135. [CrossRef] [PubMed]
26. Tack, J.; Lappalainen, J.; Diva, U.; Tummala, R.; Sostek, M. Efficacy and safety of naloxegol in patients with opioid-induced constipation and laxative-inadequate response. *United Eur. Gastroenterol. J.* **2015**, *3*, 471–480. [CrossRef]
27. Corsetti, M.; Tack, J. Naloxegol: The first orally administered, peripherally acting, mu opioid receptor antagonist, approved for the treatment of opioid-induced constipation. *Drugs Today* **2015**, *51*, 479–489. [CrossRef]
28. Corsetti, M.; Tack, J. Naloxegol, a new drug for the treatment of opioid-induced constipation. *Expert Opin. Pharmacother.* **2015**, *16*, 399–406. [CrossRef]
29. Chey, W.D.; Webster, L.; Sostek, M.; Lappalainen, J.; Barker, P.N.; Tack, J. Naloxegol for Opioid-Induced Constipation in Patients with Noncancer Pain. *N. Engl. J. Med.* **2014**, *370*, 2387–2396. [CrossRef]
30. Ozaki, A.; Kessoku, T.; Iwaki, M.; Kobayashi, T.; Yoshihara, T.; Kato, T.; Honda, Y.; Ogawa, Y.; Imajo, K.; Higurashi, T.; et al. Comparing the effectiveness of magnesium oxide and naldemedine in preventing opioid-induced constipation: A proof of concept, single institutional, two arm, open-label, phase II, randomized controlled trial: The MAGNET study. *Trials* **2020**, *21*, 1–9. [CrossRef]
31. Yamasaki, M.; Funakoshi, S.; Matsuda, S.; Imazu, T.; Takeda, Y.; Murakami, T.; Maeda, Y. Interaction of magnesium oxide with gastric acid secretion inhibitors in clinical pharmacotherapy. *Eur. J. Clin. Pharmacol.* **2014**, *70*, 921–924. [CrossRef]
32. Ibuka, H.; Ishihara, M.; Suzuki, A.; Kagaya, H.; Shimizu, M.; Kinosada, Y.; Itoh, Y. Antacid attenuates the laxative action of magnesia in cancer patients receiving opioid analgesic. *J. Pharm. Pharmacol.* **2016**, *68*, 1214–1221. [CrossRef] [PubMed]

33. Coyne, K.S.; Locasale, R.J.; Datto, C.J.; Sexton, C.C.; Yeomans, K.; Tack, J. Opioid-induced constipation in patients with chronic noncancer pain in the USA, Canada, Germany, and the UK: Descriptive analysis of baseline patient-reported outcomes and retrospective chart review. *Clin. Outcomes Res.* **2014**, *6*, 269–281. [CrossRef] [PubMed]
34. Tabbers, M.M.; DiLorenzo, C.; Berger, M.Y.; Faure, C.; Langendam, M.W.; Nurko, S.; Staiano, A.; Vandenplas, Y.; Benninga, M.A.; European Society for Pediatric Gastroenterology, Hepatology, and Nutrition; et al. Evaluation and treatment of functional constipation in infants and children: Evidence-based recommendations from ESPGHAN and NASPGHAN. *J. Pediatric Gastroenterol. Nutr.* **2014**, *58*, 258–274. [CrossRef] [PubMed]
35. Bu, L.-N.; Chang, M.-H.; Ni, Y.-H.; Chen, H.-L.; Cheng, C.-C. Lactobacillus casei rhamnosus Lcr35 in children with chronic constipation. *Pediatr. Int.* **2007**, *49*, 485–490. [CrossRef] [PubMed]
36. Kubota, M.; Ito, K.; Tomimoto, K.; Kanazaki, M.; Tsukiyama, K.; Kubota, A.; Kuroki, H.; Fujita, M.; Vandenplas, Y. Lactobacillus reuteri DSM 17938 and Magnesium Oxide in Children with Functional Chronic Constipation: A Double-Blind and Randomized Clinical Trial. *Nutrients* **2020**, *12*, 225. [CrossRef]
37. Research Society for the Diagnosis and Treatment of Chronic Constipation/Affiliated to the Japanese Society of Gastroenterology. *Evidence-Based Clinical Practice Guideline for Chronic Constipation*; Nankodo: Tokyo, Japan, 2017.
38. Bharucha, A.E.; Pemberton, J.H.; Locke, G.R. American Gastroenterological Association Technical Review on Constipation. *Gastroenterology* **2013**, *144*, 218–238. [CrossRef] [PubMed]
39. Gwee, K.; Ghoshal, U.C.; Gonlachanvit, S.; Chua, A.S.B.; Myung, S.-J.; Rajindrajith, S.; Patcharatrakul, T.; Choi, M.-G.; Wu, J.C.Y.; Chen, M.-H.; et al. Primary Care Management of Chronic Constipation in Asia: The ANMA Chronic Constipation Tool. *J. Neurogastroenterol. Motil.* **2013**, *19*, 149–160. [CrossRef]
40. Shin, J.E.; Jung, H.K.; Lee, T.H.; Jo, Y.; Lee, H.; Song, K.H.; Hong, S.N.; Lim, H.C.; Lee, S.J.; Chung, S.S.; et al. Guidelines for the Diagnosis and Treatment of Chronic Functional Constipation in Korea, 2015 Revised Edition. *J. Neurogastroenterol. Motil.* **2016**, *22*, 383–411. [CrossRef]
41. Remes-Troche, J.; Coss-Adame, E.; Lopez-Colombo, A.; Amieva-Balmori, M.; Carmona-Sánchez, R.; Guindic, L.C.; Rendón, R.F.; Escudero, O.G.; Martínez, M.G.; Chávez, M.I.; et al. The Mexican consensus on chronic constipation. *Rev. Gastroenterol. México* **2018**, *83*, 168–189. [CrossRef]
42. Bove, A.; Bellini, M.; Battaglia, E.; Bocchini, R.; Gambaccini, D.; Bove, V.; Pucciani, F.; Altomare, D.F.; Dodi, G.; Sciaudone, G.; et al. Consensus statement AIGO/SICCR diagnosis and treatment of chronic constipation and obstructed defecation (part II: Treatment). *World J. Gastroenterol.* **2012**, *18*, 4994–5013. [CrossRef]
43. Piche, T.; Dapoigny, M.; Bouteloup, C.; Chassagne, P.; Coffin, B.; Desfourneaux, V.; Fabiani, P.; Fatton, B.; Flammenbaum, M.; Jacquet, A.; et al. Recommendations for the clinical management and treatment of chronic constipation in adults. *Gastroentérol. Clin. Biol.* **2007**, *31*, 125–135. [CrossRef]
44. Andresen, V.; Enck, P.; Frieling, T.; Herold, A.; Ilgenstein, P.; Jesse, N.; Karaus, M.; Kasparek, M.; Keller, J.; Kuhlbusch-Zicklam, R.; et al. S2k guideline for chronic constipation: Definition, pathophysiology, diagnosis and therapy. *Z. Gastroenterol.* **2013**, *51*, 651–672. [PubMed]
45. American College of Gastroenterology Chronic Constipation Task Force. An evidence-based approach to the management of chronic constipation in North America. *Am. J. Gastroenterol.* **2005**, *100* (Suppl. S1), S1–S4.
46. Serra, J.; Pohl, D.; Azpiroz, F.; Chiarioni, G.; Ducrotté, P.; Gourcerol, G.; Hungin, A.P.S.; Layer, P.; Mendive, J.M.; Pfeifer, J.; et al. European society of neurogastroenterology and motility guidelines on functional constipation in adults. *Neurogastroenterol. Motil.* **2020**, *32*, e13762. [CrossRef] [PubMed]
47. Wu, J.C.Y.; Chan, A.O.; Cheung, T.K.; Kwan, A.C.; Leung, V.K.; Sze, W.C.; Tan, V.P.Y.; Ta, V.P. Consensus statements on diagnosis and management of chronic idiopathic constipation in adults in Hong Kong. *Hong Kong Med. J.* **2019**, *25*, 142–148. [CrossRef]
48. National Institute for Health and Clinical Excellence. Constipation Clinical Knowledge Summary. 2015. Available online: http://cks.nice.org.uk/constipation (accessed on 2 December 2015).
49. O'Donnell, L.J.; Virjee, J.; Heaton, K.W. Detection of pseudodiarrhoea by simple clinical assessment of intestinal transit rate. *BMJ* **1990**, *300*, 439–440. [CrossRef]
50. Brusciano, L.; Limongelli, P.; Del Genio, G.; Rossetti, G.; Sansone, S.; Healey, A.; Maffettone, V.; Napolitano, V.; Pizza, F.; Tolone, S. Clinical and instrumental parameters in patients with constipation and incontinence: Their potential implications in the functional aspects of these disorders. *Int. J. Color. Dis.* **2009**, *24*, 961–967. [CrossRef]
51. Brusciano, L.; Gambardella, C.; Tolone, S.; Del Genio, G.; Terracciano, G.; Gualtieri, G.; Di Visconte, M.S.; Docimo, L. An imaginary cuboid: Chest, abdomen, vertebral column and perineum, different parts of the same whole in the harmonic functioning of the pelvic floor. *Tech. Coloproctol.* **2019**, *23*, 603–605. [CrossRef]
52. Yoshida Pharmaceutical Co., Ltd. Drug Information Sheet of Magnesium Oxide. Available online: https://www.rad-ar.or.jp/siori/english/kekka.cgi?n=40132 (accessed on 30 December 2020).
53. Moore, J.G.; Christian, P.E.; Coleman, R.E. Gastric emptying of varying meal weight and composition in man. Evaluation by dual liquid- and solid-phase isotopic method. *Dig. Dis. Sci.* **1981**, *26*, 16–22. [CrossRef]
54. Firoz, M.; Graber, M. Bioavailability of US commercial magnesium preparations. *Magnes. Res.* **2001**, *14*, 257–262.
55. Yamaguchi, H.; Shimada, H.; Yoshita, K.; Tsubata, Y.; Ikarashi, K.; Morioka, T.; Saito, N.; Sakai, S.; Narita, I. Severe hypermagnesemia induced by magnesium oxide ingestion: A case series. *CEN Case Rep.* **2019**, *8*, 31–37. [CrossRef] [PubMed]

56. Matsuo, H.; Nakamura, K.; Nishida, A.; Kubo, K.; Nakagawa, R.; Sumida, Y. A case of hypermagnesemia accompanied by hypercalcemia induced by a magnesium laxative in a hemodialysis patient. *Nephron* **1995**, *71*, 477–478. [CrossRef] [PubMed]
57. Tatsumi, H.; Masuda, Y.; Imaizumi, H.; Kuroda, H.; Yoshida, S.-I.; Kyan, R.; Goto, K.; Asai, Y. A case of cardiopulmonary arrest caused by laxatives-induced hypermagnesemia in a patient with anorexia nervosa and chronic renal failure. *J. Anesth.* **2011**, *25*, 935–938. [CrossRef] [PubMed]
58. Qureshi, T.; Melonakos, T.K. Acute hypermagnesemia after laxative use. *Ann. Emerg. Med.* **1996**, *28*, 552–555. [CrossRef]
59. Khairi, T.; Amer, S.; Spitalewitz, S.; Alasadi, L. Severe Symptomatic Hypermagnesemia Associated with Over-the-Counter Laxatives in a Patient with Renal Failure and Sigmoid Volvulus. *Case Rep. Nephrol.* **2014**, *2014*, 1–2. [CrossRef]
60. The Japanese Ministry of Health, Labour and Welfare. Revision of Precautions Magnesium Oxide. 2015. Available online: http://www.pmda.go.jp/files/000219708.pdf (accessed on 30 December 2020).
61. Wakai, E.; Ikemura, K.; Sugimoto, H.; Iwamoto, T.; Okuda, M. Risk factors for the development of hypermagnesemia in patients prescribed magnesium oxide: A retrospective cohort study. *J. Pharm. Health Care Sci.* **2019**, *5*, 1–6. [CrossRef]
62. Horibata, K.; Tanoue, A.; Ito, M.; Takemura, Y. Relationship between renal function and serum magnesium concentration in elderly outpatients treated with magnesium oxide. *Geriatr. Gerontol. Int.* **2016**, *16*, 600–605. [CrossRef]
63. Tatsuki, M.; Miyazawa, R.; Tomomasa, T.; Ishige, T.; Nakazawa, T.; Arakawa, H. Serum magnesium concentration in children with functional constipation treated with magnesium oxide. *World J. Gastroenterol.* **2011**, *17*, 779–783. [CrossRef]
64. Lindberg, J.S.; Zobitz, M.M.; Poindexter, J.R.; Pak, C.Y. Magnesium bioavailability from magnesium citrate and magnesium oxide. *J. Am. Coll. Nutr.* **1990**, *9*, 48–55. [CrossRef]
65. Kurtz, W. Role of bile acids in the development of peptic lesions. Studies on the bile acid-binding capacity of a new kind of antacid. *Fortschr. Med.* **1991**, *109*, 592–594.
66. Moe, S.M. Disorders Involving Calcium, Phosphorus, and Magnesium. *Prim. Care Clin. Off. Pract.* **2008**, *35*, 215–237. [CrossRef] [PubMed]
67. Medarov, B.I. Milk-alkali syndrome. *Mayo Clin. Proc.* **2009**, *84*, 261–267. [CrossRef] [PubMed]
68. Hanada, S.; Iwamoto, M.; Kobayashi, N.; Ando, R.; Sasaki, S. Calcium-Alkali Syndrome Due to Vitamin D Administration and Magnesium Oxide Administration. *Am. J. Kidney Dis.* **2009**, *53*, 711–714. [CrossRef] [PubMed]
69. Yamada, T.; Nakanishi, T.; Uyama, O.; Iida, T.; Sugita, M. A case of the milk-alkali syndrome with a small amount of milk and magnesium oxide ingestion–the contribution of sustained metabolic alkalosis induced by hypertonic dehydration. *Nihon Jinzo Gakkai Shi* **1991**, *33*, 581–586. [PubMed]

Article

# Variations in Magnesium Concentration Are Associated with Increased Mortality: Study in an Unselected Population of Hospitalized Patients

Justyna Malinowska [1], Milena Małecka [2] and Olga Ciepiela [2,*]

1. Students Scientific Group of Laboratory Medicine, Medical University of Warsaw, 02-097 Warsaw, Poland; malinowska_justyna@interia.pl
2. Department of Laboratory Medicine, Medical University of Warsaw, 02-097 Warsaw, Poland; milena.malecka@wum.edu.pl
* Correspondence: olga.ciepiela@wum.edu.pl; Tel.: +48-22-599-24-05

Received: 19 May 2020; Accepted: 17 June 2020; Published: 19 June 2020

**Abstract:** Dysmagnesemia is a serious disturbance of microelement homeostasis. The aim of this study was to analyze the distribution of serum magnesium concentrations in hospitalized patients according to gender, age, and result of hospitalization. The study was conducted from February 2018 to January 2019 at the Central Clinical Hospital in Warsaw. Laboratory test results from 20,438 patients were included in this retrospective analysis. When a lower reference value 0.65 mmol/L was applied, hypermagnesemia occurred in 196 patients (1%), hypomagnesemia in 1505 patients (7%), and normomagnesemia in 18,711 patients (92%). At a lower reference value of 0.75 mmol/L, hypomagnesemia was found in 25% and normomagnesemia in 74% of patients. At a lower reference value of 0.85 mmol/L, hypomagnesemia was found in 60% and normomagnesemia in 39% of patients. Either hypo- or hyper-magnesemia was associated with increased risk of in-hospital mortality. This risk is the highest in patients with hypermagnesemia (40.1% of deaths), but also increases inversely with magnesium concentration below 0.85 mmol/L. Serum magnesium concentration was not gender-dependent, and there was a slight positive correlation with age ($p < 0.0001$, $r = 0.07$). Large fluctuations in serum magnesium level were associated with increased mortality ($p = 0.0017$). The results indicate that dysmagnesemia is associated with severe diseases and generally severe conditions. To avoid misdiagnosis, an increase of a lower cut-off for serum magnesium concentration to at least 0.75 mmol/L is suggested.

**Keywords:** cut-off value; hypomagnesemia; hospitalization; mortality; serum magnesium

## 1. Introduction

Magnesium is one of the most important minerals for maintaining microelement homeostasis. It acts as a cofactor or activator for over 800 enzymes, is essential for neuromuscular conduction, and affects blood glucose level and blood pressure [1–3]. Its absorption and excretion are regulated by the kidneys, gut and bones [4]. Magnesium is absorbed primarily in the intestine in an amount of 30–50% of daily intake (~100 mg daily, at a daily recommended intake of 320–420 mg per day for women and men, respectively). There are two pathways in which $Mg^{2+}$ is absorbed: one is a paracellular transport within the small intestine, and the other is a transcellular transport in the cecum and colon [1,4–6]. Total storage of magnesium is estimated for 2400 mg, among which 50–60% is stored in bones bound to the hydroxyapatite crystals, 25–30% is stored in muscles, and 20–25% in other soft tissues [5–7]. Daily, ~2400 mg of magnesium is filtered by the kidneys' glomeruli, and up to 99% of this microelement is reabsorbed. This results in daily excretion of magnesium by kidneys in an amount of approximately

100 mg [1,4,5,7]. Unabsorbed magnesium from daily intake is excreted with feces (approximately 270 mg per day) [6]. $Mg^{2+}$ concentration is measured routinely in serum during medical laboratory testing, however it does not provide full information about magnesium homeostasis in the body, since serum concentration constitutes ca. 1% of the total amount of the body's magnesium [6]. Despite being the second most abundant ion in the cells, its free concentration remains low (up to 1.2 mM). Nevertheless, its role in the cells metabolism is inestimable: bound to polynucleotides, ribosomes, or adenosine triphosphate (ATP), it takes part in nucleotide binding, supports variable enzymatic reaction as a crucial activator or cofactor, allows protein synthesis, and controls cell proliferation [6]. Magnesium homeostasis is connected to calcium and phosphates. Imbalances of those element concentrations result in serious disorders, i.e., arrhythmia, convulsions, and respiratory distress. Renal regulation of its serum concentration occurs through glomerular filtration, tubular reabsorption, or secretion. Under physiological conditions, calcium, phosphates, and magnesium are metabolized by adjusting the amounts excreted in urine to the amount of supply [4].

Hypomagnesemia is reported to be a frequent condition. From the clinical laboratory point of view, hypomagnesemia is defined as decreased serum magnesium concentration below reference ranges. The discussion about reference values for magnesium is still ongoing, since several studies show discrepancies between clinical symptoms of Mg deficiency and the threshold in serum magnesium concentration applied to diagnose magnesium deficiency [8–10]. Despite the fact that serum magnesium concentration is routinely used to assess body magnesium status, it has to be underlined that Mg in serum may not correspond with total body magnesium content, since it reflects only 1% of total body stores [2,6,11,12]. Serum magnesium concentration depends, among others, on the daily intake [2,11]. In the NHANES (National Health and Nutrition Examination Survey) results, 45% of the American population was found to be dietary deficient [13], and other studies also point to populational inappropriate magnesium intake [14–16]. Magnesium deficiency is associated with enhanced oxidative stress, inflammation, impaired glucose transport, reduced pancreatic insulin secretion, increased insulin resistance, and impaired endothelial function [6,17–19]. It is a part of the pathophysiology of type 2 diabetes mellitus, ischemic heart disease, hypertension, dyslipidemia, metabolic syndrome, liver diseases, migraines, and depression [17,19–27].

Hypermagnesemia is a less common condition and usually remains undetected. It occurs mostly as a result of chronic kidney disease or the misuse of magnesium-containing supplements or medicines [5]. Despite being a rare condition, extreme hypermagnesemia has serious clinical manifestation and may lead to hypotension, complete heart block, coma, and death [6].

As mentioned above, the most commonly used method as an assessment of serum magnesium concentration is a total Mg concentration assessment, which does not provide full information about body magnesium status. Approximately 2% of clinical laboratories offer measurement of ionized magnesium, which represents 55–70% of total magnesium and is biologically active [5,11]. Twenty-four-hour excretion of magnesium in urine gives reliable information about Mg metabolism but should be very carefully used in patients with renal disorders. Insufficient glomerular filtration rate would falsify results of magnesium turnover [5,7,11]. The aforementioned method is particularly useful in assessment of magnesium retention during oral or parenteral loading tests. Decreased excretion of Mg with urine suggests increased magnesium embedding to the bones, which might suggest its depletion. On the other hand, excretion exceeding 60% of magnesium load with the highest probability excludes Mg deficiency [5]. Other methods, which are not routinely and commonly used, are assessment of red blood cell magnesium concentration, soft and hard tissue tests for magnesium content, or isotope studies [5,11].

The aim of this study was to assess the frequency of total serum magnesium imbalance among unselected subjects from inpatient departments using different lower reference values (0.65, 0.75, and 0.85 mmol/L) and its association with in-hospital mortality.

## 2. Materials and Methods

We performed a retrospective analysis of serum magnesium assessment among 20,438 patients hospitalized between February 2018 and January 2019 in the Central Clinical Hospital of the Medical University of Warsaw. Among all subjects included, there were 11,537 men (56%) and 8901 women (44%). Average mortality among included subjects was estimated for 3.6%, while overall mortality in the corresponding period of time among all hospitalized patients was 2.3%. Mean age was 60.96 ± 16.90 years, median 61 years (18–99 years). Magnesium concentration was tested upon admission. The frequency of magnesium measurement was dependent on patients' condition and clinical requirements. For the analysis, the first time-measurements and measurements repeated at least 10 times were included. Peripheral blood was taken by venipuncture into tubes with clotting accelerator and left for 30 min for clotting. Then, samples were centrifuged for 10 min at 1500× $g$ and magnesium in serum was measured using a Cobas C702 analyzer (Roche). The principle of the method is based on the reaction of magnesium with xylidyl blue in alkaline solution, which produces a purple diazonium salt. Absorbance of the purple product is directly proportional to the magnesium concentration. The measuring range was 0.10–2.0 mmol/L (0.243–4.86 mg/dL), with high precision (coefficient of variation(CV) 0.7–1.3%).

All data were anonymized. The study included analysis of magnesium concentration, age, gender, and outcome of hospitalization (mortality or survival). As a retrospective analysis, this study was conducted according to rules of the Bioethical Committee of the Medical University of Warsaw.

Statistical analysis was performed using Microsoft Office Excel 2019, GraphPad Prism 6.0, and Statsoft Statistica. A nonparametric analysis was performed using the Mann–Whitney test (difference in magnesium concentration between 2 groups), one-way analysis of variance (ANOVA) (difference in magnesium concentration between age groups), the Fisher exact test (sex difference among different magnesium state groups), and the Kruskal–Wallis test (difference in magnesium concentration between 3 groups), where appropriate. The correlation between age and magnesium concentration was estimated with nonparametric Spearman correlation. Variations in magnesium concentration in patients who were tested at least 10 times were calculated as a delta (Δ) between the lowest and the highest serum magnesium concentration during one hospitalization. Normality of distribution of the results was assessed using Shapiro–Wilk and Kolmogorov–Smirnov tests. Results were considered statistically significant at $p < 0.05$.

## 3. Results

The mean serum magnesium concentration was 0.82 ± 0.13 mmol/L. According to different lower reference values of serum magnesium that are recommended in the literature [8,9,11,12], analyses were performed separately for cut-offs of 0.65, 0.75, and 0.85 mmol/L.

*3.1. Reference Values of 0.65–1.2 mmol/L*

Hypomagnesemia (Mg < 0.65 mmol/L) was found in 1505 subjects (7%), among which 840 were men (56%) and 665 were women (44%). Patients with hypomagnesemia had a mean Mg concentration of 0.57 ± 0.06 mmol/L.

Normomagnesemia (Mg 0.65–1.2 mmol/L) was found in 18,711 (92%) subjects, among which 10,580 (57%) were men and 8131 (43%) were women. Mean Mg concentration in the normomagnesemic group was 0.83 ± 0.1 mmol/L.

Hypermagnesemia (Mg >1.2 mmol/L) was found in 196 subjects (1%), among which 105 (54%) were men and 91 (46%) were women. Mean Mg concentration in this group was 1.36 ± 0.22 mmol/L. All three groups differed significantly in serum magnesium concentration, $p < 0.0001$ (the Kruskal–Wallis test) (Figure 1A).

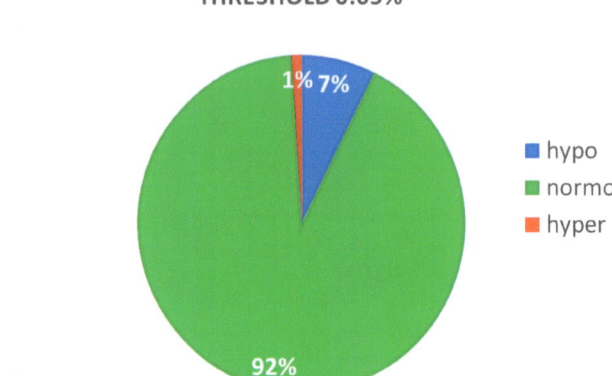
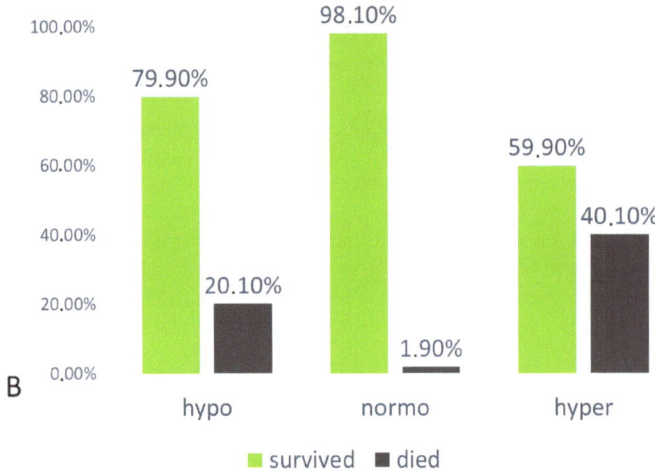

**Figure 1.** (**A**) Percentages of subjects assigned to hypo-, normo- and hyper-magnesemia groups at reference values of 0.65–1.2 mmol/L, and (**B**) in-hospital mortality ratio among separated groups. The difference between death ratio is statistically significant (Chi-square test), $p < 0.0001$.

The highest ratio of in-hospital mortality was found in the group of hypermagnesemic subjects (40.1%). In the group of hypomagnesemic patients, in-hospital mortality was significantly lower (20.1%) than in hypermagnesemic, but also significantly higher than in the normomagnesemic group (1.9%), $p < 0.0001$ (Chi-square test) (Figure 1B).

### 3.2. Reference Values 0.75–1.2 mmol/L

Hypomagnesemia (Mg < 0.75 mmol/L) was found in 5109 subjects (25%), among which 2878 were men (56%) and 2231 were women (44%). Patients with hypomagnesemia had a mean Mg concentration of 0.66 ± 0.07 mmol/L.

Normomagnesemia (Mg 0.75–1.2 mmol/L) was found in 15,183 (74%) subjects, among which 8553 (56%) were men and 6630 (44%) were women. Mean Mg concentration in the normomagnesemic group was 0.86 ± 0.08 mmol/L.

Hypermagnesemia (Mg >1.2 mmol/L) was found in 196 subjects (1%), among which 105 (54%) were men and 91 (46%) were women. Mean Mg concentration in this group was 1.36 ± 0.22 mmol/L. All three groups differed significantly in serum magnesium concentration, $p < 0.0001$ (Figure 2A).

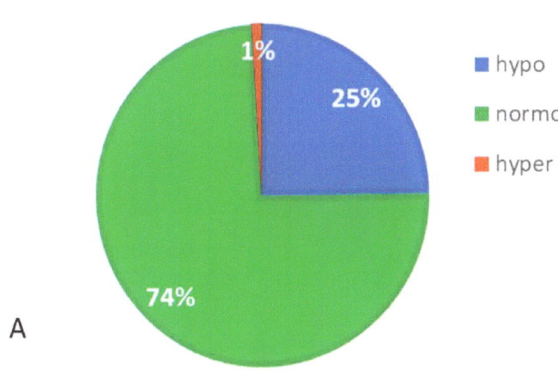

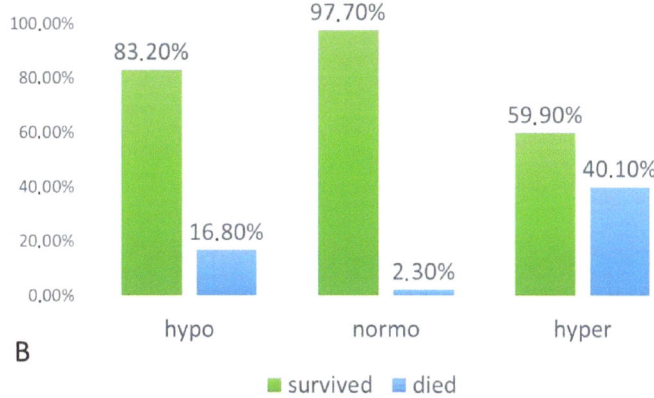

**Figure 2.** (**A**) Percentages of subjects assigned to hypo-, normo- and hyper-magnesemia groups at reference values of 0.75–1.2 mmol/L, and (**B**) in-hospital mortality ratio among separated groups. The difference between death ratio is statistically significant (Chi-square test), $p < 0.0001$.

In the group of hypomagnesemic patients, in-hospital mortality was significantly lower (16.8%) than in hypermagnesemic (40.1%), but also significantly higher than in the normomagnesemic group (2.3%), $p < 0.0001$ (Chi-square test) (Figure 2B).

## 3.3. Reference Values 0.85–1.2 mmol/L

Hypomagnesemia (Mg < 0.85 mmol/L) was found in 12,177 subjects (60%), among which 840 were men (56%) and 665 were women (44%). Patients with hypomagnesemia had a mean Mg concentration of 0.74 ± 0.08 mmol/L.

Normomagnesemia (Mg 0.85–1.2 mmol/L) was found in 8064 (39%) subjects, among which 4596 (57%) were men and 3468 (43%) were women. Mean Mg concentration in the normomagnesemic group was 0.92 ± 0.07 mmol/L.

Hypermagnesemia (Mg >1.2 mmol/L) was found in 196 subjects (1%), among which 105 (54%) were men and 91 (46%) were women. Mean Mg concentration in this group was 1.36 ± 0.22 mmol/L. All three groups differed significantly in serum magnesium concentration, $p < 0.0001$ (Figure 3A).

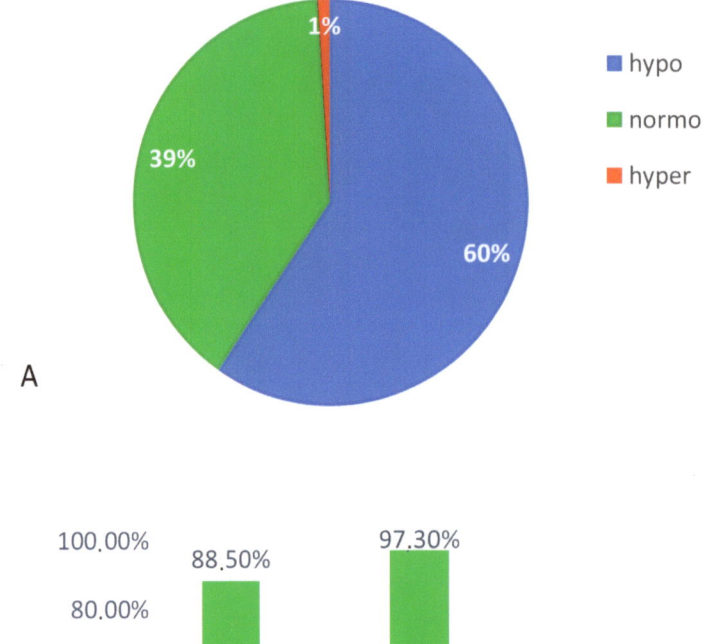

**Figure 3.** (**A**) Percentages of subjects assigned to hypo-, normo- and hyper-magnesemia groups at reference values of 0.85–1.2 mmol/L, and (**B**) in-hospital mortality ratio among separated groups. The difference between death ratio is statistically significant (Chi-square test), $p < 0.0001$.

In the group of hypomagnesemic patients, in-hospital mortality was significantly lower (11.5%) than in hypermagnesemic (40.1%), but also significantly higher than in the normomagnesemic group (2.7%), $p < 0.0001$ (Chi-square test) (Figure 3B).

There was a statistically significant difference in in-hospital mortality between hypomagnesemic groups, when using different lower reference values of serum magnesium (0.65 mmol/L, 0.75 mmol/L, and 0.85 mmol/L), with the highest ratio of mortality in the group with the lowest reference value applied (Figure 4).

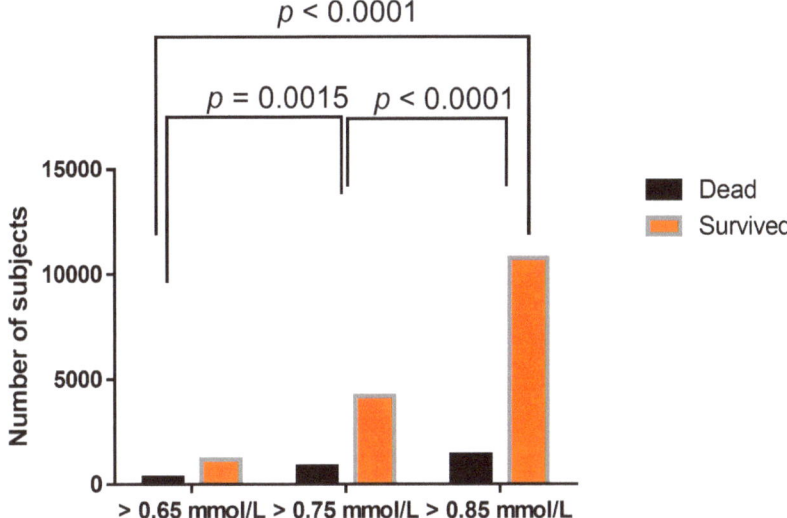

**Figure 4.** Differences in in-hospital mortality ratio between hypomagnesemic groups in dependence on lower reference value applied. Analysis was performed using Chi-square test, $p < 0.0001$.

There was no difference in magnesium concentration between men and women, either in all enrolled subjects, or in hypo-, normo-, and hyper-magnesemia groups at all reference values applied, $p > 0.05$ (the Mann–Whitney test).

There was a slight positive correlation between magnesium concentration and age of studied subjects ($r = 0.07$, $p < 0.0001$) (Figure 5A). However, differences in magnesium concentration were found between different analyzed aged groups (Figure 5B and Table 1)

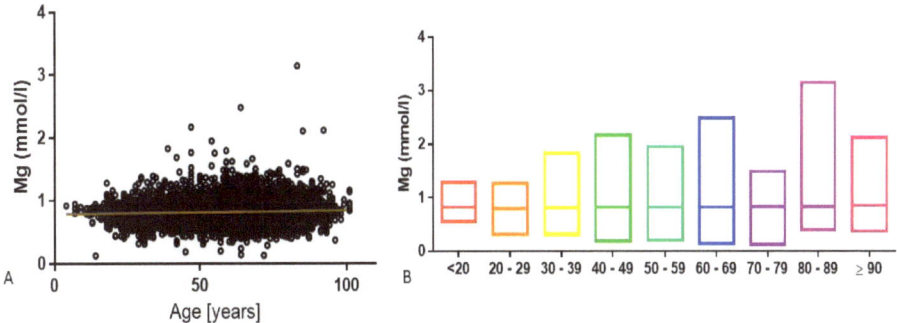

**Figure 5.** (**A**) Magnesium concentration showed a slight positive correlation with patient age ($r = 0.07$, $p < 0.0001$), assessed with nonparametric Spearman correlation. (**B**) Median and minimum/maximum values in different age groups.

**Table 1.** Statistical differences in median magnesium concentration among studied aged groups (Kruskal–Wallis test).

| Compared Aged Groups | Median Mg Values | $p$-Value |
|---|---|---|
| <20 vs ≥90 | 0.82 vs 0.85 | <0.01 |
| 20–29 vs 40–49 | 0.80 vs 0.82 | <0.001 |
| 20–29 vs 50–59 | 0.80 vs 0.82 | <0.0001 |
| 20–29 vs 60–69 | 0.80 vs 0.82 | <0.0001 |
| 20–29 vs 70–79 | 0.80 vs 0.83 | <0.0001 |
| 20–29 vs 80–89 | 0.80 vs 0.83 | <0.0001 |
| 20–29 vs ≥90 | 0.80 vs 0.85 | <0.0001 |
| 30–39 vs 50–59 | 0.81 vs 0.82 | <0.01 |
| 30–39 vs 60–69 | 0.81 vs 0.82 | <0.01 |
| 30–39 vs 70–79 | 0.81 vs 0.83 | <0.0001 |
| 30–39 vs 80–89 | 0.81 vs 0.83 | <0.0001 |
| 30–39 vs ≥90 | 0.81 vs 0.85 | <0.0001 |
| 40–49 vs 80–89 | 0.82 vs 0.83 | <0.001 |
| 40–49 vs ≥90 | 0.82 vs 0.85 | <0.01 |
| 50–59 vs 80–89 | 0.82 vs 0.83 | <0.01 |
| 50–59 vs ≥90 | 0.82 vs 0.85 | <0.01 |
| 60–69 vs 80–89 | 0.82 vs 0.83 | <0.001 |
| 60–69 vs ≥90 | 0.82 vs 0.85 | <0.01 |
| 70–79 vs ≥90 | 0.83 vs 0.85 | <0.05 |

In the study, we also analyzed variation in magnesium concentration in patients who were examined more than 10 times during one hospitalization. This group ($n = 52$) was divided into two subgroups, depending on the outcome of hospitalization: death ($n = 17$, 33%) or survival ($n = 35$, 67%). The variation in magnesium concentration was expressed as a difference between the lowest and highest serum magnesium concentration during one hospitalization ($\Delta$). In the group of patients with in-hospital mortality, variations in serum magnesium concentration were higher than in the group of subjects who survived ($0.64 \pm 0.47$ mmol/L vs $0.39 \pm 0.2$ mmol/L; $p = 0.0017$; Figure 6). We also found that the highest instability in magnesium concentration in the group of subjects with in-hospital mortality was associated with hematological malignancies ($0.94 \pm 0.79$ mmol/L, $n = 5$), septic shock ($0.6 \pm 0.1$ mmol/L, $n = 2$), and multi-organ failure ($0.6 \pm 0.16$ mmol/L, $n = 4$). In the group of patients who survived, magnesium variations were as follows: hematological malignancies $0.44 \pm 0.22$ mmol/L ($n = 14$, $p = 0.03$), septic shock $0.69 \pm 0.15$ mmol/L ($n = 4$, $p = 0.53$), and multi-organ failure $0.40 \pm 0.13$ mmol/L ($n = 5$, $p = 0.19$).

Based on different lower reference values for magnesium concentration in serum, we analyzed how many patients from the group which was tested at least 10 times were hypo-, normo- or hyper-magnesemic upon hospital admission. When setting the reference value for 0.65 mmol/L, there were 44 normomagnesemic and 8 hypomagnesemic patients, among which there was only 1 patient with hypomagnesemia who died during hospitalization (16 were normomagnesemic upon admission). If the lower reference value is set to 0.75 mmol/L, among these 52 subjects, there were 17 hypomagnesemic and 35 normomagnesemic; moreover, 4 hypomagnesemic patients did not survive hospitalization (and 13 normomagnesemic). If we use 0.85 mmol/L as a lower reference value, among 52 patients who were tested at least 10 times, there were 32 hypomagnesemic and 20 normomagnesemic upon admission. Ten patients with hypomagnesemia and seven with normomagnesemia died. None of the patients who were tested at least 10 times had hypermagnesemia upon submission (Table 2).

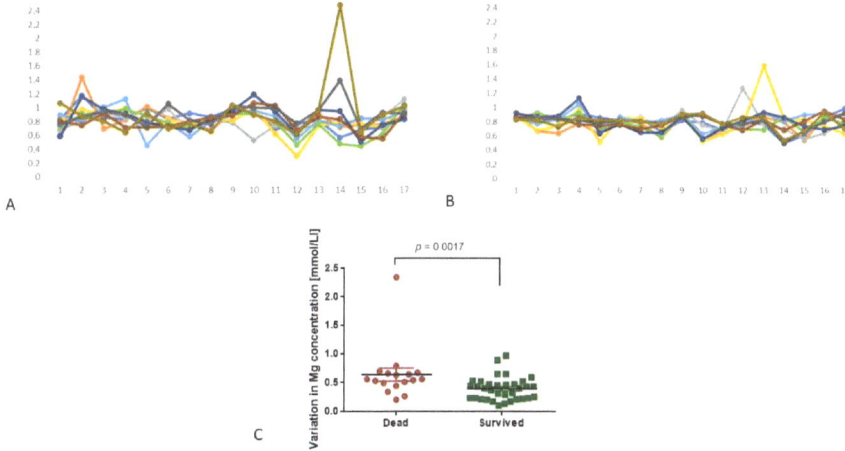

**Figure 6.** Variation in magnesium concentration in group of patients who were examined at least 10 times and whose hospitalization resulted in (**A**) death or (**B**) survival. Figures A and B present variations in magnesium concentration of ten subjects from each group, in which these fluctuations were expressed the most. The value of magnesium variation was expressed as a difference between the lowest and the highest serum Mg concentration during hospitalization. (**C**) The difference between these two groups assessed with Mann–Whitney test (death, $n = 17$ and survival, $n = 35$) was statistically significant ($p = 0.0017$).

**Table 2.** Number of patients ($n = 52$) tested at least 10 times who died ($n = 17$) or survived hospitalization ($n = 35$) with regard to their magnesium status. Data are separately presented for different lower cut-off for reference magnesium concentration in serum.

|  | 0.65–1.2 mmol/L | | 0.75–1.2 mmol/L | | 0.85–1.2 mmol/L | |
|---|---|---|---|---|---|---|
|  | Dead ($n$) | Survived ($n$) | Dead ($n$) | Survived ($n$) | Dead ($n$) | Survived ($n$) |
| Hypomagnesemia | 1 | 7 | 4 | 13 | 10 | 22 |
| Normomagnesemia | 16 | 28 | 13 | 22 | 7 | 13 |

## 4. Discussion

In the present study, disturbances in magnesium homeostasis were assessed with regard to different lower cut-offs for reference values of serum magnesium concentration: 0.65 mmol/L (applied in the authors' clinical laboratory), 0.75 mmol/L, and 0.85 mmol/L. At the lowest cut-off value, normomagnesemia was found in 92% of patients, hypomagnesemia was found in 7%, and hypermagnesemia in 1% of studied cases. An increase of the lower reference value for serum magnesium allowed to increase hypomagnesemia frequency to 25% and 60% respectively, and decrease normomagnesemia to 74% and 39%, respectively.

Laboratory relevant hypomagnesemia is recognized when total serum magnesium concentration is below the lower reference value. However, discussion concerning an adequate cut-off for normomagnesemia has been going on since the end of the 20th century. Wong et al. defined hypomagnesemia when total serum Mg concentration was below 0.6 mmol/L, and this value was obtained after assessment of serum magnesium in 341 individuals [10]. von Ehrlich [8] was considering two cut-off values: 0.70 and 0.75 mmol/L to diagnose hypomagnesemia, and found that approximately 9% of patients with a syndrome of magnesium deficiency would remain omitted in clinical practice if the lower cut-off value would be set for 0.7 mmol/L. Increasing the cut-off to 0.75 mmol/L allows to increase the ratio of appropriately diagnosed patients four times. Other studies suggest setting the lower reference value in a range of 0.65-0.95 mmol/L [12,28–33]. Nevertheless, it has to be underlined that magnesium deficiency should be assessed clinically, based on deficiency syndromes, not only based on reference values provided by local laboratories or test manufacturers. Costello et al. underline that clinical magnesium deficiency is accurately diagnosed at a cut-off value of 0.82 mmol/L, with proper urinary magnesium excretion (40–80 mg/day) [12]. Liebscher and Liebscher point out that the misdiagnosis ratio of magnesium disturbances could be decreased in a vast majority of patients (99%), if the cut-off value was 0.9 mmol/L. They based their conviction on clinical studies in patients with magnesium disorders, who expressed unspecific symptoms of Mg deficiency attributed to underlying disease [9,11].

Here, we found that the hypomagnesemia ratio strictly depends on the cut-off value of serum magnesium concentration. We found hypomagnesemia in 7%, 25%, and 60% of in-patients, using cut-off values of 0.65, 0.75, and 0.85 mmol/L in hospitalized patients. Wong et al., based on a cut-off value of 0.6 mmol/L, recognized hypomagnesemia in 11% of in-patients [10], which is similar to our data and the results of a study performed last year in Italy [29]. Our observation is in contrast to the recent study of Lorenzoni et al., who found that hypomagnesemia at a cut-off value of 0.65 mmol/L is recognized in over 58% of elderly in-patients [28]. We have obtained such a level of hypomagnesemia diagnosis at a cut-off value of 0.85 mmol/L. Our data obtained when a 0.75 mmol/L cut-off value was applied are also in line with Cheungpasitporn et al.'s results, who diagnosed hypomagnesemia (<0.7 mmol/L) in 20.2% of in-patients [31].

Literature data show that increased magnesium concentration in patients with severe conditions is associated with a high risk of mortality. Here, we also confirmed this observation. Similar to an aforementioned study performed in the Mayo Clinic, Rochester, USA [31], we assessed mortality ratio in patients regarding their serum magnesium status. Comparably, we obtained a high mortality rate in subjects of serum magnesium upon admission, with less than 0.75 mmol/L (16.8%) and 0.65 mmol/L (20.1%) [31], but in our study, significantly higher mortality was associated with hypermagnesemia upon admission (40.1%), a group that is much smaller (1%) than hypomagnesemia upon admission (7%, 25%, or 60% depending upon cutoff). Our results are in line with a study by Lorenzoni et al., who assessed a mortality rate in hypermagnesemic subjects of 45% [28]. A higher mortality rate in patients with hypermagnesemia has been also confirmed in other studies [33,34]. This phenomenon may be associated with the influence of magnesium excess on heart muscle function—arrhythmia associated with high magnesium levels may lead to death [34–36]. On the other hand, severe conditions may disturb electrolyte homeostasis and induce the release of magnesium from cells, which leads to hypermagnesemia. Moreover, it has been shown that, in general, hypermagnesemia bears a high risk of in-hospital mortality, as well as of 30-day and 12-month mortality [33]. We also observed that mortality in the group of enrolled subjects was higher than overall mortality among all hospitalized subjects in a corresponding period of time. It suggests that, firstly, patients in whom magnesium concentration is assessed upon admission are, in general, in worse condition than others, and secondly, that hypo- (regardless, defined as less than 0.65 or 0.85 mmol/L) and, what is more expressed, hyper-magnesemia, may characterize patients who are more susceptible to in-hospital mortality.

In our study, hypermagnesemia was less frequent than in other studies (1.78–12%) [5,29–31]. The differences may be associated with the selection of study group—we did not focus on patients hospitalized due to calcium–phosphate–magnesium disorders, rather we analyzed the general hospitalized population. Regardless of the reference value applied, disturbed magnesium homeostasis is an adverse prognostic factor in in-patient subjects, especially in intensive care units, thus constant and sufficient magnesium concentration might contribute to a decrease of in-hospital mortality [28,37].

Here, we did not observe significant changes in magnesium concentration with patient age—there was only a very slight correlation, which showed that magnesium concentration increases with age. A similar observation was reported by Cheungpasitporn et al. [31]. Interestingly, the mean age of patients with hypomagnesemia was 57.5 years, and with hypermagnesemia, 63 years. Similar observations were reported in other studies, where the mean age with hypomagnesemia was 60 years, and with hypermagnesemia, 65 years [31]. This phenomenon could be associated with inadequate Mg intake in elderly people or decreased kidney function, which appears with age [37]; however, we did not analyze estimated glomerular filtration rate (eGFR) in enrolled subjects and we cannot confirm this observation. Wakai et al. showed that the risk of hypermagnesemia in patients administering magnesium oxide increases in subjects who are ≥68 years old or have decreased eGFR (≤55.4 mL/min), or urea nitrogen of ≥22.4 mg/dL, and their daily administration of magnesium oxide exceeds 1650 mg [38].

Here, we showed that variations in magnesium concentration are associated with in-hospital mortality. This is in line with the observations of Rhee et al., who showed that fluctuations in magnesium concentration in dialyzed patients constitute a risk factor for increased mortality [39]. In our study, the highest variations in magnesium concentration were associated with hematological malignancies, septic shock, and multi-organ disfunction. Literature data show that both hypo- and hyper-magnesemia increase the risk of septic shock in patients who are recognized with systemic inflammatory response syndrome [40]. On the other hand, magnesium was not found as a marker that could help to predict organ failure in critically ill adult patients [41]. Here, we also found differences in magnesium concentration in a group of patients who suffered from different hematological malignancies with different hospitalization outcomes. Significantly higher values were found in subjects with in-hospital mortality. Variation in magnesium concentrations in patients suffering from acute leukemias have already been reported [42] and may be associated with either tumor lysis syndrome triggered by chemotherapy or improper parenteral nutrition. Regardless of etiology, variations in magnesium concentration in these patients may lead to increased mortality due to arrythmia and cardiotoxicity [43].

## 5. Conclusions

To conclude, it has been shown that disturbances in magnesium homeostasis accompany systemic diseases and severe conditions [39,40,43]. Either hypo- or hyper-magnesemia are associated with increased risk of mortality, and, for this reason, primarily, hypermagnesemia should be treated as a laboratory biomarker of critical value. Despite local laboratory reference values for serum magnesium concentration, its deficiency should always be diagnosed based on clinical symptoms. To avoid misdiagnosis, the increase of a lower cut-off value is suggested. High variation in magnesium levels during hospitalization may be associated with increased in-hospital mortality [40,42]; thus, magnesium levels should be routinely tested in hospitalized subjects.

**Author Contributions:** Conceptualization, O.C.; methodology, J.M. and O.C.; software, J.M. and M.M.; validation, J.M., M.M., and O.C.; formal analysis, J.M., M.M., and O.C. investigation J.M., M.M., and O.C.; resources, J.M. and O.C.; data curation, O.C.; writing—original draft preparation, J.M. and O.C. writing—review and editing, J.M., M.M., and O.C.; visualization, J.M., M.M., and O.C.; supervision, O.C.; project administration, O.C. All authors have read and agreed to the published version of the manuscript

**Funding:** This research received no external funding.

**Conflicts of Interest:** The authors declare no conflict of interest.

## References

1. Grober, U.; Schmidt, J.; Kisters, K. Magnesium in Prevention and Therapy. *Nutrients* **2015**, *7*, 8199–8226. [CrossRef] [PubMed]
2. Razzaque, M.S. Magnesium: Are We Consuming Enough? *Nutrients* **2018**, *10*, 1863. [CrossRef] [PubMed]
3. Volpe, S.L. Magnesium in disease prevention and overall health. *Adv. Nutr.* **2013**, *4*, 378S–383S. [CrossRef]
4. Blaine, J.; Chonchol, M.; Levi, M. Renal control of calcium, phosphate, and magnesium homeostasis. *Clin. J. Am. Soc. Nephrol.* **2015**, *10*, 1257–1272. [CrossRef] [PubMed]
5. Jahnen-Dechent, W.; Ketteler, M. Magnesium basics. *Clin. Kidney J.* **2012**, *5* (Suppl. 1), i3–i14. [CrossRef] [PubMed]
6. de Baaij, J.H.; Hoenderop, J.G.; Bindels, R.J. Magnesium in man: Implications for health and disease. *Physiol. Rev.* **2015**, *95*, 1–46. [CrossRef]

7. Elin, R.J. Magnesium metabolism in health and disease. *Dis. Mon.* **1988**, *34*, 161–218. [CrossRef]
8. Ehrlich, B. Magnesiummangelsyndrom in der internischen Praxis. *Magnes.-Bull.* **1997**, *19*, 29–30.
9. Liebscher, D.H.; Liebscher, D.E. About the misdiagnosis of magnesium deficiency. *J. Am. Coll. Nutr.* **2004**, *23*, 730S–731S. [CrossRef]
10. Wong, E.T.; Rude, R.K.; Singer, F.R.; Shaw, S.T., Jr. A high prevalence of hypomagnesemia and hypermagnesemia in hospitalized patients. *Am. J. Clin. Pathol.* **1983**, *79*, 348–352. [CrossRef]
11. Elin, R.J. Assessment of magnesium status for diagnosis and therapy. *Magnes. Res.* **2010**, *23*, S194–S198. [PubMed]
12. Costello, R.B.; Elin, R.J.; Rosanoff, A.; Wallace, T.C.; Guerrero-Romero, F.; Hruby, A.; Lutsey, P.L.; Nielsen, F.H.; Rodriguez-Moran, M.; Song, Y.; et al. Perspective: The Case for an Evidence-Based Reference Interval for Serum Magnesium: The Time Has Come. *Adv. Nutr.* **2016**, *7*, 977–993. [CrossRef]
13. Moshfegh Alanna, G.J.; Jaspreet, A.; Donna, R.; Randy, L. *What We Eat in America, NHANES 2005–2006: Usual Nutrient Intakes from Food and Water Compared to 1997 Dietary Reference Intakes for Vitamin D. Calcium, Phosphorus, and Magnesium*; US Department of Agriculture, Agricultural Research Service: Washington, DC, USA, 2009.
14. Elwood, P.C.; Fehily, A.M.; Ising, H.; Poor, D.J.; Pickering, J.; Kamel, F. Dietary magnesium does not predict ischaemic heart disease in the Caerphilly cohort. *Eur. J. Clin. Nutr.* **1996**, *50*, 694–697.
15. Choi, M.K.; Weaver, C.M. Daily Intake of Magnesium and its Relation to Urinary Excretion in Korean Healthy Adults Consuming Self-Selected Diets. *Biol. Trace Elem. Res.* **2017**, *176*, 105–113. [CrossRef] [PubMed]
16. McKay, J.; Ho, S.; Jane, M.; Pal, S. Overweight & obese Australian adults and micronutrient deficiency. *BMC Nutr.* **2020**, *6*, 12.
17. Del Gobbo, L.C.; Song, Y.; Poirier, P.; Dewailly, E.; Elin, R.J.; Egeland, G.M. Low serum magnesium concentrations are associated with a high prevalence of premature ventricular complexes in obese adults with type 2 diabetes. *Cardiovasc. Diabetol.* **2012**, *11*, 23. [CrossRef] [PubMed]
18. Lecube, A.; Baena-Fustegueras, J.A.; Fort, J.M.; Pelegri, D.; Hernandez, C.; Simo, R. Diabetes is the main factor accounting for hypomagnesemia in obese subjects. *PLoS ONE* **2012**, *7*, e30599. [CrossRef] [PubMed]
19. Wu, J.; Xun, P.; Tang, Q.; Cai, W.; He, K. Circulating magnesium levels and incidence of coronary heart diseases, hypertension, and type 2 diabetes mellitus: A meta-analysis of prospective cohort studies. *Nutr. J.* **2017**, *16*, 60. [CrossRef] [PubMed]
20. Fang, X.; Han, H.; Li, M.; Liang, C.; Fan, Z.; Aaseth, J.; He, J.; Montgomery, S.; Cao, Y. Dose-Response Relationship between Dietary Magnesium Intake and Risk of Type 2 Diabetes Mellitus: A Systematic Review and Meta-Regression Analysis of Prospective Cohort Studies. *Nutrients* **2016**, *8*, 739. [CrossRef] [PubMed]
21. Peters, K.E.; Chubb, S.A.; Davis, W.A.; Davis, T.M. The relationship between hypomagnesemia, metformin therapy and cardiovascular disease complicating type 2 diabetes: The Fremantle Diabetes Study. *PLoS ONE* **2013**, *8*, e74355. [CrossRef]
22. Spiga, R.; Mannino, G.C.; Mancuso, E.; Averta, C.; Paone, C.; Rubino, M.; Sciacqua, A.; Succurro, E.; Perticone, F.; Andreozzi, F.; et al. Are Circulating Mg(2+) Levels Associated with Glucose Tolerance Profiles and Incident Type 2 Diabetes? *Nutrients* **2019**, *11*, 2460. [CrossRef] [PubMed]
23. Sarrafzadegan, N.; Khosravi-Boroujeni, H.; Lotfizadeh, M.; Pourmogaddas, A.; Salehi-Abargouei, A. Magnesium status and the metabolic syndrome: A systematic review and meta-analysis. *Nutrition* **2016**, *32*, 409–417. [CrossRef]
24. Tarleton, E.K.; Littenberg, B. Magnesium intake and depression in adults. *J. Am. Board Fam. Med.* **2015**, *28*, 249–256. [CrossRef] [PubMed]
25. Liu, M.; Yang, H.; Mao, Y. Magnesium and liver disease. *Ann. Transl. Med.* **2019**, *7*, 578. [CrossRef] [PubMed]
26. Dolati, S.; Rikhtegar, R.; Mehdizadeh, A.; Yousefi, M. The Role of Magnesium in Pathophysiology and Migraine Treatment. *Biol. Trace Elem. Res.* **2019**. [CrossRef] [PubMed]
27. Altura, B.M.; Altura, B.T. New perspectives on the role of magnesium in the pathophysiology of the cardiovascular system. I. Clinical aspects. *Magnesium* **1985**, *4*, 226–244.
28. Lorenzoni, G.; Swain, S.; Lanera, C.; Florin, M.; Baldi, I.; Iliceto, S.; Gregori, D. High- and low-inpatients' serum magnesium levels are associated with in-hospital mortality in elderly patients: A neglected marker? *Aging Clin. Exp. Res.* **2020**, *32*, 407–413. [CrossRef]
29. Catalano, A.; Bellone, F.; Chila, D.; Loddo, S.; Corica, F. Magnesium disorders: Myth or facts? *Eur. J. Intern. Med.* **2019**, *70*, e22–e24. [CrossRef]

30. Ayuk, J.; Gittoes, N.J. Contemporary view of the clinical relevance of magnesium homeostasis. *Ann. Clin Biochem.* **2014**, *51 Pt. 2*, 179–188. [CrossRef]
31. Cheungpasitporn, W.; Thongprayoon, C.; Qian, Q. Dysmagnesemia in Hospitalized Patients: Prevalence and Prognostic Importance. *Mayo Clin. Proc.* **2015**, *90*, 1001–1010. [CrossRef]
32. Assadi, F. Hypomagnesemia: An evidence-based approach to clinical cases. *Iran. J. Kidney Dis.* **2010**, *4*, 13–19. [PubMed]
33. Tazmini, K.; Nymo, S.H.; Louch, W.E.; Ranhoff, A.H.; Oie, E. Electrolyte imbalances in an unselected population in an emergency department: A retrospective cohort study. *PLoS ONE* **2019**, *14*, e0215673. [CrossRef] [PubMed]
34. Naksuk, N.; Hu, T.; Krittanawong, C.; Thongprayoon, C.; Sharma, S.; Park, J.Y.; Rosenbaum, A.N.; Gaba, P.; Killu, A.M.; Sugrue, A.M.; et al. Association of Serum Magnesium on Mortality in Patients Admitted to the Intensive Cardiac Care Unit. *Am. J. Med.* **2017**, *130*, 229.e5–229.e13. [CrossRef] [PubMed]
35. Haider, D.G.; Lindner, G.; Ahmad, S.S.; Sauter, T.; Wolzt, M.; Leichtle, A.B.; Fiedler, G.; Exadaktylos, A.K.; Fuhrmann, V. Hypermagnesemia is a strong independent risk factor for mortality in critically ill patients: Results from a cross-sectional study. *Eur. J. Intern. Med.* **2015**, *26*, 504–507. [CrossRef]
36. Cascella, M.; Vaqar, S. *Hypermagnesemia*; StatPearls: Treasure Island, FL, USA, 2020.
37. Van Laecke, S. Hypomagnesemia and hypermagnesemia. *Acta Clin. Belg.* **2019**, *74*, 41–47. [CrossRef]
38. Wakai, E.; Ikemura, K.; Sugimoto, H.; Iwamoto, T.; Okuda, M. Risk factors for the development of hypermagnesemia in patients prescribed magnesium oxide: A retrospective cohort study. *J. Pharm. Health Care Sci.* **2019**, *5*, 4. [CrossRef]
39. Rhee, C.M.; Chou, J.A.; Kalantar-Zadeh, K. Dialysis Prescription and Sudden Death. *Semin. Nephrol.* **2018**, *38*, 570–581. [CrossRef]
40. Thongprayoon, C.; Cheungpasitporn, W.; Erickson, S.B. Admission hypomagnesemia linked to septic shock in patients with systemic inflammatory response syndrome. *Ren. Fail.* **2015**, *37*, 1518–1521. [CrossRef]
41. Helliksson, F.; Wernerman, J.; Wiklund, L.; Rosell, J.; Karlsson, M. The combined use of three widely available biochemical markers as predictor of organ failure in critically ill patients. *Scand. J. Clin. Lab. Investig.* **2016**, *76*, 479–485. [CrossRef] [PubMed]
42. Afridi, H.I.; Kazi, T.G.; Talpur, F.N. Correlation of Calcium and Magnesium Levels in the Biological Samples of Different Types of Acute Leukemia Children. *Biol. Trace Elem. Res.* **2018**, *186*, 395–406. [CrossRef] [PubMed]
43. British Committee for Standards in Haematology; Milligan, D.W.; Grimwade, D.; Cullis, J.O.; Bond, L.; Swirsky, D.; Craddock, C.; Kell, J.; Homewood, J.; Campbell, K.; et al. Guidelines on the management of acute myeloid leukaemia in adults. *Br. J. Haematol.* **2006**, *135*, 450–474. [CrossRef] [PubMed]

© 2020 by the authors. Licensee MDPI, Basel, Switzerland. This article is an open access article distributed under the terms and conditions of the Creative Commons Attribution (CC BY) license (http://creativecommons.org/licenses/by/4.0/).

MDPI
St. Alban-Anlage 66
4052 Basel
Switzerland
Tel. +41 61 683 77 34
Fax +41 61 302 89 18
www.mdpi.com

*Nutrients* Editorial Office
E-mail: nutrients@mdpi.com
www.mdpi.com/journal/nutrients

www.ingramcontent.com/pod-product-compliance
Lightning Source LLC
LaVergne TN
LVHW072316090526
838202LV00019B/2296